Cornelia Denz

# Optical Neural Networks

W0042903

Cornelia Denz

# Optical Neural Networks

Edited by Theo Tschudi

Die Deutsche Bibliothek – CIP-Einheitsaufnahme

**Denz, Cornelia:**
Optical neural networks: an introduction with special emphasize
on photorefractive implementations / Cornelia Denz. Ed. by
Theo Tschudi. – Braunschweig; Wiesbaden: Vieweg, 1998
    (Optics and photonics)

Anschrift Dr. Cornelia Denz
Institut für Angewandte Optik
Technische Universität Darmstadt
AG Photorefraktive Optik
Hochschulstr. 6
D-64289 Darmstadt

Alle Rechte vorbehalten

© Springer Fachmedien Wiesbaden 1998
Ursprünglich erschienen bei Friedr. Vieweg & Sohn Verlagsgesellschaft mbH, Braunschwieg/Wiebaden 1998.
Softcover reprint of the hardcover 1st edition 1998

Das Werk einschließlich aller seiner Teile ist urheberrechtlich geschützt. Jede
Verwertung außerhalb der engen Grenzen des Urheberrechtsgesetzes ist ohne
Zustimmung des Verlags unzulässig und strafbar. Das gilt insbesondere für
Vervielfältigungen, Übersetzungen, Mikroverfilmungen und die Einspeicherung
und Verarbeitung in elektronischen Systemen.

http://www.vieweg.de

Umschlaggestaltung: Klaus Birk, Wiesbaden

Gedruckt auf säurefreiem Papier

ISBN 978-3-663-12274-6    ISBN 978-3-663-12272-2 (eBook)
DOI 10.1007/978-3-663-12272-2

Light is the first of painters.
There is no object so foul
that intense light
will not make it beautiful

*- Ralph Waldo Emerson -*

# Preface

In recent years, there has been a rapid expansion in the field of nonlinear optics as well as in the field of neural computing. Up to date, no one would doubt that nonlinear optics is one of the most promising fields of realizing large neural network models due to their inherent parallelism, the use of the speed of light and their ability to process two-dimensional data arrays without carriers or transformation bottlenecks. This is the reason why so many of the interesting applications of nonlinear optics — associative memories, Hopfield networks and self-organized nets — are realized in an all optical way using nonlinear optical processing elements. Both areas attracting people from a wide variety of disciplines and judged by the proliferation of published papers, conferences, international collaborations and enterprises, more people than ever before are now involved in research and applications in these two fields. These people all bring a different background to the area, and one of the aims of this book is to provide a common ground from which new development can grow. Another aim is to explain the basic concepts of neural computation as well as its nonlinear optical realizations to an interested audience. Therefore, the book is about the whole field of optical neural network applications, covering all the major approaches and their important results. Especially, it its an introduction that develops the concepts and ideas from their simple basics through their formulation into powerful experimental neural net systems.

In writing this book, my objective was not to completely review the vast range of activity that can be actually found in the area on optical neural networks. It would burst the volume of that series to describe all concepts, models and promising systems that have been realized up to now in nonlinear optics in detail. One consequence of the quick development of the area is the growing diversity of the field which has meant that, especially for the newcomer, the underlying principles often lie hidden in the technical literature amongst a wealth of details. Therefore, this book is intended to fulfil the need for a self-contained account of the most important implementations of neural nets using nonlinear optics, fully developed from its basic concepts. However, to get at the same time an insight in the current scientific possibilities of actual optical realizations of neural networks, I have chosen one of the forthcoming areas of nonlinear optics — photorefraction — as a crucial point. Most of the examples of the last part which describes the actual implementations of neural net systems in nonlinear optics are related to that effect. Here, the basic concepts are extended to some detailed models of systems, that allow for promising applications as associative memories with large data storage, Hopfield networks or self-organizing multilayer networks.

I have tried throughout the book to assume as little as possible in the reader and have attempted to set the book out in a clear and logic order, showing the basic concepts behind each of the major approaches. No prior knowledge of nonlinear optics or neural network algorithms apart from the most simple notations and a basic mathematical and physical understanding is required of the reader. All the ideas are explained in English text first, followed in many cases by the mathematical treatments of the models. The important derivations and proofs are also included since they form a vital part of the development of the area as a whole, and indicate points at which apparently insurmountable problems were reached and finally overcome. The book is written for graduate students of physics and electric engineering, as well as for the increasing number of scientists and engineers entering the field. Especially for these persons that belong to one of the fields that are combined in that book — neural networks or nonlinear optics — the book wants to provide a self-consistent source of the basics in the other area. Moreover, the book is also directed to the established researcher who is looking for a source reference for the derivation of fundamental formulas with confidence that the notation, numerical factors and units are treated consistently.

Throughout the book, various little labels appear again and again - they contain information about the content of the section in which they appear and have the following meaning:

 This symbol has been chosen to describe a section of text that is mathematical in nature. Not all the mathematical parts of the book are indicated like this, since they are usually important to the overall understanding. However, sections with this icon at the start can be skipped over at a first reading without losing too much of the following discussion for those readers that are not familiar with these mathematics. The sections contain material that is useful, interesting and essential for a detailed understanding of the subject and therefore I would like to suggest that an effort of looking at them should be made in a second reading.

 This icon represents an algorithm for a particular model. Most often, these algorithms are summarized in a box in the chapters concerning the neural network structures. They should be looked at when they are encountered in the text, but again they may be omitted on first reading.

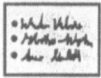 This symbol usually appears at the end of every section. Here, I have tried to compress the major ideas in a succinct summary. Most often, the summary consists of a list of items that are the most important points to bear in mind. It can be used to check that you have followed the main features of the present section.

 This finally represents some suggestions for further reading at the very end of each chapter. I have tried to summarize here the most important references, directing the attention towards specific books or overview articles that deal with a particular subject in depth. Detailed references for certain derivations throughout the book are still summarized at the end of the book, then referring to the original work and specialized scientific papers.

### What is Where: Organization of this Volume

This book is divided into three parts and has 13 chapters that are organized in such a way to retain continuity and to provide an insight in the whole area of nonlinear optical neural net realizations.

The first part, the **Background for Optical Neural Networks** contains chapter 1 and 2 and is dedicated to the background in the area of neural network models and basic concepts in nonlinear optics that is necessary to understand the following parts.

Chapter 1, the **Introduction**, contains the background to the subject. The first sections takes a more philosophical look at the differences between humans and computers. The second section describes a simplified model of the real brain. Then, learning in biological systems is described and the differences between artificial and biological neural nets are presented. Finally, the reasons and advantages of nonlinear optical implementations of neural nets are discussed.

Chapter 2, the **Principles of Neural Nets,** develops first the basic model of a simple neuron, the one layer perceptron, its learning rule, operation and features. It is followed by a description of the multilayer perceptron in the second section, showing how the one layer version can be altered to be more powerful. It covers the concepts of back-propagation, including the generalized delta rule, gradient descent, the concepts of feature extraction by hidden units and the energy landscape representation. The fourth section looks at the different paradigm of unsupervised learning, namely the Kohonen self-organizing feature networks. Section 5 deals with Hopfield networks, containing the description of that fully connected net, whereas section 6 is dedicated to its probabilistic partner, the Boltzmann machine. The problems of optimization in Boltzmann nets are described using the mean-field theory and are explained using the example of spin-glasses. At last, the adaptive resonance theory has been chosen as a fully connected system based on unsupervised learning, highlighting the differences between the architecture and approach of this system to those covered earlier.

In Chapter 3, **Basic Concepts of Nonlinear and Photorefractive Optics** are explained in the third chapter, beginning with the origins of optical nonlinearities. Starting from the constitutive relations between the polarization and the electric field, a structured development of the underlying mathematics is performed. The following sections are dedicated to a representative description of nonlinear-optical phenomena, beginning with effects arising from the quadratic polarization as the second harmonic generation and parametric amplification. Effects that arise due to the cubic polarization are treated in the next section, with special emphasize on the photorefractive effects and its various applications as two wave mixing and phase conjugation.

In the second part of the book, **Devices, Components and Subsystems** of nonlinear optical neural networks are described. Here, these components are derived in their operation apart from the implementation in large neural nets, allowing thus for a comprehensive description of their function and performance.

In Chapter 4, **Nonlinear Optical Storage and Interconnection Concepts** are discussed. A detailed section on interconnections using the principles of volume holographic storage is followed by the description of multiple image storage. Both principles are the basics for the realizations of the heart of associative memories. Here, the implementations and applications of photorefractive systems play an important role, leading to high capacity and high resolution storage units for a broad range of applications that are useful not only in neural networks realizations but in a lot of optical processing and computimg applications too.

In Chapter 5, **Nonlinear Optical Thresholding** is treated, which may be performed using a lot of different nonlinear optical principles. In this chapter, I will discuss electrooptic as well as all-optical, coherent realizations of thresholding using a wide range of different materials and devices, as liquid crystal cells, Fabry-Perot etalons or photorefractive phase conjugation.

Chapter 6, named **Further Computing Elements,** describes operations that are useful for neural nets but not necessary to realize the most simplest neural net structures. These are e.g. data reduction devices, optical logic and bistable operations and time-resolving elements. Moreover, I will describe in this chapter the methods to retain and refresh data im optical feedback system as well as the methods and problems connected with pure optical, resonant feedback.

The last part of the book describes the most important and actual **Concepts and Architectures of Optical Neural Networks,** beginning with simple models for associative memories, followed by complex multilayer feedback systems up to self-organized networks with special realization requirements. Again, this section emphasizes implementations that have been realized using photorefractive devices combined with the most common nonlinear processing techniques.

In Chapter 7, **Associative Memories** are described, because these systems were the earliest neural network structures being realized optically. Today, optical implementations of associative memories represent the largest class of neural network systems in optics. Especially volume holography with its possibility to store a large amount of data that should be associatively compared combined with optical Fourier filtering techniques are promising candidates to realize high capacity auto- or heteroassociative systems. I will start with linear holographic associative memories, followed by a variety of system realizing nonlinear associative memories mainly with photorefractive interconnection devices and optoelectronic feedback mechanisms. Moreover, I will describe resonator associative memories, where the information is stored in the eigenmodes of the resonators, and bidirectional memories in which the information can be cycled in both, forward and backward direction. This leads finally to a section describing complex multilayer associative memories.

Chapter 8 deals with **Optical Realizations of Perceptron-like Neural Networks**, simple systems based on prescribed learning. Because the perceptron algorithm is one of the most simple and popular structure in optical neural network realization, although it allows only for processing of nonlinear decision problems, a lot of realization using planar and volume holographic interconnection devices have been presented. The most interesting and promising of them are described in that chapter.

Chapter 9 is the one that introduces **Optical Realizations of Multilayer Perceptrons**. These systems show a further step into the direction of realizations of nonlinear optical neural nets. The crucial point is here the realization of error backpropagation, which may be realized optoelectronically or using nonlinear optical effects. The concept of phase conjugation is perfectly suited to match all the conditions connected with backpropagation back to the input that are necessary for a proper net with several layers.

Chapter 10 introduces **Optical Realizations of Self-Organizing Neural Networks** as Kohonen networks and implementations of winner-takes-all networks that are realized using nonlinear optics. I will describe how competitive learning may be realized using photorefractive volume holograms and optoelectronic threshold devices.

In Chapter 11, **Optical Realizations of fully connected networks as Hopfield or Boltzmann nets** and an overview over actual investigations in the field of nonlinear optical realizations of Hopfield networks are the main topics. Among the concepts that are promising here, interpattern association, an extension of the Hopfield model that relies more on the similarities on different input patterns than on their differences, will be described as well as some more complex system realizing optically simulated annealing

Chapter 12 finally describes examples of **Optical Realizations of Adaptive Resonance Theory Networks**, one of the most complex unsupervised learning networks. This network type fits perfectly with the ability of optics to process large input-pattern fields in parallel, leading to promising optoelectronic and all-optical concepts for pattern recognition.

Chapter 13 is the **Outlook** that describes shortly some very recent developments and gives a couple of hints for further reading for following the actual developments.

If the aim of the reader is to properly understand optical realizations of neural networks, I would suggest to read the whole book in order. An alternative for those who are already familiar with neural net algorithms or have some basic knowledge in nonlinear optics and for those who have a particular interest in the more recent development of the field, is to briefly read Chapters 4 to 6 in order to familiarize themselves with the foundations of the subject, followed by Chapters 7 through to 12.

Because of the vast scope of the subject and the limits of space, the contents of that book are inevitably a compromise. In order to restrict the topic to an amount that can be handled and on the same time extend the topic exemplarily for actual developments, I have chosen among the different attractive nonlinear optical effects the case of photorefractive optics for detailed explanations. I hope, however, that the choice of topics will provide an understanding of the broad principles. On the one hand, I have chosen among existing work a lot of basic early ideas that are reviewed in detail in order to give a profound insight into the mechanisms of the different realizations. On the other hand, I have given room to promising recent developments and systems in order to show the current state of the art. I consider that the material presented here represents the minimum knowledge necessary for an appreciation of key aspects of the current literature.

Nonlinear optics as well as neural net applications have developed through the efforts of countless researchers. Throughout the chapters, work of a lot of important authors is presented. It is the sum of these investigations that represent the key for the actual success of optical neural networks. I am indebted to these authors for being able to present their work here. For brevity, further references are limited to those which supplement the text. I sincerely apologize to the many authors of important works whom I have failed to cite by name.

### With the help of ...

A project like writing a book involves the collaboration of a number of people and this book has been no exception. I am greatly indebted to those people who have helped me through their support, comments and criticism to transform the inital idea to reality. I appreciate their assistance and effort, especially that of my colleagues at the Technische Hochschule Darmstadt, Institute for Light and Particle Optics, and in particular those within my research group, who have contributed freely to discussions and thus influenced many parts of that book. I am grateful to Prof.Dr. Theo Tschudi who has initialized my work on the topic of nonlinear optical neural networks. Dr. Wolfgang Balzer has contributed a lot to my understanding of neural network algorithms. Most of the work of my research group that has been presented throughout the book has mainly been successful due to Silke Aumann, Gerhard Balzer, Thilo Dellwig, Matthias Fiegler, Thorsten Heimann, Oliver Knaup, Jan Lembcke, Kai-Oliver Müller, Jürgen Petter, Torsten Rauch, Michael Schwab, Markus Sedlatschek and Jürgen Trumpfheller. Gerhard Balzer, Torsten Rauch and Markus Sedlatschek contributed a lot to work on ring resonators, volume holographic storage, and novelty filtering. Thilo Dellwig and especially Markus Sedlatschek had moreover a lot of work correcting the manuscript and looking out for typos. The graphical representation of the images was realized by Barbara Hackel and Anna Zilch.

My personal thanks must extend to Wilfried Denz, who has tolerated preoccupations, shared disappointments, read many drafts and still found time for encouragement and support. Finally, I thank Irmgard and Horst Denz for their encouragement as well as Tobias and Silas. The book is better because of them all — if there are bad parts, the blame lies with me.

**Darmstadt, September 1997**                                                **Cornelia Denz**

# Contents

# Part I

# Background for Optical Neural Networks

# Part I

# Background for Optical Neural Networks

# 1 Introduction

What have optics to do with mathematical algorithms for neural networks? In order to compute, we need to transport data from place to place, connect components together, store data and be able to switch on and off. All these features seem to be realized in an easy way by electronic components as there are wires or conducting pathways on silicon, electrical junctions, a memory and the transistor. However, for the purpose of neural networks, there are many practical difficulties in implementation. The most obvious of these is that neural networks require by their nature an *adaptive feature*, which means that they are able to distinguish between states and change their decision rules themselves as necessary. It is very difficult to implement such a feature in integrated circuit technology. Moreover, the high interconnectivity also demands a lot of effort in electronics.

Thus the question is obvious if there is another area in physics that is able to fulfil the requirements that neural network structures demand. Optics already have a long tradition in interacting with electronics, so why not having a look at how optics are able to fulfil these requirements. To get a clear answer to the question of how optics are able to implement neural network algorithms, let us first have a more detailed look into the features and requirements of neural networks. Then we will be able to look once again at the question if optics are advantageous compared to conventional electronic neural net realizations.

## 1.1 Comparing Computers and Humans

No one would doubt that human beings are more intelligent than computers. Where does this attitude originate from? Has it anything to do with the fact that we are human ourselves and do not want to think about a lumpy machine being better or more skilful than we are? Or is it because they work different from our brain in terms of the structure and operations they perform?

In some areas for instance, computers are highly superior to human brains: multiplicating two numbers with nine or ten digits is a trivial problem for computers, but it is a rather difficult problem even for the most adept person. In contrast, computers have more difficulties in finding a tree in a picture or solving crosswords.

These two tasks that are examples for all problems arising in vision and speech usually involve working out what a partial information is referring to and requires intuition and guesswork. In that case, human beings follow lines of enquiry that are not immediately obvious, but are sparked off by some recollection or idea that happens to come along. Why then can't computers do these things while they seem to be so easy for human beings, whereas they do other tasks — as addition or multiplication — in a portion of the time we need to solve them. One of the answers appear to be in the nature of their specific design and structure. Computers are designed to carry out the instructions of a mathematical algorithm one after another in an extremely rapid way. We call this a serial process. They typically carry out a few million operations every second, whereas our brain works with many, but slower units. On the average, about ten times per second an operation is performed. However, the units in the brain are arranged in a complex fashion, each connected to thousands of other units. Therefore, many of these units are working in *parallel* on many different things at once. This is an operation modus, computers can not perform: the computer is a high-speed, serial machine, in contrast to the slow, highly parallel nature of the brain. With this background, is it so surprising that the computer fails to perform in the same way as the brain? It manages tasks which suit its design very well: computing is an essentially serial activity — as is addition or multiplication — and so computers can beat the brain many times in this domain.

For vision or speech recognition in contrast, the problem is a highly parallel one, with many different and conflicting inputs triggering many different and conflicting ideas and meanings. It is in fact only the combination of all these different factors — even if they are contradictory — which enables our brains to operate in parallel easily. The conclusion that we can draw from all of these explanations is that the problems the brain is solving so easily are immensely parallel ones. They require the processing of lots of different ideas of information, all of them interacting to provide a solution. The knowledge required to solve these problems comes from many different sources, each of it having its own contribution to the final output.

The parallel design of the brain enables to represent and store this knowledge in an accessible way. It is also able to process this knowledge along with the many different stimuli it receives, due again to the parallel nature of its operation. Speed and complexity of one stimulus are not the important factors. Parallelism is what makes the brain ideally suited for all those tasks requiring high connectivity, cooperation and competition of different channels — as speech and vision. This talent is the reason, why computer specialists and designers have become interested in the structure and function of the brain, hoping to get an insight in the basic principles of its mechanism and to apply them to computer systems. Because we do not know exactly how the brain represents high-level information, we can not mimic it. But we know that it uses many slow units which are highly interconnected. Computer models of these basic systems of the brain are called neuro-computers. They should be able to represent knowledge in a parallel fashion and process it in a similar way. One way to realize these neuro-computers is to *simulate* the brain structures in a serial fashion, so that one is not forced to build new computers. A more elegant way is to implement them on parallel systems — or to modify an old proverb — to have the right architecture for the right job. Here, optics come into play. They are the most promising field for implementations of parallel neuro-systems, being in perfect agreement with the structures required for highly parallel systems.

In the following chapters, I show how the study of such real neural systems has allowed to model the parallelism that exists in the brain. Moreover, some other useful features of real neural systems have been incooperated into the new, artificial network concepts. I will give an insight in them either.

Perhaps the most important of these features is that the brain is able to learn — it can teach itself. The principle behind this mechanism is *learning by example* — it is the way in which children pick up speech, learn to eat, drink and write as well as to develop an own set of standards and morals. In contrast, conventional computers are not able to learn or to rearrange algorithms at all. Conventional computer systems usually are directed by long and complicated programs, which give specific instructions for every stage in its operation. Each step of the way towards the solution has to be worked out. It is fairly obvious that the human brain does not work this way at all, since when we come to write such programs, it takes us hours of patient and careful work to write down in detail exactly what we mean in a form that the computer can understand it. Small mistakes in that flow of instructions can cause all sorts of unexpected things to happen. They are known as bugs — making the hell out of a computer scientist's life. Indeed, these mistakes are recognized as being immensely difficult to avoid and most of the large programs have many bugs in them. Therefore, wouldn't it be a nice dream if instead of having to develop a large program with many millions of lines of instructions without bugs to do a certain task, you could simply let the computer system observe some examples of the task and — who knows — may find itself an even better way of doing it than you would? Such a system would probably have also bugs in its solution algorithms initially, so that it is occasionally doing something wrong — but it would learn from its mistakes and not repeat the error.

## 1.2   The Brain: Structure and Learning

The human brain is one of the most complicated things that scientists have studied in detail and is in the whole only poorly understood. Satisfactory answers to the fundamental questions as "What is my mind?" or "How do I think?" are still missing. Nevertheless, we have now a basic understanding of the operation of the brain. It contains approximately ten thousand millions ($10^{10}$) basic units, which are called neurons. Each neuron is connected to about ten thousand ($10^4$) others. The *neuron* is the basic unit of the brain. In comparison with computer technology it can be named a single analog logical processing unit. Generally, two main types of neurons can be distinguished. Local neuron cells that have their input and their output connect along small distances (about 100 microns) and output cells that connect different regions of the brain to each other, e.g. connect the brain to a muscle or connect sensory organs with the brain. Although the operation of the neuron is a highly complicated one, which is not fully understood on the microscopic, molecular level, the basic details are already relatively well known. Each neuron accepts many inputs, which are all added up in some fashion. If enough active inputs are received at once, then the neuron will be activated — brain scientists call this a "firing" neuron. If the number of activation inputs is too low or — what is the same — if the real, present activation level or *potential* is too low, then the neuron will remain inactive, in its "quiet" state. A representation of the basic features of the net of neurons in the brain is shown in fig. 1.1. The body of the neuron is called the soma. Attached to it are long, irregular shaped filaments, the dendrites, which act as the *connections* through which all the inputs to the neuron arrive. The cell body is able to perform more complex functions than simple addition on the inputs they receive, but for our purposes considering a simple summation is a reasonable approximation. Another type of nerve process attached to the soma is the axon, an electrically active one, which serves as the output channel of the neuron in contrast to the dendrites. Axons always appear on output cells, but are often absent from neurons in interlayer positions, which have both inputs and outputs on dendrites. The axon can be regarded as a *nonlinear thresholding device*, producing a voltage pulse or activity potential only when the testing potential within the soma rises above a certain critical threshold value. In detail, the action potential is composed of a series of rapid voltage spikes, lasting about one millisecond ($10^{-3}$ s). The mechanism of the axon is therefore an "all-or-nothing" principle, deciding whether a signal is going further on into the net or not. The coupling center of the axon with the

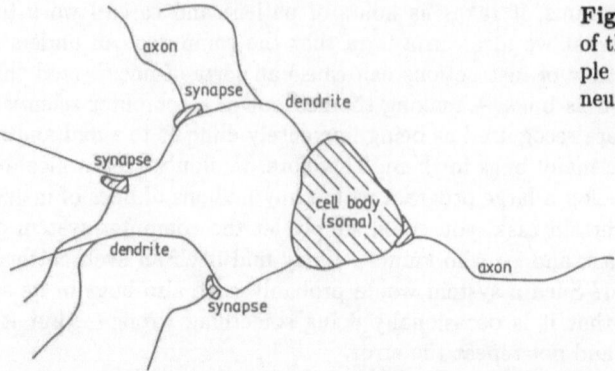

**Figure 1.1:** Representation of the basic features of a simple biological neural net of neurons in the brain.

dendrite of another cell is called the synapse. There is no direct link across the junction; rather it can be regarded as being a temporary chemical junction. The synapse releases chemical substances called neurotransmitters when its potential is raised sufficiently by the action potential. It may take e.g. the arrival of more than one activation potential before the synapse is triggered. The neurotransmitters that are released by the synapse diffuse across the gap between axon and dendrite and chemically activate gates on the dendrite, which in turn allow charged ions to flow when the gates are open. It is this flow of ions that alters the dendrite potential and provides a voltage pulse on the dendrite, which then is conducted along into the next neuron body. The number of gates that will be opened on the dendrite depends on the number of neurotransmitters released. Moreover, it may also appear that some synapses excite the dendrite they affect, while others serve to inhibit it. This corresponds to altering the local potential of the dendrite in a positive or negative way — it acts as a bipolar system.

In biological systems as the structure of the brain in fig. 1.1, learning is defined as the modifications that are made to the effective coupling between one cell and another at the synaptic junction. Biochemically, the mechanism for achieving this seems to facilitate the release of more neurotransmitters. This has the effect of opening more gates on the dendrite and thus increasing the coupling effect of two cells. The adjustment of coupling so as to favourably reinforce good interconnections is one of the most important features of artificial neural net nodes, as is the effective coupling or weighting, that occurs on connections into a neuronal brain cell. The process of learning becomes even more important when we look at the global structure or organization of the brain.

The brain is divided into different regions, each of it being responsible for different functions. The largest parts of the brain which occupy most of the interior of the skull are the cerebral hemispheres, which are layered structures. The most complex of them is the outer layer, known as the cerebral cortex, which is the seat of the higher order functions of the brain — sensory processing, interaction with an environment, awareness. In other words, it is the seat of the characteristics that we generally refer to as indicating intelligence. The cerebral cortex also has a localized structure, because different areas of the cortex fulfil different functions, such as motion control, hearing, speech and vision. For an example of the complexity of these areas let us pick out the visual part. Here, electrical stimulation of the cells can produce the sensation of light. Moreover, specific layers of neurons are sensitive to particular orientations of input stimuli, so that one layer responds maximally to horizontal lines, while another responds to vertical ones. Although much of this structure is genetically pre-determined, the orientation specific layout of the cells seems to be learnt at an early stage. This effect of environmental influence is provided by the fact that e.g. animals which are brought up in an environment of purely horizontal lines do not develop neuron structures that respond to vertical orientations. The formation of those structures is called self-organization, since there is no external teacher to guide the development of the structures. Because self-organization enables structuring and reduction of large amounts of input data, it is also an interesting process for the field of artificial neural networks. The work of T. Kohonen in that field has shown that such *feature maps* can be developed in artificial neural systems as a consequence of simple learning rules.

## 1.3   Electronic Computer Architectures and Learning Principles

Now that we have got an insight in the structure and organization of the brain, we have to compare it to the structure of electronical systems. The ability to learn is generally not possible in a computer machine. Normally, they can do only what they are programmed to do in a sequential, serial way and can not adapt to their surroundings. This is on the one hand an advantage — today's computers are by thousands of times faster at logical inference and numerical computation — but require a high exactness in programming. Without that exactness, faults may arise with incorrigible results, like that U.S. satellite that went off into space without fulfiling its planned mission because of a misplaced comma in a computer program. Therefore, learning could reduce human programming effects. In contrast to computers, humans do not store determined addresses as in random access memories. Rather, they learn by association. How these associations can be performed in electronics will be described in the next section.

As an example, let us look at how a computer would solve the problem of *classification* — which is to determine which of a number of classes is most representative for an unknown static input pattern containing $N$ input elements. In a speech recognizer the inputs might be the output envelope values from a filter bank spectral analyzer sampled at one time instant and the classes might represent different vowels. In an image classifier the input might be the gray scale level of each pixel for a picture and the classes might represent different objects. The electronic, traditional classifier contains two stages as shown in fig. 1.2a. The first computes matching scores for each class and the second selects the class with the maximum score. The inputs to the first stage are symbols representing values of the $N$ input elements. These symbols have to be entered sequentially and decoded from the external symbolic form into an internal representation, which is suited for performing arithmetic and symbolic operations in the computer's own language. An algorithm then computes a matching score for each of the $M$ classes which indicates how closely the input matches the exemplar pattern — the most representative pattern of each class. Here, to generate input patterns from each of the $M$ possible exemplars, the behaviour of the machine does not have to be so rigid and determinated as it is commonly regarded in many situations. Probabilistic models are often used to model the generation of input patterns from exemplars. The matching score in that case represents the likelihood or probability that the input pattern was generated from each of the $M$ possible exemplars. In those cases, strong assumptions are typically made concerning the underlying distribution of the input elements. The parameters of these distributions can be estimated using training data as shown schematically in fig. 1.2. If e.g. multivariant Gaussian distributions are used, the algorithms for computing matching scores become

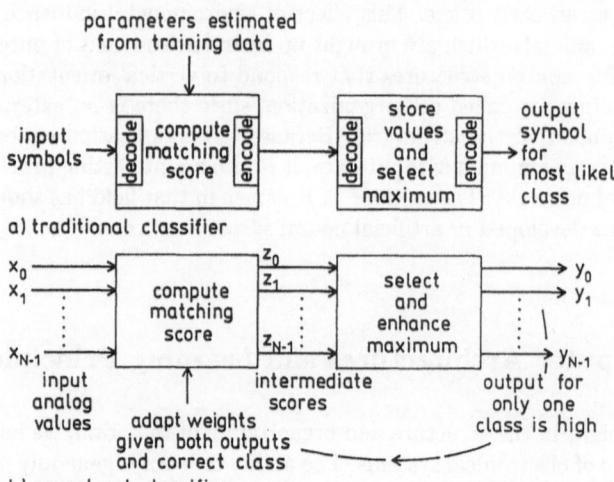

**Figure 1.2:** Block diagram
of a traditional (a) and a
neural net classifier.

very simple, showing that the quite natural feeling that complicated programs and systems are required to achieve a decisional behaviour such as a classification is misleading. Here, no complicated and deterministic program is responsible for achieving a behaviour which we still would classify as intelligent. To be transferred into the next stage, matching scores are coded back into the symbolic representation and passed sequentially forward. In the second stage, they are decoded once again and the class with the maximum score is selected. A symbolic representation of that class is then sent out to complete the classification task. Perhaps the most surprising feature of that system is that even though it is based on a lot of algorithms that complicate matching, it can learn by training data (which may be examples of right input patterns) to respond correctly to a classification task. However, at the same time it is obvious that various data, time and storage place consuming operations are necessary to perform a task which can be solved by a three pound human brain in a much easier and more elegant way.

## 1.4   Learning in Neural Networks

To get an insight in the possibilities to realize a more brain-like net, let us stay for a moment at the example of classification. How would a brain-like structure be able to solve that task? Such an adaptive neural net classifier is shown in fig. 1.2b.

Here, input values are fed in parallel to the first stage via $N$ input connections. Each connection carries an analog value which may take on two levels for binary inputs or may vary over a large range for continuously valued inputs. As in the case of the electronic version, the first stage computes matching scores and outputs them in parallel to the next stage over $M$ analog input lines. This stage reveals the most important difference to the electronic version of a classifier. Here, very simple processing elements working in parallel fulfil the task instead of a single complicated one doing the whole job serially. In the next stage, the maximum of the $M$ values is selected and enhanced. This second stage has one output for each of the $M$ classes. When the classification is complete, only that output corresponding to the most likely class will be on strongly — we may call this stage "on" or "high"; other outputs will be "off" or "low". Note that in this design, outputs exist for every class and that this multiplicity of outputs must be preserved if further processing stages are used. In the simplest classification system, these output lines may go directly to lights with labels that specify class identities. In more complicated cases, they may go to further processing states where other inputs are taken into consideration, too. Up to now, only the parallelism of the system makes it different from an electronic one. The most important advantage of the system is that if the correct class is provided, then this information can be fed back to the first stage of the classifier to adapt weights using a learning algorithm (see fig. 1.2). Adaptation will then make a correct response more likely for succeeding input patterns that are similar to the current pattern. Therefore, in classification processes adaptation is a learning process and it is again that function that makes the system well-suited for all tasks where decisions between several solutions are required.

## 1.5   The Differences: Artificial versus Biological Neural Networks

To summarize the results of the precedent chapters, we have seen that the brain is excellent at performing many of the tasks that we would like computers to perform, such as vision, speech, learning by example and so on. We have also seen that it requires a rather complicated structure to achieve electronically such a computer system which is able to solve decision problems — whereas the brain is structured in such a way as to manage to accomplish these complex tasks with an apparent minimum of effort.

This is due to its highly developed structure — that of a massively parallel system, in which many simple processing elements share the job of working out what is going on, rather than trying to make one fast node do all the work. This divison of labour has another advantage. Since many processing units or neurons are involved at any of the time, the contribution made by a single one is not too important. So if it happens for one neuron to go wrong, it is unlikely to affect the others in a significant way. This distribution of work is called *distributed processing* and allows to tolerate errors without producing nonsense at the output. Indeed, because the brain can learn, it is able to adjust to the permanent loss of one of its neurons and can bring in new ones. This ability to function with only some of the processing elements working correctly is known as *fault tolerance*. It is a vital feature of the operation of the brain because every day a few neurons die as part of the natural occurence of events. Even in the case of a continuing, severe damage, parallel distributed systems exhibit what is known in computer circles as graceful degradation, what indicates that the performance of the system slowly falls off from a high level to a reduced one, without dropping dramatically down to zero. In contrast, if a single unit in a computer — which works hard to calculate lots of things quickly enough to provide a correct output — breaks down, there is no hope of obtaining a sensible answer — a situation computer specialists sometimes obtain after uncountable hours of programming.

Computers are very different in structure, too. Rather than being composed of many millions of relatively slow, highly interconnected processing elements like the brain, they consist of one (or on modern transputer machines, maybe a few more) exceptionally fast processor, which is capable of performing many million calculations per second. This is the reason for the traditional computer's advantage in simple, repetetive actions like adding numbers, but makes it at the same time poorer at the task of processing the vast quantities of different types of data that a vision system requires. They also suffer from their serial approach instead of a parallel distributed one. As a consequence, they are intrinsically intolerant to faults. If the processor in a computer breaks, that may cause the screen of the monitor to go black, or worse, an aircraft may crash or all the lights of a city may go out — the consequences may be far-reaching and difficult to correct or even anticipate. These problems have led to the current interest in developing systems that adjust the principles developed by the evolution of the brain — that is, formulated in a boorish way, keep the unit processing cell simple, keep it joined up and have lots of it to share the load.

## 1.6   Advantages of Optical Neural Network Implementations

The question that arises immediately at that point is: which system and structure is well-suited to implement these characteristic features of the brain? Electronics seem to fail for it, because the design of integrated circuits with highly interconnected, complex data pathways between the layers of units or nodes and the implementation of adaptive weights in integrated circuit technology are both difficult to realize and extremely expensive. However, there are actually several approaches to realize weights electronically, which I want to mention at least: analog, digital and fixed weights. Analog techniques for creating modifiable weight connections include variable resistors, field emitting transistor gate voltage control and capacitive storage methods. The major drawback of most of these methods is that they require large amounts of silicon space, resulting in only a very low density of neural nodes available on a chip. Digital techniques use addressable registers to store and modify the weights. This technique is useful but is again limited by the space required for multiplying and adding units on the silicon. The third alternative avoids the problem of modifying weight values by only allowing the value of the weights be set once. The idea behind this is to learn the correct weight matrix in a simulation environment, and then load this into the chip permanently. All these methods also suffer from their difficulties in implementing the other important feature of neural networks, namely high interconnectivity. In their sum, these restrictions lead to the conclusion that in many application environments integrated circuit technology is just not suitable. Other technologies are advancing at a remarkable rate, however, and these difficulties in implementation will not hinder progress of neural network chips for too long. In contrast, optics — especially nonlinear opticcs — represent a rapidly developing information technology area that has several advantageous features for overcoming the limitations described for electronic realizations.

Optical computing uses light to transport information instead of electric signals. That approach holds three major advantages for computation in general and artificial neural networks in particular. The first is the inherently high speed achievable — data can flow at the speed of light.

Optical switches can therefore go much faster than electronic ones, where the transmission time between e.g. neurons is limited by their distance or — in the case of the high speed electronic machines where additional power is used to provide speed and elements are located close to one another — by the mechanism of heat transfer out of the system. Another advantage of optics is the high bandwidth that enables more information to be carried. While electronic communication along wires requires changing of a capacity that depends on its length, optical signals in fibers, optically integrated circuits or free

**Figure 1.3:** A lens as a simple device in optics that offers an immense interconnectivity.

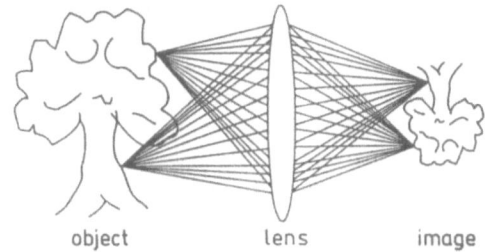

object                    lens                    image

space propagation of light do not have to change anything and are therefore inherently fast. However, for neural computing in particular, the most important reason is due to the fact that one beam of light can cross another and emerges completely inaffected by its encounter whithout distorting the information carried on it, whereas two electrical wires can not. This effect is commonly known as the *principle of superposition* in optics. It is possible because the information carrying elements of light — the photons — are uncharged particles, which do not interact with one another as readily do electrons. This opens up the potential for massive interconnections within a small place, especially in the case of highly parallel systems, where loops of connections are necessary. In the case of electronics, loops of that set will generate noise voltage spikes whenever the electromagnetic field through the loop changes. Furthermore, high frequency or fast switching pulses will cause influence or interference in neighbouring wires. Even when they are routed in silicon, a certain distance between neighbouring pathways has to be established in order to avoid interference. In contrast, signals in adjacent optical fibers or in optical interconnection channels do not interfere with each other nor do they pick up noise due to loops. The third advantage of optics is that images or arrays of light points — so-called pixels — may be handled in parallel. Consequently, a simple lens can be thought of as a powerful interconnection device, as it is schematically shown in fig. 1.3. The image it forms is a collection of rays of light reflected from different points of the object. Therefore the lens effectively connects countless of these rays from the object to the image. These reasons have led to investigations how to explore optics for computing purposes and how to realize the equivalent — and on the same time more adequate and powerful — components of electronic systems with optical devices.

Storage in optical systems can be accomplished using holograms — a sort of three dimensional photograph containing enough information to reconstruct an image of a solid object. The physical principle underlying the recording and reconstruction of these holograms will be explained further in chapter 3.3. Holograms can also be used as distributors by directing light that falls onto them in different directions depending on the inital incident angle of the beam on the hologram. The amount of information that can be stored on a hologram is high, since a single one can hold and discriminate many images. The methods of discrimination of different images stored in a hologram at the same time will be explained in detail in chapter 4. Optical switches, which use a small signal to control a larger one for producing gain — as in a transistor — can also be made optically. One approach is to affect a crystal structure with an electric or magnetic field, which alters its optical properties (in most cases the refractive index of the material) and so affect incoming light differently. These devices work at a speed of around $10^{-10}$ seconds, compared to the best electronic transistor switching times of down to $10^{-12}$ seconds. Other switching devices use nonlinear crystals that alter the amount of light that they transmit depending on the intensity of the incoming beam. Here, best optical switching times are about $10^{-14}$ seconds using e.g. the nonlinearities in Gallium Arsenide devices. Thus, these devices have the speed edge even over electronic ones (see chapter 5 and 6).

To summarize, optics have the most promising potentialities of realizing neural network models which may be able to solve the sort of problems, humans deal with so easily. Therefore, I will focus my attention in this book on optical neural network implementations, composed of linear and nonlinear optical devices for storage, switching and decision purposes. To understand these systems in detail, an insight in both areas, the functional principles of artificial neural networks and in basic concepts of optics — especially nonlinear optics — is necessary. It will be provided in the following chapters, beginning with the basic models of neural networks. Because both — optics, especially nonlinear optics, as well as neural computing — are two forthcoming, rapidly expanding fields, this book will focus on the major concepts, combined with some of the most actual examples of applications, chosen mainly out of the field of photorefractive optics. I will not be able to describe all the fascinating new concepts in both areas throughout the book, but Further Reading hints for those who want to plonge deeper into the field it are provided at the end of each chapter.

 ## Summary

The most important features of the structure of the brain are:
- The brain is a parallel operating system.
- Its processing is distributed over a large number of simple processing units.
- The basic processing unit is called neuron. About $10^{10}$ neurons are each connected to $10^4$ others.The neuron is operating by firing a pulse down the axon when sufficient input is received from the dendrites.
- The connections between neurons are realized via chemical junctions, the synapses.
- Learning increases the efficiency of a synaptic junction to get the desired net output.

Optics is ideally suited to realize large, highly-interconnected neural network models because of
- the high speed of optical information processing (speed of light).
- the high bandwidth of transferred signals.
- the high interconnectivity without a transfer or carrier medium.
- the possibility of realizing the most important devices for neural networks, interconnection units and thresholding elements in a simple and well-performing way.

 ## Further Reading

1. R. Beale, T. Jackson, *Neural Computing: An Introduction*, A. Hilger, Bristol (1992). An instructive, detailed and easy-to-read introduction into the field. Many of the concepts in the first two chapters of this book originate from that introduction.

2. P.D. Wasserman, *Neural Computing: Theory and Practice*, Chapman & Hall (1989).
   A well-written introduction with a lot of examples.
3. R.P. Lippmann, *An Introduction to Computing with Neural Nets*, IEEE ASSP Magazine, April 1987.
   An excellent and compact overview of the whole area. The mathematical algorithms in that book have been formulated following the instructions in that paper.
4. J.L. McClelland, D.E. Rumelhardt, *Parallel Distributed Processing*, Vol. 1-3, MIT Press (1986).
   This book covers the foundations and many of the current approaches and models.
5. R. Rosenblatt, A. Anderson, *Neurocomputing: Foundations of Research*, MIT Press (1988).
   The most extensive reference book of most of the major papers in that field.
6. E.R. Kandel, J.H. Schwartz, *Principles of Neural Science*, Elsevier (1985).
   A well-written overview over the structure and field of neural networks.

# 2 The Principles of Neural Networks

This chapter wants to give an overview over established ideas, algorithms and methods of neural computing. It contains the basic definitions necessary for further reading. Beginning with the model that is most equivalent to the brain's structure — the single neuron — some simple examples of binary and analog network models are presented in their structure and in their principles. These examples have been chosen because they are fairly well-suited for optical implementations. Therefore, all following examples of optical neural net architectures will refer to the basic explanations in the present chapter. For the reader who is already familiar with the principle functions of several net structures, some of the explanations and algorithms might already be well-known. Then this chapter might be briefly read to familiarize with the foundations of the subject or even skipped for a first reading.

To fill the following explanations of basic network functions with life, I have chosen again the example of the first chapter — pattern recognition and pattern classification. Pattern recognition in all forms becomes more and more the area of choice of a lot of optical scientists for promising applications of neural networks. Currently, it is a large area of computer science — most image comparisons, enhancements and modifications are done digitally on the computer — but analog optical models with coherent or incoherent laser sources, the use of liquid crystal displays as image generators and holograms as image storage devices are coming up. One example of an often required task in pattern recognition — classification of inputs to a fixed series of examples — has already been mentioned in the introductory chapter. Here, I want to look at it once again, but this time in a more general way in order to extract the principle mechanism for simple neural network systems out of it.

## 2.1 Classification of Neural Networks for Pattern Recognition

A pattern recognition system can be classified as a two stage device — the first stage is feature extraction, the second one is classification — where a feature is defined as a measure and taken on the input pattern that is to be classified. Typically, one will be looking for features that will provide a definitial characteristic of the input type. A trivial example may be the characterization of letters: to distinguish the letter P from the letter B, we would need to count the number of vertical lines and half circles in the character. In reality, feature extraction is a rather complicated problem and often the most difficult one of the recognition task. The second stage, classification, is supplied with the list of measured features. Its function is to map these input features into a classification state. Here, the classifier, which relies typically on distance metrics or probability theory, must decide which type of class category the input matches most closely.

Classification nets can perform three different tasks. First, they can identify which class best represents an input pattern. This task is inherent to the system. In real application situations, inputs are often corrupted by noise or some other process. The problem then

**Figure 2.1:** Three different forms of associative memories.

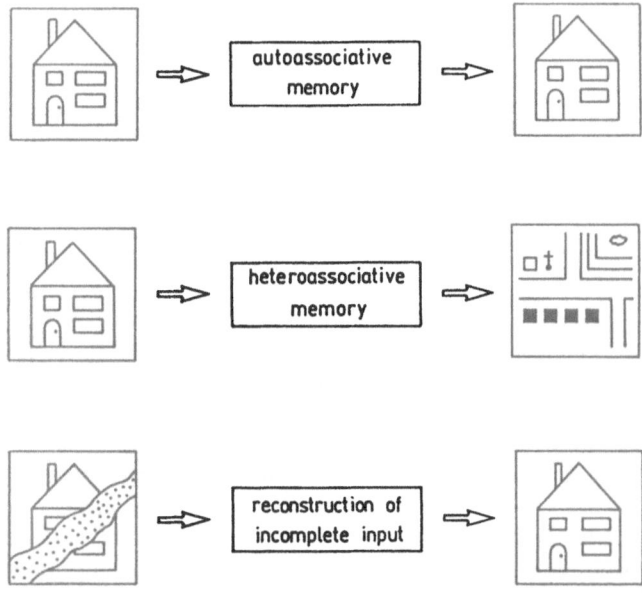

is to decide which pattern is the original one. Second, classifiers can be used as *associative memories*, where the class exemplar is required and the input pattern is used to determine which exemplar to produce. Associative memories appear familiar to us since they correspond to the way in which our own memory operates. In associative memories, the aim of the presentation of one set of input signals is to achieve the recall of another set of signals from the memory. The idea of accessing the memory on the basis of the structure of the input pattern gives rise to the name *content-addressable-memory* (CAM). Such a memory is able to recall the complete description of an object despite having only parts of the input available. This tolerance to partial input description makes associative memories useful for pattern completion tasks and for closest match recognition. An example of such a memory is bibliographic retrieval of journal references from partial information.

So far, I have discussed the association between the "input" and the "output", without reference to any particular type of input or output. In fact, there are two types of association depending on the nature of the two patterns to be associated. A recall, where a pattern is associated with itself (e.g. when an incomplete pattern in the input will result in the recall of the complete pattern) or in other words where the datum itself acts as a pointer to either itself, is called *homoassociative* or *autoassociative*. If the input pattern is taught in association with a different output pattern, the presentation of this input will cause the corresponding pattern ot appear on the output. Thus, when the datum acts as a pointer to other stored data, we have the case of a *heteroassociative memory*. These two association possibilities have been scetched in fig. 2.1 in chapter 2. All types of association are schematically shown in fig. 2.1. A third task these systems can perform is to *vector quantization* or *cluster* the inputs into a small number of clusters. These vector quantizers are used in speech and image recognition applications to compress the amount of data that must be processed further on without losing important information.

Now let us have a look at how to model the basic computing elements of an artificial neural network in order to realize one that is able to fulfil the task of classification prop-

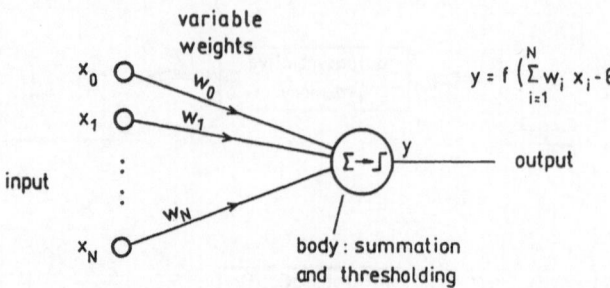

$$y = f \left( \sum_{i=1}^{N} w_i \, x_i - \theta \right)$$

**Figure 2.2:** The basic neuron. It performs a weighted summation, followed by an internal thresholding.

erly. Therefore, we have to combine the principles of our pattern recognition mechanism together with the function of our well-working biological example — the brain. The idea here is to rebuild certain properties of our real nervous system — exaggerating some features whilst others are ignored. However, I do not intend to use them as detailed copies of any real nervous system, I am only dealing with some basic functions of the brain. This fact is often emphasized in literature as the idea of computer "brains" is grabbing the imagination, making human thinking unnecessary. In the following neural network models, I am not attempting to build computer brains, nor I am trying to mimic parts of real human brains exactly — rather I am aiming to discover the properties of models that take this behaviour from extremely simplified versions of natural neural systems, usually at a massively reduced scale as well. Whereas the brain has at least $10^{10}$ neurons, each connected to $10^4$ others, we are concerned with maybe a few hundred neurons at all, connected to a few thousand input lines — later on we will see, that only optics can reach the same high connectivity as biological networks, but still then the node is only a poor copy of what the brain is able to realize.

This means that in our simplified attempt to copy the most important features of the brain, we have to add up inputs and to produce an output if the sum is greater than a certain value, known as threshold value. The easiest way is doing that straightforward, leading to the simplest artificial neuron, called a node. As shown in fig. 2.2, it performs a weighted summation of its inputs and compares this to the internally fixed threshold value. If the value is exceeded, the output of the neuron is open, if not, it stays off or gives a negative output, showing that there is only a small agreement of the input with the example. Such a basic neuron is completely characterized by its internal threshold or offset-value $\Theta$ and the sort of nonlinear mechanism that is used to perform the comparison. There are several possibilities of nonlinear functions that are able to realize a successful comparison. Examples of nonlinearities are shown in fig. 2.3. They can be bipolar, processing positive as well as negative outputs (fig. 2.3a,b) or unipolar, referring to the model of the neuron being "on" or "off" (fig. 2.3c,d). The function of these nonlinearities defines the "strictness" of the decision and the speed of learning. They may be hard limiters (fig. 2.3a), logic thresholds (fig. 2.3c) or have a sigmoidal form (fig. 2.3b, d). All complicated neuronal network models are operating with this basic node element. Their difference arises from the specific node characteristics, as e.g. the different nonlinear threshold functions, the net topology or structure and the learning rules. These rules specify an initial set of weights and indicate how weights should be adapted during use to improve the performance of the net. Both thresholding or classifying programs and training rules are topics of actual research in order to realize complex neural network structures for a variety of different problems. With these characteristics in mind, all

**Figure 2.3:** Different nonlinearities for application as decision elements in the basic neuron. a), b) are bipolar functions, c), d) are unipolar ones. a), c) are hard-limiting functions, b), d) are soft thresholders having a sigmoidal form.

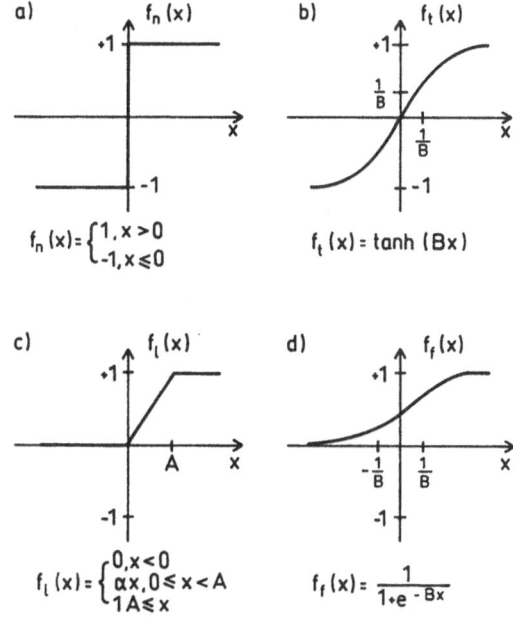

currently available neural net models can be classified in the following arrangement (see fig. 2.4). Nets can be first divided into those working with binary or continuously valued inputs. Below this, nets are divided in those trained with and without supervision. Among the nets that are trained with supervision, the Hopfield net and the perceptron can both be used as associative memories or as classifiers. They differ in the form of the net structure to obtain the correct class of the input: whereas the perceptron relies on a straightforward structure of weighting and thresholding — sometimes in multiple layers arranged one behind the other — the Hopfield net is a fully-connected network, where each node is connected to every other node in such a way that the first output is taken as the new input, which in turn produces a new output and so on. Nets trained without supervision, such as Kohonens feature map, can be used as cluster forming units. No information concerning the correct class is provided during training to these nets. That is why they are often called self-organizing systems, too.

## 2.2 Learning in Simple Neurons — the Single Layer Perceptron

To obtain a network that can do anything useful, we need a mechanism being able to *train* the network. It is this capability to learn that makes neural networks so useful for various complex tasks as pattern recognition and class discrimination. Moreover, in order to keep our model in an easy frame to understand its principle still well, we want to find a learning rule as simple as possible. Again, we can get an inspiration for that mechanism looking at real biological neural systems. Children are praised for doing something well, e.g. a mathematical problem, and are scolded for doing nonsense or dangerous activities as plugging their fingers into a wall plug. Hence, education is based on the principle that

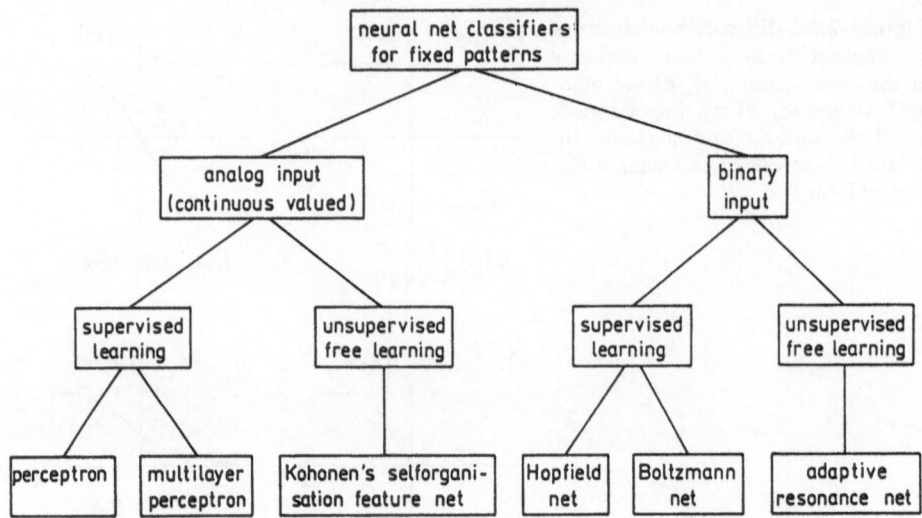

**Figure 2.4:** Classification arrangement for different neural net architectures.

good behaviour is reinforced, whilst bad behaviour is repremanded. How can we include that concept of learning within a simple design of the model neuron? The guiding principle is to allow the neuron to learn from its mistakes. If it produces an incorrect output, we want to reduce the chances of that happening again — if it comes up with the correct output, then we need do nothing.

Thus to train the network, we will begin with random weights on the neuron's input lines, corresponding to a static state in which it "knows" nothing. After the presentation of an example, we increase the weights and the active inputs when we want the output to be active and decrease them when we want the output to be inactive. One can achieve this by adding the input values to the weights when we want the output to be on and subtracting the input values from the weights when we want the output to be off. This already defines the complete learning rule. It is important to note that only those inputs which are active at the time will be affected: weights are not adjusted on input lines which do not contribute to the incorrect response, since each weight is adjusted by the value of the input on that line, $x_i$, which would be zero. This is sensible since the inactive ones do not contribute to the weighted sum. Therefore, changing them will not affect the result for the particular input in question, but may well upset what has already been learnt.

A variant of this simple, but effective learning rule has been proposed in 1949 by D. Hebb and is therefore called Hebbian learning. Its basic principle is the reinforcement of active connections only. Since this form of learning is guided by the knowledge of the result we want to achieve, it is also known as supervised learning (compare with the taxinometry of the different nets, fig. 2.4) and is the dominant one used today in learning models. This learning paradigm is summarized in the algorithm described in Box 1, which can be used to implement a perceptron network on a conventional computer by coding the step in any programming language.

The basic perceptron learning algorithm has been subject to various modifications which have the aim to make the network more robust for "wrong" decisions or learning results. One alteration is to introduce a multiplicative factor of less than one into the

---

  ## Box 1: The Perceptron Algorithm

---

**Step 1: Initialize weights and thresholds**
Set the weight $w_i(t)$, $(0 < i < N - 1)$ from input $i$ at time $t = 0$ to $w_i(0)$ and the threshold $\theta$ in the output node to small random values.

**Step 2: Present new input and desired output**
Present new continuous-valued input $x_0, x_1, \cdots, x_{N-1}$ along with the desired output d(t).

**Step 3: Calculate actual output**

$$y(t) = f_h \left( \sum_{i=0}^{N-1} w_i(t) \cdot x_i(t) - \theta \right) \qquad (2.1)$$

**Step 4: Adapt weights**

$$w_i(t + 1) = w_i(t) + a(t) \cdot x_i(t) \quad \text{for} \quad 0 \leq i \leq N - 1 \qquad (2.2)$$

with

$$a(t) = \begin{cases} 0 & \text{if output is correct} \\ +1 & \text{if output 0, should be 1 (class A)} \\ -1 & \text{if output 1, should be 0 (class B)} \end{cases} \qquad (2.3)$$

Note that the weights are unchanged if the correct decision is made by the net: $w_i(t+1) = w_i(t)$.

**Step 5: Repeat by going to step 2**

---

weight adaptation term. This has the effect of slowing down the change in the weights making the network take smaller steps towards the solution. In the algorithm, the change takes place in step 4, which may be replaced by the one shown in Box 2.

Another change was suggested by Widrow and Hoff, taking into account that the weights must be changed a lot when the weighted sum is a long way from the desired value, whilst a small alteration is sufficient when the weighted sum is close to that required to give the correct solution. The corresponding learning algorithm is called the Widrow-Hoff delta rule and calculates the difference between the actual weighted sum $y(t)$ and the required output $d(t)$ — the error $\Delta = d(t) - y(t)$. Weight adjustment is then carried out proportional to that error. This takes care of the addition or subtraction, since if the desired output is 1 and the actual one is 0, $\Delta = +1$ results in an increase of the weights. Conversely, if the desired output is 0 and the actual one is $+1$, $\Delta = -1$ and thus the weights will be decreased. Note that the weights remain unchanged if the net makes the correct decision, since $d(t) - y(t) = 0$ in that case. Again, the learning algorithm step 4 has to be modified to achieve weight adaptation via the Widrow-Hoff-rule (see Box 3).

A last alternative proposed is to use inputs that are not 0 or 1 (binary or unipolar), but are instead -1 or +1, known as bipolar. Whereas unipolar binary values mean that input lines with 0's on them are not trained, bipolar values allow all the inputs to be trained each time. This simple alteration speeds up the convergence process and is therefore preferred in large and much more complicated networks.

Although the perceptron mechanism is useful for simple nets, there are limitations

## Box 2: The Perceptron Algorithm — Version 2

$$w_i(t+1) = w_i(t) + \eta a(t) \cdot x_i(t) \quad \text{for} \quad 0 \le i \le N-1 \tag{2.4}$$

and

$$a(t) = \begin{cases} 0 & \text{if output is correct} \\ +1 & \text{if output 0, should be 1 (class A)} \\ -1 & \text{if output 1, should be 0 (class B)} \end{cases} \tag{2.5}$$

where $\eta$ is a positive gain fraction less than 1 that controls the adaptation rate.

## Box 3: The Widrow-Hoff Delta Rule

$$w_i(t+1) = w_i(t) + \eta\,[d(t) - y(t)] \cdot x_i(t) \quad \text{for} \quad 0 \le i \le N-1 \tag{2.6}$$

and

$$d(t) = \begin{cases} +1 & \text{if input from class A} \\ -1 & \text{if input from class B} \end{cases} \tag{2.7}$$

to the capabilities of the perceptron. The most important one lies in the fact that the perceptron, when used for classification, is only able to separate classes linearly into two groups. To examine this limitation in more detail, let us once again choose the example of a conventional pattern recognition task. Such a pattern recognition system can be considered as a two stage device. The first stage is feature extraction, the second one is classification. Here, a feature is defined as a measurement taken on the input pattern that is to be classified in order to provide a definite characteristic of that input type.

The classifier is supplied with the list of measured features. Its task is to map these input features onto a classification state. That means, the classifier must decide which type of class category the inputs match most closely. Classifiers typically rely on distance metrics and probability theory to do this and usually need several different measurements to be able to adequately distinguish inputs that belong to different classes or categories. Mathematically, we may describe this procedure as creating a set of features by making $n$ measurements and call the resulting algebraic notation a feature vector. The dimensionality of the vector, that is, the number of elements in it, creates an $n$-dimensional feature space. The simplest model of a feature space is the two-dimensional example, where two measurements on the pattern define the feature vector. In fig. 2.5, we have chosen as an example the task of distinguishing circles from squares. If we imagine the circles being dinosaurs and the squares beeing mice (as if it isn't rather obvious!), then we might decide that two distinctive measurements that categorize each type are height and weight. If we make a series of height and weight measurements on typical examples of each class, then we are able to plot the range of readings in a two-dimensional Euclidian plane $(x, y)$, that defines our feature space. The problem now is that there are many situations, where the decision between two classes is not as easy as in our example. In

**Figure 2.5:** A two-dimensional Euclidian feature or pattern space with linear classification decision boundary.

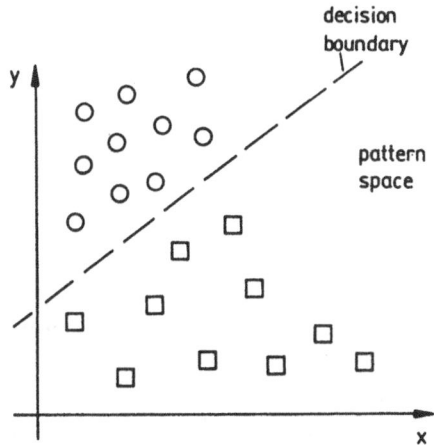

most cases, the decision boundary is not a straight line, as we may draw it in fig. 2.5, but a much more complex region boundary. Imagine e.g. some squares being located in the region of the circles as shown in fig. 2.6. There is no longer any possibility to find a linear region, where only one class can be found. Thus, these problems are called *linearly inseparable*. They are the biggest class of problems in classification. If we look once again at the perceptron convergence procedure presented in box 1, it becomes obvious that the perceptron is trying to find — in the feature space representation — the straight line that separates classes. In all problems, where the perceptron is not able to find any such line, it will fail to solve the problem. In fact, a single-layer perceptron can not solve any problem that is linearly inseparable.

When M.A. Minsky and S.A. Papert demonstrated in 1969 in their book "Perceptrons" — which contained a detailed analysis of the capabilities and limitations of perceptrons — that perceptrons could only do linearly separable problems, the scientific community regarded this as a mortal blow to the area. The majority of the scientists working in this

**Figure 2.6:** Nonlinear separable classification task in the Euclidian pattern space.

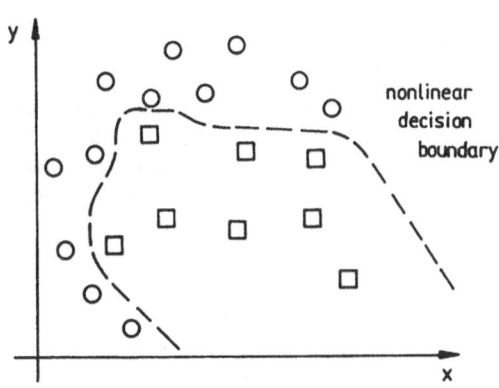

area walked away from the topic and not much happened in the area. It was until 1986 when Rumelhardt and McClelland produced an improvement that fused the perceptron idea with some modern adaptations of the learning procedure and the structure of the network to obtain solutions for the case of nonlinear problems.

 **Summary**

Although the single-layer perceptron is quite a simple model, it has shown great success. It has exhibited the features that are important for a neural classification system and has shown that it is able to distinguish between different classes of objects — provided that they are linearly separable in pattern space. The most important points are

- The perceptron is an artificial neuron.
- It computes the weighted sum of all inputs and outputs an answer, if the sum is larger than the threshold.
- Learning means adjusting the values of the weights of the system.
- Positive reinforcement (the increase of active junctions or Hebbian learning) is the learning approach most often used.

## 2.3  The Multilayer Perceptron

How may we overcome the problem of being unable to solve linearly inseparable problems with our perceptron? An initial approach would be to use more than one perceptron, each set up to identify small, linearly separable sections of the inputs and then combining their outputs into another perceptron, which would produce a final indication of the class. This method, which has been sketched for the case of the combination of two perceptrons in fig. 2.7, seems to be fine on a first examination, but investigations in detail reveal, that such an arrangement of perceptrons will be unable to learn. Therefore, there will be no way of training it to solve a nonlinear separation problem.

Each neuron in the structure still thresholds the weighted sum of its inputs and gives an output on the value 0 or 1. Because the perceptron in the second stage or layer takes as its inputs the outputs from the first layer, it does not know which of the real inputs were on or off. Since learning corresponds to strengthening the connection between active input and active output units, it is impossible to strengthen the correct parts of the network. The reason for that behaviour lies in the fact that the hidden or intermediate layers between inputs and outputs mask off the actual inputs from the output units and prevent the strengthening procedure required. The problem is not a new one in neural network science. For a long time, it is known as the credit assignment problem, since it means that the network is unable to determine which of the input weights should be increased and which should not, and so it is unable to work out which changes should be made to produce a better solution next time.

How to go on further in that dead end street? It is the character of the two-state neuron or the hard-limiting threshold function which gives us no indication of the scale by which we need to adjust the weights. Actually, weighted inputs that just turn on a neuron should

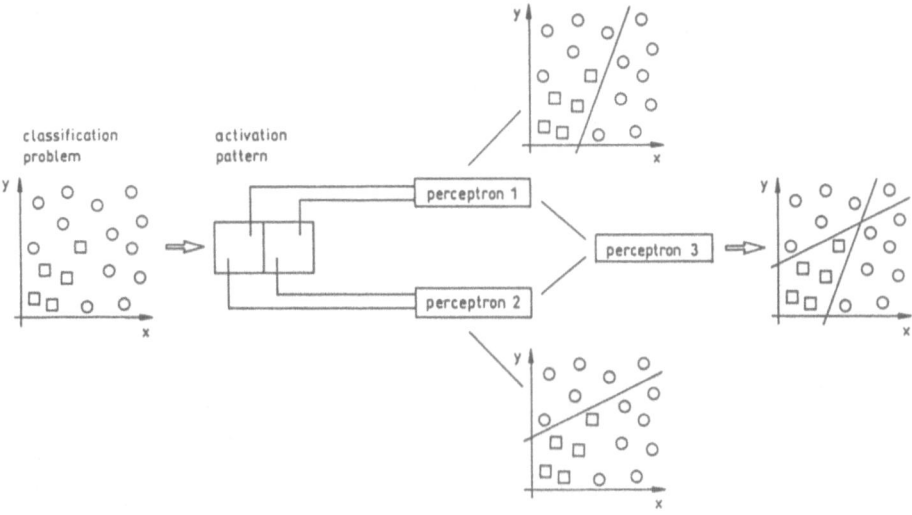

**Figure 2.7:** The combination of two perceptrons to realize a more complex linear classification task.

**Figure 2.8:** Principle effect of the nonlinear sigmoidal threshold function.

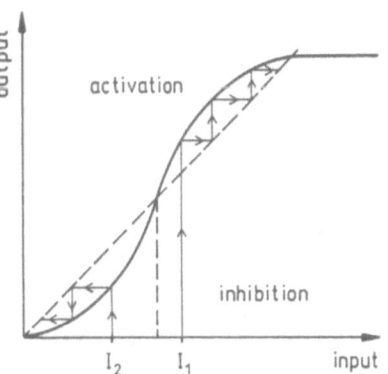

not be altered to the same extent as those in which the neuron is definitively turned on. However, because we are not able to get this information from the net, we can not make a reasonable adjustment. As a result, the hard-limiting threshold function removes the information that is needed if the network should be able to work successfully. The way out around that difficulty is to change the step function as a thresholding element into a smoothly changing nonlinearity. If we use a function that has a slowly sloping region in the middle that allows us to "see" the information on the inputs, we will be able to determine when we need to strengthen or weaken the relevant weights. A couple of possibilities for those thresholding functions have already been shown in fig. 2.3. With that new class of threshold functions, the inputs are not simply on or off, but ly within a range. Because we are using functions that approximate the step function at the extremes of its range, stable output answers are still reached, but now without losing the information that relates the output to its inputs. This behaviour is shown in fig. 2.8. We have to use such a nonlinear threshold function, since layers of perceptron units using linear functions are not more powerful than a suitably chosen single layer. In that case, a single-layer net

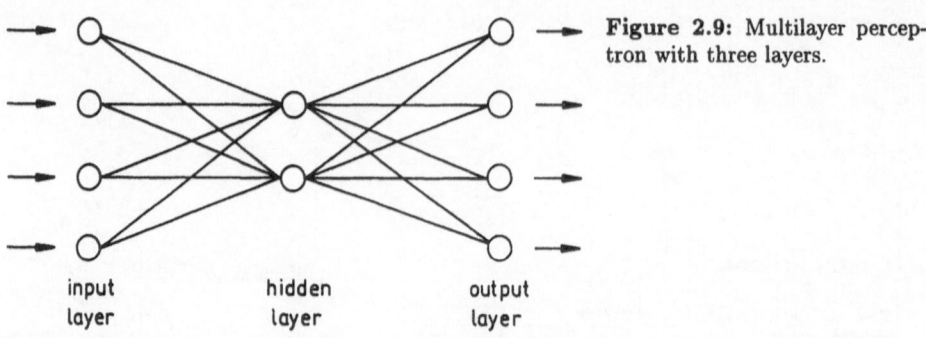

**Figure 2.9:** Multilayer perceptron with three layers.

input          hidden          output
layer          layer           layer

with appropriately chosen weights could exactly duplicate the calculations performed by any multilayer net. This is because each layer would perform a purely linear operation on its inputs, which could be condensed into one operation. To summarize, changing scale in networks by introducing more layers with linear thresholding is a linear operation, since all things are affected by an equal amount. It does not change the quality of the decision rules and therefore is not able to realize other decisions than single-layer systems. Therefore, it is the threshold function itself which has to be changed; if it becomes nonlinear by smoothing its slope, the possibility to transfer the knowledge of the inputs to the next layer becomes real and allows to solve nonlinear decision problems. As an example for such a multilayer system, a system with three layers is shown in fig. 2.9. The layers in between the output and the input layers are not directly connected to the input and the output layers and are therefore called hidden layers. Each unit in a hidden layer and in the output is like a single perceptron unit, except that the thresholding function is a smooth, most often a sigmoidal one. Because the hidden layer and the output layer are therefore the active thresholding layers, systems with three layers overall are often called two-layer perceptrons.

For comparison, the capabilities of perceptrons with one, two or three layers are illustrated in fig. 2.10. The figure indicates in its second column the types of decision regions and shows in the following columns examples of decision boundaries between classes in different problems. Whereas, as denoted already above, a single-layer perceptron forms half-plane decision regions, a two-layer perceptron can form any convex region shape in the space spanned by the inputs. A three-layer perceptron finally forms arbitrarily complex decision regions and is capable of separating any classes. The complexity of the shapes of our decision regions in a three-layer perceptron is limited by the number of nodes in the network, since these define the number of edges that we can have. The arbitrary complexity of shapes that we can create means that we never need more than three layers in a network — a statement which is known as the Kolmogorov theorem.

The operation of a multilayer perceptron is similar to that of the single-layer perceptron, in that we present a pattern to the net and calculate its response. Comparison with the desired response then enables the weights to be altered so that the network can produce a more accurate output next time. The learning rule provides the method for adjusting the weights in the network and is called the *generalized delta rule* or *backpropagation rule*. It was suggested in a first version in 1974 by Werbos. In 1982, Parker published similar results, but it took until 1986, when D.E. Rumelhardt, J.E. McClelland and R. Williams investigated and characterized different multilayer networks with the multilayer perceptron learning rule. Their book "Parallel Distributed Processing"

**Figure 2.10:** Types of decision regions that can be formed by single- and multilayer perceptrons with one and two layers of hidden units and two inputs. Shading denotes decision region for class B.

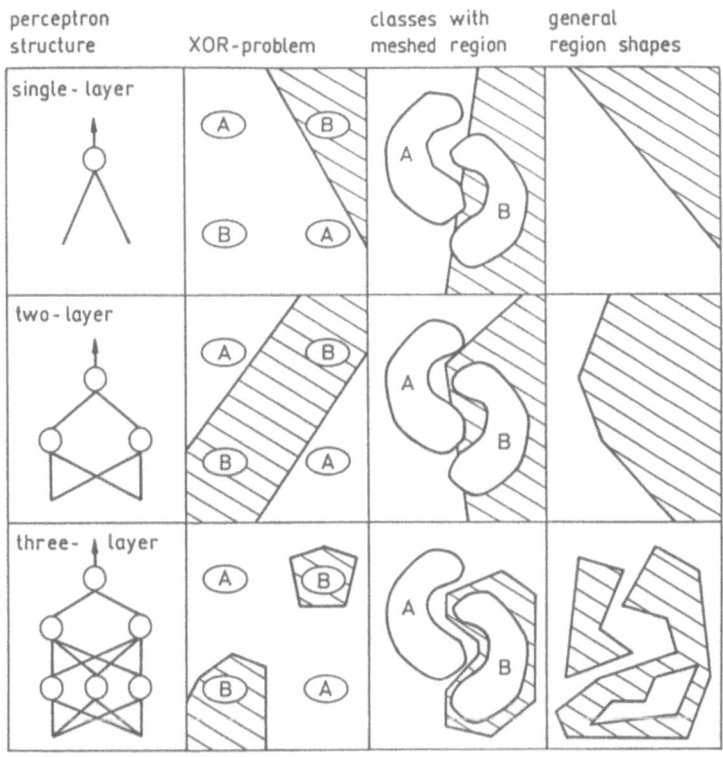

signalled the "renaissance" of the whole area of perceptron research and is still one of the most important books in the field.

### 2.3.1 Learning — the Backpropagation Algorithm

The learning rule is a little more complex than the previous one and can be explained best by looking at how the net behaves when patterns are taught to it. The untrained network will answer to any input pattern with a random output. To adjust the network, we need to define an error function that represents the difference between the network's current output and the correct, desired output that we want the network to produce. Because we prescribe the correct pattern in that algorithm, this type of learning is also called *supervised or prescribed learning*. In order to achieve a successful learning procedure, we have to approach the net output and the desired one, that is, we want to continuosly reduce the value of this error function. This is achieved by adjusting the weights on the links between the units. The generalized delta rule does this by calculating the value of the error function for that particular input and then backpropagating (hence the name!) it from one layer to the previous one. Each unit of the net has its weights adjusted that way so that it reduces the value of the error function; for units on the output layers, this comparison with the desired output is relatively simple, but for units in the intermediate, the middle layers, the adjustment is not so obvious.

The mathematics of the generalized delta rule — as it is summarized in box 4 — have to guarantee, that a particular node should be adjusted in direct proportion to the error in the units to which it is connected. In the following, I will exemplarily derive the mathematics of the multilayer perceptron for a large range of learning procedures. Let me denote $E_p$ as the error function for a pattern $p_i$, $t_{pj}$ the target output for pattern $p$ on node $j$, whilst $o_{pj}$ represents the actual output at the node and $w_{ij}$ is the weight from node $i$ to node $j$.

---

  ## Box 4: The Backpropagation Training Algorithm

---

**Step 1: Initialize weights and offsets**
Set all weights and node offsets to small random values

**Step 2: Present input and desired output**
Present a continuous valued input vector $x_0, x_1, \cdots, x_{N-1}$ and specify the desired outputs $t_0, t_1, \cdots, t_{N-1}$. If the net is used as a classifier, all desired outputs are typically set to zero except for that corresponding to the class the input is from. That desired output is 1.

**Step 3: Calculate actual outputs**
Use the sigmoid nonlinearity

$$f(\alpha) = \frac{1}{1 + e^{(\alpha - \theta)}} \tag{2.8}$$

to calculate outputs $o_0, o_1, \cdots, o_{N-1}$.

**Step 4: Adapt weights**
Use a recursive algorithm starting at the output nodes and working back to the first hidden layer. Adjust weights by

$$w_{ij}(t+1) = w_{ij}(t) + \eta \delta_j x'_i \tag{2.9}$$

where $w_{ij}(t)$ is the weight from hidden node $i$ or from an input to node $j$ at time $t$, $x'_i$ is either the output of node $i$ or is an input, $\eta$ is a gain term, $\delta_j$ is an error term for node $j$.
If node $i$ is an output node, then

$$\delta_i = o_i \cdot (1 - o_i) \cdot (t_i - o_i) \tag{2.10}$$

where $t_i$ is the desired output of node $i$ and $o_i$ is the actual output.
If node $j$ is an internal hidden node, then

$$\delta_j = x'_j \cdot (1 - x'_j) \cdot \sum_{k>j} \delta_k w_{jk} \tag{2.11}$$

where $k$ is over all nodes in the layers above node $j$.
Internal node thresholds are adapted in a similar manner by assuming they are connection weights on links from auxiliary constant-valued inputs. Convergence is sometimes faster if a momentum term is added and weight changes are smoothed by

$$w_{ij}(t+1) = w_{ij}(t) + \eta \delta_i x'_i + \alpha(w_{ij}(t) - w_{ij}(t-1)) \tag{2.12}$$

where $0 < \alpha < 1$.

**Step 5: Repeat by going to step 2**

---

Then the error function can be defined to be proportional to the square of the difference between the actual and the target output for all patterns to be learnt:

$$E_p = \frac{1}{2} \sum_j (t_{pj} - o_{pj})^2 \tag{2.13}$$

The activity of each unit $j$ for pattern $p$, which will be named $\alpha$ for the case of the multilayer perceptron, can be written as a weighted sum

$$\alpha_{pj} = \sum_i w_{ij} o_{pi} \tag{2.14}$$

as in the single-layer perceptron learning algorithm (see boxes 1, 2 and 3). Here, the output from each unit $j$ is the threshold function $f_j$ activity on the weighted sum

$$o_{pj} = f_j(\alpha_{pj}). \tag{2.15}$$

The change in error can be described as a function of the change in the net inputs for a unit as

$$-\frac{\partial E_p}{\partial \alpha_{pj}} = \delta_{pj} \tag{2.16}$$

Using the chain rule,

$$\frac{\partial E_p}{\partial w_{ij}} = \frac{\partial E_p}{\partial \alpha_{pj}} \frac{\partial \alpha_{pj}}{\partial w_{ij}} \tag{2.17}$$

substituting its second term in eq. (2.14) and using $\frac{\partial w_{kj}}{\partial w_{ij}} = 0$ except when $k = i$ (then it equals 1), gives

$$\frac{\partial \alpha_{ij}}{\partial w_{ij}} = \frac{\partial}{\partial w_{ij}} \sum_k w_{kj} o_{pk} = \sum_k \frac{\partial w_{jk}}{\partial w_{ij}} o_{pk} = o_{pi} \tag{2.18}$$

Thus, we are able to rewrite eq. (2.16) as

$$-\frac{\partial E_p}{\partial w_{ij}} = \delta_{pj} o_{pi} \tag{2.19}$$

That equation shows that decreasing the value of $E_p$ means nothing else than making the weight changes proportional to $\delta_{pj} o_{pi}$, e.g. in the following way:

$$\Delta_p w_{ij} = \eta \delta_{pj} o_{pi} \tag{2.20}$$

Now we need to know $\delta_{pj}$ for each of the units of the layer, because then we are able to decrease $E$. Using eq. (2.16) and the chain rule, we can write

$$\delta_{pj} = -\frac{\partial E_p}{\partial \alpha_{pj}} = -\frac{\partial E_p}{\partial o_{pj}} \frac{\partial o_{pj}}{\partial \alpha_{pj}} \tag{2.21}$$

The second term of eq. (2.21) gives with eq. (2.15)

$$\frac{\partial o_{pj}}{\partial \alpha_{pj}} = f_j'(\alpha_{pj}). \tag{2.22}$$

The first term in eq. (2.21) can be evaluated with eq. (2.13). By differentiating $E_p$ with respect to $o_{pj}$, we obtain

$$\frac{\partial E_p}{\partial o_{pj}} = -(t_{pj} - o_{pj}) \tag{2.23}$$

Thus we get

$$\delta_{pj} = f'_j(\alpha_{pj})(t_{pj} - o_{pj}) \tag{2.24}$$

This form is useful for the output units, since the target and the output are both available. However, for hidden units, the targets are not known, and therefore, we should rewrite eq. (2.24) in a form, where the change in the error function is described with respect to the weights in the network. So if $j$ is not an output unit, we can write by the help of the chain rule again:

$$\frac{\partial E_p}{\partial o_{pj}} = \sum_k \frac{\partial E_p}{\partial \alpha_{pk}} \frac{\partial \alpha_{pk}}{\partial o_{pj}} \tag{2.25}$$

$$= \sum_k \frac{\partial E_p}{\partial \alpha_{pk}} \frac{\partial}{\partial o_{pj}} \sum_i w_{ik} o_{pi} \tag{2.26}$$

$$= -\sum_k \delta_{pk} w_{jk} \tag{2.27}$$

By the help of this equation and noticing that the sum drops out since the partial difference is zero except for only one value (compare to eq. (2.18)), we finally obtain for the error function $\delta_{pj}$

$$\delta_{pj} = f'_j(\alpha_{pj}) \sum_k \delta_{pk} w_{jk} \tag{2.28}$$

The function in eq. (2.28) is proportional to the errors $\delta_{pk}$ in subsequent nets, so the error has to be calculated in the output unit first and then passed back through the net to the earlier stages to allow them for altering their connection weights. It is because of this passing back of the error value that the network is being referred to as backpropagation network.

Eqs. (2.16) and (2.24) show clearly that the backpropagation training algorithm is an iterative gradient algorithm designed to minimize the mean square error between two layers. Therefore, it requires continuous, differentiable nonlinearities. A function which is quite like the step function and fulfils that requirement is the sigmoid or sigmoidal logistic function (see fig. 2.3d)

$$f(\alpha) = \frac{1}{1 + e^{(\alpha - \Theta)}} \tag{2.29}$$

Here, $\Theta$ is the so-called "threshold", a positive constant that controls the "spread" of the function.

## 2.3.2  The Energy Landscape

To get a deeper insight in the generalized delta rule, we need a method of visualizing what is going on in the network. The backpropagation function error the multilayer network produces can be interpreted as an energy function. This energy function represents the amount of energy, by which the output of the net differs from the required output. Since the output of the net is related to the weights between the nets and the input applied, the energy is a function of the weights and inputs to the network.

If we consider a very odd network in which we can vary only one weight, this would lead
to a graph of the energy function which may look like the one in fig. 2.11. If we now extend
our thinking so that we can vary two weights, we will have a three-dimensional graph with
two weight axes — the energy landscape. If we add more and more weights, we finally
achieve a multidimensional energy function. However, most of us have difficulties to image
more than three dimensions, because we are not able to draw them. Our understanding
of these higher-dimensional cases is helped greatly by the analogies that we can visualize
easily in the three-dimensional situation. Therefore, I will stay for my explanations at the
three-dimensional situation, always having in mind that the situation of real networks
has many more dimensions.

The energy surface in three dimensions is a rippling landscape of hills and valleys, wells
and mountains, with points of minimum energy corresponding to the wells and maximum
energy found on the peaks. The generalized delta rule now aims to minimize the error
function $E$ by adjusting the weights in the network in such a way that they correspond to
those at which the energy surface is lowest. It does this by a method known as gradient
descent, where changes are made in the steepest downward direction. This is guaranteed
to find a solution in cases where the energy landscape is simple. Each possible solution is
represented as a hollow or a basin in the landscape. These basins of attraction represent
the solutions to the values of the weights that produce the correct output from a given
input. Remember that these basins are actually multi-dimensional, but we can only draw
(or even imagine) them in three dimensions. Learning in such a landscape can easiest be
imagined and visualized by the analogy to a large, stretchy rubber sheet that is initially
flat. The basins of attraction are formed by placing heavy balls on the sheet: the sheet
deforms downwards creating a well. The bottom of the well represents the low energy
solution that the network has learnt.

Many features associated with multilayer perceptrons can be understood easiest if
they are considered in terms of the energy landscape. It is especially useful to look at the
difficulties or problems that are associated with learning in multilayer perceptrons and
finally form the restrictions of that neural network model.

### 2.3.3  Learning Difficulties

Occasionally, the network may settle into a stable solution that does not provide the
correct output — the energy function is in a local minimum as shown in fig. 2.11.

This implies that in every direction in which the network could move, the energy is
higher than at the current position. It may be that there is only a slight "hip" to cross
before reaching an actual deeper minimum, but the network has no way of knowing this,
because learning is accomplished by following the energy function down in the steepest
direction, until it reaches the bottom of a well. Here, there is no direction to move in
order to reduce energy any more. Consequently, a slight temporary rise in energy would
be required to reach a deeper minimum. To minimize these occurencies, there are several
alternative approaches:

- **Lowering the gain term**
  If the rate at which the weights are altered is progressively decreased, then the
  gradient descent algorithm is able to achieve a better solution. If the gain term $\eta$
  is made larger to begin with, large steps are taken across the weights and energy
  space towards the solution. As the gain is decreased, the network weights settle into
  a minimum energy configuration without overshooting the stable position, as the

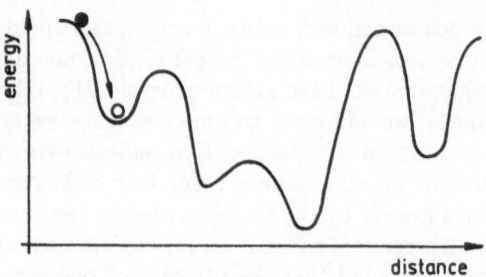

Figure 2.11: Reaching of a local minimum in the energy function in one dimension.

gradient descent takes smaller downhill steps. This approach enables the network to bypass local minima at first, and settle finally in some deeper minimum without oscillating wildly.

- **Addition of internal nodes**
  Local minima can be thought to occur when some disjoint classes are classified as being the same — the corresponding disjoint regions in the energy landscape are then clustered into one. This fault can be minimized in its occurence by adding more units to the corresponding layer, allowing a better recording of inputs and lowering thus the occurence of local minima. Another solution with a similar effect is to use multiple starts with different random weights and a low gain in order to adapt weights in a better way.

- **Addition of a momentum term**
  The weight changes can be modified by adding some "momentum". This is done by introducing an extra term into the weight adaptation equation as shown in the last line in box 4. This extra term then will produce a large change in the weight if the changes are currently large and will decrease as the changes become less. Thus the network is less likely to get stuck in local minima early on, since the momentum term will "kick" the changes over local increases in the energy function and help the weights to follow the overall downhill direction. Beneath the advantage of favouriting the downhill direction, the momentum term is of great help in speeding up the convergence along shallow gradients. The mathematical expression of such a momentum term is given in eq. (2.12).

- **Addition of noise**
  A method similar to the addition of momentum is the addition of random noise, which perturbs the gradient descent algorithm from the line of the steepest descent. This slight "shaking" is often enough to knock the system out of a local minimum and has the big advantage, that it takes very little extra computing time.

All these methods have the overall aim to enable the network to escape from the local hollows in the energy landscape and move into some deeper well that represents a better or the desired result. This concept has also revealed to be important for even all those sorts of neural networks, where classification problems need to be resolved. We will discuss this point once again in the following chapters when we look at Hopfield and Boltzmann nets. In these nets, that are based on the idea of full interconnectivity of all nodes with all others, the same problem of settling down in local minima arises and similar solutions to overcome that problem will be discussed. In terms of the energy representation, all these approaches can be named by their feature of lowering the energy function and therefore lowering or annealing the temperature variable in the energy function in such a way that local minima are avoided — a procedure that is often called *simulated annealing*.

Several models have been developed from the backpropagation algorithm to improve its performance. Among them, the feed-forward model of the Kanerva associative memory (see Further Reading section at the end of that chapter), which is very similar to the Marr and Albus models (see Further Reading) of the cerebellum, have become important for optical neural net implementations. Here, we just want to shortly mention the main features of these models.

These three models (henceforth referred to as MAK) have similar features distinguishing them from three-layer backpropagation nets as well as from mean field theory networks (see section 2.6). The input-to-hidden weights are randomly chosen and fixed, and there are many more units in the hidden layer than in the other two layers. Patterns in the hidden layer can thus be regarded as expanded, sparsely coded versions of the input patterns. The three MAK models have minor difference regarding connections. First, the Marr and Kanerva models have different connectivities between the input and hidden layers. These layers are fully connected in the Kanerva model, but in the Marr model only four to five input units are connected to each hidden unit. We will see in chapter 4 that this limited connectivity makes the Marr model especially attractive for optical implementation since in this case the input devices to the network (as e.g. spatial light modulators) can be fully used without restricting the number of interconnections in the neuronal layer. Second, while the Marr and Kanerva models set the hidden-to-output weights according to the simple Hebbian learning rule (see section 2.2), the Albus model learns those weights via iterative least mean squares, a supervised learning procedure. Without hidden units, the algorithm of backpropagation would also reduce to least mean squares.

The backpropagation algorithm has been successfully used already for a long time for a number of deterministic problems such as the XOR-problem (see e.g. the paper of R.P. Lippmann in Further Reading), on problems related to speech synthesis (see e.g. [1]) and on problems related to visual pattern recognition (see e.g. [2]). With the modifications mentioned above, the multilayer perceptron algorithm has found to perform well and to find good solutions to the problems posed. Therefore, it has become meanwhile one of the most often used algorithms in visual and speech recognition problems.

### 2.3.4 Second- and Higher-Order Perceptron Algorithms

Neural networks carry out a complex mapping from an input space to an output space. The simplest network, the perceptron, was already discussed in section 2.2 and consists of a weighted sum of the inputs followed by thresholding. In considering the limitations of this simple network, especially its inability to solve nonlinearly separable problems, M.A. Minsky and S.A. Papert (see Further Reading) introduced the idea of order. The order $k$ of a network can be intuitively understood as the ability of the network to calculate correlation terms of the form $x_1 x_2 x_3 \cdots x_k$, where $x_i$ are the input vectors. They showed that many — or indeed most — interesting problems are of order greater than one, and thus could not be solved by the simple first-order perceptron.

There are two possible solutions to this problem. The first is to cascade the output of a first-order network into a second network to form a multilayer structure, combined with nonlinear thresholding. This approach is the one described extensively in the present chapter and has found great popularity with the development of the backpropagation algorithm, which permits the training of the intermediate hidden layers in a simple and successful way. However, we will see when discussing optical implementations of the

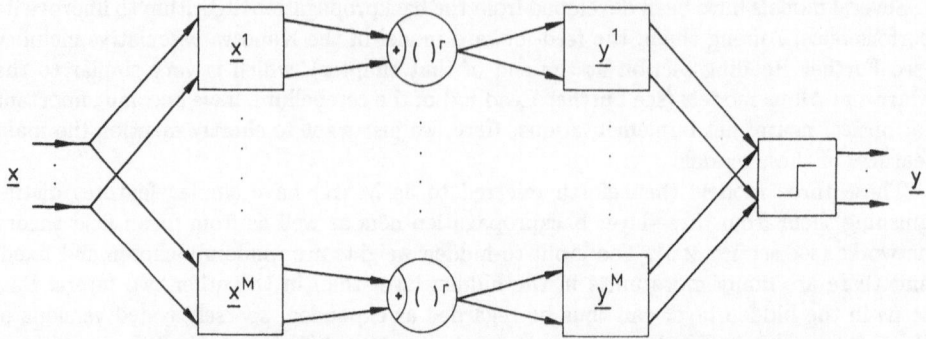

**Figure 2.12:** Outer product associative memory of the $r$th order.

backpropagation algorithm, that difficulties may arise in realizing the partial derivatives of the thresholding function.

The alternative approach is to calculate the *higher-order correlations* and to use them as input to the network, such that the output $y_i$ is obtained by

$$y_i = f\left(w_i^0 + \sum_{j=1}^{N} w_{ij}^1 x_j + \sum_{j=1}^{N} \sum_{k=j+1}^{N} w_{ijk}^2 x_j x_k + \cdots\right) \tag{2.30}$$

where $f()$ is a threshold function, and $w_i^1$, $w_{ij}^1$, $w_{ijk}^2$ are the interconnection weights of different orders. This principle is shown in fig. 2.12 for an outer product associative memory. In more detail, a second order neural network functions essentially as a linear perceptron in which the inputs are the summed correlations along each diagonal of the interconnection matrix. Thus the system can not distinguish between inputs that are composed of the same set of correlations. This feature — the extra degree of freedom arising from the second-order weights — can be used to implement translation invariance in the network. That two-stage breakdown of a second-order neural network into a correlation stage, followed by a first-order perceptron function, is shown in fig. 2.13. D. Psaltis and coworkers showed that the quadratic associative memory is equivalent to such a linear discriminant system with a square-law nonlinearity on the correlation plane [3]. The problem of these higher-order networks is that they suffer from a combinatoric explosion in the number of interconnections. However, if the order of the network is restricted to the minimum necessary to the problem at hand, this need not be an insurmountable problem. A great advantage of the single-layer structure of such networks is that both the input and the output layers are directly accessible and there are no hidden layers, which permits simple training rules to be used, in contrast to the more complex and lengthy backpropagation algorithms. The increased degrees of freedom introduced by higher-order terms can be used to increase the storage capacity of a network. Moreover, in higher-order networks the complexity lies in the interconnection rather than in the processing, and therein lies their suitability especially to optical implementations.

### 2.3.5 Local Backpropagation Learning

The conventional backpropagation algorithm is a nonlocal learning rule. If we change any of of the weights in the first layer of such a network (increase or decrease) the effect

**Figure 2.13:** Structure (upper scetch) and two-stage breakdown of a second-order network (lower scetch, both after [4]).

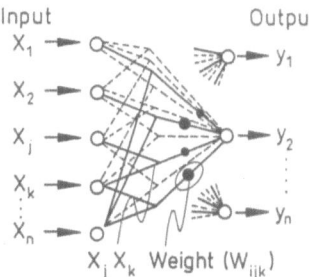

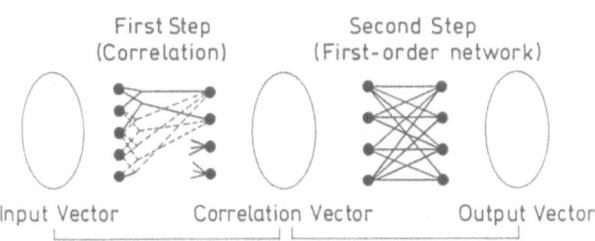

of this change on the output of the network would depend on the value and the sign of the weights of the second layer. Backpropagation is a steepest-descent method and, therefore, attempts to decrease the output error as quickly as possible. To accomplish this, we require knowledge of the second-layer weights in order to change the weights in the first layer. From an implementation point of view, this complicates matters since the information must be communicated in both directions and requires bidirectional neurons with a different functionality in the forward and the backward directions.

As an alternative solution to the problem, it is possible to use an algorithms for training two-layer optical neural networks, in which the weight updates are calculated from the signal at the input of each connection, the signal at the other end of the same connection and a global scalar error signal. The advantage of this algorithm is that it can be implemented with signals that are locally available, which is a simplification especially for optical system.

*Mathematics of Local Learning*

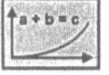 In order to get an intiuitive feel of how the local learning algorithm is working, consider the network of fig. 2.14 with only a single output neuron. Suppose that the output of that neuron is a binary one, taking only the values +1 or -1. The weights of the second layer should be trained by using the same procedure as in error backpropagation. For the weights of the first layer, if the output is wrong for a particular input, it is possible to correct it by adjusting the weights of the first layer to produce the negative of the current response of the hidden layer. Thus, the first layer can be treated as a single-layer net with one of the exisisting algorithms for training perceptron single-layer nets. I will show below that it is possible to select this training

algorithm in such a way to guarantee that the output error will decrease at each iteration. Following the explanation of Y. Qiao and D. Psaltis [5], let the number of neurons for the input, first and second layers of the two layer network be $N_0$, $N_1$, $N_2$, respectively. The inputs to the neurons of the $n$th layer are then

$$\alpha_j^{(n)} = \sum_{i=1}^{N_{\eta-1}} w_{ij}^{(n)} o_i^{(n-1)}, \tag{2.31}$$

where $w_{ij}^{(n)}$ is the weight of the interconnection between the $j$ th neuron in the $n$th layer and the $i$th neuron in the previous layer, and $o_i^{(n)}$ is the output of the $i$th neuron in the $n$th layer. The signal at the input layer is denoted by $o_i^{(0)}$. The first and the second layers of neurons perform a soft thresholding operation on their inputs to produce the outputs

$$o_j^{(n)} = f\left[\alpha_j^{(n)}\right] \tag{2.32}$$

where the function $f$ is chosen to be the bipolar sigmoidal function $f(\alpha) = \tanh(\alpha)$. The desired response that corresponds to the input $o_i^{(0)}$ is a vector $t$ whose elements are binary. The output error of the network is measured by the energy function

$$E = \sum_{k=1}^{N_2} \left\{ (1 + t_k) \cdot \ln\left(\frac{1 + t_k}{1 + o_k^{(2)}}\right) + (1 - t_k) \cdot \ln\left(\frac{1 - t_k}{1 - o_k^{(2)}}\right) \right\}. \tag{2.33}$$

Eq. (2.33) has its global minimum at $E = 0$ and reaches this minimum only if the network output is the same as the desired response. Y. Qiao and D. Psaltis [5] used this form of error function instead of the more commly used quadratic error function because this function is appropriate for the problem of recognizing handwritten zip-code digits. However, the algorithm may also work with the quadratic error function.

The backpropagation rule changes the weights by means of the gradient descent, which implies that

$$\Delta w_{kj}^{(2)} \propto -\frac{\partial E}{\partial w_{kj}^{(2)}} = 2\delta_k o_j^{(1)}$$

$$\Delta w_{ji}^{(1)} \propto -\frac{\partial E}{\partial w_{ji}^{(1)}} = 2\left(1 - (o_j^{(1)})^2\right) \cdot o_i^{(0)} \sum_{k=1}^{N_2} \delta_k w_{kj}^{(2)}, \tag{2.34}$$

where $\delta_k = t_k - o_k^{(2)}$ is the output error signal. The nonlocal nature of the backpropagation algorithm is due to the $\sum_{k=1}^{N_2} \delta_k w_{kj}^{(2)}$ factor, which contains the values of the weights in the second layer.

In order to derive the nonlocal equivalent, let $\gamma_k = \delta_k s_k^{(2)}$. Then $\gamma_k$ is positive if and only if the sign of the $k$th output unit matches the sign of the $t_k$. For example let us assume that $t_k = -1$ and $o_k^{(2)} > 0$. Then we get $\delta_k < 0$. The input signal to the $k$ th output unit $s_k^{(2)}$ has the same sign as the output of that unit and is, therefore, also positive. Thus, for this case, we achieve $\gamma_k < 0$. If we define $\gamma \equiv \sum_{k=1}^{N_2} \cdot \gamma_k$, which will be positive if the sign of most of the output units matches the target sign and will be negative if the reverse is true, we are able to construct a learning rule for the first layer of weights by using $\gamma$ as a performance metric. $\gamma$ is a scalar quantity that can be calculated

**Figure 2.14:** Schematic diagram of a feed-forward two-layer neural network used for local learning (after [5]).

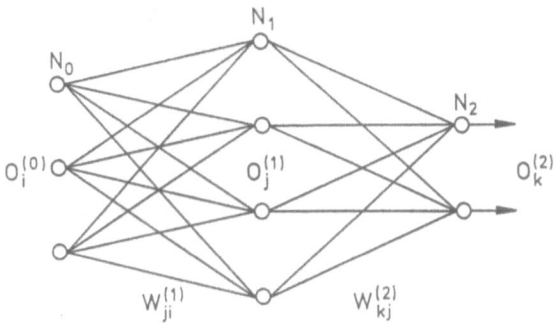

from the signals that are available at the output layer of the network. Thus, the basic idea of the new network function is to modify the first layer of weights so that we reinforce the production of the current hidden layer response if $\gamma$ is positive or reinforce the production of the negative of the current hidden layer if $\gamma$ is negative. From section 2.2 you remember that the simple Hebbian learning rule

$$\Delta w_{ji}^{(1)} \propto o_j^{(1)} o_i^{(1)} \tag{2.35}$$

reinforces the reproduction of the current response when presented with the same input. Therefore, if we want to reinforce the negative of the current response when the output is incorrect, then we can simply adopt an anti-Hebbian rule by multiplying the right-hand side of the Hebbian rule by $\gamma$. This idea led Y. Qiao and D. Psaltis [5] to the following anti-Hebbian local learning algorithm for the first layer:

$$\Delta w_{ji}^{(1)} \propto \frac{\gamma o_j^{(1)} o_i^{(0)}}{[1 - (o_j^{(1)})^2]}. \tag{2.36}$$

The denominator in eq. (2.36) is an additional term that is needed to guarantee that this learning rule always decreases the overall error at the output. Thus, even though the anti-Hebbain local learning algorithm is not a steepest descent rule, it still is a descent rule.

 ## Summary

The multilayer perceptron that utilizes a backpropagation algorithm is able to solve all classification problems because of its layered structure and the implementation of nonlinear threshold functions. The most important features of multilayer perceptrons are:

- Each layer in a multilayer perceptron has perceptron-like units.
- Thresholding is performed with a continuously differentiable function, most often a sigmoid one.
- Learning takes place in a feedforward, supervised way.
- The net is trained by the backpropagation algorithm (the generalized delta rule) that trains the network by passing errors back through the net.

- A net with three layers can represent any pattern classification.
- The most instructive description of the multilayer perceptron is the energy landscape.
- Learning difficulties as settling down in local minima may be overcome by various approaches as lowering the gain term, addition of internal nodes, a momentum term or noise.

Beneath the multilayer perceptron, higher-order neural networks allow for the solution of the same class of problems by utilizing the higher-order correlations as input to the network. Higher-order networks are

- generically translation invariant
- symmetrically weighted networks that are able to implement bipolar weighting
- of the form of single-layer networks, having no hidden layers with both the input and the output, but a square law or higher-order nonlinearity
- complex in the form of the interconnection, but easy as simple-layer perceptrons in the processing.

As an alternative to nonlocal backpropagation, it is possible to use a local learning rule, in which the weight updates are calculated from the signal at the input of each connection, the signal at the other end of the same connection and a global scalar error signal. For that purpose, an anti-Hebbian learning rule has to be applied, resulting still in a descent rule, although not performing a steepest descent to the solution.

## 2.4   Self-Organizing Feature Networks

Up to now I have looked at algorithms that relied on supervised learning techniques. This type of learning relies on an external training response — the desired output response — being available for each input from the training class. Although these techniques allow successful applications in almost all speech and pattern recognition problems, it is sometimes desired that a network itself could form its own classification of the training data — especially if large amounts of data have to be classified or clustered into classes without losing the essential input data information.

The basic principle of these feature nets is that the placement of neurons has a certain order and reflects some characteristics of the external stimulus or input. This sort of *self-organization* has its biological background in the structure of the auditary cortex region. In the auditory pathway, it is possible to distinguish a spatial arrangement of the neurons which reflects the frequency response of the auditory system. This phonotopic organization traces an almost logarithmic scale of frequency: low frequencies will generate responses at one end of the cortex region, high frequencies at the opposite extreme. Although much of the low-level organization of the brain is genetically pre-determined, it is likely that some of the organization at higher levels is created during learning by algorithms which take use of these ways of self-organization.

The ideas of self-organization were proposed as early as 1973 by C. van der Malzburg. Among the algorithms that have been proposed to realize self-organizing network structures, competitive learning is the simplest one. It allows to produce useful classifications of statistically clustered input distributions without the aid of a teacher. Because it can be regarded as the basis of more robust and sophisticated techniques such as adaptive resonance (see section 2.7) , generalized competitive learning as it was proposed by B. Kosko

**Figure 2.15:** Competitive learning network with $N$ inputs and $M$ outputs, divided into two competitive winner-takes-all patches with $m_1 = 3$ units in one and $m_2 = 2$ units in the other patch (after [6]).

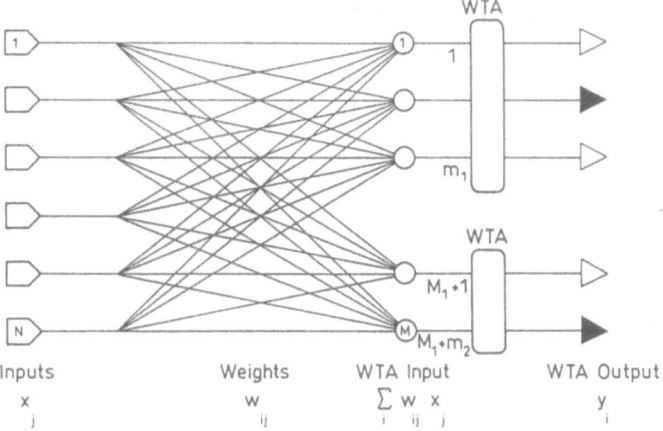

or self-organizing feature maps as they have been introduced by T. Kohonen (see Further Reading), I will begin the description of self-organizing feature maps with a more general explanation of competitive learning.

### 2.4.1  Competitive Learning

The key system components required by a competitive learing network are "winner-takes-all" neurons and adaptive outer product learning of the weights. These components are common to almost all other algorithms for unsupervised, self-organized learning as well, and it is the realization of these components that enables the construction of several different self-organizing networks. I will show in the following chapters when introducing optical models, that the development of techniques for optically implementing competitive learning networks is a critical first step in developing the capabilities required for other more sophisticated unsupervised learing algorithms.

A competitive learning network consists of an input layer of units that is adaptively interconnected to a collection of competitive patches in the output layer. Upon each pattern presentation, the interconnections to the neurons that received the largest inputs in each competitive patch are strengthened, whereas all of the other connections are slightly weakened. This simple unsupervised learning rule leads to the emerge of units tuned to respond to particular pattern clusters.

A diagram of a competitive learning network is shown in fig. 2.15. When an $N$-dimensional pattern of activity $C_j$ is presented to the input layer, the input is weighted and interconnected by the $M \times N$ adaptive weight matrix $w_{ij}$. This produces an array of $M$ inner products between the input pattern $\vec{x}$ and the rows of the weights matrix (weight vectors $\vec{w}_i$), which are the presynaptic inputs $p_i$ to each competitive patch:

$$p_i = \sum_i^N w_{ij}x_j = \vec{w}_i \cdot \vec{x} \tag{2.37}$$

The $M$ output neurons are grouped into $K$ competitive patches, where the $k$th patch contains $M_k$ competitive units, and $M = \sum_{k=1}^K m_k$. The competitive patches can be stacked lexicographically so that the rows of the weight matrix with $i \in 1, \ldots, m_1$ lead into the first patch that contains $M_1$ units, and the rows $j \in M_k + 1, \ldots, M_k + m_k$ (where

$M_k = \sum_i^{k-1} m_i$) lead into the $k$th patch. Each patch forms an independent winner-takes-all network, with the unit receiving the largest presynaptic inner-product input becoming the winner and signaling subsequent layers with a value of 1. All the outputs of the other units in that patch are suppressed to 0, so that the output is

$$y_i = \begin{cases} 1 & \text{for } p_i \geq p_{i'} \\ 0 & \text{otherwise} \end{cases} \qquad (i' \in M_k + 1, \ldots, M_k + m_k) \qquad (2.38)$$

The competitive learning rule reinforces the connections between the winning units and the input pattern. With these definitions of the input-pattern vector $x_j$ and the stacked array of winner-takes-all patches as vector $y_i$, the competitive learning rule can be formulated conveniently as an outer-product learning rule:

$$\Delta w_{ij} = \eta x_j y_i \qquad (2.39)$$

where $\eta$ is the learning rate. This simple learning algorithm reinforces the connections leading into all the winning nodes on each pattern presentation, and it leaves the losing nodes unaffected. The winning unit had the largest presynaptic inner-product input, indicating that its weight vector was the closest (in the inner product sense) to the input pattern vector. The input pattern vector is added to this weight vector as a learning perturbation, making it even more similar to that particular input pattern. The next time that a similar input pattern is presented to the network, it is even more likely that the same unit will win the competition and reinforce its connections. Cycling through an appropriately clustered statistical distribution of input pattern leads to a motion of the weight vectors towards the means of the different pattern clusters, which tends to tune the winning units to certain patterns or groups of patterns.

### Normalization of the Weight Vectors

The outer-product competitive learning rule is simple, but is has a severe disadvantage, originating from the statistical distribution of input patterns. Input patterns that have a large total activity $|\vec{x}|$ or those that appear more often in the input distribution can cause a particular unit in each competitive patch to win repeatedly. This drives the weight vectors leading to this unit towards a solution with a large norm $|\vec{w}_i|$ so that they may begin to win for all of the other pattern classes as well, preventing the evolution of a useful classifier.

To prevent this effect, it is possible to normalize the weight vectors. One possibility, proposed by C. von der Malsburg [7], is to project them back onto the surface of an $N$-dimensional hypersphere after every learning cycle. Thus the inner products between the normalized weight vectors and the input activity pattern become proportional to the norm of the input pattern vector (times the cosine of the angle between the input and the weight vectors). Because the first term is common to all of the units, the winner-takes-all decision becomes based on a pattern-similarity measure determined on the basis of angle exclusively. The weight vectors learn and evolve constrained to the surface of an $N$-dimensional hypersphere, so that winning units have weight vectors pointing towards the mean of the pattern clusters. The method is satisfactory operating — which means sufficient to stabilize competitive learning — even for approximate weight-vector normalization, where the weight evolution is constrained to a hypercube instead of a hypersphere [8].

An alternative to weight-vector normalization that is easier to implement optically is input-pattern normalization, first introduced by S. Grossberg [9]. Examples of normalized patterns include unipolar binary (0,1) patterns with the same number of ones, bipolar binary (-1,1) patterns, arbitrary analog pattern vectors divided by their arbitrary vector norm $x_j/|\vec{x}|$, and analog patterns represented as phase modulations, e.g. $\exp(i\pi x_j)$.

Phase modulation and bipolar binary modulation are especially attractive for optical implementations since they are readily implemented using exisiting liquid-crystal devices, so-called spatial light modulators. Both, spatial light modulators and phase modulation will play an important role when we realize all-optical neural network architectures (see section 4.1 and section 4.4).

When patterns are presented in this way optically with equal frequency, the use of normalized input-pattern distributions is less likely to result in a single unit in each class dominating the learning.

*Weight Forgetting*

In unsupervised learning, the patterns are often derived randomly from a fixed distribution. However, a sequence of unlucky pattern presentations can lead to a single neuron in a particular competitive patch winning a succession of competitions from patterns in different classes. Even with pattern normalization, it is possible for the corresponding weight vector to grow much larger than any of the others and to dominate all subsequent decisions. In addition, the weight vectors can grow without bound unless some weight saturation is incorporated.

Here, a simple weight forgetting term that decreases all the weights on each pattern presentation tends to produce an approximate weight-vector normalization at steady state [8]. This results in a resonably robust competitive learning rule

$$\Delta w_{ij} = \eta \frac{x_j}{|\vec{x}|} y_i - \alpha w_{ij} \tag{2.40}$$

where $\alpha$ is the small weight-forgetting constant.

Again, we will see later on that optical implementations as e.g. photorefractive volume holograms are well-suited to realize that feature of weight forgetting, because reading out optically stored data may result in an automatic erasure of previously written weight vectors. This feature unique to optical volume holographic memories allowing to read, write and adapt information is regarded as a rather disadvantageous one in many applications for supervised learning procedures, but can be exploited for self-organization in a very useful way.

The combination of pattern normalization, weight forgetting, and uniform a priori pattern-cluster probabilities produces a robust unsupervised competitive learning algorithm for stationary input distributions. When learning has stabilized on the average, the weight vectors tend to be normalized with a length proportional to $\eta/\alpha$ if all classes are equally likely and equally correlated. Correlations between the classes and unequal class probabilities can produce weight vectors with various lengths, but the algorithm often converges to useful classification results for these cases as well.

*Selection of the Initial Weight Matrix and the Problem of Noise*

In a noise-free deterministic implementation of competitive learning, the initial state of the weight matrix determines much about the evolution of the classifier. For an initial weight matrix with all elements equal, all presynaptic inputs are equal, no units win the competition, and no learning occurs.

Although in a real hardware system with analog noise in the winner-take-all units, a winner is picked at random and learning can thus be initiated, even for quite small learning rates, the winning nodes on the first cycle might capture all of the classes if the patterns are not sufficiently uncorrelated. This can be avoided by making sure that the increase in presynaptic input owing to the first epoch of training is less than the noise level in the neurons — although this small learning rate leads to slow training.

Alternatively, the initial weight matrix can be filled with random weights, and the neurons can be assumed to be noiseless. This tends to distribute the initial weight vectors uniformly in weight space with the same expected value for the length. Consequently, none will tend to be favoured initially when competing for normalized input patterns, and good classifications are likely to result. However, poor choice of the random initial weight matrix can still lead to useless classification if one unit tends to dominate for each of the input-pattern classes. To reduce the probability of useless classifiers evolving during competitive learning, it is possible to replicate each competitive patch several times, reducing the chances that they will be all unlucky and produce useless classification results. For small learning rates, noiseless neurons, and uncorrelated noiseless patterns, the final categorization produced by competitive learning is dominated by the initial projections of the patterns upon the random weights. Since the random initial weights are distributed uniformly and the patterns are normalized, all initial winner-takes-all classifications are equally likely.

To examine the likelihood of a useful classifier evolving from the random initial weight matrix, consider an input with $J$ patterns being categorized by $m$ winner-take-all nodes. The probability of an initial classification with $j_i$ of the patterns being associated with arbitrary permutations of the $m$ nodes, where $\sum_i^m j_i = J$, is obtained by [6]

$$p(j_i, \cdots, j_m) = \frac{J!m!}{m^J} \prod_{l=1}^{J+1} c_i! \prod_{i=1}^{m} j_i! \qquad (2.41)$$

where $c_1$ is the redundancy factor of the $l$th value taken by $j$, $0 < c_l \leq m - 1$, and $\sum_i^{J+1} c_l = m$. For example consider competitive learning problem with $J = 4$ patterns and a competitive patch that contains $m = 4$ units. There are a total of $m^J = 4^4 = 256$ possible separations of these patterns by the four units. The most probably outcome is that one unit wins for two patterns, two units selectively respond to one pattern and one unit remains unresponsive. K. Wagner and T.M. Slagle have shown with computer simulations that this is indeed the most likely outcome of noiseless competitive learning [6]. By adding more units than patterns, the most likely outcome is that each pattern is associated with an individual unit. Fewer units than patterns tend to force the grouping of the data into natural clusters. Competitive learning in the presence of noise evolves the probabilities of a particular classification but the estimates in eq. (2.41) obtain a worst-case boundary on the evolution of useless classifiers with all patterns associated with one winning unit, $p(0, \cdots, m) = m^{1-J}$.

For these reasons, it is advantageous to have numerous competitive patches with a wide distribution of the number of nodes per patch if the number of classes is unknown.

Small competitive patches produce coarse classifications that may generalize as well, but several patches of the same size may be necessary to ensure that at least one generates useful categorizations. Large numbers of units in a patch in contrast permit clustering on a finer scale, and multiple copies of these larger patches may not be required.

When the input patterns are noisy, such as deterministic patterns corrupted with Gaussian noise, then there is effectively an infinite number of patterns, and clustering is inevitable, as is desired. The expected steady-state values of the weight vectors are pointed towards the means of each pattern cluster, and all patterns in that cluster are represented by an individual winning node. Noise at the presynaptic inputs to the winner-take-all nodes have an effect similar to that of the noise in the patterns.

*Problems of Learning*

If the pattern clusters are correlated with each other, some of the initial random weight vectors may point towards the group of patterns, while others may be nearly orthogonal. Units can learn only when they win in a competition, so these initially orthogonal units never win, and they never participate in learning. Adding noise dynamically to the weight matrix could potentially help to alleviate this problem for low learning rates, since it can permit losing units to wander into the subspace spanned by the patterns.

Another technique, called leaky learning, can draw the weight vectors away from regions of low pattern density so that they can participate in the competitive learning process [8]. Leaky learning is implemented by setting the outputs of the losing units to a small positive value, rather than to zero, as before. This is again an almost unavoidable situation in optics, because of the limited contrast ratio in the optical storage devices. Outer product learning then adds a small version of the input pattern to the weight vectors of the losing units. This causes a slow movement of the losing weight vectors towards the region in weight space corresponding to a higher probability of the input patterns. As these weight vectors migrate towards the region of weight space containing the pattern clusters, they can potentially start winning and can acquire a cluster of their own to represent the group.

*Complex Weights and Activity*

Bipolar weigths, inputs and activity for competitive learning become important when we think of potential optical applications, because the information-bearing variables in a coherent optical system are complex. Thus, to transport amplitude and phase in an optical neural network, modifications to obtain bipolar or complex-valued weights in the competitive learning algorithm have to be introduced.

Complex pattern inputs $\hat{x}_j$ can be multiplied by complex weights $\hat{w}_{ij}$ and can be linearly combined, but the presynaptic input is modulus squared by the optical detectors, as e.g. photodetectors

$$P_i = |\sum_j < N\hat{w}_{ij}\hat{x}_j > |^2 \tag{2.42}$$

Ideally, unipolar input patterns result in all excitatory weights, producing positive linearly combined inputs only, so the monotonic modulus-squaring operation does not affect the winner-takes-all decision, and competitive learning is not affected. An interesting aspect of this modulus-squared presynaptic input for bipolar (or complex) input patterns with inverted contrast is that they produce an equally strong response. Thus a natural class invariance is built in for contrast reversed patterns. However, typical optical disturbances as the presence of complex noise in the hologram and undesired phase shifts between the beams may affect the behaviour of the algorithm. The complex outer-product learning rule operates as before, but there is no driving term for the imaginary part of the weights unless complex inputs are used or noise is added to the system. The resulting complex noisy pattern-normalized competitive learning rule with weight erasure is obtained by the evolution of both the real and the imaginary parts of the weights

$$\frac{dw_{ij}^r}{dt} = \eta \cos\phi \frac{\hat{x}_j}{|\vec{x}|} y_i - \alpha w_{ij}^r + \sigma^r,$$  (2.43)

$$\frac{dw_{ij}^i}{dt} = \eta \sin\phi \frac{\hat{x}_j}{|\vec{x}|} y_i - \alpha w_{ij}^i + \sigma^i.$$  (2.44)

In this expression, $y_i$ is the winner-takes-all result of operating in the modulus-squared presynaptic inputs $P_i$, $\phi$ is a time-varying phase instability, and $\sigma^r$ or $\sigma^i$ are the correlated real and imaginary noise contributions.

With proper phase stabilization so that $\phi(t) \approx 0$, the competitive learning rule driving term affects primarily only the real part of the weights. The imaginary parts of the weights are driven only by the noise but are erased simultaneously by the forgetting term. As the learning rule cycles through an entire training set, there is, on the average, a consistent driving term pushing the real parts of the weights towards the desired solution, while the noise driven imaginary parts are attracted back to the origin through weight forgetting. Simulations indicate that the imaginary parts of the weights do not adversely affect the learning as long as the forgetting term can dominate the noise term [6].

### 2.4.2  Kohonen's Self-Organizing Feature Network

The most popular self-organizing algorithm derived from competitive learning which produces spatial mapping to model complex data structures internally has been presented for use in speech representation by T. Kohonen in 1984 and was called by him *self-organization feature map*. It is this model that I want to pick out from the models existing for self-organizing networks. T. Kohonen used a technique known as vector organization to perform data compression on the vectors to be stored in this network. Moreover, it allows the network to store data in such a way that spatial or topological relationships in the training data are maintained and represented in a meaningful way.

The structure of a typical self-organizing network is shown in a simplified version in fig. 2.16. It is a one-layer-two-dimensional Kohonen network. The most obvious difference to the former models of multilayer perceptron structure is that the neurons are not arranged in layers — as input, hidden and output layer — but on a flat grid. All inputs are connected to every node in the network. The output nodes are arranged on a two dimensional grid and are extensively interconnected with many local connections. Feedback is therefore restricted to lateral interconnections between neighbouring nodes. This concept of local neighbourhood can be described mathematically with a mexican-hat activation function as it is shown in fig. 2.17. In that representation, nodes that are physically close to an active one have the strongest links to it. In contrast, those at a certain distance

# Box 5: Algorithm for Self-Organizing Feature Maps

**Step 1: Initialize weights**
Initialize weights from N inputs to the M output nodes (see fig. 2.16) to small random values.
Set the initial radius of the neighbourhood as shown in fig. 2.18.

**Step 2: Present new input**

**Step 3: Compute distance to all nodes**
Compute distances $d_j$ between the input and each output node j using

$$d_j = \sum_{i=0}^{N-1} (x_i(t) - w_{ij}(t))^2 \tag{2.45}$$

where $x_i(t)$ is the input to node i at time t and $w_{ij}(t)$ is the weight from input node i to output
node j at time t.

**Step 4: Select output node with minimum distance**
Select node $j^*$ as that output node with minimum $d_j$.

**Step 5: Update weights to Node $j^*$ and all neighbours**
Update weights to Node $j^*$ and all nodes in the neighbourhood defined by $NE_{j^*}$ as shown in
fig.2.18. New weights are

$$w_{ij}(t+1) = w_{ij}(t) + \eta(t) \cdot (x_i(t) - w_{ij}(t)) \tag{2.46}$$

with the term $\eta(t)$ beeing a gain term $(0 \leq \eta(t) \leq 1)$ that decreases in time.

**Step 6: Repeat by going to step 2**

---

defined by the mexican hat function are actually switched to inhibitory links. Note that
there is no separate output layer — each of the nodes in the grid is itself an output node.

To teach or train the system, continuously valued input vectors are presented sequentially in time without specifying the desired output. Where inputs match the node vectors, that area of the map is selectively optimized to represent an average of the training
data for that class. After enough input vectors have been presented, weights will specify
cluster or vector centers that sample the input space in such a way that the point density
function of these centers tends to approximate the probability density function of the input vectors. Thus the grid settles from a randomly organized set of nodes into a feature
map that has local representations and is self-organized in such a way that topologically
close nodes are sensitive to inputs that are physically similar. Output nodes will thus be
ordered in a natural manner.

The algorithm that forms feature maps (see Box 5) requires a neighbourhood to be
defined around each node. It is naturally very simple: it finds first the closest matching
unit to a training input, and then increases the similarity of that unit and those in the
neighbouring proximity (as defined by the appropriate function as in fig. 2.17) relative
to the input. Then it slowly decreases the neighbourhood in size as shown in fig. 2.18 in
order to minimize the number of connections between nodes.

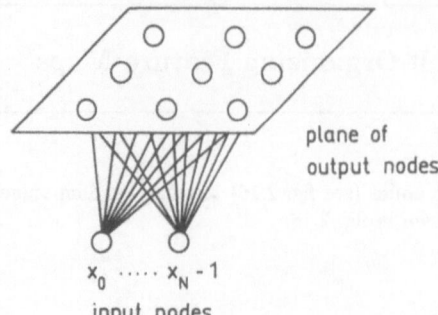

plane of
output nodes

$x_0$ ..... $x_N - 1$

input nodes

**Figure 2.16:** Two-dimensional array of output nodes used to form feature maps after Kohonen. Every input is connected to every output node via a variable connection weight.

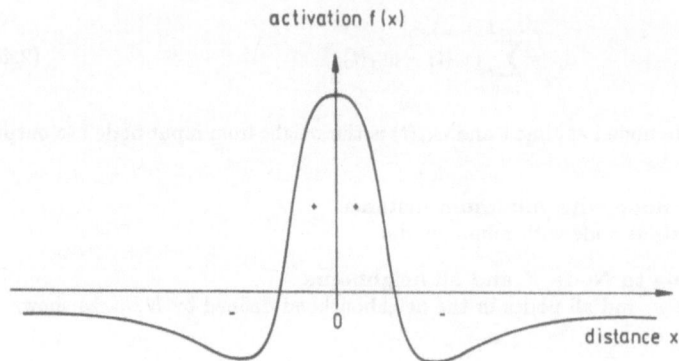

activation f(x)

0

distance x

**Figure 2.17:** Mexican hat activation function to define a local neighbourhood as it is used in self-organizing feature maps.

Weights between input and output nodes are initially set to small random values and an input is presented. The distance between the input and all nodes is computed as shown. If the weight vectors are normalized to have constant length (which is possible because the sums of the squared weights from all inputs to each output are identical), the node with the minimum distance can be found by using the net of fig. 2.16 to form the dot product of the input and the weights. The selection required in step 4 then turns into a problem of finding the node with a maximum value. This node can be selected e.g. using extensive lateral inhibition. Once this node is selected, weights to it and to other nodes in its neighbourhood are modified to make these nodes more responsive to the current input. This process is repeated for further inputs. Finally, weights eventually converge and are fixed after the gain term in step 5 is reduced to zero.

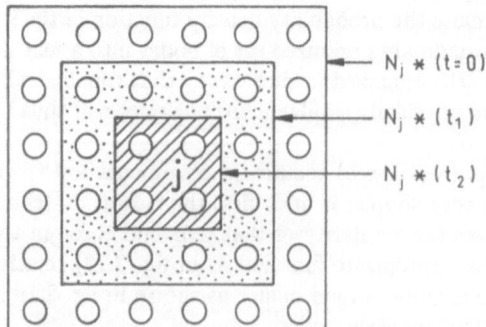

$N_j * (t = 0)$

$N_j * (t_1)$

$N_j * (t_2)$

**Figure 2.18:** Topological neighbourhoods at different times as feature maps are formed. The neighbourhood starts large and slowly decreases in size over time. In this example, $0 < t_1 < t_2$.

In conclusion, the two central issues to adaptive self-organizing learning in a Kohonen network are the weight adaptation process and the concept of topological neighbourhood of nodes. Both of these ideas are very different from the neural concepts I have discussed so far. These features are the ones that make Kohonens feature nets so well-suited for data compression and organization of large amounts of data without losing information.

 **Summary**

Self-organizing networks are completely different from the nets I have discussed up to now. Among them, competitive learning networks are the most simple self-organizing networks. Their main features are:

- The key components are winner-takes-all neurons and adaptive outer product learning of the weights.
- The net consists of an input layer of units that are adaptively interconnected to a collection of competitive patches — each being a winner-takes-all network — in the output layer.
- Only the interconnections that receive the largest inputs in each competitive patch are strengthened, the rest is weakened.

The self-organizing feature map or Kohonen map is a more complex net with the following features:

- Kohonen nets are self-organizing, with similar inputs mapped to nearby nodes.
- All nodes of a Kohonen net are located in one two-dimensional layer.
- The algorithm requires a neighbourhood to be defined around each node. It finds first the closest matching unit to a training input, and then increases the similarity of that unit and those in the neighbouring proximity Then it slowly decreases the neighbourhood in size in order to minimize the number of connections between nodes.
- Lateral excitation and inhibition in the neighborhood proximity is applied using a "Mexican hat"-function.

## 2.5   Hopfield Networks

Hopfield nets are the most popular and often mentioned networks if someone is looking for an introduction to neural network structures. In contrast to the nets presented up to now, Hopfield nets are normally used with two-state inputs, which may be binary unipolar $(0, 1)$ or bipolar $(-1, +1)$. They are most appropriate when an exact binary representation is possible — as e.g. with black and white images whose input elements are values of pixel brightness. They consist of a number of nodes containing hard-limiting nonlinearities, each of it connected to every other node. Therefore, Hopfield nets are *fully-connected* networks and thus have an inherent feedback mechanism as sketched in fig. 2.19. As a result, there is no obvious input or output node — each node is equal to each other! This is the major feature of the Hopfield net and it is the one which

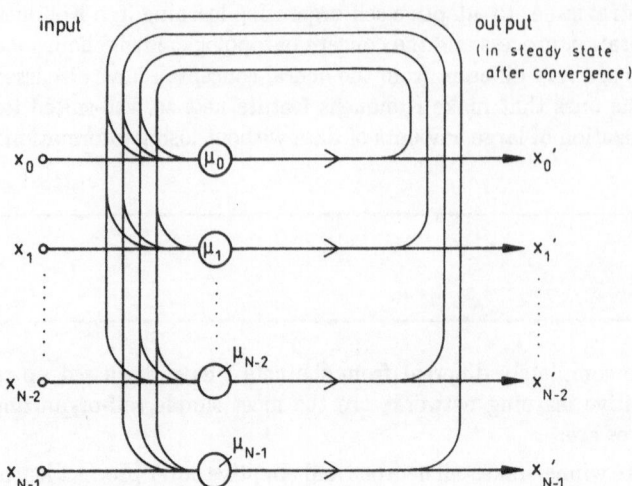

**Figure 2.19:** Schematic representation of a fully-connected Hopfield net.

distinguishes it completely from the nets I have discussed so far. Moreover, the net is *symmetrically-weighted*, since the weights on the link from one node to another are the same in both directions. As a consequence, a symmetric, zero-diagonal weight matrix is assumed in Hopfield nets. This symmetry is central to the operation of the network. Even slight deviations from it can give rise to instabilities which may end up with networks that do not settle into any final state. The difference in architecture means that the network operates also in a different way: inputs to the network are applied to all nodes at once. Then the net is left alone and proceeds to cycle through a succession of states by setting each output as the new input until it converges to a stable solution. Here, stability is defined as the situation, when the values of the nodes no longer alter. When the net has reached such a stable, steady state, its output is taken to be the value of all the nodes.

The process of finding the steady state can be considered as a competition process in which in the initial state a lot of different values try to affect each other. This is likely to be unstable, since one value may be trying to turn other nodes on, while another one is trying to turn them off. As the net cycles through a succession of states, it is trying to reach a compromise between all the values in the network. Therefore, the final steady state may be regarded as the "best compromise" solution that the network is able to find. In this state, the number of inputs trying to switch a unit "on" or "off" is equal. Thus the net remains in its stable state. The mathematical algorithm belonging to that chapter is shown in box 6.

First, weights are set using the given recipe from exemplar patterns for all classes. When this teaching stage has finished, an unknown pattern is imposed in the net by forcing the output nodes of the net to match the representation of that unknown pattern. Following this initialization, the net is then allowed to iterate freely in discrete time steps until it reaches a steady state. The net is considered to have converged when the outputs change no longer on successive iterations. J.J. Hopfield could prove in his derivation of the mathematical principles of these nets, that such a solution is possible, if the weights of the net are symmetric [10]. Hopfield also demonstrated that the net converges when graded nonlinearities as the sigmoid function are used and that the net can be used as an associative memory. In that case, the net output after convergence is used directly as the complete restored memory. Even autoassociation of a pattern is possible in Hopfield nets. It means here, that the presentation of a corrupt input pattern will result in the reproduction of the perfect pattern as the output — resulting in a content-addressable memory. If the Hopfield net should be used as a classifier, the output after convergence has to be compared to the $M$ exemplars to determine if it matches an exemplar exactly. If it does, the output is that class whose example matches the output pattern — if not, a "no match" result occurs.

 **Box 6: Hopfield net algorithm**

**Step 1: Assign connection weights**

$$w_{ij} = \begin{cases} \sum_{s=0}^{M-1} x_i^s x_j^s & i \neq j \\ 0 & i = j, \, 0 \leq i \,, j \leq M - 1 \end{cases} \qquad (2.47)$$

where $w_{ij}$ is the connection weight from node i to node j and $x_i^s$ (which can be -1 or +1) is element i of the exemplar for class s.

**Step 2: Initialize with unknown input pattern**

$$\mu_i(0) = x_i \qquad 0 \leq i \leq N - 1 \qquad (2.48)$$

where $\mu_i(t)$ is the output of node $i$ at time $t$ and $x_i$ (+1 or -1) is element $i$ of the input pattern.

**Step 3: Iterate until convergence**

$$\mu_j(t+1) = f_h \left[ \sum_{i=0}^{N-1} w_{ij} \mu_i(t) \right] \qquad 0 \leq j \leq N - 1 \qquad (2.49)$$

where $f_h$ is the hard limiting nonlinearity.
The process is repeated until node outputs remain unchanged with further iterations. The node outputs $\mu_j(\text{stable}) = x_j^s$ then represent the exemplar pattern of class $s$ that best matches the unknown input.

**Step 4: Repeat by going to step 2.**

### 2.5.1   The Energy Representation Revisited

We have already used successfully the energy landscape model to describe the behaviour of perceptrons. Since it provides a visual analogy that allows us to form an intuitive view of what happens in a net during learning and recall, why not trying to use it in the same way to visualize the processes in a Hopfield net? As in the case for perceptrons, the energy landscape has hollows (stable states or basins of attraction) that represent the patterns stored in the network. An unknown input represents a particular point — which can be imagined as a ball in the three-dimensional situation — in the landscape. As the network iterates its way towards the solution, this ball moves through the landscape towards one of the hollows.

We are able to express this behaviour in more detail, if we look at it mathematically. For the perceptron, the energy function was $E = \frac{1}{2} \sum (w_{pj} - o_i)^2$, but this expression depends on the knowledge of the required output as well as the actual output of the net. In contrast, in the Hopfield net, which steps its way towards the solution, the required intermediate steps are not known. Thus the energy function of the Hopfield model has to be more applicable to this architecture. However, it would be sensible to retain some of the features of the perceptron energy function: it should be large for large errors and small for small errors. The weight values in the network must affect the energy, as must the actual patterns presented. In analogy to the perceptron, we may express such a suitable energy function mathematically as

$$E = -\frac{1}{2} \sum_i \sum_{j \neq i} w_{ij} x_i x_j \quad + \sum_i x_i \Theta_i \tag{2.50}$$

where $w_{ij}$ represents the weight between node $i$ and node $j$ of the network, $x_i$ represents the output from node $i$ and the threshold value of node $i$ is represented by $\Theta_i$. Because the output is fed back into the net, each output at any time represents the next set of inputs. So both the weights and the inputs are explicitly represented in that energy function. The weights in the network contain in turn the pattern information so that finally all patterns are included in this energy function.

Now let us have a look at the physical content of eq. 2.50. Nodes are not directly connected to themselves, so the term $w_{ii}$ has to be zero. Since the connections need to be symmetric for a stable operation (see above), we postulate $w_{ij} = w_{ji}$. Having defined the error function, we are able to describe the mechanism of storing and recalling a pattern: if we make our pattern occupy the low points in the energy landscape, then we can perform gradient descent on the energy function in order to end up in one of these minima, which will represent our solution.

### Storing Patterns

 How are we now able to store patterns in a Hopfield net? Storing in this system corresponds to lowering the energy function as much as possible for a particular pattern so that it occupies a minimum point in the energy landscape. However, we also want to leave any previously stored patterns in their hollows, so that adding new patterns does not destroy all the previous information. Because the weight matrix contains the information about the stored patterns, we have to find an expression for the weight values that will produce a minimum in the energy function. In terms of the energy function, we have to minimize eq. 2.50 for a particular pattern $s$ that has a set of input elements $(x_o, x_1, \cdots, x_{n-1})$. We want each term to be negative, thus requiring that $\sum_i x_i \Theta_i$ is negative. This can be achieved e.g. by setting $\Theta_i$ to the opposite sign of $x_i$ for a particular pattern. However, a different pattern will have different values of $x_i$ and then the threshold term may well increase the value of $E$. In order to avoid this, the best solution is to set the threshold to zero, which will neither increase nor decrease the energy function for any pattern.

Now let us split the weight matrix into two parts, one representing the effects of all the patterns except the $s$th one — which will be denoted by $w'_{ij}$ — and a second one which is the contribution made by the $s$th pattern alone, shown as $w^s_{ij}$. This results in the new energy function

$$
\begin{aligned}
E &= -\frac{1}{2}\sum_i\sum_{j\neq i} w'_{ij} x_i x_j \\
&\quad -\frac{1}{2}\sum_i\sum_{j\neq i} w^s_{ij} x^s_i x^s_j \\
&= E_{\text{all except } s} + E_s
\end{aligned}
\tag{2.51}
$$

where I have separated the contribution made to the energy function from the $s$th pattern. Eq. (2.51) can be thought of as viewing the energy as a "signal" plus a "noise" term; the "signal" is the energy due to the pattern $s$, whilst the "noise" is due to the contributions from all the other patterns. Because storing means to make the energy function as small as possible and because "noise" can not be altered much, we have to look how to minimize the contribution to the energy function from the signal term. Making

$$
E_s = -\frac{1}{2}\sum_i\sum_{j\neq i} w^s_{ij} x^s_i x^s_j
\tag{2.52}
$$

as small as possible corresponds to making

$$
\sum_i\sum_{j\neq i} w^s_{ij} x^s_i x^s_j
\tag{2.53}
$$

as large as possible, due to the minus sign in eq. 2.52. Now, because the elements in $x_i$ are either +1 or -1 (+1 or 0), we would get into trouble if we try to maximize this equation. However, $x^2_i$ is always positive, and therefore it would be much easier to make the energy term dependent on $x^2_i x^2_j$. We can do this most simply by equating

$$
\sum_i\sum_{j\neq i} w^s_{ij} x_i x_j = \sum_i\sum_{j\neq i} x^2_i x^2_j
\tag{2.54}
$$

Thus all we have to do is set the weight term to be

$$
w^s_{ij} = x^s_i x^s_j.
\tag{2.55}
$$

In order to calculate the weight values for all the patterns, we sum this equation over all patterns to get an expression for the total weight set between nodes as

$$
w_{ij} = \sum_s w^s_{ij} = \sum_s x^s_i x^s_j
\tag{2.56}
$$

Comparing this to the first step of the Hopfield net algorithm as summarized in Box 6, we find that it is identical to our result of storage — so step 1 really does store all the initial patterns in the network! Referring to eq. (2.52), altering the $w_{ij}$ each time will alter the value of $E_{\text{all except } s}$ somewhat. Therefore, adding a pattern does disrupt the previous state to some extent, which is unavoidable. To summarize, the Hopfield net has no iterative learning algorithm as all patterns are simply stored by lowering their energies. The network has no hidden units either and thus is unable to encode the data.

*Recall*

 Having stored the pattern, a method is required to recall them. Compared once again with the algorithm of the perceptron, we have to find a method to perform gradient descent on the energy function. Considering eq. (2.50), we need to calculate the contribution that a particular node's value makes to the energy, and then we can cycle around the net, reducing each node's contribution until the energy value is at a minimum. For that purpose, let us first express the energy function in two parts, splitting off the contribution made by the $k$th node.

$$
\begin{aligned}
E \;=\; & -\frac{1}{2}\sum_{i\neq k}\sum_{j\neq k} w_{ij}x_i x_j + \sum_{i\neq k} x_i \Theta_i \\
& -\frac{1}{2}x_k \sum_j x_j w_{kj} - \frac{1}{2}x_k \sum_i x_i w_{ik} + x_k \Theta_k
\end{aligned}
\tag{2.57}
$$

Now we have to calculate the change in energy $\Delta E = E_2 - E_1$ caused by the state change $\Delta x_k = x_{k_2} - x_{k_1}$. We will find from eq. (2.57) that this energy difference is

$$
\Delta E = -\frac{1}{2}\left[ \Delta x_k \sum_j x_j w_{kj} + \Delta x_k \sum_i x_i w_{ik} \right] + \Delta x_j k \Theta_k
\tag{2.58}
$$

The first two terms of eq. (2.57) are unaffected by the alteration of neuron $k$. So they remain unchanged and cancel out. Since the matrix $w_{ij}$ is symmetric, we are allowed to interchange the indices in eq. (2.58). This simplifies the expression to

$$
\Delta E = -\Delta x_k \left[ \sum_j x_j w_{kj} - \Theta_k \right]
\tag{2.59}
$$

where $\sum_j x_j w_{kj}$ is the weighted sum of the input node $k$ and $\Theta_k$ is the threshold of unit $k$. Because the only outputs of the nodes are $\pm 1$ $(0,1)$ in a binary net as the Hopfield net, decreasing $\Delta E_k$ means giving out a $+1$ if the weighted sum is greater than zero and $-1$ if it is less than zero, since this will always serve to reduce the value of $\Delta E_k$. If we compare this to the update function for nodes in a Hopfield net:

$$
\sum_{i\neq k} w_{ij}x_i = \begin{cases} > 0 & \text{then } x_i \to +1 \\ = 0 & \text{then return to previous state} \\ < 0 & \text{then } x_i \to -1 \end{cases}
\tag{2.60}
$$

we see that it performs exactly this operation and therefore implements a gradient descent in $E$. To summarize, this algorithm allows the recall of the patterns from the net by cycling through a succession of states, each of which has a lower or equal energy level than the previous one. The process stops, when the net has found its way into an energy minimum and thus has produced the target pattern.

There are actually two slightly different methods of performing update. The first one is to carry out update on all nodes simultaneously, where the values on the network are temporarily frozen and all the nodes then compute their next state. Consequently, the update is performed in one single time step across the entire network. That is the reason for calling this updating mechanism synchronous updating. The alternative approach, called asynchronous updating, occurs when a node is chosen at random and updates its output according to the input it is receiving. This processing is then repeated until all nodes are updates. The change in output of one node affects in that case the state of the system as a whole and can therefore affect the next node's change. This involves that the order in which the nodes are updated becomes important. It affects the behaviour of the network to a certain extent. The effects are evident in the recall state, since the random nature of choice of the next node to be updated alters the sequence of patterns that the network evolves through. In contrast, with synchronous updating, all nodes are updated together and so the intermediate pattern does not alter. The asynchronous updating therefore adds some uncertainty into the way from the input to the final steady state. However, both methods rely on the same general characteristics of the network and therefore the choice of the updating mechansim is rarely an important factor for the performance of the network.

*Limitations of the Hopfield Network*

The Hopfield net has two major limitations when used as an associative memory. First, the number of patterns that can be stored and accurately recalled is severely limited. If too many patterns are stored, the net may converge to a novel, spurious pattern different from all exemplar patterns. In terms of the energy landscape, there has been sufficient interference between patterns to form intermediate local minimum states that were not taught to the network. These states are also known als metastable states. Hopfield himself showed that this may occur when the number of classes is more than 0.15 times the number of input elements or nodes in the net [10, 11].

A second limitation of the Hopfield net is that an exemplar pattern will be unstable if it shares many bits in common with another exemplar pattern. It may happen that an exemplar pattern is applied at time zero and the net converges to some other exemplar. Again, the main reason of this behaviour is the tendency of the net to converge into a local minimum. The more the exemplar patterns are similar, the more the local minima or holes are close together and consequently the probability rises that the systems ends up in a local minimum.

Beneath these restrictions, the Hopfield net and its concept of global interconnectivity revealed to be very useful in many situations. Therefore, improvements of the Hopfield net are a main aim of investigations since the early days of neural network research. The main point in these improvements is always to find a method which allows the net to escape from these local hollows and move into some deeper well that represents a better result.

### 2.5.2 Interpattern Association

As in several neural network algorithms, in the Hopfield model the similarities among patterns are used to recognize a certain pattern. Thus, the interconnection weight matrix is constructed by correlating the elements within each pattern, but ignoring the relationships among the reference patterns. However, in all cases where the exemplar patterns are quite similar, the probability of the net to settle down in local mimina and end up in a wrong solution rises dramatically. For these cases, a net that relies rather on the differences than on the similarities among patterns to perform pattern recognition seems to be more adapted. This idea has been lead to the *interpattern association model*, in

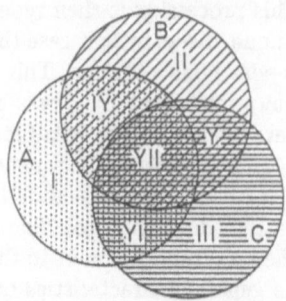

**Figure 2.20:** Common and special areas of three reference patterns (after [12]).

which the basic logical operations are used to determine the interpattern association (i.e. association between the reference patterns). Thus, the information obtained from special areas in the network (i.e. areas occupied by one pattern) are emphasized and considered to be more important than those from common or overlapped areas. The model first determines the common and special features among the reference patterns by applying a logical operation. For that purpose, simple logical rules are applied to construct a tristate interconnection in the network. There are a couple of logical operations possible to realize this comparison and I will present shortly the most promising of these algorithms. Fig. 2.20 shows an example of a reference set that consists of three overlapping patterns $A$, $B$, and $C$ in the pattern space. These patterns can be divided into seven subspaces. I, II, and III are the special areas of pattern $A$, $B$, and $C$, respectively. IV, V, and VI are the common areas of $A$ and $B$, $B$ and $C$, $C$ and $A$, respectively. VII finally is the common area of $A$, $B$, and $C$. The rest can be defined as an empty space $\Phi$.

With this information, we can start building the interconnections for a single-layer neural network. To illustrate the concept of interpattern association, in this example, let us map each pixel in the reference pattern space to each input/output neuron. When an input neuron in area VII is on (i.e. the input of this neuron is 1), it implies that there is an input but not whether it belongs to any pattern. Thus this neuron can only excite the output neurons within area VII and has no connection with the ouptut neurons in other areas. When an input neuron in area V is on, it implies that the input pattern is not $A$ but not whether it belongs to pattern $B$ or $C$. Thus this neuron must excite the output neurons in areas V and VII but inhibit the output neurons in area I. Similar rules can be constructed for input neurons in area IV or VI. For the case when an input neuron is on in area I, it implies that pattern $A$ appears at the input end. Therefore, this neuron can excite all the output neurons in pattern $A$ (i.e. areas I, IV, VI, and VII), but inhibit the output neurons in areas II, III, and V. Similarly, for the input neurons in area II or III, they must excite output neurons in area $B$ or $C$ and inhibit neurons in areas I, III, and VI or areas I, II, and IV, respectively. The descriptive logic of such a net can be summarized in the following *interpattern association rule A*: assume that a subarea $X$ is represented by

$$X = P \wedge \bar{Q} \tag{2.61}$$

where $P = p_1 \wedge p_2 \wedge \cdots \wedge p_n$, and $Q = q_1 \vee q_2 \vee \cdots \vee q_m$, $p_1, p_2, \cdots, p_n$ and $q_1, q_2, \cdots, q_m$ are reference patterns, $n$ and $m$ are two positive integers, and $n + m = M$ is the total number of reference patterns. The input neurons in area $X$ must excite (i.e. have positive connections with) all the output neurons in area $P$, inhibit (i.e. have negative connections with) all output neurons in area $Q \wedge \bar{P}'$ where $P'$ is defined by $P' = p_1 \vee p_2 \vee \cdots \vee p_n$

and have no connection with the output neurons in the remaining areas. For simplicity, the connection strenghts (i.e. weights) are defined as 1 for positive connections, $-1$ for negative connections, and 0 for no connection. Thus the interpattern association neural network can be constructed in a simple three-state structure. If bipolar weights are an undesired feature of the network, a unipolar interconnection weight matrix can also be realized using the interpattern association. By a search of the redundant interconnection links which is in detail described in [13], a method can be found to remove all negative links. Fortunately, the reduction of the links to the necessary ones results in a reduction in the noise performance, resulting in a better performance of the unipolar model, while preserving the storage capacity of the bipolar interpattern association model.

To illustrate further the construction of such an interpattern associative net, let us assume $A$, $B$, and $C$ as three 2 x 2 array patterns as shown in fig. 2.21, (a), (b) and (c), respectively. The resulting neural net applying the rule derived above to the three reference patterns is illustrated in fig. 2.21, (d). This is a one-layer network with four input neurons and four output neurons. Each neuron is matched to one pixel of the reference patterns. For example, the first input neuron correspond to pixel 1 of the input pattern and excites only the first output neuron. The second input neuron (the corresponding pixel belongs to patterns $A$ and $B$), excites both the first and second output neurons, while inhibiting the fourth output neuron, which belongs to the special area of pattern $C$. It is interesting to note that the weight matrix in that case becomes a 4-D matrix, which can be partitioned into a 2-D submatrix array, as illustrated in fig. 2.21, (e). This weight matrix can be divided into four blocks, each corresponding to one output neuron. The four elements in one block represent the four neurons in the input end. For example, since all four elements in the upper left block have a value of 1, it indicates that either one of the four input neurons can excite the first output neuron. In the upper right block, as another example, the first and the third elements are 0, the second element has the value 1, the fourth is equal to $-1$. From this example we can derive that the first and the third input neurons have no connection to the second ouptut neuron, the second input neuron excites the second output neuron and the fourth input neuron inhibits the second output neuron. To simplify the algebraic operations, an equivalent rule can be developed. If $D_{ij}$ is a 2-D matrix corresponding to the matrix in fig. 2.21, with $i$ and $j$ being the row and column numbers respectively, and $d_i$ is the number of patterns that are bright at the $i$th pixel, we get

$$d_i = \sum_{l=1}^{M} D_{l,i}. \tag{2.62}$$

The sum of the product of columns $i$ and $j$ is

$$k_{ij} = \sum_{l=1}^{M} D_{l,i} D_{l,j} \tag{2.63}$$

Then we can construct an interpattern association neural network by applying the following *interpattern association rule B*:

- If $k_{ij} = \min(d_i, d_j)$:
  when $d_i < d_j$, pixel $i$ must excite pixel $j$, but pixel $j$ must not excite pixel $i$;
  when $d_i = d_j$, pixels $i$ and $j$ must excite each other;
  when $d_i > d_j$, pixel $j$ must excite pixel $i$, but pixel $i$ must not excite pixel $j$.

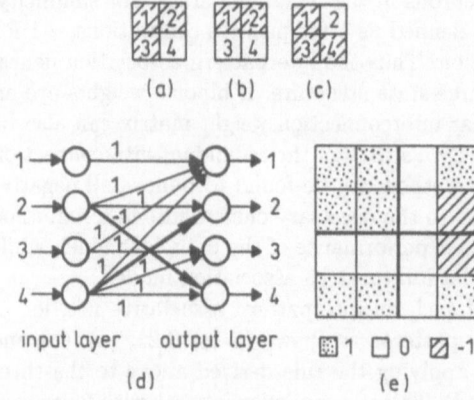

**Figure 2.21:** Example of construction of an interpattern association neural network: three 2 x 2 reference patterns $A$ (a), $B$ (b), $C$ (c), the corresponding tristate neural network (d), and the interconnection weight matrix (e) (after [12]).

- If $0 < k_{ij} < \min(d_j, d_j)$:
  pixels have no connection to each other.
- If $k_{ij} = 0$:
  when $d_i \neq 0$ and $d_j \neq 0$, pixels $i$ and $j$ must inhibit each other; when $d_i = 0$ and / or $d_j = 0$, pixels $i$ and $j$ must have no connection to each other.

Note that these rules are simple and straightforward, which makes the algorithm suitable for optical implementations as well as for computer calculations. Moreover, they are completely equivalent to the formerly mentioned algebraic rule A.

F.T.S. Yu and coworkers [12] could show in 1990, that the interpattern association model is superior to the Hopfield model in computer simulations with similar reference patterns. The latter becomes unstable if the weights of the patterns are similar, because the definition of the threshold value then becomes difficult. Moreover, since the first developments of the interpattern association model, several optical realizations of the model have been proposed and realized using ferroelectric spatial light modulators as input and feedback devices and lenslet arrays as the interconnection matrices. More details on these systems will be discussed in sections 7 and 11.

 ## Summary

Compared to the perceptron algorithms, Hopfield nets have no specific input and output. The consequences arising from that structure are:
- Hopfield nets are symmetric, fully-interconnected systems.
- The system iterates through its states towards a stable state which then represents the solution.
- The system acts as an autoassociative memory.

Problems of the Hopfield net are
- The number of patterns that can be stored and accurately recalled is limited. Metastable or intermediate local minimum states may occur when the number of classes is more than 0.15 times the number of the input elements or nodes in the net.

- An exemplar pattern will be unstable if it shares many bits in common with another exemplar pattern because of the tendency of the net to converge into local minima.

An important improvement of the Hopfield network is the interpattern association model. Its significant differences to the Hopfield net are:

- Instead of constructing the interconnecting matrix by correlating the elements within each pattern and rely on the similarities among patters, it relies rather on the differences among patterns.
- Informations obtained from special areas are emphasized and considered to be more important than those from common or overlapped areas.
- If the exemplar patterns are similar, the interpattern association model is superior to the Hopfield model.

## 2.6  The Boltzmann Machine

In the energy landscape representation, it is easy to imagine that — if the solution to the inputs is represented as a small ball — giving the ball some supplementary (intrinsic or thermal) energy will allow it to randomly walk around in the potential wells and probably escape from a local minimum. However, this "shaking" of nominally stable situations, which I have already discussed in different versions to help multilayer perceptrons to find a way out of the local minima, needs to be done rather carefully, as a violent shaking can move the solution as well away from a stable point as towards it and a soft shaking may result in nothing. Therefore, the best method is to provide a lot of energy (in form of thermal noise) at first and then slowly reduce the amount as the network makes its way towards a global solution.

The idea is similar to that in metallurgy, where the low energy state of a metal is produced by melting it, then slowly reducing its temperature. This annealing of a metal ensures that it reaches a stable, low energy configuration and is the example which gave the name to that novel method to reach a stable output state in a Hopfield neural net — *simulated annealing*.

Using that analogy, a "high temperature" is first simulated resulting in a lot of thermal noise, then "temperature" is slowly lowered so that the amount of thermal noise decreases. To obtain these features, our novel net — which is called Boltzmann machine after the Boltzmann-distribution of states that underlies a thermal annealing process — has a probabilistic update rule. Each node in the network calculates which state it should switch into to reduce the energy as in the Hopfield net, but instead of just switching, it changes to that state depending on the values of the probability function. Therefore, the net is sometimes enabled to jump into higher energy states instead of lower ones, which allows local minima to be escaped. The probability function is chosen in such a way that the likelihood of changing the state of a unit depends on the reduction in the overall energy — if this reduction is great, switching becomes likely, but if there is not a great deal to be gained by changing, it becomes more uncertain. It has also a parameter for varying its "temperature" — at high temperatures, jumps to higher energy states are much more likely to occur than at lower temperatures. As the temperature is lowered, the probability of assuming the correct low energy state approaches one, and the network is said to have reached the so-called *thermal equilibrium*.

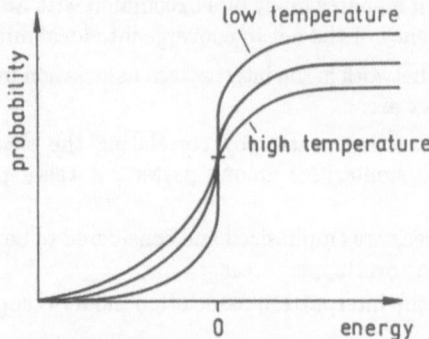

**Figure 2.22:** Probability function $1/[1 + \exp(-\Delta E_K/T)]$ for different temperatures showing that the probability of a transition to a higher energy state is greater at higher temperatures than it is at lower ones.

### 2.6.1  Mathematical Formalism

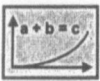

Mathematically, simulated annealing can be expressed as follows. Each unit in the network computes an energy gap given by

$$\Delta E_k = \sum_i w_{ki}x_i - \Theta_k \tag{2.64}$$

and switches into the state of lower energy according to the probabilistic update rule with

$$p_k = \frac{1}{1 + e^{-\Delta E_k/T}} \tag{2.65}$$

Therefore, the network can settle into one of a large number of global energy states, the distribution of which is given by the *Boltzmann distribution*. If $P_\alpha$ is the probability of the network to settle into some global energy state $E_\alpha$, then the Boltzmann distribution has the form:

$$P_\alpha = ke^{-E_\alpha/T} \tag{2.66}$$

Calling $P_\beta$ the probability of a state with energy $E_\beta$ being larger than $E_\alpha$, then

$$\begin{aligned} \frac{P_\beta}{P_\alpha} &= \frac{e^{-E_\alpha/T}}{e^{-E_\beta/T}} \\ &= e^{-(E_\alpha - E_\beta)/T} > 1 \end{aligned} \tag{2.67}$$

and therefore $P_\alpha > P_\beta$. This means that as the network approaches thermal equilibrium, lower energy states are more probable, dependent only on their relative energy. The setting into equilibrium is strongly dependent on the system's temperature, as illustrated in fig. 2.22. At high temperatures, the net reaches equilibrium quickly, but good global energy states are not much more likely to occur than poor ones. In that temperature regions, escaping from local minima states via higher energy states is as likely to occur as transitions from lower minima to higher energy states.

As the temperature is lowered, the probability of escaping from a higher energy minimum to a lower one falls, but the probability of travelling in the reverse direction falls even faster. Thus more low energy states are reached. Eventually, the system settles down at low temperatures in thermal equilibrium — which means that the output probabilities of the states become constant, not the values of the states themselves.

Naturally, this description of simulated annealing is an oversimplification — the energy landscape is a high-dimensional space and therefore the energy basins between states are usually massively degenerated. This implies that there are many ways of going from one state to another, the number of which increases exponentially with the amount of thermal noise added to the system.

The rate at which the temperature is decreased — mathematically achieved by adjusting the steepness of the sigmoid function as shown in fig. 2.22 — is important, since this affects the opportunities that the network has to find a globally optimum solution. If the temperature is lowered too quickly, the net does not have enough opportunities to escape from local minima. In contrast, if the temperature is lowered very slowly, the network can escape from local minima but will take a long time to converge to a final solution. At high temperatures, the net moves into high energy states easily and the overall energy of the system is high. At the other extreme, low temperatures mean that transitions to higher energy states are extremely rare, and the net will tend to stay in its current state of relatively low energy. However, the transition between these two extremes is not a gentle one, since there is a period during the lowering of the temperature when the transition from higher to lower energy minima occurs much more often than transitions in the opposite direction. It is during that period, that the overall energy of the network decreases most rapidly, and so the time spent in this transition region should be as long as possible. However, the actual problem is that determining the phase transition point in practice is rather difficult.

In physics, this behaviour is known as *phase transition* in substances as they cool and change from one state to the other: there is a critical temperature (melting or boiling point) where the state of the system suddenly changes from a high overall energy to a much lower one. With this analogy in mind, it is important that for fastest convergence to a good global minimum the temperature is decreased in such a way as to spend most of the time in the phase transition part of the system.

### 2.6.2   Learning

Learning takes place in two phases in Boltzmann machines. The network is fully-connected, but an arbitrary choice is made concerning the question which units are to be input units and which are to be output units. In the first phase, input and output units are clamped to their correct values and the net is allowed to cycle through its states with the temperature being gradually lowered until the hidden units reach thermal equilibrium. Weights connecting two units that are both "on" are then incremented — a method which is called *Hebbian or incremental learning*.

In the second phase, only the inputs are clamped to their correct values. The hidden and output units are left free. After a run as before weights between any two units that are "on" are decremented *(decremental learning)*. Thus the first phase — the incremental one — reinforces connections, while the second, the decremental one "unlearns" poor associations If the net produces the same output in both phases, learning of the correct response has been accomplished, because the same weights will be incremented and decremented by the same amount, thus cancelling out the effect of each other. In contrast, if the output e.g. of the decremental phase is not matched, some of the weights will be left on, whilst others will be turned off. After a period of time, learning will take place in such a way that only the weights between units that produce the correct output will have been left on. In contrast to the structure of multilayer perceptrons, the distinction

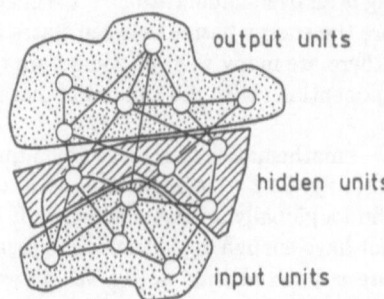

**Figure 2.23:** Visualization of the division of a fully-connected Boltzmann net into layers. For clarity, not all connections between units have been drawn.

output units

hidden units

input units

between layers is less clear in Boltzmann nets: it is fully-interconnected with the output units connected back up to the input units and the hidden units. An idea of how this arrangement could be visualized is given by fig. 2.23. The learning procedure with its two phases, the incremental and the decremental stage, is summarized in Box 7.

### 2.6.3  Optimization: Mean Field Theory

One of the disadvantages of the simulated annealing process is that the probability of switching into a state is calculated by summing the weighted outputs and subtracting the threshold of all the other units in the network. Because these units are also changing their output over time, we ought to calculate the probability based on the average output of the other units. Unfortunately, this calculation is rather time consuming. The problem can be simplified if the binary state of a unit is replaced by a real number representing the probability of that unit being in the "on" state and if we use this to estimate its average effect on the unit in question. The similarity to the *mean field theory* in physics — where the average effect of different fields acting on a body is approximated by the effect of the average of the different fields — has led to the name of the method. Mathematically, the approximated probability can be written as

$$p_i = G\left(\frac{1}{T}\sum_j <x_j> w_{ij}\right) \tag{2.70}$$

with the sigmoid function

$$G(x) = 1/(1 + e^{-x}) \tag{2.71}$$

and $<x_i>$ being the mean value of $x_i$.

Naturally, in using the mean field approximation, we introduce errors into the system and consequently the Boltzmann learning procedure is no longer strictly correct. However, the system still works and it does it much faster than before.

J.J. Hopfield made a great breakthrough in the understanding of the behaviour of Hopfield and Boltzmann nets by demonstrating that their behaviour could be expressed in terms of the energy function, and that the energy function itself was similar to the energy function encountered in the world of spin glass theory in physics. The instructiveness of the model has led me to include it into this book, although it is a completely non-optical one. Although I will not go further into detail than to demonstrate the mathematical

**Box 7: Algorithm for Boltzmann Machine Learning**

**Phase 1: incremental**

**Step 1: Clamp the input and output units to their correct values**

**Step 2: Let the net cycle through its states**
Calculate the energy of a state

$$\Delta E_k = \sum_i w_{ki} x_i - \Theta_k \qquad (2.68)$$

with $0 \leq i \leq n - 1$. Then switch to a lower energy state with the probability

$$p_k = \frac{1}{1 + e^{-\Delta E_k / T}} \qquad (2.69)$$

Reduce T until the output is stable.

**Step 3: Increment the weight between two units if they are both on (Hebbian learning)**

**Phase 2: decremental**

**Step 1: Clamp the input only, leave the output and the hidden units free**

**Step 2: Let the net reach thermal equilibrium again as in Phase 1**

**Step 3: Decrement the weights between both units if they are both on**

**Step 4 (both phases): Repeat until the weights are stable**

---

analogy between the Boltzmann net and spin glass models, this step is an important one since it shows that the techniques of statistical mechanics, as they are used in spin glass theory, can be applied in the analysis of highly-connected networks.

### 2.6.4 A Non-Optical Model: Spin Glasses

Spin glasses are magnetic alloys that show a new magnetic behaviour in a certain region of concentration: below a certain threshold temperature, the magnetic moments — which can be imagined as particles each having a particular *spin* — freeze into accidentally chosen directions. This magnetism is not a spontaneous one, but appears only if an external field is switched on for a short time. Because the spin glass behaviour appears in many different materials, there are only two major explanations for that behaviour. Normally, the spins of the particles make them want to be aligned in a common direction. However, there are usually additional forces trying to align the particles differently, such as an external field or localized effects due to surrounding particles. Depending on the material, the spins want to be aligned in a specific pattern — e.g. in one common parallel direction as in the case of ferromagnetism or in an antiparallel one as in antiferromagnets (see fig. 2.24). If both forces begin to act simultaneously in a material on different points the decision what direction to take becomes difficult for the spin. the spin particles now

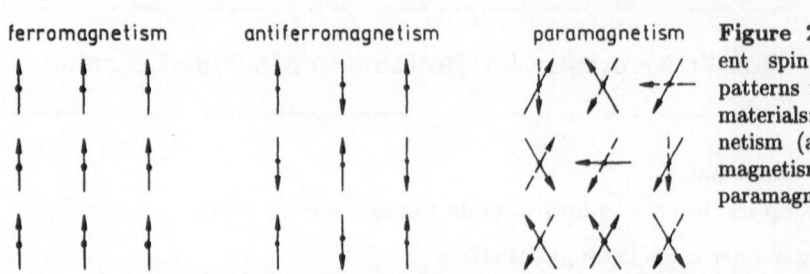

**Figure 2.24:** Different spin orientation patterns for magnetic materials: ferromagnetism (a), antiferromagnetism (b) and paramagnetism (c).

receive competing ordering instructions, which they can not fulfil simultaneously. They suffer from that situation which lets attraction and repelling act on the same time on one spin — a situation that physicists named *frustrated*. Already a very small model of spin glasses shows the decision frustration for the spin in an instructive way. Try to arrange four spins with three parallel aligning forces and one antiparallel aligning force — you will not achieve a situation that satisfies all four forces. There will be at best three of the spinforces being satisfied, the fourth one is always frustrated.

The surprising analogy to the Boltzmann machine can be seen if we consider the form of the energy function of spin glasses. The two states of the model neuron, "on" and "off", can be represented in spin glass theory using the Edwards-Anderson-model, where the states $s_i = +1$ or -1 can be interpreted as "spin up" or "spin down" states, the so-called Ising-spins. The coupling between two spins is explained by the term $j_{ij}$, which is in analogy to the weights in our neural network models.

The behaviour of the spin glass system is described by the energy function

$$H = -\frac{1}{2} \sum_i \sum_{j \neq i} j_{ij} s_i s_j + h \sum_i s_i \tag{2.72}$$

where the first term corresponds to the interaction between pairs of particles and the second to locally induced effects on single particles.

A comparison with eq. (2.50) shows, that eq. (2.72) has exactly the same representation as the Hopfield model. Again, the matrix $j_{ij}$ is symmetric ($j_{ij} = j_{ji}$). The condition for stability against a single spin flip from $s_i$ to $-s_i$, which will increase the energy of the system, is

$$s_i = \text{sign} \left[ \sum_j j_{ij} s_i s_j - h_i \right] \tag{2.73}$$

Comparison with e.g. eq. (2.52) shows that the conditions for stability are comparable as well if we consider as the nonlinear function in our neural network the Heaviside function. Then, the output at the next time interval can be given by

$$x_i = \text{sign} \left[ \sum_j w_{ij} x_j - \Theta_i \right] \tag{2.74}$$

Out of these analogies, the use of spin glass theory in the analysis of Hopfield networks has been very successful. Let us here just outline shortly the most important results:
One can imagine e.g. that stable states of spin glass configurations represent stored patterns of the system. If a spin glass system should be used as an associative memory,

recalling a pattern means to find a minimum energy configuration to a given input. In analogy to Boltzmann machines, there are metastable states other than the stored patterns and mixture states which overlap several of the stored prototypes. In addition, there are spin glass states, which bear little relation to the stored states and can therefore be considered as spurious for memory retrieval. Reliable pattern retrieval is possible in spin glass systems up to a storage density of about 0.15 stored patterns per unit number of nodes as in Hopfield nets (refer to section 2.5). As in Boltzmann nets, the effect of temperature on the system acts like noise, and for low values can smoothen the energy surface and eliminate metastable states, but for high values no retrieval solution, only spin glass ones, are found.

Now that we have found the analogy between spin glasses and Boltzmann machines, let us have a look at the more general analogy or suitability of Boltzmann machines for problems, where situations of *constraint satisfaction* involving partial frustration or so-called optimization problems have to be solved. More details on spin-glasses can be found e.g. in [14].

### 2.6.5 Optimization Problems

In general, the Boltzmann machine produces solutions that are equivalent to minima in the energy function. If it should be used as a content-addressable memory, we only have to ensure that the patterns to be stored occupy the minima in the energy function. In other words, we have to find an optimum output for the given energy function, since the network converges to some minimum value. Therefore it is possible that the same network design is able to optimize different problems, especially a number of constraint satisfaction problems, where some constraints fix the conditions under which an optimum solution has to be found. If we can express the constraints of a problem that we want to satisfy in terms of a suitable energy problem, then our Boltzmann network will produce a solution to that function that minimizes the energy.

One of the most important and well-studied problems of this optimization class is the *travelling salesman problem* or TSP problem. Its contents is the following: imagine you are a travelling salesperson for a company and have to visit all the cities in your area without visiting any city more than once. The cities are different distances apart and the problem you face is to decide the shortest route to take. For a detailed explanation to these optimization problems, refer e.g. to the textbook of by R. Beale and T. Jackson (see Further Reading of chapter 1).

Another class of problems are cost problems, e.g. if we have to minimize the cost of transporting goods and the cost is porportional to the distance the goods have to be moved. The energy function will have to be large when the distances involved are large and small when the journeys are short. Minimizing this will then correspond to minimizing the transport costs. In physics, constraint satisfactory problems arise for the construction of many different devices, e.g. for the arrangement of magnets in particle accelerator configurations.

Therefore, the concept of the Boltzmann net is extremely useful not only for neural network architectures, but for all problems, where constraints or conditions can be formulated in terms of an energy function, which has to be minimized in order to find an optimum solution.

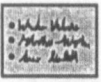 **Summary**

The Boltzmann machine can be viewed as a Hopfield net with an update rule to avoid trapping in local minima of the energy function. The points to note concerning these systems are:

- The update rule for adapting the weights is a probabilistic one.
- To assist convergence to a global minimum, simulated annealing (high temperature falling to a low one) is used.
- The Boltzmann machine is able to solve constraint satisfaction and optimization problems.

An important improvement of the Boltzmann algorithm is the mean field theory. Its features are:

- To speed up the process, the binary state of a unit is replaced by a real number representing the probability of that unit being in the "on" or "off" state.
- The average effect on the unit is estimated.

## 2.7   Adaptive Resonance Theory Networks

In the development of the adaptive resonance theory — hereafter referred to as ART — G. Carpenter and S. Grossberg designed in 1987 a net which forms clusters and is trained, in contrast to the Hopfield and Boltzmann nets, without supervision. It is the youngest among the networks I presented up to now. The major aim of this net is to establish stability in a self-organizing system. This concept arised from a problem often occuring in globally-connected networks, the *stability-plasticity* dilemma. This basic problem is based on the fact that networks are not able to learn new information on the top of old. In a multilayer perceptron network, for example, trying to add a new training vector to an already trained network may have the catastrophic side effect of destroying all the previous learning by interfering with the weight values. Because of the training time of hours or even days in computer simulations of neural nets, this is a serious limitation of a system.

The major feature of ART therefore is the ability to switch between learning states where the internal parameters of the network can be modified (*plastic states*) and stable states or fixed classification sets without detriment to any previous learning. Moreover, the network has many behavioural type properties, such as sensitivity to context — from where stems the notion of *adaptation* — that enables the network to discriminate between irrelevant information or information that is repeatedly shown in the network.

The ART network is, compared to the other network types I have discussed so far, a rather specific one. It relies much more on details of architecture than other neural network paradigms. Unlike the fairly homogeneous layers of the multilayer perceptron or Kohonen networks, every layer of the network has different functions. Moreover, there are external parts that influence the layers and control the data flow through the network.

Out of this structure, I will explain here first the way how the architecture is implemented before going on to describe the operation of the network during learning and classification.

**Figure 2.25:** Architecture of a net based on the Adaptive Resonance Theory by S. Grossberg and G. Carpenter.

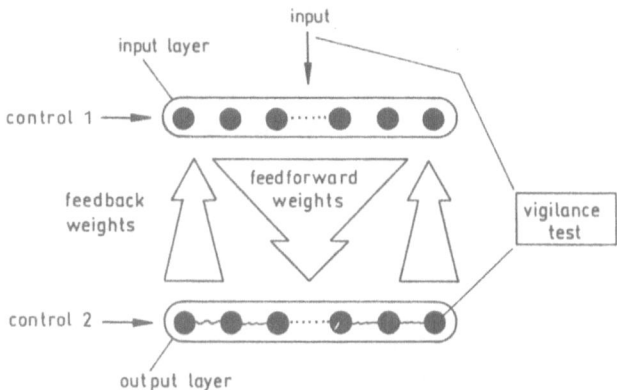

The heart of the net is the implementation of a *leader algorithm* which selects the first input as the exemplar for the first cluster. The next input is compared to the first cluster exemplar. It "follows" the leader and is clustered with the first if the distance to it is less than a certain threshold value. Otherwise it is the exemplar for a new cluster. This process is repeated for all following inputs, thus allowing the number of clusters to grow with time — depending on the threshold and the distance rules to compare input and cluster exemplars. The architecture of the net is shown schematically in fig. 2.25 and has two layers. An input or comparison layer and the output or recognition layer. These layers are connected together with extensive use of feedback in both directions, often called *memory traces*: the ART network has feedforward weight vectors from the input layer to the output layer and feedback weight vectors from the output to the input layer. These weight vectors or *long term memory traces* are named in the following text with $W$ (feedforward or *adaptive filter*) and $T$ (feedback, represents the stored exemplar vectors). For each layer, there are also logic control signals or *attentional gain control channels* in the terminolgy of Grossberg that control the data flow through the layers at each stage of the operating cycle — we will call them control-1 and control-2. Moreover, between the input and the output layer, there is a reset circuit. It is actually responsible for comparing the inputs to a vigilance threshold that determines whether a new class pattern should be created for an input pattern.

The description above is naturally a rather short and simplified version of the basic ART architecture, having substituted most of S. Grossberg's terminology by terms that seemed more instructive to me. I have included whatever I feel adds to a fuller understanding of the working principles of the network, but have reduced the description to a fairly basis level. The points to note of the ART architecture are the extensive feedback connections in both forward and feedback directions, the separate functions of each layer and the external control signals.

In operation, one has to distinguish several different phases for learning and classification. The most striking difference from the nets we have seen up to now is that the continually modified input vector is passed forward and backward between the layers in a cyclic process. This gives rise to the model's name *Resonance Theory*. We will describe the action of that network in terms of the behaviour at the separate layers for each of the four most important phases in the ART operation: the initialization phase, the recognition phase, the comparison phase, and the search phase.

### 2.7.1   The Initialization Phase

For initialization, the two control signals which direct the data flow through the network during the various learning and classification phases, and the weight vectors have to be defined. Control-1 determines the course of the data flow for the input layer — its binary value switches the first layer of nodes between its two modes: input and comparison. Control-1 has state "one" whenever a valid input is presented to the nework and is forced to zero if any node in the recognition layer is active. Control-2 is a simple one — its binary value enables or disables the nodes in the recognition layer. It is "one" for any valid input pattern, but zero after a failed vigilance test. To initialize the weight vectors $W$ and $T$, feedback weight links are all set to binary one, which means that every node in the output is initially connected to every node in the input via a feedback link. The feedforward links are set to a constant real value by

$$b_{ij}(0) = (1 + N)^{-1} \qquad (2.75)$$

where $0 \leq i \leq N - 1$ with the number of imput nodes $N$, $0 \leq j \leq M - 1$. The feedback links are set to one:

$$t_{ij}(0) = 1 \qquad (2.76)$$

At last, the vigilance-threshold $\rho$ which indicates how close an input must be to a stored exemplar to match, has to be set to a value in the range of $0 < \rho < 1$.

### 2.7.2   The Recognition Phase

In the recognition phase, the input vector is matched against the classification represented at each node in the output layer. Therefore, this layer was called a *category representation field* by S. Grossberg. To achieve this, the nodes in the input layer each have three inputs: a component of the input vector, $x_i$, the feedback signal from the output layer and the control-1 signal. The rule that governs data flow is the "two-thirds-rule" suggested by S. Grossberg and G. Carpenter: if any two inputs to a node are active then a "one" is output from the node; otherwise the node is held at zero output. At first, the input vector is compared to the exemplar at each node. As in Kohonen networks, each weight vector at each recognition node can be thought of as a *stored template* or exemplar class pattern. The best match is found via the dot product of the input vector and a node's weight vector. Therefore, the node with closest weight vector to the input will yield the largest or best result. Because several nodes in the recognition layer may respond with a high level of activation, the lateral inhibition between the nodes comes into play, turning off each node except the maximum response node. This node will inflict the largest inhibitory effect on the other nodes, so that although all the nodes are actually trying to turn each other off, it will be the maximum response node that dominates the effect. Each node has also positive feedback to itself to reinforce its own output value. The combination of these effects of reinforcement and lateral inhibition will ensure that only one node remains significantly active in the layer.

The winning node finally passes its stored class pattern back to the comparison layer. Because the exemplar is actually stored as a binary weight vector in the feedback links to the input layer, we can see that the exemplar can be passed to the comparison layer by simply mapping the winning nodes' activation through the feedback weights to the input layer.

### 2.7.3   The Comparison Phase

Two vectors are now present at the input layer for the comparison phase. On one input of each node the input vector is clamped and on the second input the exemplar vector from the recognition layer is clamped. The third input — the control-1 signal — is zero for the duration of this phase because the recognition layer has a fully active node. Thus the two-thirds-rule results in an AND-operation for the exemplar and the input vector to produce a new vector on the output of the comparison layer. We will call it the comparison vector $Z$. This vector is then passed to the reset circuit along with the current input vector to accomplish the vigilance test.

This test of similarity of the input vector and the comparison vector against the vigilance threshold is simply a ratio count of the number of ones in both the input and the comparison vector. Mathematically, it can be described with the ratio $S$:

$$S = \frac{\sum_{i=0}^{N-1} t_{ij} x_i}{\sum_{i=0}^{N-1} x_i} \tag{2.77}$$

The test now consists in testing whether

$$S > \rho \tag{2.78}$$

If the ratio is below the threshold, then this implies that the network has not yet found the correct exemplar and has to enter the search phase. Because the comparison layer represents a comparison of similar features, it was named a *feature representation field* in Grossberg's original ART model.

### 2.7.4   The Search Phase

In the search phase, the network is trying to find a new matching vector in the recognition layer for the current input vector. To achieve this, the present active output node is first disabled and its output is "zeroed". As a result, the node does not enter the best-match comparison any further and the control-1 signal is forced to zero. Afterwards, the input vector is reapplied to the recognition layer and the best-match comparison starts once again as described above. This process is repeated, consecutively disabling nodes in the output layer, until a node is found in the recognition layer that matches the input within the limits of the vigilance threshold. If there is no such node, the network decides to declare the input vector as an unknown class and allocate it to a previously unassigned node in the output layer.

This completes the working description of the various states of the ART network, showing clearly the resonant way in which the input vector is "bounced" back and forth between the input and the output layers before it finds a stable state. The ART network has much more complexity than the network algorithms discussed previously, but it can be described mathematically — as suggested by R.P. Lippmann (see Further Reading of chapter 1) — in a rather simple fashion, which is presented in Box 8.

## Box 8: The Carpenter/Grossberg ART Net Algorithm

**Step 1: Initialization**

$$t_{ij} = 1 \tag{2.79}$$

$$b_{ij} = \frac{1}{1+N} \tag{2.80}$$

$$0 \le i \le N-1, \qquad 0 \le j \le M-1 \tag{2.81}$$

Set $\rho$ with $0 \le \rho \le 1$, where $b_{ij}(t)$ is the bottom up and $t_{ij}$ is the top down connection weight between input node i and output node j at time t. These weights define the exemplar specified by output node j. $\rho$ is the vigilance threshold which indicates how close an input must be to be a stored exemplar to match.

**Step 2: Apply new input**

**Step 3: Compute matching scores**

$$\mu_j = \sum_{i=0}^{N-1} b_{ij}(t) x_i \qquad 0 \le j \le M-1 \tag{2.82}$$

where $\mu_j$ is the output of output node j and $x_i$ is element i of the input which can be 0 or 1.

**Step 4: Select best matching exemplar**

$$\mu_j^* = i_{max}\,[\mu_j] \tag{2.83}$$

This is performed using extensive lateral inhibition.

**Step 5: Vigilance test**

$$\| X \| = \sum_{i=0}^{N-1} x_i \tag{2.84}$$

$$\| T \cdot X \| = \sum_{i=0}^{N-1} t_{ij^*} \cdot x_i \tag{2.85}$$

$$S = \frac{\| T \cdot X \|}{\| X \|} = \begin{cases} \le \rho & \text{go to step 6} \\ \ge \rho & \text{go to step 7} \end{cases} \tag{2.86}$$

**Step 6: Disable best matching exemplar**

The output of the best matching node selected in Step 4 is temporarily set to zero and no longer takes part in the maximization of Step 4. Then go to Step 3.

**Step 7: Adapt best matching exemplar**

$$t_{ij^*}(t+1) = t_{ij^*}v(t) \cdot x_i \tag{2.87}$$

$$b_{ij^*}(t+1) = \frac{t_{ij^*}v(t) \cdot x_i}{0.5 + \sum_{i=0}^{N-1} t_{ij^*} \cdot x_i} \tag{2.88}$$

**Step 8: Repeat by going to step 2**

(First enable any nodes disabled in step 6)

### 2.7.5    Learning and Training of the ART Network

The training cycle for the ART network has a rather different learning philosophy compared to other neural net paradigms. The learning algorithm is optimized to enable the net to re-enter the training mode at any time and to incorporate new training data.

There are two main learning schemes for ART, which are described as *fast learning*, where the weights in the feedforward path are set to their optimum values in a very few learning cycles, and *slow learning*, which forces the weights to adapt slowly over many training cycles.

ART is the most sensitive system discussed up to now. It is very sensitive to variations in its network parameters during the training cycle. The most critical parameter is the vigilance threshold, which can alter a lot the performance of the network. The initialization of the feedforward weight vectors is important as well — if they are not set to low or small values, any vector with a high value will invariably win the best match comparison at the recognition phase. This would result in all the input vectors being assigned to just one output node — which would be a rather poor categorization of the input! The vigilance threshold parameter also has to be chosen very carefully. If it is low ($< 0.4$, [15]), we will obtain a low resolution classification process, creating fewer class types. Conversely, a high vigilance threshold (up to 1) will produce a very fine resolution classification, meaning that even slight variations between the input patterns will force a new class exemplar. In the worst case, each input pattern will then have its own class. The optimal solution would be to vary the vigilance value dynamically during the training process. A low initial value would quickly assign the coarse clustering of the input patterns and increasing this later in the training cycle may optimize the classification. Grossberg describes the possibility of modifying the vigilance value during training as a "punishment environment" — referring to the fact that an erroneous classification should be "punished" by a negative reinforcement, which must be administrated by an external circuit that monitors the response of the network. In that case, the network is already using a sort of reinforced learning instead of an unsupervised one.

### *The Feedforward Weights*

In the learning algorithm of ART, a process called *self-scaling* of the feedforward real-valued weight vector $W$ takes place. The effect of this process is as critical to the classification performance as the vigilance parameter, since it makes a step towards distinguishing noise from the signal in an input vector. The scaling process is governed by the equation for adapting the weights in the feedforward path:

$$W_{ij} = \frac{LZ_{ij}}{(L - 1 + \sum_k Z_k)} \tag{2.89}$$

The term $Z_k$ in the denominator is equal to the number of active bits in the binary comparison vector. Consequently, all the weight components $Z_{ij}$ are "normalized" by the active bit count of ones in the comparison vector. This causes the comparison vectors with a higher number of bits set to one to produce smaller weight values than those with a comparatively low active bit count. Thus that scaling prevents any vector that is a subset of another being classified in the same category. The consequence of this is that two vectors that share common features, but are in different classes, can still be distinguished.

| Input | Recognition layer | Control-1 | Result by 2/3 rule |
|-------|-------------------|-----------|--------------------|
| 1     | 1                 | 1         | 1                  |
| 1     | 1                 | 0         | 1                  |
| 0     | 1                 | 0         | 0                  |

**Table 2.1:** The two-thirds-rule and its function in the ART network

Let us now have a look at how the net passes through its stages in a training cycle. We assume a network with three input nodes, an arbitrary number of e.g. fifteen output nodes and suppose that it is initialized (control-1 = 0; control-2 = 0; output layer: all zero). The weight vectors are set to their starting states and the feedforward weights are set to a value determined by

$$b_{ij} = \frac{1}{1 + N} \tag{2.90}$$

with the dimensionality $N$ of the input vector. The exemplar patterns, which are stored in the feedback weights, are all set to binary one. When the input vector $X_i$ is applied to the input layer, the two-thirds-rule allows to AND the input vector with the control-1 signal and thus the input vector is passed unchanged through to the next, the recognition layer. Here, the input is matched against the feedforward vectors at each node by calculating the dot product of the input and the weight vectors. Because all feedforward weights have been initialized to the same starting value, it will be an arbitrary choice as to which is selected as the best-match. The "winner-node" now passes its stored exemplar back to the input layer, and the control-1 signal is forced back to zero. Consequently, the input layer has now three inputs and the two-thirds-rule has now to work in "real". The different outputs of the layer as they result from the two-thirds-rule are shown in table 2.7.

The comparison vector (1,1,0) and the input vector (1,1,0) are now both passed to the reset circuit for the vigilance test. The similarity ratio $S$ for the two vectors is evaluated — here, we get for identical vectors simply $S = 1 : 1$ — and then compared to the vigilance threshold (which we could assume e.g. to be 0.8). In our case, the similarity ratio is above the threshold and the input vector is assumed to be correctly classified. Once the vigilance test is passed, the winning node weight vector is updated to incorporate the features of the input vector. This is done by ANDing the old exemplar vector with the current input:

$$T_{ij,new} = T_{ij,old} \cap X_i \tag{2.91}$$

With $X_i = (1,1,1)$ and $T_{ij,old} = (1,1,0)$ we achieve for the new exemplar $T_{ij,new} = (1,1,0)$. The input vector $X_i$ is now stored as a class type at the node in the recognition layer.

If we now apply another training input to the network, e.g. $X_2 = (1,0,1)$ and recalculate the matching scores at the recognition layer, we find that the node assigned to the $X_1$ input will be the winning one. This is because its feedforward weight values are much larger than those of the other, yet unassigned nodes. As a result, the exemplar for class 1 (input $X_1$), will be passed, erroneously, to the comparison layer with the input $X_2$. However, if we trace the exemplar through to the reset circuit as before, we obtain

$$S = \frac{\sum T_{ij} X_i}{\sum X_i} \tag{2.92}$$

which results in our example in $S = 0.5$ and is thus lower than the vigilance threshold value $\rho$. Thus the net considers the classification as wrong and enters the rest phase. This means that node one will be disabled for the duration of the present input, the recognition layer is set to all zero and the vector is reapplied to the recognition phase without node one. Then the classification will proceed as for the first input and $X_2$ will be assigned to an unused output node.

One of the important features of ART is — beneath its full exploitation of parallel processing— its fast learning time compared to the iterative convergence procedures proposed for most other networks. This is mostly due to the fact that none of the weight values are modified at all until the search process has halted and one node has been selected. The other important advantage is that its classification stage remains open to adaptation in the event of new information being applied to the network — an unknown input will be assigned to a new class in the output (recognition) layer. The limit of this process is only given by the number of nodes available in the recognition layer.

A further development of the ART concept is the construction of ART 2, where both binary and/or real valued inputs are possible. The criticism on ART concerns its poor results for noisy input conditions, the use of non-distributed coding of data and the implausible architecture of the network which made it difficult for physicists to implement or simulate it in real experimental systems. However, the different concept of ART and the ability of staying stable in a changing evironment, e.g. when new training data enter to a set of trained nodes, makes it extremely interesting for a lot of applications in large data amount nets which are often submit to changes.

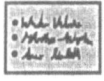 **Summary**

The ART network has many significant differences from the other neural paradigms discussed up to now. The most important are

- ART is based on cognitive and behavioural concepts as they are lend from psychological research.
- ART is an unsupervised, vector-clustering, competitive learning net algorithm.
- ART is able to deal with the stability-plasticity dilemma of learning in a changing environment.
- ART uses extensive feedback between input and output layers.
- ART may be implemented for both real or binary inputs.
- ART is mathematically fully-described by nonlinear differential equations.

 **Further Reading**

1. D. Hebb, *The Organization of Behaviour*, J. Wiley & Sons (1949).
   The original ideas of Hebb concerning learning by reinforcement with special emphasize on learning in simple neurons and the one-layer perceptron.

2. F. Rosenblatt, *The perceptron: a probabilistic model for information storage and organization in the brain*, Psych. Rev. 65 (1958), p. 386.
   One of the original publications on the beginning of the perceptron model.

3. M.A. Minsky, S.A. Papert, *Perceptrons. An introduction to Computational Geometry*, MIT Press (1969) and *Perceptrons. Expanded edition*, MIT Press (1988).
   Concepts and criticism on single-layer perceptrons, with a lot of mathematical derivations and proofs.

4. P. Kanerva, *Sparse Distributed Memory*, MIT Press, Bradford Books, Cambridge (Mass.) (1986)
   An overview over models that are derived from the backpropagation paradigm.

5. J.S. Albus, *Brains, Behavior and Robotics*, McGraw-Hill, New York (1981)
   A summary of models that originate from the multilayer perceptron idea.

6. T. Kohonen, *Self-Organization and Associative Memory*, Springer Verlag, Berlin (1984).
   An introduction to concepts of associative memory and discussion of self-organization principles.

7. T. Kohonen, *Adaptive, associative, and self-organizing functions in neural computing*, Appl. Opt. 26 (1987), p. 4910.
   This paper describes certain adaptive and cooperative functions encountered in neural networks, including biological and mathematical models of description.

8. H. Ritter, T. Martinetz and K. Schulten, *Neural computing and self-organizing maps*, Addison-Wesley, Reading, Massachusetts (1992).
   Self-organization is prescribed for neural computing, including the formation of feature nets and applications for computational tasks.

9. J.J. Hopfield, D.W. Tank, *Computing with neural circuits: a model*, Science 233 (1986), p. 633.
   A well-written description of the Hopfield model.

10. G.E. Hinton and T.J. Sejnowski, *Learning and relearning in Boltzmann machines*, in J.L. McClelland, D.E. Rumelhardt, *Parallel Distributed Processing*, Vol. 1, MIT Press (1986), chapter 7.
    This chapter of the famous *Parallel distributed Processing* book gives an excellent overview over the principles of Boltzmann machines.

11. G.A. Carpenter, S. Grossberg, *The ART of Adaptive Pattern Recognition*, IEEE Computer **21** (1988), no. 3.
    A well-structured introduction to ART and a big source for further references.

# 3 Basic Concepts of Nonlinear and Photorefractive Optics

Nonlinear optics are not part of our everyday experience. Their discovery and investigation became possible only after the invention of strong energy sources, especially the laser. Although the concept of laser radiation was already formulated by A. Einstein at the beginning of the century, it was not realized and developed until the early sixties. Often, the discovery of second-harmonic generation in 1961 by Franken et al. is taken as the beginning of the field of nonlinear optics, which dates only shortly after the invention of the first laser by Maiman in 1960. Therefore, the field of nonlinear optics is a rather new one and still in the state of expansion.

In general, optics treat the interaction of light with matter. This interaction is described by the index of refraction $n$, and the absorption coefficient $\alpha$. At relatively low light intensities as they normally occur in nature or are caused by conventional light sources, these optical properties are independent of the intensity of illumination. If light waves penetrate and pass in that case through a medium, reflection, refraction, absorption and the velocity of the waves are constants of the medium, depending only on the material characteristics and not on the illumination intensity. This implies two major principles of linear optics: The *principle of superposition* and the *conservation of frequency*. The principle of superposition relies on the fact that waves pass and penetrate a medium without any interaction between them. In other words, a light wave travels in a medium, independent from the fact whether there is another, second light wave or not. That independent superposition of different waves is e.g. the most important condition for application of the Fourier-decomposition, where an arbitrary light wave is divided into a sum of contributions to that light wave with different frequencies or the spectral analysis of the light. The conservation of frequency means that for every interaction of light with matter no new frequencies can appear and no frequency components of the incoming light can disappear. The frequency of a light wave remains constant inside and outside a material.

These two principles are the basis of all the optical properties of matter that are familiar to us through our visual sense — as diffraction, refraction, interference, scattering and absorption. However, if the illumination is made sufficiently intense, the optical properties begin to depend on intensity and other characteristics of light. In that case, the light waves may begin to interact with each other as well as with the medium — leaving the medium in some cases severely changed behind after a single pass through it.

This is the area of nonlinear optics. It is nonlinear in the sense that nonlinear optical phenomena occur when the response of a material system to an applied optical field depends in a nonlinear manner upon the strength of the optical field. For example, second-harmonic generaton occurs as a result of the part of the atomic response that depends quadratically on the strength of the applied optical field. Consequently, the intensity of the light generated at the second-harmonic frequency tends to increase as the square of the intensity of the applied laser light.

The behaviour which is produced by nonlinear optics provides on the one side insight into the structure and properties of matter, on the other hand it can be utilized to great

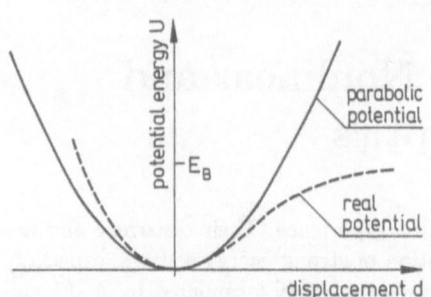

**Figure 3.1:** Comparison of an ideal potential (parabolic curvature) with the typical form of a real potential.

effect in the construction of nonlinear optical devices and techniques for applications in optical information and signal processing, telecommunication, optical display devices or — as in that book — for applications in optical neural network models. The basis of these devices is mostly given by the fact that an information carrying light wave — one can imagine e.g. an image with a grey scale representation — changes the medium's characteristics — e.g. the index of refraction — proportional to the information of the light wave. Thus, the light wave's content is visible or "stored" in the medium even when the wave has passed through and there must be possibilities to recall and process the information which is now available in the medium independent from its source — the travelling light beam. The nonlinear effects thus allow us to manipulate, store or process a given light wave.

## 3.1   Origins of Optical Nonlinearities

In the simplest model of nonlinear optical behaviour, we can think of a material as a collection of electrons and ion cores. When an electric field is applied, the charges begin to move: the positive charges tend to move in the direction of the field, whilst the negative ones move the opposite way. In conductors, some of the charged particles are free to move through the material for as long as the electric field is applied, giving thus rise to a flow of electrical current. In dielectric materials, the situation reveals to be different. Here, the charged particles are electrically bound together. Therefore, the force on the charges induced by the outer electric field causes a transitory motion of the charges: they are displaced slightly from their usual position — which results in a collection of induced electric dipole moments. In other words, the effect of the field on a dielectric medium is to induce a polarization.

As light waves consist of electric and magnetic fields which vary sinusoidally at frequencies of about $10^{13}$ to $10^{17}$ Hz, the motion of the charged particles in a dielectric medium in response to an optical electric field is oscillatory: they form oscillating dipoles. Because the effect of the optical magnetic field on the particles is much weaker and the positively charged particles — the ion cores — have a much greater mass than the electrons, it is the motion of the electrons in an electric field that is most significant for a certain material. The response of an electron to the electric field can be imagined with a simple mechanical analogy: that of a particle in an anharmonic potential well. In fig. 3.1, such an ideal potential well is compared to the real potential structure for an electrically bound electron. The differences to such an ideal potential well become only important for higher

**Figure 3.2:** Real and imaginary part of the electric susceptibility $\chi$ which represents the refractive index and absorption of the medium respectively.

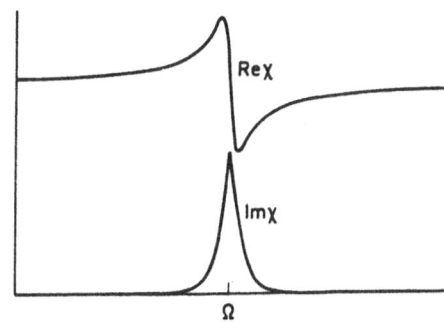

oscillation energies. Thus, the position of the electron varies in response to the electric field $E(t)$ in a manner governed by the equation of motion for an oscillator:

$$m\left[\frac{d^2\vec{x}}{dt^2} + 2\Gamma\frac{d\vec{x}}{dt} + \Omega^2\vec{x} - (\xi^{(2)}\vec{x}^2 + \xi^{(3)}\vec{x}^3 + \cdots)\right] = -e\vec{E} \tag{3.1}$$

where $\vec{x}$ is the displacement from the mean position, $\Omega$ is the resonance frequency and $\Gamma$ is a damping constant. $\xi^{(2)}\vec{x}^2 + \xi^{(3)}\vec{x}^3 + \cdots$ is the anharmonic term representing the nonlinear interaction in the oscillation. The term on the right-hand side of eq. (3.1) represents the external force on the electron by the applied field which drives the oscillation.

For the moment, we may ignore the anharmonic term $\xi^{(2)}\vec{x}^2 + \xi^{(3)}\vec{x}^3 + \cdots$ and consider the harmonic response, which has the form:

$$\vec{E}(t) = \vec{E}_0 \cos(\omega t) = \frac{1}{2}\vec{E}_0\left[e^{-i\omega t} + e^{+i\omega t}\right]. \tag{3.2}$$

Substituting eq. (3.2) in eq. (3.1) gives a linear equation with the solution

$$\vec{x} = -\frac{e\vec{E}_0}{2m} \cdot \frac{e^{-i\omega t}}{\Omega^2 - 2i\Gamma\omega - \omega^2} + c.c. \tag{3.3}$$

where $c.c.$ denotes the complex conjugate.

If there are $N$ such electric dipoles per volume, the polarization induced in the medium is

$$\vec{P} = -Ne\vec{x} = \frac{1}{2}\epsilon_0\chi\vec{E}_0 e^{-i\omega t} + c.c. \tag{3.4}$$

where we have expressed the linear dependence of the polarization $\vec{P}$ on the field $\vec{E}$ in terms of the susceptibility $\chi$

$$\chi = \frac{Ne^2}{\epsilon_0 m} \cdot \frac{1}{\Omega^2 - 2i\Gamma\omega - \omega^2} \tag{3.5}$$

and $\epsilon_0$ is the free-space permittivity. The electric dipoles (and with them the polarization) therefore oscillate at the same frequency as the incident optical field. They radiate into the medium and modify the way in which the waves propagate. Since the electric displacement is $\vec{D} = \epsilon_0\vec{E} + \vec{P}$, we see that the dielectric constant is $\epsilon = \epsilon_0\epsilon_r$ with $\epsilon_r = 1 + \chi$ and the refractive index becomes $n = \text{Re}(\sqrt{1+\chi})$. Losses in the medium are taken into account by the imaginary part of $\chi$, which includes the component of $\vec{P}$ being quadratic with the field. Both real and imaginary part of $\chi$ are sketched in fig. 3.2, showing the familiar linear optical behaviour in a medium.

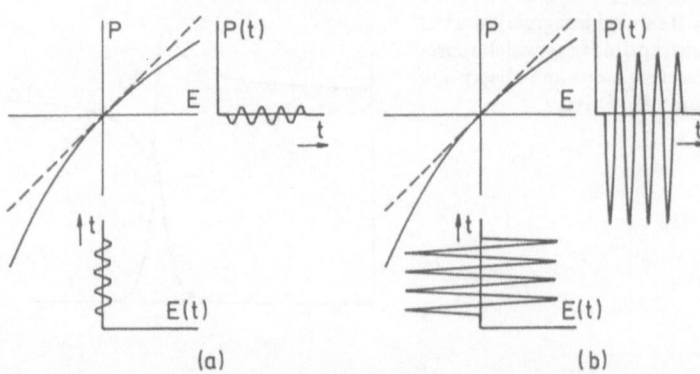

**Figure 3.3:** The effect of a nonlinear dependence of the polarization $P$ on the electric field $E$ compared to the linear dependence. For small input fields (a), $P$ does not deviate significantly from a linear dependence (dashed line). At larger fields (b), $P$ has a distorted waveform which contains significant components at harmonic frequencies.

However, in physics the linear dependence of one quantity on another is almost always an approximation, which is valid only in a limited range of values. In our case, the motion of charged particles in a dielectric medium can be considered to be linear with the applied field only if the displacement $\vec{x}$ is small. With conventional light sources having fields about $|E| = 1$ V/cm, we achieve displacements about $|\vec{x}| \approx 10^{-16}$ cm which is small compared to the atomic scale of approximately $10^{-8}$ cm. Therefore, the "ideal" potential remains valid for conventional light.

The situation changes completely for stronger fields, as they are created by lasers. The displacement can grow large enough to observe the nonlinearities in the input-output characteristic (see fig. 3.3). In the mathematical representation (see eq. (3.1)), the non-linearity is accounted for by terms which represent an additional anharmonic restoring force depending on quadratic and higher terms in $\vec{x}$: $\xi^{(2)}\vec{x}^2 + \xi^{(3)}\vec{x}^3 + \cdots$ with the constants $\xi^{(2)}, \xi^{(3)}, \cdots$. Graphically, the behaviour is illustrated in fig. 3.3. An anharmonic response gives rise to an induced polarization which can be considered to be either linear (to a good approximation) or significantly nonlinear, depending on the magnitude of the applied field. A spectral analysis of the polarization wave in the nonlinear case reveals that, in addition to the major component oscillating at the input frequency $\omega_i$, it contains significant contributions oscillating at the harmonic frequencies $2\omega$, $3\omega$, $\cdots$ and a component at zero frequency. Although this is the analogue to the well-known harmonic distortion of signals in an electric circuit, whose response is not perfectly linear, the important difference in electromagnetic theory is that an oscillatory electric dipole limits a radiation field to the frequency of the oscillators. Therefore, the component of the polarization that oscillates at the second-harmonic frequency $2\omega$ can radiate a field at $2\omega$. This is the famous process of *second harmonic generation*.

When the anharmonic terms are included, there is no longer an exact solution for the equation of motion (eq. (3.1)). The polarization is then a highly complex function of the fields and can be expressed by:

$$\vec{P} = \epsilon\chi(\vec{E}) \cdot \vec{E} \tag{3.6}$$

The electric susceptibility and therefore also the index of refraction are now a function of the electric field of the light. However, provided that the anharmonic terms are small

compared to the harmonic one, we can solve eq. (3.1) to successive orders of approximation by expressing $\chi$ as a power series in $\vec{E}$. Equivalently, we can express the polarization
$\vec{P}$ in a Taylor-series in the form

$$\vec{P} = \epsilon(\chi^{(1)}\vec{E} + \chi^{(2)}\vec{E}^2 + \chi^{(3)}\vec{E}^3 + \cdots) \tag{3.7}$$

Here, $\chi^{(1)}$ describes the linear susceptibility we discussed previously, and the quantities
$\chi^{(2)}$, $\chi^{(3)}$, $\cdots$ are called the nonlinear susceptibilities of the medium. Writing down the
power series expression (eq.(3.7)) for a realistic medium, we have to take into account that
the vectorial relation between the polarization $\vec{P}$ and the electric field $\vec{E}$ is not as easy as
the equation suggests. Only in the case of isotropic materials (glass, polymers, liquids and
gases) and cubic crystals, the susceptibility $\chi$ is a scalar. In that case, the medium has no
outstanding direction and the velocity of light is independent of the state of polarization
of the light. Here, $\vec{E}$ and $\vec{P}$ point in the same direction. For anisotropic media (e.g.
birefringent crystals), the velocity of the light in the material depends on the state of
polarization of the incoming light and on the crystal orientation. As a consequence, the
polarization $\vec{P}$ and the index of refraction $n$ become dependent on the direction of $\vec{E}$ and
$\vec{P}$. Consequently, $\vec{E}$ and $\vec{P}$ point no longer in the same direction — in other words, the
displacement of the elastically bound electrons does not occur in the same direction as the
driving force. In the vectorial equation, each of the three components of the polarization
depends on the three components of the electric field. Therefore, there are nine constants
$\chi_{ij}^{(1)}$ $(i, j = 1, 2, 3)$ of the susceptibility necessary to describe the relation between $\vec{P}$
and $\vec{E}$ sufficiently. This sort of expression is called a tensor, which is the general, linear
complimentation of two vectors. The situation gets even more complicated if we want to
describe the relation between the polarization and the quadratic term of the electric field.
Here, each component of the polarization has to be combined with the combinations of
$E_i \cdot E_j$, e.g. $E_x \cdot E_x, E_x \cdot E_y, E_y \cdot E_z$ — which involves 27 constants $\chi_{ijk}$ to be defined. The
symbol for all $\chi_{ijk}$ is $\chi^{(2)}$ and defines a third order tensor which describes the quadratic
combination between two vectors. However, depending on the symmetry of the materials,
not all of these susceptibility tensor elements may be different from zero, because they
are not all independent of each other and may act in such a way to compensate their
relative values.

To get a feeling which intensities we need to realize any nonlinear effect experimentally,
let us consider how large the incident optical field must be to allow atoms and molecules
to reveal their nonlinear properties. From the above discussion it is apparent that for a
significant nonlinearity arising from the anharmonic motion of the electrons, we require
an incident field which is not entirely negligible in comparison with the internal field $E_a$
which binds together the electrons with the ion cores. Because $E_a$ is typically about $3 \cdot 10^{10}$
$Vm^{-1}$, we would need an incident intensity of $10^{14}$ $Wcm^{-2}$ to obtain an optical field of
such a magnitude. These intensities may be achieved by focusing powerful picosecond-
duration laser-pulses from mode locked lasers.

Fortunately however, these high intensities are not necessary for the observation of
many nonlinear optical effects, because the assembly of induced dipoles may oscillate coherently — which means that a definite phase relationship between them exists — and the
fields they radiate may then add constructively to produce a much higher total intensity.
In nonlinear optics, this condition of constructive interference is known as *phase matching*. The intensity required to observe some nonlinear processes can be reduced further
by many orders of magnitude by choosing one or more of the optical frequencies of the
external field in such a way that they lie close to a resonant frequency of the oscillating

dipoles. This is called *resonance enhancement*. In nonlinear optics, resonance enhancement can be used on the one hand to operate nonlinear processes and devices effectively at low power levels, thus increasing their range of use and efficiency. On the other hand, resonant nonlinear phenomena provide the basis for nonlinear spectroscopy, because the observation of these effects provides information about the structure of matter that is not accessible using conventional optical spectroscopy.

| quadratic effects | $\chi^{(2)}$ | polarization |
|---|---|---|
| frequency doubling | real | $\vec{P}^{(2)}(2\omega) = \epsilon_0 \cdot \chi^{(2)} \cdot \vec{E}^2(\omega)$ |
| linear electrooptic effect (Pockels effect 1893) | real | $\vec{P}^{(2)}(\omega) = \epsilon_0 \cdot \chi^{(2)} \cdot \vec{E}_1(0) \cdot \vec{E}_2(\omega)$ |
| linear Stark effect | imaginary | $\vec{P}^{(2)}(\omega) = \epsilon_0 \cdot \chi^{(2)} \cdot \vec{E}_1(0) \cdot \vec{E}_2(\omega)$ |
| parametric amplification (1962) | real | $\vec{P}^{(2)}(\omega_1 \pm \omega_2) = \epsilon_0 \cdot \chi^{(2)} \cdot \vec{E}_1(\omega_1) \cdot \vec{E}_2(\omega_2)$ |
| Faraday effect (1845) | complex | $\vec{P}^{(2)}(\omega) = \epsilon_0 \cdot \chi^{(2)} \cdot \vec{E}_1(\omega) \cdot \vec{H}_2(0)$ |
| optical rectification (1962) | real | $\vec{P}^{(2)}(0) = \epsilon_0 \cdot \chi^{(2)} \cdot <\vec{E}^2(\omega)>$ |

| cubic effects | $\chi^{(3)}$ | polarization |
|---|---|---|
| frequency tripling (1962) | real | $\vec{P}^{(3)}(3\omega) = \epsilon_0 \cdot \chi^{(3)} \cdot \vec{E}^3(\omega)$ |
| self-focusing (1964), thermal effects | real | $\vec{P}^{(3)}(\omega) = \epsilon_0 \cdot \chi^{(3)} \cdot \vec{E}(\omega) \cdot <\vec{E}^2(\omega)>$ |
| saturable absorption, saturable amplification | imaginary negative | $\vec{P}^{(3)}(\omega) = \epsilon_0 \cdot \chi^{(3)} \cdot \vec{E}(\omega) \cdot <\vec{E}^2(\omega)>$ |
| two-photon-absorption (1961) | imaginary positive | $\vec{P}^{(3)}(\omega) = \epsilon_0 \cdot \chi^{(3)} \cdot \vec{E}(\omega) \cdot <\vec{E}^2(\omega)>$ |
| quadratic Stark effect | imaginary | $\vec{P}^{(3)}(\omega) = \epsilon_0 \cdot \chi^{(3)} \cdot \vec{E}_1(\omega) \cdot \vec{E}_2^2(0)$ |
| electrical Kerr effect (1875) | real | $\vec{P}^{(3)}(\omega) = \epsilon_0 \cdot \chi^{(3)} \cdot \vec{E}_1(\omega) \cdot \vec{E}_2^2(0)$ |
| frequency doubling in an external field (1962) | real  real | $\vec{P}^{(3)}(2\omega) = \epsilon_0 \cdot \chi^{(3)} \cdot \vec{E}_1^2(\omega) \cdot \vec{E}_2(0)$  $\vec{P}^{(3)}(2\omega) = \epsilon_0 \cdot \chi^{(3)} \cdot \vec{E}_1^2(\omega) \cdot \vec{H}_2(0)$ |
| Cotton-Mouton effect (1907) | real | $\vec{P}^{(3)}(\omega) = \epsilon_0 \cdot \chi^{(3)} \cdot \vec{E}_1(\omega) \cdot \vec{H}_2^2(0)$ |
| induced Brillouin-, Raman-, Rayleigh- scattering (1962/1964) | imaginary negative | $\vec{P}^{(3)}(\omega \pm \Omega) = \epsilon_0 \cdot \chi^{(3)} \cdot \vec{E}_1(\omega \pm \Omega) \cdot \vec{E}_2^2(\omega)$ |

**Table 3.1:** Examples for quadratic and cubic nonlinear effects. The right column gives the polarization $\vec{P}^{(2)}$ resp. $\vec{P}^{(3)}$, depending on the frequency at which the polarization or the electric field are oscillating. $\vec{E}(0)$ or $\vec{H}(0)$ are representing dc-fields. $<\vec{E}^2(\omega)>$ is the temporal average, i.e. the effect is caused by the average radiation intensity. The coefficients $\chi^{(2)}$ and $\chi^{(3)}$ are the nonlinear susceptibilities as defined by eq. (3.7).

An overview over important quadratic and cubic nonlinear effects is given in table 3.1. In the following chapters, I will shortly review some of the main nonlinear optical phenomena that may occur. I have chosen especially those which are of particular interest for applications and simulations of neural network structures.

## 3.2 Quadratic Effects Arising from the Second Order Nonlinear Polarization

As I have shown in the previous chapter, the effect of nonlinear dependence of the polarization $\vec{P}$ on the electric field $\vec{E}$ gives rise to a polarization which has a distorted waveform, containing significant components of harmonic frequencies of the incoming light. As a consequence, the second order polarization

$$\vec{P}^{(2)} = \epsilon_0 \chi^{(2)}(\vec{E})\vec{E}^2 \tag{3.8}$$

basically causes all effects which are mixing phenomena, involving the generation of sum and difference frequencies and second-harmonic generation, depending on how many frequencies are contained in the incoming beam.

### 3.2.1 Second Harmonic Generation

The most important of these second order polarization effects, the second harmonic generation, giving rise to $P_{NL}(2\omega)$, can be generated in a number of crystals as e.g. Potassiumdihydrogenphosphate (KDP), if only a single input frequency is operating. It has found many applications in the field of doubling of strong laser frequencies that can be brought into the visible region by second harmonic generation.

At a first glance, the effect of a coherent optical field on the medium is to induce an assembly of electric dipoles which oscillate coherently. Such an assembly of oscillating dipoles comprises a macroscopic polarization and therefore we are just concerned with the component of the polarization which oscillates at $2\omega$. This may be interpreted as to give rise to the radiation of an electric field $\vec{E}_2(2\omega)$ with the double frequency $2\omega$ compared to the incoming light. However, if we look somewhat deeper into the details of the mechanism, there are a lot of difficulties that arise before a second harmonic wave can be created. The phase of the induced polarization component is not constant throughout the medium. Instead, fronts of constant phase move through the medium in the propagation direction with a phase velocity $c_1 = \omega/k_\omega$. This gives rise to the idea of a polarization "wave" (although the polarization does not actually move in space). The second harmonic light wave, on the other hand, travels with a phase velocity $c_2 = c_0/n_2 = 2\omega/k_{2\omega}$. The phase-matching condition $\Delta k = 2k_\omega - k2\omega = 0$ is satisfied when these two phase velocities are equal. Because in most cases, $n_2(2\omega)$ is larger than $n_1(\omega)$ — which is called *normal dispersion* — $c_2$ is smaller than $c_1$ and the second harmonic is much slower in the medium than the creating polarization $\vec{P}_{NL}$. As a result, $\vec{P}_{NL}$ and $\vec{E}_2$ are no longer in phase with increasing crystal length and the contributions of the second harmonic which have been created at different areas in the crystal are no longer able to superimpose in phase. Although the radiance intensity of the second harmonic wave rises first, it drops again down to zero after a certain length in the crystal because of these

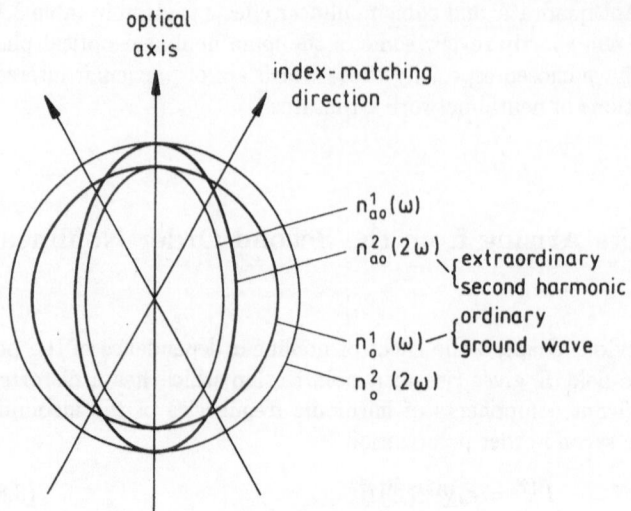

**Figure 3.4:** Dependence of the ordinary (o) and extraordinary (ao) refractive indices on the angle $\theta$ of the incident light beam relative to the optical axis of the uniaxial, birefringent crystal KDP as an example. The figure is symmetric around the optical-axis.

phase mismatching effects. Afterwards, when the second harmonic radiance intensity has reached its zero level, the process can start again. Therefore, the crystal can exhibit a periodically varying radiance intensity of the second harmonic wave, but destroys the possibility to create an intense light beam of double frequency.

How can we then be able to take use of the second harmonic generation? Theoretically, we have to prevent the crystal from realizing the conditions for destructive interference or, in other words, we have to take care that a situation for constructive interference is possible, as it is when the condition $c_1(\omega) = c_2(2\omega)$ or $n_1(\omega) = n_2(2\omega)$ is fulfilled. This method is indeed used very commonly to obtain intensive frequency doubled light and is called *phase or index matching*. However, for all crystals the index of refraction rises with frequency (normal dispersion), making the fulfilling of the phase matching condition impossible in reality. Does this mean that there is no way to get a second harmonic wave out of a real crystal? Fortunately, physicists have found a tricky method to circumvent the crystal's characteristics. The trick is to take advantage of dispersion and birefringence at the same time. I will explain this method for the case of KDP, an optical uniaxial crystal with two indices of refraction, the ordinary $n_o$ and the extraordinary $n_{ao}$. The idea of the method is that the ordinary polarized incoming light beam of frequency $\omega$ creates a nonlinear polarization $\vec{P}_{NL}(2\omega)$, which is parallel to the optical axis. Therefore, the electric field of the second harmonic $\vec{E}_1(2\omega)$, which is radiated by that polarization, has the same polarization parallel to the optical axis — and is hence an extraordinary beam in the crystal. Thus an ordinary ray of the basis frequency creates an extraordinary ray at the double frequency. The incoming wave and its second harmonic are consequently polarized perpendicular to each other. In KDP, infrared light with the wavelength $\lambda = 1,06\,\mu m$ generates a blue line at $\lambda = 0,53\,\mu m$. Now, we can take advantage of the fact that the index of refraction $n_{ao,2}$ is dependent on the angle of the incident light beam. Consequently, there is a special angular configuration relative to the optical axis, at which $n_{o,1}(\omega) = n_{ao,2}(2\omega)$. If the incoming light beam travels at this angle relative to the c-axis through the crystal, the velocities for the incoming wave, the polarization wave and the second harmonic wave are the same. This situation is sketched in fig. 3.4. There is no phase mismatch between the polarization wave and the second harmonic one so that all contributions to the second harmonic wave interfere in phase. Finally, the intensity of the second harmonic wave rises with the crystal thickness until the whole power of the ground wave is converted into the second harmonic wave. In real crystals, efficiencies of about 50 - 60% can be experimentally achieved, provided that the crystal has a large nonlinearity and that index matching is possible in an angular range which is easily accessible experimentally.

Now that I have discussed the principle of index matching, one may ask whether there is also the possibility to obtain third and fourth harmonics in that way. Theoretically, the same matching principle may be applied for that generation of higher order harmonics, but the efficiency drops dramatically with the nonlinear degree of $\chi$ — third harmonic generations are possible with efficiencies of about $\eta \approx 10^{-5}$. To get acceptable results at the output of the crystal in that case, one has to enhance the incoming light intensities as much as that the damage threshold of the material may be approached. Out of that reason, the favourite method of obtaining higher order harmonics is to double the frequency of a frequency doubled beam once again. For KDP, the blue light may be reconverted into ultra-violet light with $\lambda = 265$ nm with an efficiency of about $\eta \approx 10^{-3}$. More nonlinear effects that are caused by the second order polarization term arise when the applied optical field contains two frequencies. The simplest case of mixing occurs when one of the frequencies is zero, e.g. when we send an optical wave through the medium in the presence of a d.c. electric field. This may give rise to different effects depending on which parameter of the material the external field acts.

### 3.2.2   Pockels, Stark and Faraday Effect

To produce the Pockels effect, the second order polarization has to contain a term that is proportional to the product of the optical and the applied d.c. field. Hence, the refractive index at the optical frequency depends on the d.c. field. This is the linear electrooptic or Pockels effect, which is widely used in optical modulators. A similar effect is the linear Stark-effect, where it is again the external d.c. field that alters the material properties. In contrast to the Pockels effect, the external d.c. field induces here a shift in the energy levels of the atoms. Thus, not only the refractive index, but both, refractive index $n$ and absorption coefficient $\alpha$ are changed. In the case of a negligible change in $n$, but a large change in $\alpha$, the susceptibility $\chi^{(2)}$ becomes imaginary.

Although these two electrooptic effects are well described within the framework of nonlinear optics, they are exceptions to my opening remark, that nonlinear effects require intense illumination for their observation. Indeed, the discovery of the Pockels effect in 1893 predated the invention of the laser by exactly 70 years. The same is true for the magnetooptic or Faraday effect, which was found in 1845 and which is based on the interaction of a magnetic d.c. field with matter. Again, the index of refraction and the absorption are altered via a shift in the atomic energy levels. Compared to the Stark effect, the change in the absorption is small in real magneto-optic materials, leading to approximately real susceptibilities $\chi^{(2)}$ and therefore to almost pure refractive index changes.

### 3.2.3  Parametric Amplification

If two or more intensive light waves are superimposed in the crystal, a wide range of interesting frequency mixing phenomena may occur. Perhaps the most interesting one arising from the second order polarization of that kind is the parametric amplification, especially by photorefraction. It has an essential meaning for a variety of optical neural network models, in particular for data storage and comparison. Therefore, it is the key nonlinear effect for the applications described in this book and thus will be treated in a separate section, section 3.4, in more detail. Here, I will describe the basic features of parametric amplification in the frame of its relation to other nonlinear optical polarization effects.

Parametric amplification occurs when we send a small optical signal at frequency $\omega_s$ through the medium in the presence of a powerful optical field — called the pump wave — at a higher frequency $\omega_p$. The pump and the signal beat together to produce a field at the difference frequency $\omega_1 = \omega_p - \omega_s$ — which is called the idler field. It is proportional to the product of the pump and signal fields. Then a second beating action takes place. This time, the idler beats with the pump to produce a term in the polarization at the difference frequency $\omega_p - \omega_1 = \omega_s$, which is proportional to the product of the signal field and the pump intensity (pump field squared). Thus, as a result of that double beating action, there is an extra term in the total polarization which is linear in the signal field. In its effect on the signal field, this extra term is equivalent to changing $\chi^{(1)}$ by an amount proportional to the pump intensity. The effect is called a "parametric" one, because the pump field may be regarded as modulating the linear susceptibility — the parameter $\chi^{(1)}$ — at the pump frequency.

The parametric interaction is particularly strong when it is phase-matched. This is equivalent to ensure that the propagation constants $k_s$, $k_1$ and $k_p$ at the frequencies $\omega_s$, $\omega_1$ and $\omega_p$ respectively, satisfy the condition $k_s + k_1 = k_p$. In that case, power is removed from the pump wave to the benefit of both the signal and idler waves, which are consequently amplified. Naturally, this effect is extremely useful for optical devices such as amplifiers and oscillators as well as for neural network models. It can be exploited in photorefractive materials using dynamic holography. Because this quadratic nonlinear effect has an essential meaning for applications to optical neural networks, we will consider the fundamentals of these two-wave mixing phenomena in a separate section, section 3.4, in more detail.

There is also a special class of materials, where $\chi^{(2)}$ vanishes: in crystals with inversion symmetry, opposite directions are completely equivalent and so the polarization must change its sign when the crystal's electric field is reversed. Hence there can be no even powers of the field in the expansion of the polarization. In these media, the first nonlinear term in the polarization is the cubic term, which I will now consider.

---

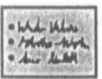 ## Summary

---

Nonlinear optics arise due to an electric field that is incident on a material via a strong light beam. It induces an oscillation of the electron that is no longer harmonic. Thus, the oscillation amplitude as well as the polarization contains higher harmonic contributions.

- In second harmonic generation, an ordinarily polarized ray of the basic frequency creates an extraordinarily polarized ray at the double frequency in the medium. A special angular configuration has to be chosen, where the indices of refraction of these two beams are equal. Then, the velocities of both waves are equal, no phase mismatch occurs between the polarization wave and the second harmonic one and all contributions to the second harmonic wave interfere in phase.
- Third, fourth and fifth harmonic generation may also be realized applying the same matching principle, but the efficiency drops dramatically with the degree of nonlinearity.
- If a d.c. electric field is applied to a medium, the refractive index of the incident optical field frequency depends on it. If the dependence is linear, it gives rise to the linear electrooptic or Pockels effect.
- If the same d.c. electric field changes the energy levels of the atoms too, inducing thus a change in the absorption characteristics, the linear Stark effect takes place.
- The interaction of a d.c. magnetic field with a light field also changes the energy levels of the atoms as in the case of the Stark effect and is described by the Faraday effect.
- If a small optical signal beam is sent through a medium in the presence of an intensive light wave, the pump, having another frequency, both waves may beat to produce a field at the difference frequency. This wave in turn beats the pump wave to produce a term in the polarization of their difference frequency — an effect that is called parametric amplification.

## 3.3 Cubic Effects Arising from the Third Order Nonlinear Polarization

The third order polarization

$$\vec{P}^{(3)} = \epsilon_0 \chi^{(3)} (\vec{E}) \cdot \vec{E}^3 \tag{3.9}$$

gives rise to third harmonic generation and related mixing phenomena. Because third harmonic generation relies on the same principle as second harmonic generation, which was introduced in the previous section, and because its efficiency is in many materials too low to allow for a wide range of applications, the next most interesting effect which is used for optical applications is the quadratic electrooptic or d.c. Kerr effect.

### 3.3.1 The Kerr Effect

In principle, the Kerr effect takes place if an external electric field aligns the permanent dipoles of the molecules of a medium and changes thus the index of refraction of it. Here, an optical wave propagates through the medium in the presence of a d.c. field. The medium itself "sees" along the averaged intensity of the quick oscillations of the light field and experiences therefore the effect of an optical wave that propagates through the medium in the presence of a d.c. field.

The d.c. Kerr effect is used in many fast acting optical shutters. One of them is illustrated in its principle action in fig. 3.5. The heart of the setup is a cuvette with

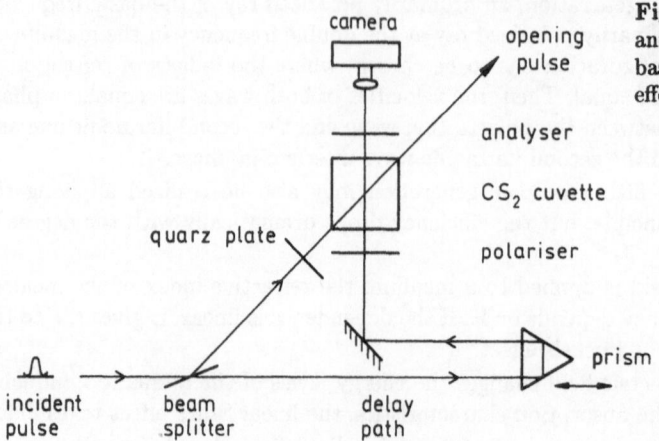

**Figure 3.5:** Principle mechanism of a fast optical shutter based on the nonlinear Kerr effect.

carbonsulfide ($CS_2$), often called a Kerr cell too. It is placed between two crossed polarizers. Therefore, the cell shuts off the light beam or closes the light path in the "out" state. The cell can be made partially transparent — the pathway may be opened — if an electric field (this might be an optical beam too) is applied to the $CS_2$ cell. In that case, the molecules of the cell align in the direction of the field, causing $CS_2$ to become birefringent. Thus the polarization direction of the light will follow the path of the molecules in the cell and will be turned. Consequently, the light will be able to pass — at least partially — through the analyzer. If the field is switched off again, the forced alignment of the molecules disappears and the cell blocks the light once again. Because the alignment of the molecules is a rather quick process, switching times of $10^{-12}$ s can be achieved in these cells, being thus much faster than the quickest mechanical shutters. An important application of such a shutter is the photography of a short pulse by itself. For that purpose, the incoming pulse is split into an opening pulse and a passing one. The passing pulse solely arrives at the detection plane provided that both pulses meet in the cell. As illustrated in fig. 3.5, a variable delay path is used to measure the outcoming intensity at the dectector in dependence on the delay-time. In that way, the duration of the pulse can be calculated and the pulse can be photographed. That method of pulse autocorrelation is widely spread especially for beam analysis.

Perhaps the most important of all third order processes is the intensity-dependent refractive index: an optical field passing through the nonlinear medium induces a third order polarization which is proportional to the third power of the field. In its effect on the wave, this term is equivalent to changing the effective value of $\chi^{(1)}$ to $\chi^{(1)} + \chi^{(3)} \vec{E}^2$ — in other words, the refractive index is changed by an amount proportional to the optical intensity. Physically, this effect is identical to the Kerr effect I have just described. The only difference lies in the fact that for the nonlinearities produced by the Kerr effect an external electric field is responsible, whereas in the case of the intensity-dependent refractive index it is the field of the light wave itself that influences the dipole arrangement of the molecules.

**Figure 3.6:** A symmetric, bell-like intensity distribution of a laser beam causes a slight change in the refractive index in a medium in such a way that this change acts as a focusing device on the laser beam itself.

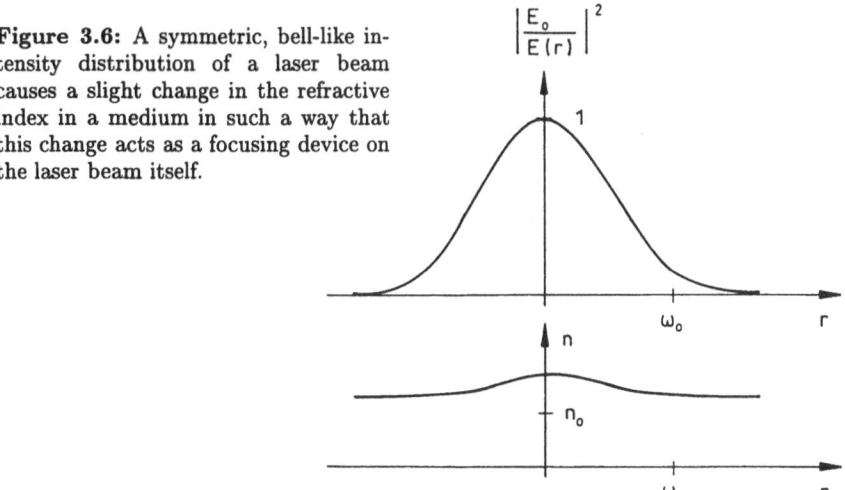

### 3.3.2 The Intensity-Dependent Refractive Index

The intensity-dependent refractive index is, in many cases, the key nonlinear effect used in optical switching and signal processing. Among them, self-focusing of a laser beam, self-phase and frequency modulation, soliton pulse propagation and especially phase-conjugate reflection are of special importance.

Here, I want to show only two examples for these applications: the nonlinear optical lens and the bistable Fabry-Perot etalon. The effect of phase conjugation will be treated in detail in chapter 3.5. If an intensive laser beam with a Gaussian intensity distribution as shown in fig. 3.6 illuminates a medium, the intensity-dependent refractive index is no longer uniform over the beam diameter. In the middle of the beam, where the intensity of the Gaussian beam profile is highest, the index of refraction is larger than in the border regions. Thus, beamparts at the border of the profile pass quicker through the medium than the beam components in the middle. As a consequence, the radiation field will focus itself, the medium acts as a convex, focusing lens. Because the self-focusing effect gives rise to a higher electric field in the central region of the beam, the index of refraction still rises more and the focusing effect grows even stronger. Finally, the process enhances itself until the damage threshold of the material is reached.

The second important device is a Fabry-Perot etalon for bistable switching purposes. In linear optics, a Fabry-Perot etalon, which consists mainly of two mirrors, aligned in opposite to each other and acts on the basis of multiple beam interference, has the property that it blocks an incident monochromatic beam, unless the wavelength of the light satisfies the resonance condition $2nd = p \cdot \lambda$ (with the index of refraction $n$ between the two mirrors, the integer $p$ and the length of the etalon $d$). Then, a high transmission is obtained. As in the case of the Kerr cell, let us assume that the space between the mirrors of a Fabry-Perot etalon exhibits a nonlinear refractive index. We suppose also that, at low incident intensities, the wavelength $\lambda$ and all other parameters fail to satisfy the resonance condition. Therefore, the transmitted light beam is blocked ("off state"). If the intensity is now increased gradually, the refractive index $n$ changes. When a value of $n$ is reached, which satisfies the resonance condition (point $I = x_0$ in fig. 3.7), the transmitted beam is switched "on". A more detailed analysis of the operation of that

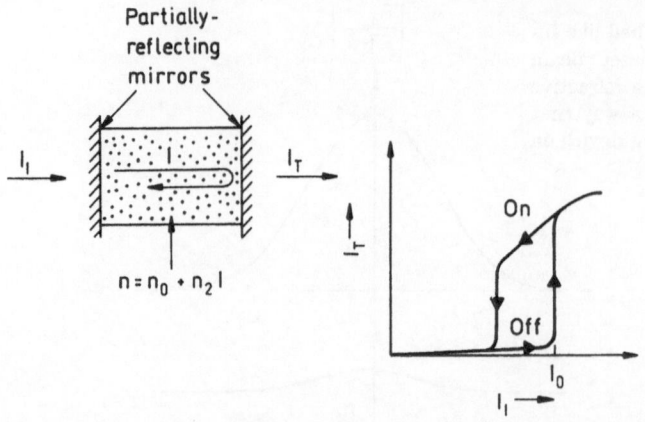

**Figure 3.7:** Optical switching behaviour is exhibited by a Fabry-Perot etalon that contains a material having a refractive index $n$ which depends on the optical intensity $I$. The hysteresis loop in the curve of transmitted intensity $I_T$ versus input intensity $I_I$ implies bistable "on" and "off" states for some values of $I_I$.

nonlinear etalon shows that under certain circumstances it can exhibit a "latching" action — a bistability — rather like an electronic bistable logic circuit. The important point is that the device operation is controlled by the light beam itself: it is an-all optical switch.

Beneath these important effects for applications of nonlinear optics, the third order nonlinear susceptibility gives rise to saturable absorption or amplification. At sufficiently high irradiance, the density of occupation of the energy levels begins to change and absorption or amplification changes, too. In that case, the change in the refractive index can be neglected and $\chi^{(3)}$ becomes imaginary. Moreover, the effect decreases with increasing irradiance. At last, there is also the possibility of a quadratic Stark effect, which has the same consequence as the linear Stark effect, but is proportional to the square of the electric field strength. All of the nonlinear effects considered up to now are essentially classic. However, the third order polarization also gives rise to some effects which are of quantum mechanical nature. They will be shortly discussed in the next section.

### 3.3.3 Two-Photon Absorption and Stimulated Raman Scattering

These effects arise when a weak signal wave at frequency $\omega_s$ propagates through a medium in the presence of a strong pump wave at frequency $\omega_p$ and their frequencies are chosen in such a way that $\omega_p \pm \omega_s = \Omega$. Moreover, $\Omega$ should be a transition frequency of the medium. In the case $\omega_p + \omega_s = \Omega$, the transition can take place through the simultaneous absorption of a pump photon and a signal photon. Thus we have a two-photon absorption process which produces signal attenuation even though the signal frequency itself is not equal to a transition frequency of the medium. In other words, the imaginary part of $\chi^{(1)}$, which describes linear absorption, arises here from a photon absorption process and, as shown in fig. 3.2, it vanishes unless the signal frequency is close to a transition frequency of the medium.

In the other case, when $\omega_p - \omega_s = \Omega$, the transition can take place via simultaneous absorption of a pump photon and emission of a signal photon. This is the stimulated Raman scattering effect which results in amplification of the input signal.

Both, the two-photon absorption and the stimulated Raman effect arise from the term in the third order polarization which is proportional to the product of the signal field and the pump intensity. In its effect on the signal wave, this term is equivalent to changing $\chi^{(1)}$ by an amount proportional to the pump intensity.

### 3.3.4  Four-Wave Mixing

Four-wave mixing refers to the third-order nonlinear process with four interacting electromagnetic waves. In cubic or isotropic media, three different cases can be considered. If we have as first case three pump fields, the signal can be greatly enhanced if phase matching ($\Delta \vec{k} = 0$) is obeyed. It can be achieved in an infinite number of ways by properly adjusting the directions of propagation of the three pump waves. Which arrangement is preferred often depends on practical considerations, such as optimal beam overlapping length and better spatial discrimination against scattering background. Second, the output field is assumed to be in the same mode as one of the input fields. Then, the input field should experience gain or loss induced by the nonlinear wave interaction. At last, if we select two strong waves acting as the pump fields, two counterpropagating weak waves can get amplified, we achieve backward parametric amplification and oscillation. This case resembles the parametric amplification case of section 3.2, except that two pump fields instead of one are used here. The two weak waves are the signal and the idler respectively. The solution then is essentially the same as that described in section 3.2. Assuming perfect phase matching, which can be achieved easily in this case, and negligible pump depletion, we find that an oscillation between the signal and the idler may occur. For special frequency conditions — e.g. if all four waves have the same frequency — this may result in the construction of a phase-conjugate wave of the signal. Phase conjugation is defined as the process in which the phase of the output wave is complex-conjugate to the phase of an input wave. In other words, the process reverses the phase of the input. If the phase-conjugate output propagates in the backward direction with respect to the corresponding input wave, then it can be used to correct aberration due to phase distortions experienced by an input wave.

In the case of degenerate four-wave mixing, where all waves have the same frequency, the creation of the phase-conjugate signal can be understood from the following physical picture: Two or three input waves interfere and form either a static grating or a moving grating with an oscillation frequency $2\omega$. The third input wave is scattered by the grating to yield the output wave. In many cases where $2\omega$ is away from resonance, the contribution from the static gratings should dominate. With three input waves, three different static gratings are formed.

The grating formed by the $\vec{k}_p$ (pump) and the $\vec{k}_s$ (signal) waves scatters the $\vec{k}_p'$ wave to yield outputs as $\vec{k}_o = \vec{k}_p' \pm (\vec{k}_p - \vec{k}_s)$. The one formed by the $\vec{k}_p'$ and the $\vec{k}_s$ waves scatters the $\vec{k}_p$ wave to yield the outputs as $\vec{k}_o = \vec{k}_p \pm (\vec{k}_p' - \vec{k}_s)$. The one formed by the $\vec{k}_p'$ and the $\vec{k}_p$ waves scatters the $\vec{k}_s$ wave to yield the outputs as $\vec{k}_o = \vec{k}_s \pm (\vec{k}_p' - \vec{k}_p)$. For the special case of $\vec{k}_p' = -\vec{k}_p$, a degenerated case of counterpropagating pump waves is obtained, resulting in a generation of a wave that is counterpropagating to the input signal wave. Altogether, three output waves with different wave vectors, $\vec{k}_o = \vec{k}_p + \vec{k}_p' - \vec{k}_s$, $\vec{k}_p - \vec{k}_p' + \vec{k}_s$, and $-\vec{k}_p + \vec{k}_p' + \vec{k}_s$ are expected to be generated. However, we realize that since $|\vec{k}_o|$, in general, is not equal to $\omega \epsilon^{1/2}/c$, the generation of the three output waves may not be phase-matched. Consider, e.g. the case with $\vec{k}_p' = -\vec{k}_p$ (counterpropagating pump waves). The output waves are expected to have $\vec{k}_o = -\vec{k}_s$ and $\vec{k}_s \pm 2\vec{k}_p$. However, only the generation of the output at $\vec{k}_o = -\vec{k}_s$ is always phase-matched, whereas the other two cases are not phase-matched. Thus usually only the output at $\vec{k}_o = -\vec{k}_s$ needs to be considered.

It is interesting to see the connection between this case of four-wave mixing and holog-

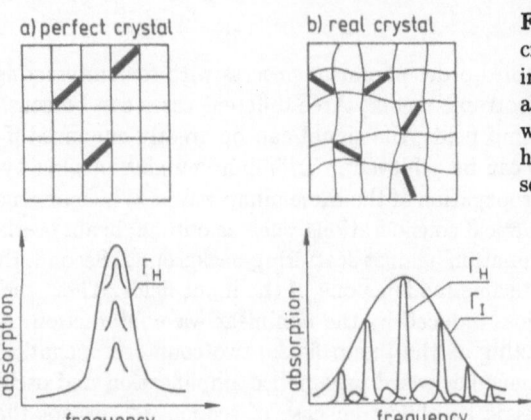

**Figure 3.8:** Comparison of an ideal crystal with a regular grating, having a homogeneous linewidth of its absorption line (a) and a real crystal with deformations that result in inhomogeneous broadening of the absorption line (b).

raphy. In both cases, the output wave ($\vec{k}_o$), arising from scattering of one or two pump waves ($\vec{k}_p$ or $\vec{k}_p'$) off the interferogram formed by the object wave ($\vec{k}_s$) and the other pump wave ($\vec{k}_p$ or $\vec{k}_p'$) retraces back the path of the object wave $\vec{k}_o = -\vec{k}_s$. Because an object can be represented by a group of $\vec{k}_s$ waves, we see that an image of the object can be reconstructed by the corresponding output waves. We will discuss this case of degenerate four-wave mixing in photorefractive media in detail in section 3.5.

### 3.3.5 Photon Echo and Spectral Hole Burning

Another class of three-wave interaction may take place, if a material with inhomogeneous broadening of the absorption line is exposed to a strong, short light pulse. Whereas homogeneous broadening is a result of incoherent interaction of the excited states of a medium as collisions, radiation decay, or coupling of vibrational, rotational, or electronic interactions, inhomogeneous broadening arises in gases because of the Doppler effect or in solids because of local variations of the crystal field, i.e. local strain. Because of the difference in local strain experienced by each ion from its environment in the crystal, many narrow homogeneous lines, with shifted center frequencies, superimpose. As a result, the energy spectrum associated with a given electronic transition is inhomogeneously broadened. In fig. refpic:hom-inhom) the difference between the homogeneous linewidth of a perfect crystal and the inhomogeneous one of a real crystal, where the grating is deformed, are shown. To simplify our discussion on the principle function of both effects, consider the storage of one data pulse, which occurs before the writing pulse (laser pulses with nanosecond pulse duration, 10 GHz linewidth and energies about 50 $\mu$J). After the interaction with the data pulse at a certain time $t_1$, a portion of the ions is resonantly excited by the laser pulse. Because the process is coherent, each excited ion memorizes its initial phase (determined by the laser radiation). Thus the temporal information is converted into spectral information by the crystal. However, because of the presence of inhomogeneous broadening, normally each excited ion oscillates at its own resonant frequency which may differ from the laser frequency, resulting in a dephasing of the oscillators and damping of the macroscopic polarization. When the write pulse arrives at $t_2$, it interferes with the effect of the first data pulse through laser-material interactions: if the dipoles of the excited ions are in phase with the writing pulse, the interference is constructive;

otherwise it is destructive. This interference, similar to the spatial interference of two coherent beams in holography, contains the complete information of the combined laser pulses. If the storage material would be a simple two-level system, the storage time would be determined by the lifetime of the excited state. The ground- and the excited state populations will then recombine and no spatial information is retained. Thus, data storage would not be possible. The crucial point thus is to prevent the ground- and excited state populations from recombining in order to preserve the structure that contains the information about the data pulse.

In spectral hole burning, excited-state atoms or molecules are made to undergo permanent photochemical processes at a a particular frequency within the inhomogeneous line, thereby preventing their return to the ground state. Therefore, persistent spectral hole burning allows for long-time storage, because the gratings are photochemically recorded at different optical frequencies within the sharp inhomogeneously broadened absorption band of solid-state materials (such as molecules in solid-state solution, color centers in crystalline hosts or a dye molecule in a low-temperature matrix).That interaction depletes the population of that subset while leaving the remainder of the line unchanged, thus burning a hole in the inhomogeneous line profile. When the presence or absence of a hole at designated frequencies corresponds to a single binary storage unit, the frequency domain can be used for information storage with the storage density increasing by a factor defined by the ratio of the inhomogeneous to the homogeneous linewidth. This procedure will be described in detail in section 4.1.

In photon echo processes, the entire storage-retrieval process is realized in the time domain and involves the use of a write pulse and a read pulse in addition to the data pulses to be stored. In rare-earth solids as one example of photon echo materials, the recombination of the ground- and the excited state population — which I have discussed above as a critical effect to destroy information storage in excited electronic states — can be prevented by the presence of hyperfine structures in the ground state. Excited ions are then likely to decay to hyperfine sublevels different from the one at which the excitation was originated from. Thus the spectrally modulated population distribution generated by the data and write pulses can persist in the hyperfine states for a longer time. A maximal storage time of several hours has been demonstrated [16, 17, 18]. To recall stored information from the hyperfine sublevels after having written the data pulse at $t_1$ and the write pulse at $t_2$, we have to re-excite the crystal by the read pulse at $t_3$. This pulse induces coherent radiation (echoes) from the crystal at $t_e = t_3 + (t_2 - t_1)$. An echo signal will be generated if the excitation pulses induce both a delayed rephasing of atoms at each spatial location and a global phase relation between locally rephased atoms, which allows the echo field radiated at various locations throughout the sample to interfere constructively. The latter requirement is equivalent to the well-known phase-matching condition of nonlinear optics. In gas-phase samples, both conditions depend on laser-excitation pulse directions, and a phase-matched sample will not necessarily rephase locally. In solid samples, local rephasing is independent of excitation direction. The temporal profile of the echo is proportional to the Fourier transform of the spectrally modulated ground state population distribution. The technique of photon echo storage and amplification uses the ability of atomic transitions to store the Fourier spectrum of laser phase structures as a population distribution in the inhomogeneous absorption profile. Thus it is possible to store temporally and spatially encoded laser pulses. The holographic geometry of backward stimulated echo also permits high-speed optical processing of temporal and spatial data and — if one uses three excitation pulses as in photorefractive materials — phase conjugation of the incoming images [16, 17, 19].

The nonlinear effects mentioned in the last two sections are the most important and widely studied ones, because they promise various applications in optical computing and neural network architectures. Although processes involving fifth, seventh or yet higher order interactions can still be observed, especially second and third order effects as photorefractive beam coupling, four-wave mixing, spectral hole burning and photon echo processes will be found at many places throughout that book when implementing special components of optical neural networks.

 ## Summary

The most important cubic effects arising from the third order polarization are the Kerr effect, the intensity-dependent refractive index and with it self-focusing and -defocusing, two-photon absorption, stimulated Raman scattering, four-wave mixing that can be derived from the photorefractive effect, persistent hole burning, and photon echo effects.

- The Kerr effect takes place if an external electric d.c. field aligns the permanent dipoles of the molecules of a medium and changes thus its index of refraction.
- The intensity-dependent refractive index is based on the fact, that the refractive index changes with the intensity of the incoming light. Self-focusing arises if a Gaussian-shaped beam profile changes the index of refraction more in the middle than in the border regions of the crystal. As a consequence, the radiation field will focus itself, the medium acts as a convex, focusing lens. Self-defocusing can be achieved by adapting the incoming beam profile in order to realize a concave, defocusing lens.
- If the frequency $\omega$ of the optical field is far away from the resonance frequency $\Omega$, no absorption takes place. Here, two photons of equal or different frequencies may be absorbed simultaneously and achieve a transition, provided that the frequency sum equal the transition frequency: ($\omega_1 + \omega_2 = \Omega$).
- In the case when the two optical fields have frequencies that can be subtracted to give the resonance frequency of the medium ($\omega_p - \omega_s = \Omega$), stimulated Raman scattering takes place by simultaneous absorption of a pump photon and emission of a signal photon. The effect results also in amplification of the input signal.
- If three waves overlap in a third order isotropic nonlinear medium, the signal can be greatly enhanced if phase matching is obeyed. Moreover, if we select two strong waves acting as the pump fields, backward parametric amplification may be obtained.
- For special frequency conditions, if e.g. all four waves have the same frequency and the pump waves are counterpropagating, phase conjugation may occur, reversing the phase of an input wave and allowing for corrections of phase distortions and aberrations.
- In materials with inhomogeneously broadened absorption lines, excited atoms or molecules may undergo permanent photochemical processes, that deplete the population of a certain frequency in the line, thus burning a hole in the inhomogeneous line profile.

- The technique of photon echo uses the ability of these atomic transitions to store the Fourier spectrum of laser phase structures as a population distribution in the inhomogeneously broadened absorption profile, using a first data pulse to excite and a second write pulse to interferometrically store the information. This information may be recalled by addressing the medium with a third read pulse.
- By backward stimulated photon echo, phase conjugation of the data pulse is possible.

## 3.4 Photorefractive Beam Coupling

Although most of the nonlinear effects of second and third order nonlinearities presented in the last chapter have been exploited in detail since the invention of the laser, there is one which has developed in recent years even more rapidly: the field of parametric amplification and phase conjugation in electrooptic and photorefractive materials. Among them, especially the photorefractive effect, defined as a photo-induced change of the indices of refraction by electrooptic effects, has become the nonlinear optical mechanism of choice for optical image processing. This is primarily because the effect allows effective interaction between light waves, leading to a series of nonlinear effects in which light waves can be controlled by other optical beams. Up to now, a series of all-optical devices such as e.g. optical phase conjugators, light amplificators and dynamic holographic memories have been constructed based on the photorefractive effect. These applications combined with the interest the mechanism currently attracts for neural network applications are the main reasons why I will focus my attention on them in this and the following chapters.

The principle of the photorefractive mechanism — which will be discussed later on in detail — can be summarized as follows: the interaction of a signal and a pump wave in the photorefractive medium leads to an intensity or energy interference grating, creating thus periodically alternating zones of high and low energy. In the regions with a high energy level, the particles that are mainly responsible for the process — impurities at energy levels between the conduction and the valence band of the ferroelectric crystal — are liberated from their bound states and are allowed to move in the conduction band. If they recombine at free impurity places, they cause a change in the charge carrier density of the material and thus a change in the refractive index via the electrooptic effect. Because — as one may know from conventional holography — the interference grating carries the whole information of the signal, the same is true for the interference-modulated refractive index and thus the effect can be considered as a signal storage mechanism.

The nonlinearities in this process are much larger than in "classical" nonlinear optics, which makes use of the nonlinear polarizability of the band electrons, since the polarizability of the lattice is also employed. In materials with large lattice polarizabilities as e.g. ferroelectrics, the nonlinearities are large enough to be substantial even at milliwatt continuous-wave power levels. Moreover, the effect can easily be applied to real-time parallel processing in bulk materials, especially in solid state crystals. No reticulation of pixels nor other complex material processing is required to use bulk crystals as storage or processing devices for optical beams even with a complex two-dimensional structure. The information processing capacity is determined only by the diffraction of the optics and the three-dimensional nature of the optical beam interaction within the material's volume.

The photorefractive effect has a particularly interesting history, which is characteristic for the discovery of a lot of physical effects. Originally, the photorefractive effect was discovered more than 30 years ago as an undesirable optical damage effect in nonlinear and electrooptic crystals [20]. The usefulness of crystals with large electrooptic coefficients such as Lithiumniobate (LiNbO$_3$) was believed to be limited by changes of the refractive index induced by the incoming laser light, because the index change gave rise to decollimation, scattering of laser beams and noise. Therefore, devices as modulators and frequency doublers were severely limited in their performance through the *optical damage effect* as it was named by scientists. It was not until some years later that materials exhibiting this damaging effect were discovered to be advantageous as means for holographic recording and storage [21]. From that beginning up to now, the photorefractive effect has been observed in a number of different crystals. The major ones used for applications in dynamic and stationary three-dimensional holography are Lithiumniobate (LiNbO$_3$), Lithiumtantalate (LiTaO$_3$), KTN (KTa$_x$Nb$_{1-x}$O$_3$), Bariumtitanate (BaTiO$_3$), BSO (Bi$_{12}$SiO$_{20}$), BGO (Bi$_{12}$GeO$_{20}$), and Potassiumniobate (KNbO$_3$). Up to now, a lot of other materials have been showed to be photorefractive, among them semiconcductors as InP or GaAs, photopolymers, organic materials as bacteriorhodopsine, and certain classes of doped fullerenes and zeolithes.

Especially in the last years, the photorefractive effect has revealed to play an important role in optical data storage. The reason for that impressing development can be found on the one side in the advantages that volume holographic materials offer for optical data storage devices, especially high parallelism in combination with the global communication features that arise because all three dimensions in space can be used and exploited simultaneously. On the other hand, the photorefractive effect and with it the possibility for real time optical storage and recall is easily accessible for low energy regimes because of the high sensitivity of photorefractive materials. The energy required to obtain a noticeable photorefractive effect in e.g. BSO or KNbO$_3$ is close to the sensitivity of silver-halogenide layers. The variety of photorefractive materials that are available up to now allows their appropriate choice for a lot of special application purposes — e.g. high nonlinearities, high amplification rates or short time constants.

The gratings in photorefractive materials are mostly recorded with visible cw-laser light (Ar-, Kr-, He-Cd-, He-Ne-lasers). The storage times of the recorded holograms or grating decay times range from milliseconds for KNbO$_3$ to days or months in BaTiO$_3$, depending on the dopant material and years for LiNbO$_3$ if a fixing process is applied. The holographic gratings can be erased by uniform illumination or heating when the recombined impurity particles are liberated again and thus "forget" the information of the signal which was stored by their fixed position.

The detailed mechanism of the photorefractive effect can be explained by the interaction of two coherent waves in the photorefractive crystalline medium. The interference pattern produces a stationary intensity distribution throughout the crystal volume and thus an equivalent energy distribution. In regions with high energy density, charges (electrons or holes) originating from impurity states in the crystal are generated, which try to migrate to the regions of lower energy density. This migration may have three different origins. First, it may result from diffusion, indicating that the charges move towards a place of lower charge concentration due to their thermal mobility. Second, it can occur due the action of an external field which causes a drift mechanism or as a third possibility due to the photovoltaic effect — a drift effect caused by an internal electric field which was in turn created by the interference light pattern itself. The special thing about the

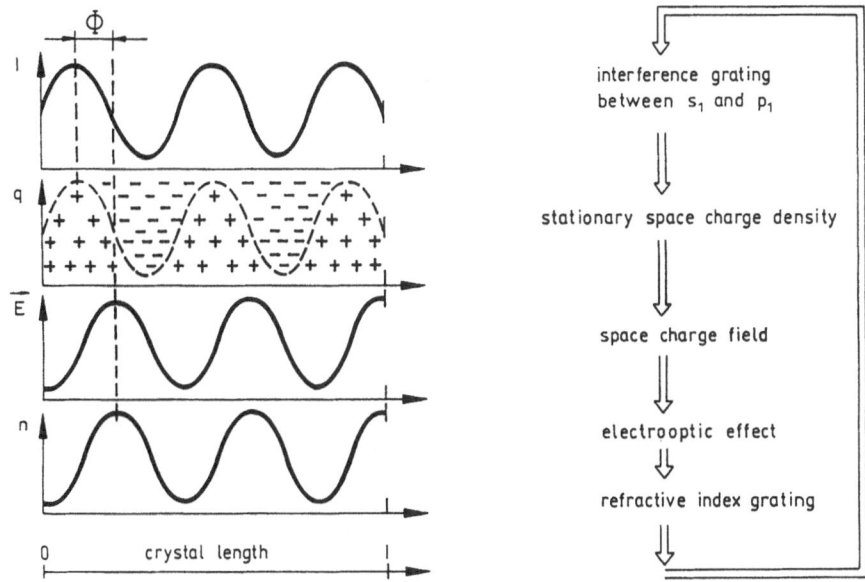

**Figure 3.9:** The principle steps of the photorefractive beam coupling mechanism.

photovoltaic effect is, that when photoelectrons are excited into the charge transfer band, they migrate along a preferential direction which is determined by the direction of the polar axis of the crystal. Finally, the electrons (or holes) are subsequently retrapped by free places in the lattice of the crystal, empty donors (or acceptors), creating thus a novel distribution of the charges. Because of the movement of the charges, there is a spatial difference in the excitation rate of the ionized donors resulting from the generation of the free electrons (or in the excitation rate of the acceptors for holes) and the trapping rate of the electrons or holes, resulting in a spatially modulated electrical charge. Consequently, an electrical field builds up, the *space charge field*, which is out of phase with the initial interference grating. This field in turn modulates the refractive index via the linear electrooptic effect. Depending on the dominant transport mechanism, an overall phase shift between the intensity modulation and the refractive index grating arises, which is generally time- and space dependent. The whole process of creation of the refractive index modulation is illustrated in fig. 3.9 for the case of a diffusion dominated material.

Because the modulation of the refractive index in a volume material acts at the same time as a diffraction grating for the incoming beams, a special situation arises, which characterizes the photorefractive effect: the index modulation itself creates a new redistribution of phase and intensity of the interference field, which in turn modulates the spatial distribution of the index of refraction. Thus the two interfering beams are diffracted by the index grating they created by themselves, leading to a complex, dynamical relationship between the interference grating and the resulting refractive index grating — an effect that is known as *dynamic holography* in nonlinear optics. As a matter of fact, the effect thus causes a high degree of interaction or coupling between the two waves in the crystal volume and gives therefore rise to a highly nonlinear behaviour which is called *self-diffraction*. The most striking effect of this self-diffraction is that an energy redistribution between the beams occurs, which may — for the condition of an asymmetric crystal field (one outstanding crystal axis) — lead to an amplification of one

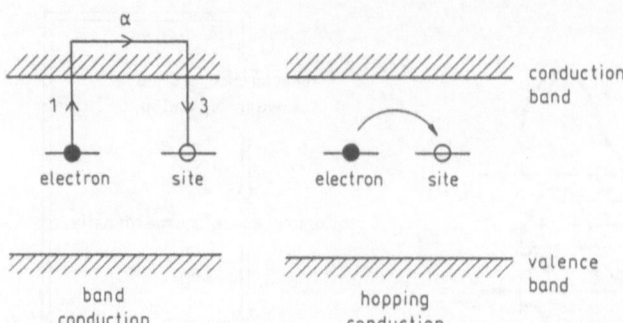

**Figure 3.10:** The creation of a photorefractive index grating: the band transport model (left) versus the hopping model (right).

of the two interacting beams. This is due to a constructive, phase-matched interference of the transmitted beam and the part of the beam diffracted by the refractive index grating. At the same time, the other beam is attenuated or even depleted, due to destructive, 180° phase-shifted interference of the transmitted beam and the diffracted part of the other beam. Both, the value and the direction of that energy exchange depend on the geometry of the incident beams relative to the crystal orientation and on different crystal parameters, e.g. the strength of the electrooptic coefficients in the susceptibility tensor.

The amplification is largest in the case of a stationary refractive index grating as it is achieved in the case of a diffusion dominated charge transport mechanism and a phase shift of $\Phi = 90°$, because the diffraction at this grating gives an additionally phase shift of 90° and leads thus as well to in-phase, constructive interference of the two beams (amplification of the signal beam) as well as 180°, destructive interference of the two beams (depletion of the pump beam).

In the following paragraph, some equations which are essential for the understanding of the photorefractive effect are given. All nonlinear processes derived from the photorefractive effect can be traced back to these fundamental equations. Therefore, they are discussed up to a certain degree in detail. However, for specific, crystal-related equations and derivations of specific transport mechanism explanations, the reader may refer to the excellent introduction book on photorefractive optics by P. Yeh.

### 3.4.1 The Band Transport Model

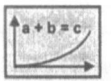 The physical description of the charge transport mechanism relies basically on the band transport model, which was first developed by N.V. Kukhtarev and coworkers [22]. The principle of that model is shown in fig. 3.10 as a three step process. In the first step, electrons or holes are liberated from donors or traps via photoionization. These charges may be provided by lattice defects or dislocations or — in the case of nominally undoped crystals — by small traces of impurities. In most of the ferroelectrics that have a pronounced photorefractive effect, ferrum impurities are the most important donor and trapping centers. After that liberation process, charge migration due to diffusion, drift or the photovoltaic effect (photoinduced stimulation) takes place as the second step in the valence or conduction band, leading the charges to areas of lower energy and charge density. At these places, they finally recombine with empty donors or traps and form the space charge field as already described above. In many photorefractive materials, it turned out that simultaneous electron and hole conductivity

is responsible for the creation of the space charge field. This process explains the build-up of the space charge field in most of the commonly used ferroelectric crystals and was the basis for our first, simple description of what happens during the photorefractive grating build-up in the previous section. However, in some cases with short time constants for the build-up of the grating or large energy band gaps between the valence and the conduction band, the *charge-hopping model* proposed by R. Jaura, T.J. Hall and P.D. Foote [23, 24] is more appropriate. It substitutes the three-step process of Kukhtarev's model by a one-step mechanism of thermal or photoinduced tunneling of excited states. Both models are trying to explain the basis of the mechanism leading to the creation of the refractive index grating and thus naturally simplify the real situation of light-particle interaction in photorefractive materials. If one looks deeper into the mechanism of refractive index creation, both explanations describe special cases of simple trap distributions, whereas in reality the distribution of traps is much more complex because of the different nature of different dopants or impurity ions. Moreover, the existence of traps near the band edge of the conduction band — *shallow traps* — or the valence band — *deep traps* — suggests, that processes of higher order with two or more transitions among traps may take place [25, 26, 27, 28]. Fortunately, the detailed knowledge of the transport mechanism is not important for the application of photorefractive materials and therefore I will restrict myself to the case of one dominating sort of charge carriers. In fact, it is this case which is one of the most frequent ones in many photorefractive materials as $BaTiO_3$ and $LiNbO_3$ [27].

We start from the rate equations for electrons and holes — as they are commomly known in semiconductor physics — which arise because an effective current density $j$ acts on the crystal when it is illuminated with an interference pattern of two interacting, coherent beams. That current density is composed of the three elements arising from the well-known different transport effects — drift in an electrical field, diffusion and the bulk photovoltaic effect. If we assume a photorefractive material of the hole-conductivity type, the current density of the electrons can be neglected. In that case the resulting current density of the holes is:

$$\vec{j}_p = \vec{j}_{\text{drift}} + \vec{j}_{\text{pv}} + \vec{j}_{\text{diff}} \tag{3.10}$$
$$= ep\mu_p(\vec{E}_0 + \vec{E}_{sc}) - e\vec{a}p - eD_p\vec{\nabla}p$$

where $D_p = \mu_p k_b T/e$ is the diffusion constant, $\mu_p$ the mobility of the holes, $\vec{E}_0$ the external field and $\vec{E}_{sc}$ the space charge field. $\vec{a}$ represents the dependence of the bulk photovoltaic effect on its direction, $p$ is the density of the holes and $e$ is the elementary charge unit. As a consequence of our assumption of a hole-conducting material, the electron number density may be neglected too. Adding the assumption that all donors are ionized, the space charge density depends only on the number density of the donors, $N_D$, and the ionized acceptors $N_{A-}$:

$$\rho = e(p + N_D - N_{A-}) \tag{3.11}$$

What we are now looking for is an expression for the change of the concentration of holes in time — that is to say we need the rate equation for electrones or holes. This change depends on one side on the change of the current density $j$, on the other hand on the rate of free carrier generation $G$ and recombination $R$. The generation of holes is proportional to the density of non-ionized acceptors ($N_A - N_{A-}$) and to the probability for thermal and photonic excitation. In a similar way, the recombination rate increases with the hole density $p$ and the number of recombination centers:

$$G = (s_p I_0 + \beta_p)(N_A - N_{A-}) \tag{3.12}$$

$$R = \gamma_{Rp} N_{A-p} \tag{3.13}$$

In these equations, $s_p$ is the constant for photoinduced excitation, $\beta_p$ the constant for thermal excitation, $I_0$ is the total incoming intensity, $N_A$ represents the total number density of the donors and $\gamma_{Rp}$ is the recombination coefficient for linear recombination.

If we imagine insulators as photorefractive materials — again $BaTiO_3$ and $LiNbO_3$ are the main representants of that class — we may neglect the thermal generation of charges relative to the intensity-dependent, photoinduced excitation. In that case, we arrive at the following continuity equation for the change of the concentration of the holes:

$$\frac{\partial p}{\partial t} + \frac{1}{e}\vec{\nabla}\vec{j}_p = G - R \tag{3.14}$$

The change of the density of the ionized acceptors depends for non-mobile charge carrriers only on the difference of the generation and recombination rate

$$\frac{\partial N_{A-}}{\partial t} = G - R \tag{3.15}$$

These two equations, combined with Gauss's law

$$\epsilon\epsilon_0 \vec{\nabla}\vec{E} = \rho = e(p + N_D - N_{A-}) \tag{3.16}$$

where $\epsilon_0$ and $\epsilon$ are the vacuum or material-dependent dielectric constants respectively, and the condition for stationary electromagnetic fields

$$\vec{\nabla} \times \vec{E} = 0 \tag{3.17}$$

are the fundamental *Kukhtarev equations* [22] which describe the propagation of the light-waves in the medium as well as the particular material properties of the photorefractive effect completely.

These equations are the basis for all further derivations of the different beam coupling processes. as two-wave mixing or four-wave mixing and all configurations of phase conjugation in photorefractive crystals.

### 3.4.2 Mathematical Formulation of Beam Coupling

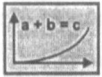 The interference of the two incoming waves, as shown in fig. 3.11, the signal wave

$$\vec{S}_0(\vec{r}, t) = \vec{e} \cdot s_0(\vec{r}) \cdot e^{i\vec{k}_s\vec{r}} \cdot e^{-i\omega t} + c.c. \tag{3.18}$$

and the pump wave

$$\vec{P}_0(\vec{r}, t) = \vec{e} \cdot p_0(\vec{r}) \cdot e^{i\vec{k}_p\vec{r}} \cdot e^{-i\omega t} + c.c. \tag{3.19}$$

with the incident angles $\alpha_1$ and $\alpha_2$ create a volume hologram in the interaction region of the crystal, which has the following intensity distribution:

$$I(\vec{r}) = I_0 \left\{ 1 + m \cos((\vec{k}_p - \vec{k}_s)\vec{r}) \right\} \tag{3.20}$$

This intensity modulation is determined via the modulation depth

**Figure 3.11:** The principle of two-wave mixing of an intense pump wave $p$ (incident angle $\alpha_2$) and an information bearing signal $s$ (incident angle $\alpha_1$). $\vec{K}_1 = \vec{k}_p - \vec{k}_s$ is the grating vector, $\vec{c}$ is the crystal's c-axis, and $\varphi_1$ is the orientation of the grating vector relative to the c-axis.

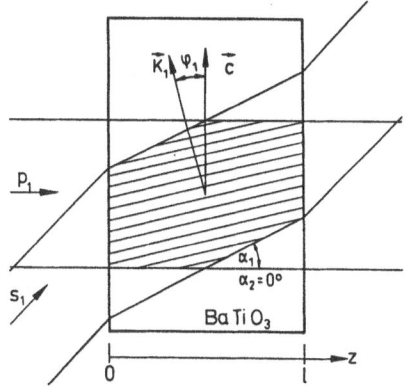

$$m = \frac{ps^* + p^*s}{p^2 + s^2},\qquad(3.21)$$

the grating vector $\vec{K} = \vec{k}_p - \vec{k}_s$ and the total incoming intensity $I_0 = |s_0|^2 + |p_0|^2$.

For the case of both waves entering the crystal from the same side, a *transmission hologram* is formed with the wave vector

$$\vec{K}_1 = \frac{4\pi n \sin((\alpha_1 - \alpha_2)/2}{\lambda}\qquad(3.22)$$

Here, $\alpha_1$ and $\alpha_2$ denote the incident angle of the signal and the pump wave respectively. In contrast, a *reflection grating* is formed if the beams enter the medium from opposite sides. This grating is described by the wave number

$$\vec{K}_2 = \frac{4\pi n \cos((\alpha_1 - \alpha_2)/2)}{\lambda}\qquad(3.23)$$

The incident angles $\alpha_1$ and $\alpha_p$ of the beams then define the orientation of these wave vectors relative to the c-axis of the crystal:

$$\varphi_1 = \frac{\alpha_1 + \alpha_2}{2}\qquad(3.24)$$

With that knowledge, we may now calculate the value of the space charge field from the charge density:

$$E_{sc} = \frac{1}{\epsilon \epsilon_0} \int \rho(\vec{r})\, e\, dr\qquad(3.25)$$

The integral in eq. (3.25) can be calculated if we assume that quadratic recombination — can be neglected and that a small modulation depth is operating. This requirement corresponds to the case of small contributions of high diffraction orders in the decomposition of the diffraction pattern into a Fourier-series or — in other words — to the assumption of a quasi-linear modulation. That simplification is justified because of the selectivity of the Bragg-condition in volume materials. This condition gives the minimum angular distance of two beams that are required to write two different, separable gratings with the same pump wave. As we will see later on, in chapter 4.4, the consequences of the Bragg-selectivity for multiple data storage are even more restrictive. With these conditions in mind, the space charge field is equal to

$$E_{sc} = \frac{\gamma_{Rp}N_{A-}}{\gamma_{Rp}N_{A-} + s_p I} \; m \; \frac{E_q \, (E_d + i(E_0 + E_{pv}))}{E_0 + E_{pv} + i(E_d + E_q)} \tag{3.26}$$

with the maximum space charge field $E_q$, which depends on the acceptor number density of the specific crystal, the bulk photovoltaic field $E_{pv}$ and the diffusion field $E_d$:

$$E_q = \frac{e(N_A - N_{A-})}{\epsilon \epsilon_0 |\vec{K}|} \tag{3.27}$$

$$E_{pv} = \chi \alpha I \sigma^{-1} \tag{3.28}$$

$$E_d = \frac{k_B T}{e} |\vec{K}| \tag{3.29}$$

Here, $\chi$ is the anisotopic tensor of the charge transport, $\alpha$ is the absorption coefficient, $\sigma = \sigma_d + \alpha f(I)$ represents the conductivity ($\sigma_d$ = dark conductivity) and $E_0$ the externally applied field.

As I already mentioned above, the amount of energy transfer between the two beams depends critically on the phase shift between the interference grating and the refractive index grating. This phase shift in turn is determined by the ratio of the imaginary contribution of the space charge field (diffusion), which characterizes the maximum energy coupling, and the real contribution (drift), which is responsible for phase coupling:

$$\tan \Phi = \frac{\text{Im}(E_{sc})}{\text{Re}(E_{sc})} = \frac{E_d E_q + E_d^2 + (E_0 + E_{pv})^2}{E_d(E_0 + E_{pv})} \tag{3.30}$$

Thus, to achieve a stationary beam coupling situation, one has to adjust the optimum phase shift at $\Phi = 90°$, e.g. by using the crystal without any external or photovoltaic field or by adjusting these two fields in such a way that they compensate and fulfil the equation $E_0 + E_{pv} = 0$. For that special case, the space charge field is given by:

$$E_{sc} = -im \frac{E_d E_q}{E_d + E_q} \tag{3.31}$$

which reaches again its maximum value for $E_d = E_q$ or a pure diffusion dominated refractive index modulation.

Now that we have calculated the space charge field, the variation of the refractive index results from the action of the linear electrooptic effect. It can be calculated via the induced change in polarization. As already discussed in chapter 3.2, the linear electrooptic effect is described by a polarization contribution of second order:

$$P_k^{nl} = 2 \sum_{i,j} \epsilon_0 \cdot \chi_{ijk} \cdot E_i(\vec{K}, \omega = 0) \cdot E_j(\vec{K}, \omega = 0) \tag{3.32}$$

Thus the polarization is determined by the field of the incident light waves $E_i(\vec{K}, \omega)$ and the stationary space charge field $E_j(\vec{K}, \omega = 0)$ — which in turn was generated by the interference of the incident waves.

The second order tensor of the susceptibility in eq. (3.32) depends on the indices of refraction and the electrooptic coefficients of the material:

$$\chi_{ijk} = -\frac{1}{2} n_i^2 n_j^2 r_{ijk} \tag{3.33}$$

Finally, we arrive at the index of refraction by substituting the space charge field in the wave equation:

$$\Delta \vec{E} = \mu_0 \left\{ \epsilon_0 \frac{\partial^2 \vec{E}}{\partial t^2} + \frac{\partial^2 \vec{P}}{\partial t^2} \right\} \tag{3.34}$$

$$\Delta n = -\frac{1}{2} n_0^3 \cdot r_{\text{eff}} \cdot E_{sc}^{ij} \tag{3.35}$$

with $i = 1$ for the pump wave and $i = 2$ for the signal wave. The notation of the waves with $i = 1, \cdots, n$ has been chosen to allow the extension to more than two interacting waves, as it is the case for complex image structures. Here, the effective electrooptic coefficient $r_{\text{eff}}$ has been introduced, pointing into the direction of the grating vector $\vec{K}$ to take the anisotropy of the susceptibility into account. For extraordinarily polarized interacting waves, $r_{\text{eff}}$ gives [29]:

$$r_{\text{eff}}^{ij} = \frac{1}{n_0 n_e^3} \cos\frac{\alpha_i + \alpha_j}{2} \cdot \left\{ n_0^4 r_{13} \sin \alpha_i \sin \alpha_j + \right. \tag{3.36}$$

$$\left. + 2n_0^2 n_e^2 r_{42} \sin^2 \left( \frac{\alpha_i + \alpha_j}{2} \right) + n_e^4 r_{33} \cos \alpha_i \cos \alpha_j \right\}$$

For ordinary polarization we obtain:

$$r_{\text{eff}}^{ij} = r_{13} \cos \left( \frac{\alpha_i + \alpha_j}{2} \right) \tag{3.37}$$

Depending on the crystal chosen for beam coupling experiments, the largest electrooptic coefficient can be selected by changing the polarization direction of the incident light. Finally, it is the refractive index given by eq. (3.35) that acts as a new diffraction grating for the incoming beams, offering thus the possibility to exploit a variety of beam coupling effects for optical processing applications. The most important and successfully applied one of these effects, two-wave mixing, which was named after the number of beams involved in the coupling process, will be discussed next.

### 3.4.3 Two-Beam Coupling and Amplification

The theory of coupled waves was completely derived for the first time by H. Kogelnik and coworkers at Bell Laboratories in the United States of America [30]. He suggested, that for the solution of the scalar wave equation (eq. (3.34)) we may assume that the field varies only slowly in time, allowing thus to neglect the second order derivatives in the wave equation. This approximation is often used in nonlinear optics and is named *slowly varying field approximation*. For that case which corresponds exactly to the case of photorefractive materials, he derived the amplitudes of the signal and pump waves by combining eq. (3.34) and eq. (3.26) or eq. (3.31) with eq. (3.35) respectively (refer also to fig. 3.11):

$$\frac{ds_1}{dz} = \frac{\gamma_1}{I_0} |p|^2 s_1 - \alpha s_1 \tag{3.38}$$

$$\frac{dp^*}{dz} = -\frac{\gamma_1}{I_{0s}} \frac{\cos \alpha_1}{\cos \alpha_2} |s_1|^2 p^* - \alpha p^* \tag{3.39}$$

with the *coupling constant*

$$\gamma_i = \frac{i\omega\Delta n_{pi}e^{i\varphi}}{2cn_0\cos\alpha_i} \qquad (3.40)$$

which summarizes the effects that arise from the incoming beam geometry, the intensity relation between the interacting beams and the crystal parameters. For diffusion dominated crystals as e.g. $BaTiO_3$, $\gamma_i$ becomes real because of the phase difference of $\Phi = \pi/2$ between the interference and the refractive index grating.

Eqs. (3.38) and (3.39) represent a set of differential equations, which describes completely the behaviour of the beams propagating through the crystal. If we assume energy conservation for all beam coupling processes, which indicates that although the energy is exchanged between the beams, there is the same overall incoming and outgoing light intensity (we neglect side effects like scattering), we arrive at

$$|s_{1,0}|^2 + |p_0|^2 = |s_1(z)|^2 + |p(z)|^2 \qquad (3.41)$$

where $z$ is the variable of the propagation length inside the crystal. If we assume furthermore that no depletion of the pump wave supplying the system with enery occurs, we can solve that system of equations and obtain (setting $\gamma_1 =: \gamma$) [31]

$$\frac{|s_1(z)|^2}{|s_{1,0}|^2} = e^{(2(\gamma-\alpha)l)}\frac{I_0}{|p_0|^2 + |s_{1,0}|^2\exp(2\gamma z)} \qquad (3.42)$$

If we choose the direction of the energy exchange between the two beams in such a way that the signal beam is amplified at the output of the crystal, we are able to evaluate the amplification $V$ which is defined as the ratio of the signal intensity with and without coupling with the pump wave to be

$$V = \frac{|s_{1,l}|^2}{|s_{1,0}|^2} = \exp(\alpha l)\frac{1+r}{\exp(-2\gamma l)+r} \qquad (3.43)$$

with $r = |s_{1,0}|^2/|p_0|^2$ representing the signal-to-pump-intensity ratio of the incoming beams.

As eq. (3.43) shows, the amplification is mainly determined by the ratio of the incoming beams and the coupling constant, which in turn depends mostly on the beam geometry and the electrooptic constants.

The case of largest amplification we may imagine is achieved for negligible absorption ($\alpha = 0$) and non-depleted pump-waves ($r \ll 1$). Then the amplification depends exponentially on the coupling coefficient and the crystal thickness: $V = \exp(2\gamma l)$.

### 3.4.4   Multiple Beam Coupling

What happens if we increase the number of signal beams incident upon the crystal? This question is a rather important one for all image processing applications since in most real situations there are always more than two beams interacting in the crystal. Imagine e.g. an image which should be stored in the photorefractive crystal instead of a laser signal. If the crystal is positioned in the Fourier plane of a lens, all beamlets originating from the image transparency impinge on the crystal from slightly different directions. This situation is sketched in fig. 3.4.4. Thus the situation is much more complex than the one in two-beam coupling. Here, a lot of signal beams interact with one pump beam, all of them trying to produce an index grating or to become amplified. Because of the complex

**Figure 3.12:** Principle geometry of multiple beam coupling using one pump wave $p$ and many signal waves $s_i$. $\alpha_i$ and $\alpha_p$ are the incident angles in the crystal, $\vec{c}$ is the crystal axis.

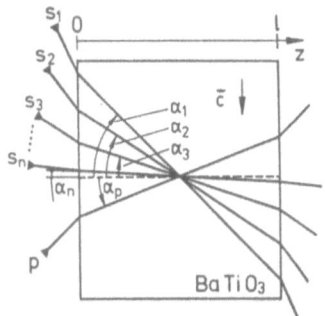

geometry and intensity configurations of the set of incident beams, the signals begin to compete with each other for amplification. A complex scenario of interaction, dominated by processes of competition and coupling, takes place.

*Mathematical Formulation of Multiple Beam Coupling*

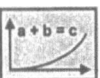 In order to understand the basic mechanism of multiple beam coupling, we may rely on the fact that every image can be divided via a series of Fourier components into a weighted sum of plane waves. These plane waves may have different wave vectors and are considered to interact in a region in the crystal, where all beams overlap completely. Thus all beams involved in the coupling process are interacting at the same time, whereas regions where only some of the waves overlap have only small contributions to the overall energy exchange and are neglected. This assumption is valid for crystals being small compared to the width of the area of interaction of the beams, which occurs for crystal thicknesses in the range of the beam diameter of the incoming waves.

The application of the *pinciple of superposition* stating that superimposed waves add linearly in their amplitude with an interference term for coherent waves, gives for $N$ linearly polarized signal waves interacting with a single pump wave — an intensity distribution which can be regarded as the linear superposition of sinusoidal gratings, each generated by a single pair of waves:

$$I(\vec{r}) = I_0 \cdot \left\{ 1 + \sum_i m_{pi} \cdot \cos((\vec{k_p} - \vec{k_i})\vec{r}) + \sum_i \sum_{j,j>i} m_{ij} \cdot \cos((\vec{k_i} - \vec{k_j})\vec{r}) \right\} \qquad (3.44)$$

with the total incoming intensity $I_0 = |p_0|^2 + \sum_{i=1}^{N} |s_{i,0}|^2$ and the modulation depths $m_{pi} = p \cdot s_i^*/I_0 + c.c.$ for pump-signal interactions and $m_{ij} = s_i \cdot s_j^*/I_0 + c.c.$, $(i \neq j)$ for intersignal interactions.

That intensity distribution gives rise to a refractive index modulation in the same way as in the case for two-wave interaction (compare with eq. (3.31)). It is again a sum of the different components of the modulation terms:

$$\Delta n = \frac{1}{2I_0} \sum_i p^* s_i \Delta n_{pi} \exp\left\{ i(\vec{k_i} - \vec{k_p})\vec{r} \right\} \qquad (3.45)$$

$$+ \frac{1}{2I_0} \sum_{i,j,i<j} s_i^* s_j \Delta n_{ij} \exp\left\{ i(\vec{k_i} - \vec{k_j})\vec{r} \right\} + c.c.$$

Here, $\Delta n_{pi} = r_{eff}^{pi} \cdot E_{sc}^{pi}$ and $\Delta n_{ij} = r_{eff}^{ij} \cdot E_{sc}^{ij}$ characterize the spatial holograms of the wave pairs $P$, $S_i$ and $S_i$, $S_j$ respectively. It is important to note that there are two different contributions to all coupling effects: One case is the signal-pump interaction where one signal is coupled with the pump wave. The other case is the signal-to-signal or intersignal interaction, where two different signals react, one of them playing the role of the energy supplier, the other one the role of the amplified wave. For the next combination of signals, these roles may be completely different, e.g. the former energy supplying wave may be amplified by another signal that is now playing the role of a supplier.

Consequently, the whole interaction scheme for all beams involved in the coupling process is rather complicated, leading to a set of $N + 1$ differential equations for $N$ signals and one pump beam [32, 33]:

$$\frac{\partial s_j}{\partial z} = \frac{s_j}{I_0} \left\{ \gamma_{pj} \mid p \mid^2 + \sum_{i=1, i<j}^{N} \gamma_{ij} \mid s_i \mid^2 - \sum_{i=1, i>j}^{N} \gamma_{ij} \mid s_i \mid^2 \right\} \qquad (3.46)$$

$$\frac{\partial p^*}{\partial z} = -\frac{p^*}{I_0} \left\{ \sum_{j=1}^{N} \frac{\cos \alpha_j}{\cos \alpha_p} \gamma_{pj} \mid s_j \mid^2 \right\} \qquad (i, j = 1, 2, \cdots, N) \qquad (3.47)$$

with the coupling constants (compare to eq. (3.40))

$$\gamma_{pj} = \omega \Delta n_{pj} / 2cn_o \cos \alpha_j \qquad (3.48)$$

and

$$\gamma_{ij} = \omega \Delta n_{ij} / 2cn_o \cos \alpha_j. \qquad (3.49)$$

Eqs. (3.46) and (3.47) state that the energy exchange of one signal results from the coupling with a pump wave (coupling constant $\gamma_{pj}$) as well as from the coupling with all other signals (coupling constant $\gamma_{ij}$). Thus, for $N$ signal waves it becomes difficult to get an overall view of all possible interactions.

Therefore, I will now discuss quantitatively only the example of three interacting beams — two signal waves and one pump wave — the behaviour. This case can be mathematically expressed in the following equations

$$\frac{\partial s_1}{\partial z} = \frac{1}{I_0} \left\{ \gamma_{p1} \mid p \mid^2 - \gamma_{21} \mid s_2 \mid^2 \right\} \cdot s_1 \qquad (3.50)$$

$$\frac{\partial s_2}{\partial z} = \frac{1}{I_0} \left\{ \gamma_{p2} \mid p \mid^2 + \gamma_{12} \mid s_1 \mid^2 \right\} \cdot s_2 \qquad (3.51)$$

$$\frac{\partial p^*}{\partial z} = -\frac{1}{I_0} \left\{ \frac{\cos \alpha_1}{\cos \alpha_p} \cdot \gamma_{p1} \mid s_1 \mid^2 + \frac{\cos \alpha_2}{\cos \alpha_p} \cdot \gamma_{p2} \mid s_2 \mid^2 \right\} \cdot p^*, \qquad (3.52)$$

giving rise to a variety of different beam interaction scenarios:
- The direction of the energy transfer depends on the signal geometry. The strength and the direction of the coupling process can be adjusted by changing the incoming angles of the signals.
- If we choose the angular configurations properly, we may achieve the situation that the signal with the largest starting intensity always gets the largest energy enhancement. This strange or — expressed in sociologic terms — "asocial" behaviour is interesting for all applications where a certain number of signals have to be suppressed at the expense of one right or desired signal. These applications are important in all neural network decision concepts. Therefore, I will discuss it further in chapter 5, where threshold devices are presented.

- The opposite case can also be adjusted. Then the lowest signal gets the largest energy increment, allowing thus to enhance weak signals relatively more compared to the stronger ones.
- It is possible to switch between these two situations by adjusting a single parameter — the incoming beam geometry or the angle of the pump wave.

That behaviour has been first investigated theoretically and experimentally by the author and her coworkers [32, 33]. Meanwhile several groups have affirmed and extended these results to different materials as SBN [34] and BSO [35, 36], including drift-dominated charge transport, to explanations on image amplification [37], diffraction efficiency [38] and beam fanning [39]. N.A. Vainos and R.W. Eason extended the results on four-wave mixing and phase-conjugation, thus allowing to control multi-wave phase conjugation [40]. In all these systems, the magnitude of interaction (or cross talk, if we consider these effects as non-desired in optical interconnection applications) can be varied or avoided by changing the geometry of interaction and the intensity ratio of the interacting beams. Thus, multiple beam coupling has become a powerful means for interconnection realizations and intensity switching for all image processing and neural network simulation applications.

### 3.4.5   Reflexive Two-Beam Coupling

Up to now, I considered the interaction of multiple beams, one of it carrying the information (signal) and the other one serving as a pump beam. It is also possible to let the information-bearing beam interact with a copy of itself. This can be done by imprinting first the image information on the beam, then splitting it by a beam splitter and redirecting both beams into the photorefractive crystal. This configuration was first realized by D.Z. Anderson and coworkers and named *reflexive beam coupling* [41].

Let us assume that the information carried by the beam consists of two or more spatially and temporally distinct signals. For example, the beam may consist of several images that appear at different times, or of images that are imposed upon distinct carrier frequencies, or of a collection of temporal signals carried by different spatial patterns. In this context, reflexive beam coupling can be considered as a special case of multiple beam coupling.

The basic idea of reflexive coupling is that the refractive index grating formed in a photorefractive medium by the reflexive interaction of an optical beam containing a number of temporal signals with itself consists of a superposition of components, each one due to the self-interaction of just one of the temporal signals. Even though each temporal signal interacts only with itself, there is an indirect coupling between different temporal signals that is due to the shared spatial components of the induced index grating. The induced grating through shared gratings allows signals to influence each other even though they may have carrier frequencies separated by more than the response bandwidth of the photorefractive medium or are present at different times. The result of this indirect coupling is that the fractional intensity and the spatial overlap of different temporal signals are modified at the output ports. The dispersion of the fractional intensities of the temporal signals carried by the optical beam determines the statistical information content of the beam. The statistical information is not conserved by reflexive interaction, and, depending on the choice of the output port, the statistical information may be selectively increased or decreased. Another application of reflexive coupling is to selectively manipulate individual signals. In particular, it is possible to spatially orthogonalize a pair of temporal signals that initially have a finite spatial overlap.

A special case of practical importance arises when the temporal signals are also spatially orthogonal. Then, they interact only with their own gratings — due to Bragg matching — and do not interact with all other grating components. On the other hand the modulation depth of each grating component is reduced by the presence of additional spatially orthogonal signals. Therefore, it is possible to manipulate spatially orthogonal signals in a multisignal beam on the basis of their intensity. Thus, if one signal is intially strongest, it also writes the strongest grating and experiences the largest energy coupling. This is similar to the behaviour in multiple beam coupling described above and can be used as a method for beam cleanup by enhancing the strongest signal and suppressing all weaker ones. By utilizing this effect in several different ways it is moreover possible to enhance strong signals, extract weak signals from a strong background, or equalize the intensity of several unequal signals. This type of intensity-based manipulation is very useful for bistable behaviours, feature extraction and contrast enhancement. Its theoretical details are described in [41] and its applications will be discussed again in chapter 6 and chapter 10.1.

In a more general description, reflexive coupling allows temporally and spatially orthogonal signals to be transformed into new temporally and spatially orthogonal signals with different temporal and spatial basis functions. In this way quite general transformations of the spatiotemporal nature of an optical beam are possible.

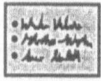 ## Summary

The photorefractive effect is based on photo-induced changes — induced e.g. by interference fringes — of the indices of refraction by the electrooptic effect. The principle mechanism can be summarized as follows:

- The interaction of e.g. a signal and a pump wave leads to a periodically modulated intensity or energy interference grating, causing electrons of impurity sites to be liberated, They migrate and recombine due to drift, diffusion or the bulk photovoltaic effect.
- While recombining, the carriers cause a change in the charge carrier density of the material.
- The induced space charge field in turn causes a change in the refractive index via the linear electrooptic effect (band transport model).
- The modulated refractive index carries the whole information of the signal, thus allowing to store the signal's information content.
- Self-diffraction of the interacting beams at the refractive index grating leads to dynamic holography, allowing for energy exchange between the beams involved.

Because the photorefractive effect allows to write gratings in a medium that represent the whole information of the interacting beams as in conventional holography, it is of immense importance for neural net applications. The main applications of the photorefractive effect for storage and neural processing purposes are:

- Two-wave mixing allows for amplification of weak, information bearing signal beams in the presence of a strong pump beam.

- If multiple signal beams are to be amplified, the strength and the direction of the coupling process can be adjusted by changing the incoming angles of the signals. Thus, the signal with the largest starting intensity or the weakest signal always gets the largest energy enhancement, depending on the configuration.
- Reflexive beam coupling — where a signal interacts with a copy of itself — allows to enhance strong signals, extract weak signals from a strong background, or equalize the intensity of several unequal beams.

## 3.5 Photorefractive Phase Conjugation

Up to now, I have discussed photorefractive beam coupling configurations where all interacting beams pass from one side through the crystal (transmission grating). A similar situation takes place if the signal and the pump wave impinge from opposite sides on the crystal. In that case, a reflection grating is formed, which allows as well as the formerly discussed transmission grating for energy exchange and amplification.

**Figure 3.13:** The principle of four-wave mixing and optical phase conjugation (transmission geometry) using two counter-propagating pump waves $p_1$ (incident angle $\alpha_2$) and $p_2$ and the signal wave $s_1$ (incident angle $\alpha_1$). $\vec{K}_1$ is the grating vector, $\vec{c}$ is the crystal c-axis, and $\varphi_1$ is the orientation of the grating vector relative to the c-axis.

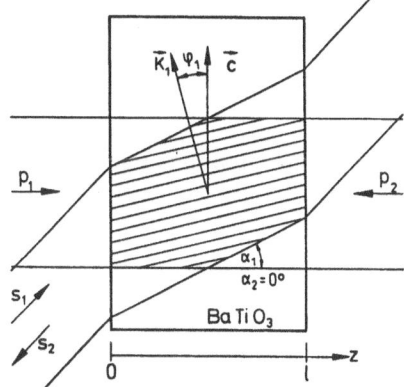

However, what will happen if we add another pump wave to the system, entering from the opposite side compared to the first pump wave in a counterparallel way? This situation is shown in fig. 3.13. In that case, we may describe the interaction process as the phase-matched scattering of the backward pump wave at the transmission grating which has been created via two-beam coupling by the forward pump and the signal. During that process, a wave travelling into the opposite direction of the incoming signal is created, which has a phase shift of $\pi$ relative to the incoming signal. This phase shift results from the sum of the phase shift of $\pi/2 = 90°$ between the interference and the refractive index grating and the phase shift of $\pi/2$ resulting from the reflection on the transmission grating. Thus that new wave is *phase-conjugate* relative to the incoming signal.

What is special about phase conjugation? The new wave retraces exactly the same way the signal has taken through the crystal. Therefore, we may characterize this process as a sort of mirror reflection. The particular interesting thing about phase-conjugate reflection is, compared to an ordinary mirror, that at each point of the beam path the

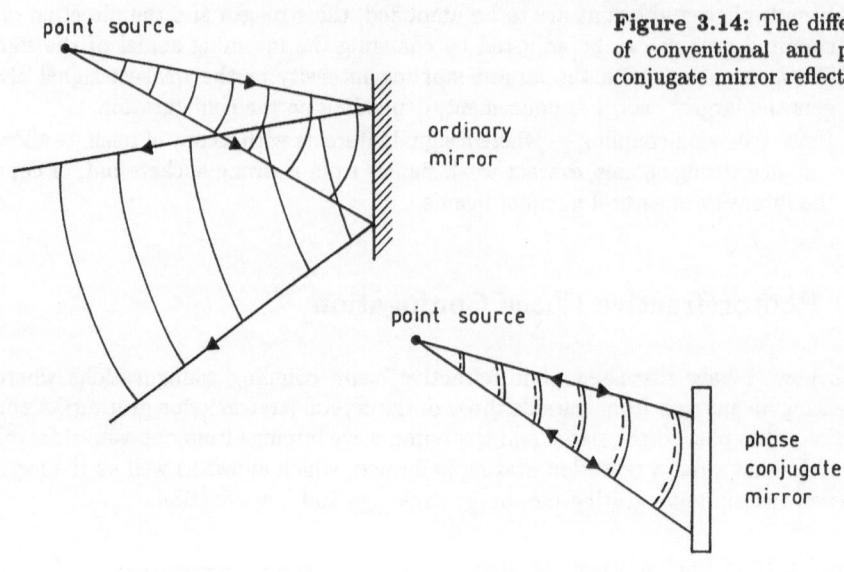

**Figure 3.14:** The difference of conventional and phase-conjugate mirror reflection.

exact reconstruction of the original phase front takes place, leading back to the point of origin of the signal (see fig. 3.14).

This effect is most impressive if you imagine looking at yourself in a phase-conjugate mirror. In an ordinary mirror, you will see everything in front of the mirror — and consequently at least your face or parts of your body. But what will be the picture of the phase-conjugate mirror, when each beam is passing back exactly the way it has taken before being reflected by the phase-conjugate mirror? Because the only rays coming back exactly to the pupil of your eye are those having started from there, you will see only these two spots in the phase-conjugate mirror. Thus the image of yourself are two spots — the reflection of your pupils' light. Some scientists even consider phase conjugation as a time reversal process, because the spatial part of the signal remains the same as before, whereas the sign of the time variable $t$ has been reversed. The applications of that mechanism which is unique in nonlinear optics and which can be realized in a lot of nonlinear materials — gases, liquids and crystals — becomes obvious if one imagines beam pathways with a lot of disturbances on it. In that case, the returning beam experiences all sorts of aberrations, distortions and disturbances the beam was subjected to on its way to the phase-conjugating medium once again, but this time in the other direction, with the phase turned about $\pi$ or 180°. This situation is illustrated in fig. 3.15. Thus, all wavefront changes during the forward way are corrected when the wave is coming back. Consequently, we get a completely noise-corrected and aberration-free signal. Moreover, as in two-wave mixing, the process allows for an amplification of the signal — provided that we use again a photorefractive material with diffusion dominated charge transport mechanisms or, which is equivalent, a stationary space charge field. So we achieve a completely corrected and amplified signal return from a phase-conjugate mirror based on photorefractive four-wave mixing.

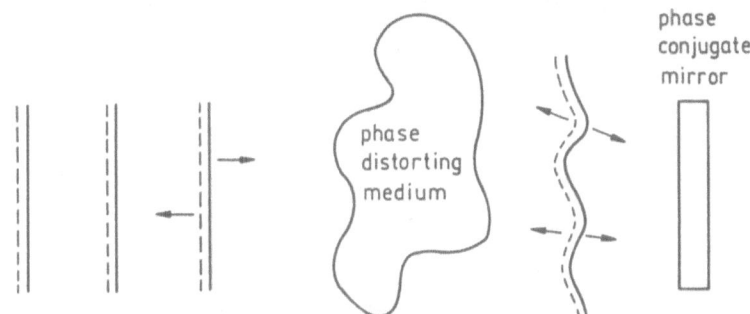

**Figure 3.15:** The correction of arbitrary phase distortions by a phase-conjugate mirror.

### 3.5.1 Mathematical Formulation of Four-Wave Mixing

 The theoretical analysis of that process is rather complicated, because instead of the easy explanation at the beginning of that chapter, we have a highly nonlinear interaction of many different gratings created in the crystal. Beneath the transmission grating, which describes the interaction of the beams $s_1$ and $p_1$ or, from the opposite direction, $s_2$ and $p_2$, reflection gratings ($s_1$ and $p_2$, $s_2$ and $p_1$) arise, each of it creating refractive index gratings via the electrooptic effect with different grating constants and modulation depths.

If we name the grating vectors $K_I$ to $K_{IV}$ with

$$
\begin{aligned}
K_I &= k_{s_1} - k_{p_1} = k_{p_2} - k_{s_2} \\
K_{II} &= k_{p_1} - k_{s_2} = k_{s_1} - k_{p_2} \\
K_{III} &= 2k_{p_2} \\
K_{IV} &= 2k_{s_1}
\end{aligned}
\tag{3.53}
$$

and if we derive the different refractive indices $\Delta n_I, \Delta n_{II}, \Delta n_{III}$ and $\Delta n_{IV}$ from eq. (3.35) by substituting the appropriate incident angles in $r_{\text{eff}}$ and the phase differences $\varphi_I$ to $\varphi_{IV}$ from eq. (3.30), we obtain a resulting refractive index modulation which represents the sum of all these contributions.

$$
\begin{aligned}
\Delta n &= \Delta n_0 + \frac{\Delta n_I}{2I_0}(s_1 p_1^* + s_2^* p_2)\, e^{i(\vec{k}_I \vec{r} + \varphi_I)} + c.c. \\[2mm]
&+ \frac{\Delta n_{II}}{2I_0}(s_2^* p_1 + s_2 p_1^*)\, e^{i(\vec{k}_{II}\vec{r} + \varphi_{II})} + c.c. \\[2mm]
&+ \frac{\Delta n_{III}}{I_0}(p_1 p_2^*)\, e^{i(\vec{k}_{III}\vec{r} + \varphi_{III})} + c.c. \\[2mm]
&+ \frac{\Delta n_{IV}}{I_0}(s_1 s_2^*)\, e^{i(\vec{k}_{IV}\vec{r} + \varphi_{IV})} + c.c.
\end{aligned}
\tag{3.54}
$$

where $I_0 = |p_{1,0}|^2 + |p_{2,l}|^2 + |s_{1,0}|^2 + |s_{2,0}|^2$ is the total intensity for the case of four-beam interaction.

Now, the spatial change of the waves throughout the crystal can be described by the following, rather complicated-looking expressions [42] — again derived using the slowly varying field approximation. They are, however, based on the same formalism as in the case of the differential equations for two-beam interaction. It is only the contribution of the high number of interference gratings that raises the number of possible interactions.

$$\frac{2c}{\omega}\cos\alpha_p\frac{\partial p_1}{\partial z} = -\frac{i\Delta n_I e^{i\varphi_I}}{I_0}(s_1^* p_1 + s_2 p_2^*)s_1 \tag{3.55}$$
$$-\frac{i\Delta n_{II} e^{i\varphi_{II}}}{I_0}(s_2^* p_1 + s_1 p_2^*)s_2$$
$$-\frac{i\Delta n_{III} e^{i\varphi_{III}}}{I_0}(p_2^* p_1)p_2 - \frac{2c}{\omega}\alpha\cos\alpha_p p_1$$

$$\frac{2c}{\omega}\cos\alpha_p\frac{\partial p_2}{\partial z} = +\frac{i\Delta n_I e^{-i\varphi_I}}{I_0}(-s_1 p_1^* + s_2^* p_2)s_2 \tag{3.56}$$
$$+\frac{i\Delta n_{II} e^{-i\varphi_{II}}}{I_0}(s_2 p_1^* + s_1^* p_2)s_1$$
$$+\frac{i\Delta n_{III} e^{-i\varphi_{III}}}{I_0}(p_2 p_1^*)p_1 + \frac{2c}{\omega}\alpha\cos\alpha_p p_2$$

$$\frac{2c}{\omega}\cos\alpha_1\frac{\partial s_2}{\partial z} = +\frac{i\Delta n_I e^{-i\varphi_I}}{I_0}(s_1^* p_1 + s_2 p_2^*)p_2 \tag{3.57}$$
$$+\frac{i\Delta n_{II} e^{-i\varphi_{II}}}{I_0}(s_2 p_1^* + s_1^* p_2)p_1$$
$$+\frac{i\Delta n_{IV} e^{-i\varphi_{IV}}}{I_0}(s_2 s_1^*)s_1 + \frac{2c}{\omega}\alpha\cos\alpha_1 s_2$$

$$\frac{2c}{\omega}\cos\alpha_1\frac{\partial s_1}{\partial z} = -\frac{i\Delta n_I e^{i\varphi_I}}{I_0}(-s_1 p_1^* + s_2^* p_2)p_1 \tag{3.58}$$
$$-\frac{i\Delta n_{II} e^{i\varphi_{II}}}{I_0}(s_2^* p_1 + s_1 p_2^*)p_2$$
$$-\frac{i\Delta n_{IV} e^{i\varphi_{IV}}}{I_0}(s_2^* s_1)s_2 - \frac{2c}{\omega}\alpha\cos\alpha_1 s_1$$

This set of differential equations can not be solved analytically for the global case of arbitrary four-wave interaction, but only for some special cases. Fortunately, they are the ones that are especially interesting for our applications in neural networks and therefore I will introduce them in the following sections.

The simplest one is the case of dominating transmission gratings, undepleted pump waves ($|p_1|^2$, $|p_2|^2 \gg |s_1|^2$, $|s_2|^2$) and negligible absorption. The solution of the corresponding, simplified set of differential equations, where we may set ($\partial p_1/\partial z, \partial p_2/\partial z = 0$) and $I_0 = |p_1(0)|^2 + |p_2(l)|^2$ is valid for most of the geometries that are usually exploited in four-wave mixing configurations and describes the real situation correctly. If we define the phase-conjugate reflectivity $R$ as the ratio of the phase-conjugate to the incoming signal and the intensity ratio of the pump waves to be $r = |p_{2,l}|^2/|p_{1,0}|^2$, we get:

$$R = \left(\frac{\sinh\left(\gamma l/2\right)}{\cosh\left(\gamma l/2 + \ln r/2\right)}\right)^2 \tag{3.59}$$

In that case, only the coupling constant $\gamma$ and the ratio of the pump beam intensities $r$ are responsible for the values of the reflectivity. That behaviour is shown in fig. 3.16. Maximum reflectivity can be achieved if $r = e^{-\gamma l \sin\varphi}$, or, for the case of diffusion-dominated materials, with $\varphi = \pi$, $r = e^{-\gamma l}$, $R_{\max}(r = e^{\gamma l}) = \sinh^2(\frac{\gamma l}{2})$ Thus, the backward pump wave has to be much smaller than the forward one to obtain high amplification factors in four-wave mixing.

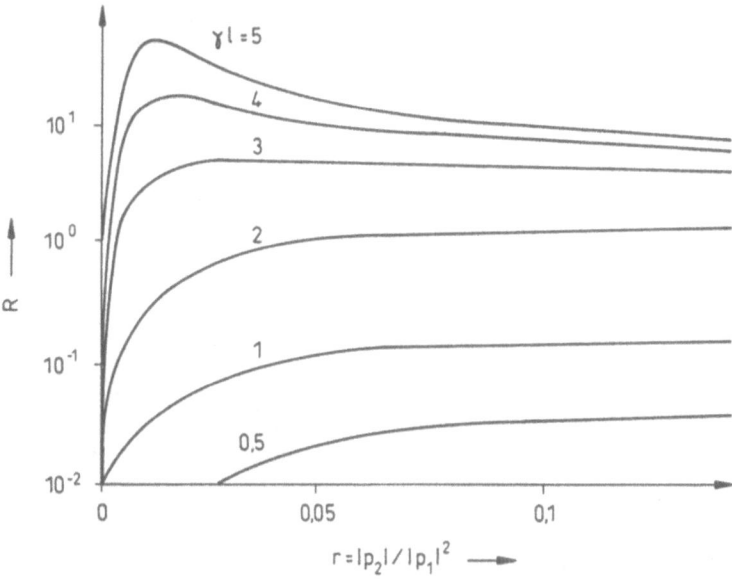

**Figure 3.16:** Dependance of the reflectivity of photorefractive four-wave mixing on the intensity ratio $r$ of the pump waves.

Up to now, I have considered the simple case of a dominating transmission grating to understand the principle mechanism of phase conjugation. In reality, the complex coupling of the different refractive index gratings in the crystal gives rise to complex energy transfer effects. Especially a competition behaviour is observed, which may lead to non-linear dynamic effects. They become important when a phase conjugate signal is used in a feedback or oscillation system, because in this case interaction and coupling effects of transverse beam distributions lead to effects like mode beating, mode competition, or optical turbulence. We will discuss these effects in more detail in chapter 6, section 6.5. For applications in optical image processing systems, these effects have to be minimized by adjusting the special operation condition — e.g. low intensities or particular geometrical configurations. Or, there is even the possibility to exploit these nonlinear effects — preferably effects in the transverse dimension relative to the propagation direction as mode competition — for storage or thresholding of information in such a way that the elementary modes of the system are used as basic eigenstates of the system. In that case, each input information is distributed between the modes of the system and can be considered as being stored in the eigenmodes of the feedback system.

### 3.5.2 Self-Pumped Phase Conjugation

The phase-conjugate replica of a complex image can be generated by the type of four-wave mixing I have described in the previous section. However, the quality of the phase-conjugate beam generated by that method is critically dependent on the alignment, intensities and phase profiles of the pump waves. For example, the phase of the backward pump wave can be used to adjust or manipulate the output phase of the phase conjugate signal. In other words, any "defect" in the phase front of the pump waves degrades the

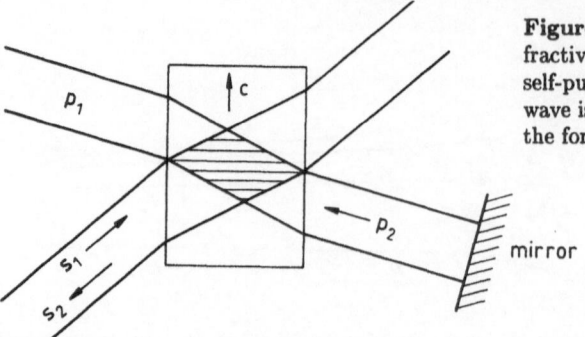

**Figure 3.17:** Creation of photorefractive phase conjugation by semi-self-pumping. The second pump wave is created by retro-reflection of the forward pump wave.

fidelity of the conjugation process. To overcome that problem, one may use an ordinary mirror behind the crystal to create the readout pump wave as a reflection from the incoming one (see fig. 3.17). Unfortunately, here again disturbances in the writing pump beam, which may even be duplicated by the ordinary mirror, degrade the fidelity of the phase-conjugate replay. However, there exists a third method of generating the necessary pump wave in an accurate way — the *self-pumped phase conjugation* process, where both pump waves are generated by the incident beam itself. Due to a complicated interaction process [43, 44, 45], the incoming beam is split or fanned off and then internally reflected from the surfaces of the crystal in such a way that the pump waves are generated. Although the fanning process, which is based on scattering phenomena, is purely statistical, the self-generated pump waves create the phase-conjugate wave in a configuration, that tends to optimize itself. That process is illustrated in fig. 3.18.

Self-pumped phase conjugators have been demonstrated in several photorefractive materials using a variety of geometries that fulfil the condition to realize the two counter-propagating pump waves. The big advantages of self-pumped phase conjugators for implementation in optical neural net systems are the self-aligning properties of the system in combination with exact phase front correction, allowing for various system configurations without any need to control the pathway of the interacting beams. In contrast to the mechanism of phase conjugation by externally-pumped four-wave mixing, which is based on the creation of transmission phase gratings that store the input beam information in the crystal volume, that configuration has a completely different grating storage mechanism. In conventional four-wave mixing, each grating represents a special interaction or weight of the coupling of two interacting beams, as already discussed for the case of multiple beam coupling (see section 3.4). Controversely, in self-pumped configurations, each weight is distributed among a large number of angularly and spatially varying gratings. Consequently, there are no limitations of the storage capacity due to the angular selectivity of Bragg-degeneracy of the material — a limitation that may become severe when using photorefractive materials as optical high-storage interconnection devices in neural networks. I will come back to this point in the following chapter on components and devices for optical neural networks, especially in section 4.3. Therefore, an optimum storage capacity may be achieved in the crystal volume.

What makes self-pumped beam coupling so different from conventional two- and four-wave mixing processes that such a different storage principle is operating? Although the exact process of self-pumped phase-conjugate generation is not yet entirely understood, there are two main processes responsible for its creation: beam fanning and internal, degenerate wave mixing. In the most commonly used model introduced by J.F. Lam and

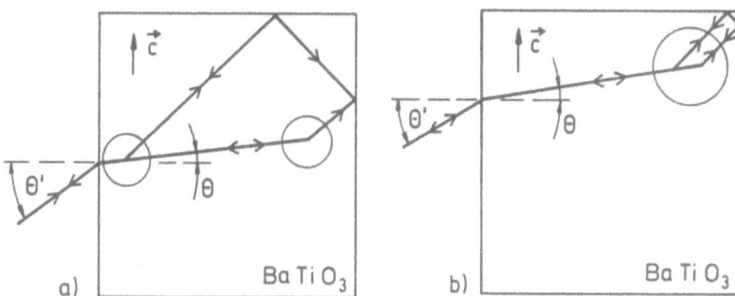

**Figure 3.18:** Principle geometry of self-pumped phase conjugation created by coupling of two four-wave mixing centers and influence of the incident beam location on the relative configuration of the two coupling centers. Coupling centers are far from each other (a) or close together (b). The latter situation results in strong nonlinear dynamic effects.

coworkers [43], Rayleigh scattering processes take place at small crystal inhomogeneities, resulting in low-amplitude scattered optical noise. Thus, some light components may be directed out of the central beam and those that are retroreflected interfere with it. In that way, a refractive index grating is created via the linear electrooptic effect in analogy to two-beam coupling configurations. Because two-beam coupling is dependent on the relative position of the interacting beams relative to the crystal's c-axis, parts of the beam in the angular range between the optical axis and the incident central beam are selectively amplified. The rest of the scattered light is attenuated. The amplified beams then write new gratings and the process cascades. Finally, the electrooptic tensor of the crystal and the angular dependence of the amplification, which gives a maximum for a certain angle between incident and scattered beam, define the preferred direction of amplified beam deflection. The net effect is that the input beam literally fans in the crystal as it writes a set of spatially and angularly distributed gratings. This directed scattering of light out of the central beam in a defined region is called the process of *beam fanning*, taking place in almost all high-gain photorefractive materials. By that process, new beam parts are generated, which are reflected at the crystal's surfaces in such a way, that they are redirected into the region of their origin and thus couple back to the incident beam.

*Mathematical Formulation of Self-Pumped Phase-Conjugation*

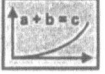 That interaction can be interpreted after a model proposed by J. Feinberg [24] as a difference mixing process. In detail, two centers are created, in which two- and four-wave mixing takes place and where the phase-conjugate replica of the incoming beam is generated. The complicated interaction of these two centers has been described by several groups [46, 47, 48, 49, 50, 51] and has been solved for the stationary case by M. Cronin-Golomb et al. [52]. All these derivations are based on the assumption that for the case of self-pumped phase conjugation, all grating distributions to beam coupling have about an equal strengths. Therefore, no simplifications in the beam coupling equations concerning neglection of weaker gratings are allowed. Consequently, the full description of photorefractive four-wave mixing has to be used for each of the four-wave mixing centers in self-pumped phase conjugation. Moreover, the energy

transfer between two coupled centers of four-wave mixing introduces a competitive behaviour, that leads naturally to dynamic effects. Therefore, the temporal dependance of the wave equations can no longer be neglected in order to achieve a complete spatio-temporal description of the self-pumped phase conjugation. The resulting set of four coupled differential equations for the case of a slowly varying field is given by:

$$(\frac{\partial}{\partial z} + \frac{d}{dt})p_1 = q_1 s_1 - q_2 s_2 - q_3 p_2 \tag{3.60}$$

$$(\frac{\partial}{\partial z} - \frac{d}{dt})p_2 = q_1^* s_1 - q_2^* s_2 - q_3^* p_2 \tag{3.61}$$

$$(\frac{\partial}{\partial z} - \frac{d}{dt})s_2 = -q_1 p_2 - q_2^* p_1 - q_4 s_1 \tag{3.62}$$

$$(\frac{\partial}{\partial z} + \frac{d}{dt})s_1 = q_1^* p_1 - q_2 p_2 - q_4^* s_2 \tag{3.63}$$

The refractive index grating strengths $q_i$ play the role of coupling the interference grating with the material parameters:

$$q_1 = \frac{-i\pi\Delta n_I}{\lambda I_0}e^{i\Phi_I}(p_1 s_1^* + p_2^* s_2) = \frac{i\pi n_e^3}{2\lambda}r_{\text{eff}}^{(1)}E_{sc}^{(1)} \tag{3.64}$$

$$q_2 = \frac{-i\pi\Delta n_{II}}{\lambda I_0}e^{i\Phi_{II}}(s_1 p_2^* + p_1^* s_2) = \frac{i\pi n_e^3}{2\lambda}r_{\text{eff}}^{(2)}E_{sc}^{(2)} \tag{3.65}$$

$$q_3 = \frac{-i\pi\Delta n_{III}}{\lambda I_0}e^{i\Phi_{III}}(p_1 p_2^*) = \frac{i\pi n_e^3}{2\lambda}r_{\text{eff}}^{(3)}E_{sc}^{(3)} \tag{3.66}$$

$$q_4 = \frac{-i\pi\Delta n_{IV}}{\lambda I_0}e^{i\Phi_{IV}}(s_2 s_1^*) = \frac{i\pi n_e^3}{2\lambda}r_{\text{eff}}^{(4)}E_{sc}^{(3)} \tag{3.67}$$

Each contribution $E_{sc}^{(i)}$ of the space charge field has the following temporal behaviour:

$$\frac{dE_{sc}}{dt} + \frac{1}{\tau}\frac{E_m + E_d + E_d \cdot E_m/E_q}{E_m + E_d}E_{sc} = \frac{i}{I_0\tau}\frac{E_d E_m}{E_m + E_d}IF(A_i) \tag{3.68}$$

where IF($A_i$) are the interference terms of the wave pairs as e.g. $(p_1 s_1^* + p_2^* s_2)$, $\tau$ is the dielectric time contant of a single grating, and $E_m = j(N_A - N_{A-})/\mu\mu_0 \mid \vec{K} \mid$ is the complete drift contribution of the space charge field. If we focus again our attention to diffusion-dominated photorefractive materials ($E_m \approx E_q \ll E_d$), the time dependance of the refractive index modulation simplifies to:

$$\frac{d\Delta n(z,t)}{dt} = -\frac{\Delta n}{\tau}\left\{1 - \frac{i}{\cos\alpha} \cdot \frac{IF(A_i)}{I_0}\right\} \tag{3.69}$$

Thus, eqs. (3.64) - (3.67) and (3.68) give four coupled differential equations to define the temporal behaviour of $Q_j$ [51]:

$$\left(\frac{d}{dt} + E^{[1]}\right)q_1 = -\gamma^{[1]}\frac{p_1 s_1^* + p_2^* s_2}{I_0} \tag{3.70}$$

$$\left(\frac{d}{dt} + E^{[2]}\right)q_2 = -\gamma^{[2]}\frac{s_1 p_2^* + p_1^* s_2}{I_0} \tag{3.71}$$

$$\left(\frac{d}{dt} + E^{[3]}\right)q_3 = -\gamma^{[3]}\frac{p_1 p_2^*}{I_0} \tag{3.72}$$

$$\left(\frac{d}{dt} + E^{[4]}\right)q_4 = -\gamma^{[4]}\frac{s_2 s_1^*}{I_0} \tag{3.73}$$

where the contributaions of the fields are given by

$$E^{[i]} = \frac{E_m^{(i)} + E_d^{(i)} + E_d^{(i)} \cdot E_m^{(i)}/E_q^{(i)}}{E_m^{(i)} + E_d^{(i)}}, \qquad (i = 1, \cdots 4) \tag{3.74}$$

and the coupling constants of the different gratings are

$$\gamma^{[i]} = \frac{n_e^3 r_{\text{eff}}^{(i)} E_d^{(i)} E_m^{(i)}}{4\lambda \tau_i \cos \alpha_i (E_m^{(i)} + E_d^{(i)})}, \qquad (i = 1, \cdots 4) \tag{3.75}$$

(compare to eqs. (3.35), (3.40), (3.48, and (3.49)). Eqs. (3.60) - (3.63) and (3.64) - (3.67) can only be solved numerically, showing that in a couple of different situations the dynamics of self-pumped phase-conjugation exhibits irregular fluctuations, bistable behaviour or even a deterministic choatic structure [51, 53, 54]. It should be noticed that even without completely overlooking the structure of these equations, the complicated coupling structure of the wave mixing centers in self-pumped phase conjugators described above makes them naturally very sensible for unstable, irregular or even chaotic behaviour, which in turn may be a source of disturbances in all sorts of devices that may severely limit their performance. In a series of detailed investigations of that irregular dynamics, the author and her coworkers have found that, depending on the position of the incoming beam relative to the y-axis of the crystal (refer to fig. 3.18), the distance of the coupling centers and therefore the complexity of their interaction changes, leading to different chaotic dynamic scenarios. In these experiments, I observed the predicted stationary behaviour as well as oscillations, pulsations, intermittencies and purely chaotic phenomena [46, 47], which have been meanwhile observed by other authors too (e.g. [55]).

In the following chapters, I will use the nonlinear optical effects, mainly beam coupling and phase conjugation in order to realize the different components of optical neural networks. In that way, effects as grating formation to realize interconnection weights, amplification to realize gain and losses for inhibitory and excitatory stimuli as well as phase conjugation to implement optical backpropagation without experiencing disturbances due to beam divergence and phase deformation noise will play an important role. Nonlinear dynamic effects, as they have been named in the present and the last section, will play only a role when they affect the performance of a system - on the one hand as disturbances that have to be controlled during network operation or on the other hand as a means to realize stable energy states or energy basins in a system similar to those we have get to known in the first chapter. Finally, they may be used in optical neural network implementations in order to realize unsupervised search for states in an arbitrary, namely chaotic manner as well as information generation for network learning. Moreover, chaos can be used for sampling, in the complementary functions of trial and memory or it can be used in neural network simulations to allow the possibility of avoiding local minima by approriate use of the flexibility of chaotic dynamics.

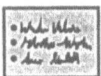 ## Summary

Photorefractive phase conjugation occurs when two counterpropagating pump waves interact with the signal beam. The main features of that process are

- The interaction process can be described as the phase matched scattering of the backward pump wave on the transmission grating.
- A new signal is created, which has a phase shift of $\pi$, thus being phase-conjugated relative to the incoming signal.
- Phase-conjugate beams retrace exactly the same way the incident beam has prescribed. It is a mirror retroreflection that exactly reconstructs the original phase fronts in each point, thus allowing for the correction of all phase aberrations in the beam pathway.
- The effect of phase conjugation via four-wave mixing with two strong pump waves allows for amplification of the incoming signal as well.
- Self-pumped phase conjugation is a completely self-aligning process. Here, the pump waves necessary to create the phase-conjugate wave are generated internally in the crystal via asymmetric scattering processes (beam fanning).

 **Further Reading**

1. D.L. Mills, *Nonlinear optics — basic concepts*, Springer Verlag, Berlin (1991).
   A short, concise overview over the principle phenomena — well suited for beginners.
2. H.M. Gibbs, G. Khinovc, N. Peyghambarian, *Nonlinear Photonics*, Springer Series in Photonics and Electronics (1990).
   A good introduction to the field with explanations of the basic principles.
3. P.N. Butcher, D. Cotter, *The elements of nonlinear optics*, Cambridge studies in nonlinear optics, Vol.9 (1991).
   A good overview over nonlinear optical phenomena, with an excellent introduction to the mathematical notation of nonlinear optics. Many interesting examples of applications of nonlinear optical devices are given.
4. Y. Shen, *The principles of nonlinear optics*, Wiley & Sons (1984).
   A well-written and comprehensive introduction to nonlinear optics.
5. P. Yeh, *Introduction to photorefractive nonlinear optics*, Wiley & Sons (1993).
   An excellent, complete introduction to all important properties of photorefraction and phase conjugation — well-written and easy to understand.
6. B.Y. Zel'dovic, N.F. Pilipisky, V.V. Shkunov, *Principles of Phase Conjugation*, Springer series in optical sciences (1988).
   A good overview over the different mechanisms to realize optical phase conjugation.
7. P. Günter, J.P. Huignard, Eds., *Photorefractive Materials and their applications*, Vol. I, II, Springer Series Topics in Applied Physics, vol. 61, vol. 62 (1987/1989).
   The first comprehensive summary of photorefractive material properties, effects and devices as feedback systems and memories.
8. R.A. Fisher, Ed., *Optical phase conjugation*, Academic Press, New York (1986).
   A basic textbook introducing in the field of optical phase conjugation with special emphasize on photorefractive four-wave mixing.
9. M.P. Petrov, S.I. Stepanov, A.V. Khomenko, *Photorefractive Crystals in Coherent Optical Systems*, Springer Verlag, Berlin (1991).
   An easy-to-read introduction into the basic applications of photorefractive optics, including a lot of examples of dynamic holography applications.

# Part II

# Devices, Components and Subsystems

# 4  Nonlinear Optical Storage and Interconnection Concepts

Now that we have got an insight in the phenomena of nonlinear optics as well as in the principles of neural networks, we have to combine them in order to derive ideas, concepts and requirements for the realization of all-optical neural networks. We have to discuss, which of the processes explained in the preceeding chapter are well-suited for a certain task that arises in the neural network concepts I have presented in chapter 2. Therefore, the present and the following chapters introduce the actual concepts of how to realize the main components necessary to build up all-optical models of neural networks.

My aim is not to present a complete review of all possible optical neural network implementations. I will focus my attention to all-optical realizations, that use the inherent advantages of nonlinear optics — parallelism, global interconnectivity, high processing speed and easy implementation of nonlinear operations. Because it would burst the volume of this book to describe completely all the concepts and promising systems that have been realized up to now in nonlinear optics, I will focus my attention to one of the forthcoming areas of nonlinear optics — photorefraction — as a crucial example technique. However, several interesting examples of realizations using other nonlinear optical phenomena are also mentioned in the following chapters. For the reader who becomes more interested in these fields, a comprehensive reference list at the end of the chapters will help to find more information about these concepts.

The results of the following three chapters will enable us to construct a series of simple neural networks with incoherent or coherent all-optical processing that allow the simulation and implementation of almost all neural net models that have been presented in the second chapter. Among them, models for one- and multilayer perceptrons as well as self-organized nets and Boltzmann structures are found. Especially the area of pattern recognition using associative memories — mostly based on Hopfield-like neural network structures — is a quickly expanding one, showing that optics offer immense possibilities for realizations of complicated multilayer nets in that area.

The most important devices for even the simplest neural network structure are storage or interconnection devices, threshold elements and logic operators. Moreover, the way of using feedback and the way of realizing several layers of a system are topics of the present part of the book, because they are important if one wants to put all the components mentioned above together to a well-working optical neural net. I have chosen out of the number of possible devices those which seem to be promising for implementations in a lot of different neural nets. Most of the concepts described here will be found again in the third part of the book, where I will put the components together to build simple, but efficiently architectured optical models of neural networks.

## 4.1   Interconnection Devices

Interconnecting a network input to an output or a hidden layer is one of the major purposes in implementations of neural networks. For these interconnection or storage units, in turn, parallel processing and global interconnection capabilities are essential.

*Parallel processing* is an inherent feature of nonlinear optical systems and certainly a highly desirable property. However, its importance as an advantageous feature compared to electronics is not the greatest one, since there is no fundamental limitation to the degree of parallelism that can be simulated and achieved electronically. The electronic realization of parallelism is based on the idea to arrange in parallel $N$ serial processors and distribute tasks between them. The gain in speed of such a system is then given by a factor $N/\ln N$ [56]. Thus, an electronic computer with e.g. 32 processors gains a factor of 9 in speed, thus being much faster than a system having only a single processor. Although the optical idea of parallelism — process a two-dimensional plane of usually about 10.000 x 10.000 $\approx 10^8$ information units or pixels in parallel — is a beautiful concept, that gains a lot more in speed because of this restriction, there are projects in progress to implement electronic systems with hundreds of thousands of electronic parallel processing elements — being still in competition with conventional optical processing capabilities.

*Global communication capability* on the other hand is a property of optics that is obviously very difficult to duplicate electronically. One of the reasons that optics can provide global communication is the fact that optical systems may be configured in three dimensions. For instance, optics can be used to interconnect a large number of processing units in a plane with the light propagating in the third dimension. The interconnection pattern itself will then be specified externally to the plane of processors. Thus, neurons may be implemented in two dimensions and the interconnections are stored in the third dimension. Consequently, if we want to build up neural nets that will have comparative features or even advantages over electronics, it is very desirable to find ways to do computation in such a manner that the communication capability is the prominent, most important component of the neural computation process.

In optics, a multiple interconnection device is defined as an element, which connects an optical input vector (or matrix) $I$ to an optical output vector (or matrix) $O$. The input vector may be provided by a one- or two-dimensional array of sources, e.g. fibers, lasers or spatial light modulators. The output vector is usually an array of fibers for nets with several layers or an array of detectors. In two-way or bidirectional links, as they are used in backpropagation or error-driven nets, the two vectors have the same role. Each vector element represents a station or a node, consisting of a transmitter and a receiver. Because every two-dimensional image consists of a lot of pixels (or smallest resolution points) which can be regarded as independent interaction channels with continuous spatial transition regions of their grey levels, the borderline between point-to-point interconnection coupling and image coupling is not a straight one. To make a distinction from pure image storage, in the first part of the chapter I want to look at small interconnection systems with discrete, separated and independent interacting pixels or beamlets as it is the case for example for 3 x 3 points in the input plane that need to be connected to 3 x 3 different ones in the output plane or the connection of one single input point with many others, typically several thousands at the output port. Tasks of that kind arise e.g. in telecommunication, where a line of independent pixels or information bits has to be connected to one or several others.

In general, the devices used for interconnections are all-optical or hybrid electrooptical

**Figure 4.1:** Four different methods of interconnecting optical inputs to optical outputs: broad-casting (fan-out) (a), combination (fan-in) (b), point-to-point (c) and crossbar switching (d) operations (after [57]).

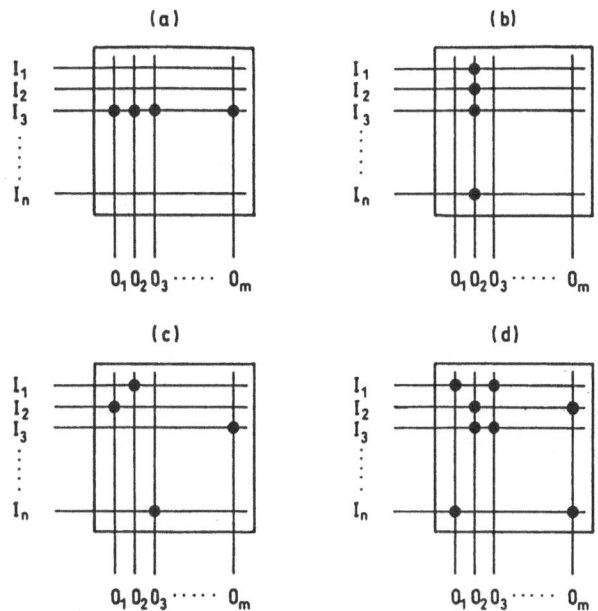

ones. The media in which the optical signals can be linked may be free space (using optical or holographic lenses), optical fibers, integrated optical waveguides or optical plane or volume holograms. For all of them, several different interconnecting methods have been proposed. The most commonly used methods are described in fig. 4.1. These are broadcasting or fan-out (a), combining or fan-in (b), point-to-point (c) and crossbar switching (d) operations.

The devices most often discussed to connect in- and outputs of a net can be divided into two-dimensional (plane) or three-dimensional (volume) interconnection media or memories. Among the existing materials and techniques for two-dimensional image storage, I want to mention optical digital storage in disks, spatial light modulators, acousto-optic modulators, and planar holograms. Three-dimensional volume interconnection devices offer on the one hand a larger interconnection capability, taking advantage of the third dimension. On the other hand, they are well-suited for the purposes of weight adaptation, because most of the nonlinear writing mechanisms allow for erasable memories with the ability of weight correction. Among these volume storage techniques, volume holography in photorefractive crystals, spectral hole burning, stimulated photon echo and two-photon effects are the most promising ones.

### 4.1.1   Planar Interconnection Devices

*Light Modulators*

Spatial light modulators (SLM's) are devices that can modulate some properties of an optical wavefront, such as its amplitude or intensity by absorption, its phase by the refractive index or its polarization by rotation. These devices form an important part of optical realizations of neural networks as well as information processors since they serve in most cases as input transducers and signal converters, as well as performing several basic processing operations on the optical wavefront. The information-bearing signals can be electrical or optical, leading to two major classes of spatial light modulators: electrically and optically addressable ones.

An electrically addressed SLM is usually constructed with a pixel structure. Thus, the local optical wavefront is modulated at each individual pixel by electric signals. The advantage of such an electrically addressed SLM is its capability of interfacing with electronic and optical systems. However, this is also the reason for its pixelation and the bottleneck in information speed. This type of modulation is most often used as a data input device, because it allows to introduce electronic data in a parallel way into an optical processing system. In contrast, an optically addressed SLM consists of a continuous structure for performing the modulation function. The addressing optical image produces an electric charge distribution over the modulation material that generates secondary effects for electrooptic modulation. In this sense, an optically addressed SLM can be described as a combination of a detector and a modulator.

In the following, I will describe shortly the main function principles of different modulators, and then give some comments on their potential for applications in optical interconnection devices.

**Acoustooptic modulators**  When an acoustic wave is launched into an optical medium, it generates a refractive index wave that behaves like a sinusoidal phase grating. An incident laser beam passing through this grating diffracts the laser beam into several orders. The angular position of the selected diffraction order (e.g. the first order) is linear proportional to the acoustic frequency, so that the higher the frequency, the larger the diffraction angle. The intensity of the diffracted light is proportional to the power of the acoustic wave, so that the intensity can be modulated too.

A variety of different acoustooptic materials are used depending on the paser parameters such as wavelength, polarization, and power density. For the visible and near-infrared region, the modulators are usually made from flint glass, tellurium oxide ($TeO_2$) or fused quartz. At the infrared region, germanium is employed. Lithium niobate ($LiNbO_3$) and gallium phosphide (GaP) are used for high-frequency signal processing devices.

There are several physical constraints on the types of signals waveforms that can serve as input and the way in which acoustooptic modulators can be used. For example, acoustic waves emitted by the transduces into the medium usually can not be seen if they are viewed by conventional optics. Moreover, the electric driving signals must general be bandpass in nature and have frequencies in the range of 1 MHz to 1 GHz. The limit of the rise and fall time of the acoustooptic modulator is the transit time of the acoustic wave propagation across the optical beam. A typical rise time for a 1 mm diameter laser beam is around 150 ns. Also, in the acoustooptic interaction, the laser beam frequency is shifted by an amount equal to the acoustic frequency.

**Magnetooptic modulators**  A magnetooptic spatial light modulator (MOSLM) is a two-dimensional electronically addressed SLM. It is based on the magnetooptic Faraday effect (see section 3.2). The MOSLM consists of a square grid of magnetically bistable pixels that can be used to modulate incident polarized light by the Faraday effect. The device can be electrically switched so that object patterns can be written by a computer input. Thus this device can function as a programmable spatial light modulator for applications in optical interconnection and processing as logic operations.

When linear polarized light is incident in such a device, the axis of polarization of the transmitted light will be rotated by 45° clockwise for a magnetic state. The plane of polarization is rotated counterclockwise by 45° for opposite magnetic state. The magnetization state of the pixel can be changed by sending current to its two adjoining drive lines. An analyzer can convert the polarization rotation into a useful input for an optical processor. If the analyzer is set at the direction making a 45° angle with the original light polarization, intensity modulation will be obtained. Alternatively, the analyzer can be set at the direction parallel to the original light polarization to generate phase modulation, which is desired for the formation of a phase-only filter. The MOSLM has a storage capability, since the magnetization state of the substance remains a stable state. The state can be switched by current. However, this current can generate heat, due to Ohmic losses, which in turn may limit the performance of the MOSLM. The switching speed of the magnetic domain itself in these devices can be very fast — of the order of tens of nanoseconds. A big disadvantage, however, of this devices is their poor transmittance. A bright pixel may only have 5% transmittance.

**Pockels optical modulators**   A Pockels readout optical modulator (PROM) is a two-dimensional optically addressed SLM that is based on the Pockels effect (see section 3.2). Because under the Pockels effect, the crystal can rotate the polarization plane of the light and because the polarization rotation can be controlled by the applied electric field, there are two common configurations referred to as transverse and longitudinal, depending on whether the electric field is applied perpendicular or parallel to the direction of propagation, respectively. PROMs may be fabricated from various electrooptic crystals such as ZnS, ZnSe, and BSO. The construction of the most common PROM device is based on a electrooptic crystal wafer that is sandwiched between two transparent electrodes and separated from them by an insulator. The crystal wafer is oriented in such a way that the field applied between the electrodes produces a longitudinal electrooptic effect. An applied dc voltage with an erase light pulse is used to create mobile carriers that cause the voltage $V_0$ in the active crystal to decay to zero. The polarity of the applied voltage is brought to zero and then reversed. When a total voltage of $2V_0$ appears across the crystal, the device is exposed to the illumination pattern of blue light. Subsequently, the voltage in the area exposed to the bright part of the input pattern decays because of the optically created mobile carriers, but the voltage in the dark area remains unchanged, thus converting the intensity pattern into a voltage pattern. The readout can be performed using a red linearly polarized light. For BSO, the crystal is 200 times more sensitive in the blue region that in the red region. Therefore, reading with a red light source in real time does not produce significant voltage decay over a period of time.

The readout is performed by reflection. In this readout mode, the area of the crystal where the voltage across it has not been affected by the input light intensity acts like a half-wave retardation plate. The angle of polarization of the linearly polarized laser light input reflected by such an area is therefore roated by 90°. Thus the light reflected by the area corresponding to the bright region has a polarization perpendicular to that of the dark region. The reflected light is then passed through a polarizer, and the amplitude of the transmitted light is attenuated according to the polarization of the input.

The disadvantages of the device are based on the fact that its function is volatile. Any extended exposure to high-intensity readout light causes substantial decay of the crystal voltage, thus reducing the read amplitude of the device. Therefore, the device cannot be read over an extended period of time under strong illumination. Hoever, PROM can be used for incoherent-to-coherent conversion, amplitude and phase modulation as well as optical parallel logic operation.

**Microchannel light modulator**   The microchannel plate spatial light modulator (MSLM) is another two-dimensional optically addressed SLM that has frequently been used in optical implementations of neural networks as well as in real-time optical signal processing and computing. It consists of a photocathode, a microchannel plate, an accelerating mesh electrode, and an electrooptic crystal plate of $LiNbO_3$ that bears, on its inner side, a high-resistivity dielectric mirror to isolate the readout from the write-in side. All this components are sealed in a vacuum tube.

The basic operation of an MSLM is the following. The write-in light (coherent or incoherent image) incident on the photocathode generates a photoelectron image, which is multiplied to about $10^5$ times by the microchannel plate, accelerated by the mesh electrode, and deposited on the dielectric mirror of the $LiNbO_3$ crystal plate. The process of electron multiplication is similar to that in an image intensifier tube. The resulting charge distribution, in combination with the biasing voltage, creates a spatially varying electric field within the crystal plate in the direction of the optical axis. This field in turn modulates the refractive indices of the crystal plate. Because of the birefringence property of the $LiNbO_3$ crystal, its refractive indices in the horizontal and vertical planes are modulated differently. Thus, after passing the crystal plane twice, the readout light, which was originally polarized in a plane bisecting the $X$ and $Y$ axes of the crystal plate, will have a relative phase retardation between its $x$ and $y$ components. The higher the charge density, the greater the phase retardation. If an analyzer is inserted in the readout light path, a coherent image that is proportional to the input image can be obtained . Generally speaking, the MSLM is not only an incoherent-to-coherent converter, but it can be applied as a wavelength converter, as an input or output transducer, and to image plane or Fourier plane processing. An important feature of an MSLM is that the device can hold an image memory (charge distribution) for a period of time. Since the output image is produced by the charge distribution, the MSLM can perform addition and subtraction by superimposing the charge distributions generated by two input images. Subtraction is performed by reversing the polarity of the biasing voltage. In addition, optical thresholding is possible by varying the biasing voltage [58]. Typically, a commercially available MSLM has a resolution of 20 lines/mm, a contrast ratio of more than 1000:1, an input sensitivity of 30 $nJ/cm^2$, a writing time response of 10 ms, an erasing time of about 20 ms and a storage time of several days at a total input window of 15 mm.

**Liquid crystal light valve**   Another example of an optically addressed SLM is the liquid crystal light valve (LCLV) [59]. Most of you are certainly familiar with liquid crystal displays, that have found their way into portable laboratory equipment, kitchen appliances, automobile dashboards and many video game displays. Most of these applications use the liquid crystal in a twisted nematic cell — a segment of the cell is either opaque or transparent, depending on the magnitude of applied voltage.

The LCLV combines the property of the twisted nematic cell in the off state with the property of electrically tunable birefringence to control transmittance over a wide continous range in the on state. For that purpose, a thin layer of nematic liquid crystal is sandwiched between two transparent electrode-coated glass plates. The electrode surfaces are treated (often rubbed) to give a preferred direction of alignment for the liquid crystal molecules. The plates are arranged in such a way that, with no electric voltage applied across the layer of liquid crystal molecules, the layer is twisted continuously by 90°. A polarizer and an analyzer are placed in front of and behind the sandwiched cell, respectively. The direction in which the light passing through the polarizer is polarized must be the same as the direction of molecular alignment at the front electrode surface. As the light passes through the twisted liquid crystal layer, its direction of polarization is also twisted by 90°. If the direction of polarization of the analyzer is made perpendicular to the direction of molecular alignment at the back electrode surface, the polarized light will not exit the cell, and the device will be in its dark state.

When a voltage is applied across the twisted nematic liquid crystal layer, the molecules tend to align themselves in the direction of the applied electric field — that is, perpendicular to the electrode surfaces, resulting in splay and bend of the molecules Thus, if the polarizer and the analyzer have parallel directions of polarization, the polarized light will pass unaffected through the liquid crystal layer and the analyzer, which is the on state for a liquid crystal alphanumerical display. Hoever, this on state is not exhibited in a LCLV.

The LCLV takes advantage of the pure birefringence of the liquid crystal material in order to modulate the output laser beam while in the on state. An incoherent image is focused onto the photoconductor layer to gate the applied alternating voltage to the liquid crystal layer in response to the input intensity at every point in the input space. Laser light illuminating the back of the LCLV is reflected back, but modulated by the birefringence of the liquid crystal layer at every point. In order to achieve this effect, the molecular alignment has a 45° twist between the two surfaces of the liquid crystal layer. The dielectric mirror plays an important role by providing optical isolation between the input incoherent light and the coherent readout beam. As with the MOSLM, a phase filter can be generated by properly orienting the analyzer.

A very similar optically addressable device are ferroelectric liquid crystal spatial light modulators.Here, liquid crystals with two bistable states under applied voltage are used instead of the twisted nematic ones, giving bistable switching behaviour for binary spatial light modulation and again the possibility for phase-only modulation.

In addition to incoherent-to-coherent conversion and other coherent processsing applications, the LCLV is also used as as wavelength converter. Commercially available LCLV have resolutions of about 60 lines/mm. Because of the slow response time of the liquid crystals, the readout time is about 10 ms. The required write energy is typically 6 $\mu J/cm^2$.

**Liquid crystal television** Although originally produced as televisions, projection TVs, or video camera view finder systems, commercially available liquid crystal televisions (LCTVs) are now widely spread as devices for electronically addressed SLMs .The basis structure is a substrate filled with a twisted nematic liquid crystal , each pixel individually controlled by electronic signals, sandwiched between two polarizing sheets, one acting as the polarizer and the other one as the analyzer. A plastic diffuser provides diffused illumination on one side and a clear plastic window protects the device from dust on the opposite side. If the plastic diffuser, the window and the two polarization sheets are removed carefully from these devices, the remaining cell can be used as a SLM in a similar way as the LCLV: When no electric field is applied to the LCTV, the plane of polarization for linearly polarized light is rotated through 90° by the twisted liquid crystal molecules. Thus no light can be transmitted trough the analyzer. Under an applied electric field, however, the twist and the tilt of the molecules are altered, and the liquid crystal molecules attempt to align parallel with the applied field, which results in partial transmission of light through the analyzer. As the electric field increases further, all the liquid crystal molecules align in the direction of the applied field. The molecules do not affect the plane of polarization, so all the light passes throught the analyzer. In general, an analyzer parallel with (0°) and orthogonal to (90°) the polarizer will produce a positive and negative image, respectively. If the analyzer is set at 45°, a phase modulation can be obtained. Varying the applied voltage at each liquid crystal pixel allows light transmission to be varied. If the applied voltages of the LCTV screen are generated from a computer, a video camera or a TV receiver, the LCTV can produce grey-scale images.

The major advantages of the LCTV are its low cost and programmibility, which make the device a very practical SLM for many optical signal processing, computing and neural network applications [60, 61, 62]. .

**Phase-only liquid crystal spatial light modulators**  The most promising liquid crystal device has revealed to be the phase-only liquid crystal spatial light modulator. which is light-efficient and allows nearly continuous phase modulation. Although the liquid crystal TV or the LCLV can be tuned to realize in a certain parameter region pure phase modulation, a modulator that enables phase modulation in a continous way over the whole voltage range has to be prepared in a special way. The panel consists of a thin layer of nematic liquid crystal material sandwiched between two glass plates on which the transparent pixel electrodes are deposited. As in conventional liquid crystal panels (see above), the orientation of the liquid crystal molecules is determined by rubbing the inside surfaces of the glass plates before assembly. However, in that panel the rubbing directions are the same for both front and rear glass plates, resulting in liquid crystal molecules being aligned in the same direction throughout the layer (planar parallel aligned molecules). In this case, application of an electric field results in phase modulation, but polarization effects are negligible. Each pixel in the panel is provided with its own nonlinear diode ring (active matrix technology) which holds the electric field across the liquid crystal layer essentially constant between successive frame scans.

This device is able to perform interconnection schemes as fan-out, nearest neighbour- or next neighbour interconnects, and perfect shuffle interconnections. The diffraction efficiencies in all these cases are close to 15% [63]. Moreover, phase-only liquid crystal spatial light modulators may be used to realize correlation systems and optical pattern recognition ( see [64] and references therein).

**Spatial light rebroadcasters**  Most SLM's do not have the ability of long-term memory. Therefore, the electronic or optical signal for addressing the electronic SLM or the optical SLM during writing must be present when reading the device. In contrast, a spatial light rebroadcaster (SLR) is an optically addressable long-term memory device [56, 65]. It can store the weights for a neural network on a two-dimensional surface and permit them to be individually increased by writing with one wavelength (e.g. 514 nm, argon-ion laser source) and individually decreased by using a different wavelength (e.g. infrared laser source).This is an excellent match to the perceptron algorithm, because it enables bipolar weight modifications in a pure optical way. Moreover, the SLR has the further advantage that the number and the size of its pixels may be changed in a definite way at will, because of its continuous surface.

**Deformable mirror device**   The deformable mirror device (DMD) is another example for an electronically addressed SLM. It involves mechanical deformation of the modulator in response to an electrostatic force. The modulating elements of the DMD are tiny metal mirrors, arranged in two-dimensional grids. These mirrors are fabricated to from the cantilever beam structure over an underlying silicon address structure. The mirror and the underlying address structure form an air-gap capacitor that is typically 2 to 3 $\mu m$ wide. The addressing structure, which is a transistor array, allows a prescribed amount of charge to be deposited below each mirror. The amount of deflection of the mirror is determined by electrostatic attraction. In contrast to other SLMs, this particular SLM modulates the phase or the optical path length of the light beam. However, intensity modulation can be obtained by using a schlieren optical or phase contrast system, where the phase information is tranferred into an amplitude modulation. More details about this new device that has been recently fabricated by Texas Instruments can be found in [66]

In all the interconnection systems based on these SLM principle, the possibility to adjust the connection weights depends strongly on the achievable transmission versus applied voltage or applied intensity characteristic. If it is smooth enough over a reasonable voltage or input intensity range, an adequate performance of the network may be realized, but if the characteristic is steep in a small voltage or intensity range, no exact adjustment of the interconnection weights and therefore no reliable output answers can be expected. Because the smoothness may be limited both by the physics underlying the SLM's operation and their driving parameters, the use of these devices leads unavoidably to a network in which the weights are discretized. Imagine e.g. the driving electronics that control the voltage at each SLM pixel being digital and therefore having a limited accuracy. Although in general, discretization would not be a disadvantage if it could have a sufficient resolution, it is connected with the number of weight levels determined by the number of pixels over which the transmittance of the device is discretized. Considering these factors, about 100 weight levels can be achieved in SLM interconnection devices at the present time, consequently resulting in a device with a poor learning performance.

This disadvantage of SLM and AOM systems becomes serious in the case of all-optical net realizations that rely on energy minimization processes to find the optimum solution as the Boltzmann machine or multilayer perceptrons. It is the main reason why I will not emphasize in that book realizations of network interconnections that utilize spatial light modulators. Consequently, only a selected number of concepts concerning SLM interconnection technology and their applications in simple neural networks will be presented throughout the book. However, almost all optical neural network realizations use these devices for data input, because they represent the ideal transfer device between optical and electronical data. An exhaustive review of present SLM technology for neural net applications and some promising network concepts based on it are given in [67], detailed investigations on SLM applications are found e.g. in [68, 69, 70].

*Optical Disks*

In optical disks, which are well-known as digital computer storage devices, information is stored across a planar surface in a digital manner using a string of pits. The recorded signal is encoded in the length of the pit and the spacing of the pits along the track. The most important device is the reading or pickup system of the optical disk. The readout light is focused onto the surface, reflected and transmitted by a beam splitter into the

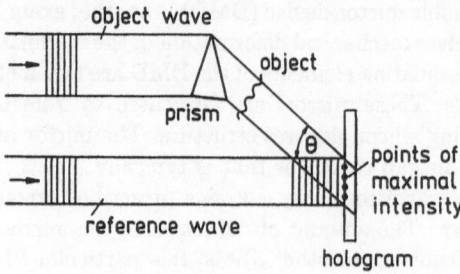

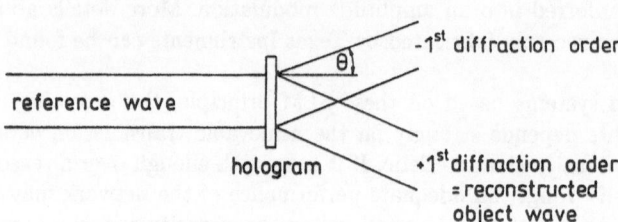

**Figure 4.2:** Recording and reconstruction processes of holograms of transparent objects with a plane reference wave.

detector. If there is no pit, the light will be fully reflected to the detector. When there is a pit, the focused beam will cover both pit and the surrounding land, which are both highly reflectant. The depth of the pit and the surrounding land is made such that the phase difference between the reflected light from a pit and a land is $\pi$. Consequently, destructive interference occurs for a pit, resulting in less light at the detector. When such a disk is read out in parallel, it can be considered as a pixelated spatial light modulator, having the pixel not in Cartesian coordinates, but in a polar format.

Due to its serial processing nature, that storage device is not able to provide parallel access. In addition its access time grows with increasing capacity. Therefore, they are not well-suited for parallel neural net processing, where a high degree of parallelism in interconnection and processing (e.g. thresholding) is required. However, we will see in section 7, that combinations of optical disks with other storage media may result in a successful combination for realizing neural network structures.

### Planar Holograms

Holography allows for the complete storage of an information bearing beam. That is, amplitude as well as phase are conserved. These conditions can be supplied by letting the information carrying beam or the object beam interfere with a non-modulated plane wave, the reference. The interference pattern built by these beams is then the code carrying the amplitude as well as the phase information of the object. In classical holography, this interference pattern is then registrated on the recording medium. The image may be recalled by addressing the storage device solely with the reference wave. The registration and reconstruction process for the case of a planar holographic plate is shown in fig. 4.2. These planar holograms — as films or layers on substrates (e.g. dichromated gelatine or silver halide on glass) realized by optical interferometric exposures or as computer generated holograms — are widely spread and easy to fabricate and implement. However, they suffer from the lack of possible modification: once stored, the information can not be changed or adapted. This may be useful for neural net configurations that use prescribed learning that is finished before the operation of the net begins. Unfortunately, adaptivity is at the same time the feature that is absolutely required for processes with adapted learning as Boltzmann machines or self-organized systems. In these cases, planar holograms with fixed grating registration are not suitable at all. The second big disadvantage of these holograms is that their storage operation is severely limited because there are only two dimensions available for storage of data. Therefore, planar holograms are well-suited for small nets built with prescribed learning procedures or for associative memories that have to recall a limited number of classes, but they do not fit for all error-driven or self-organized systems.

With the development of computer generated holograms, a lot of different systems using prescribed learning in planar holograms have been proposed and set up for pattern classification or association. In most of these implementations of prescribed learning, the step of weight adaptation before the operational state of the net for implementations of e.g. perceptron neural networks is done electronically. The trained weights are then transferred into the hologram that is used in a setup to perform the operation of the network (see e.g. [71, 72]). In such a way, optical matrix-matrix as well as matrix-vector multiplication may be achieved.

For optical neural networks implemented with these computer generated planar holograms, the space-bandwidth product of the hologram is a major consideration. This number gives an estimate about how many spatial frequencies a system is able to process in a certain spatial region without degradation or losses. Off-axis holograms can be fabricated with a single binary transmission mask. However, the carrier frequency greatly increases the space-bandwidth product. In contrast, on-axis holograms use a lower space-bandwidth product to encode interconnections but require a multilevel phase transmission profile, which may cause errors during the fabrication process.

An interesting example of how the ability to learn may be implemented even with such a planar hologram using a computer generated hologram matrix and a dot liquid crystal display as a shutter array has been proposed by A. Kajiki and coworkers [73]. In their configuration, the required weight modification can be performed by selecting special computer generated holograms in the hologram matrix and by superimposing the reconstructed images from the selected hologram. Although this principle results only in small neural nets — a system with five outputs has been experimentally realized — it allows at least in a limited way for the realization of adaptation as the most important feature for complex neural net structures.

*Electron-Trapping Materials*

Electron trapping materials can emit different output photons that correlate spatially in intensity with the input photons. Although an electron trap material does not modulate the phase, amplitude and polarization of the wavefront passing through it, it is similar to spatial light modulators in that they both transfer and/or process information from a spatial format to an optical carrier. At the same time, electron trap materials may store information and recall them when addressed with a read beam, thus being able to realize optical memories and interconnection devices.

Electron trap materials are stimulable phosphors consisting of IIA - VIB compounds (alkaline-earth chalcogenides) with two specific rare-earth dopants added. An example of an electron trapping material is SrS:Eu,Sm. Although the precise details of the luminescence mechanisms are not known, an empirical model can be given to explain the mechanism. In an electron trapping material, both the ground and excited states of each impurity exists within the band gap of the wide-band-gap host material ($\sim$ 4 eV). Visible light excites an electron in the ground state of $Eu^{2+}$ into its excited state. Some of the electrons at this higher energy level of the $Eu^{2+}$ tunnel to the lower $Sm^{3+}$ ions, where they remain trapped until stimulated by infrared light. The $Sm^{2+}$ ions so formed are thermally stable deep traps of about 1.2 eV. Such a material has to be heated up to about 450°C before the electrons are freed. Upon stimulation with infrared light (e.g. 1064 nm from a Nd:YAG laser), trapped electrons are released from the $Sm^{2+}$ ions. These released electrons then tunnel back to the Eu ions, resulting in the characteristic $Eu^{2+}$ emission when the electrons return to the ground state.

The operation of electron trapping materials can be summarized as follows. Visible light (e.g. at 488 nm) excites electrons from their ground level into an electron trapping level so that optical information can be stored. The information is retrieved by returning the trapped electrons to the ground state, with an emission at 620-630 nm (orange to red light). The retrieval can be controlled, since the recombination of electrons from the trapping band to the ground level requires infrared light (e.g., 1064 nm).

In addition to storage, electron trapping materials are interesting as spatial light modulators capable of performing multiplication, addition, and subtraction within a dynamic range covering four orders of magnitude. For that purpose, the devices consists of an electron trapping thin film coated on a transparent substrate. An exhaustive treatment of applications of electron trapping materials is given in [74].

### 4.1.2  Volume Interconnection Devices

*Photorefractive Volume Holograms*

The principle of photorefractive two- and four-wave mixing has already been explained in section 3.4. Using appropriate schemes to allow multiple gratings to superimpose independently in a volume, a large number of interconnections can be realized in photorefractive crystals. Moreover, when image storage or page-oriented storage in a volume is required, two-dimensional images may be independently stored and addressed in the volume.

Experimentally, many authors have proved high storage capacities in photorefractive volume materials [75, 76, 77, 78, 79]. The largest storage capacity achieved up to now is more than 150.000 two-dimensional pages each having about $10^5$ bits per page (or $5 \cdot 10^8$ interconnections), achieved as page-oriented storage using angularly multiplexed reference waves in photorefractive $LiNbO_3$, combined with spatial multiplexing over 16 locations [80]. Thus, photorefractive memories are well-suited candidates to realize the interconnection or storage unit in neural network implementations. I will describe several different experimental approaches to hologram multiplexing and realization of interconnection devices using photorefractive volume holography in the following sections.

*Spectral Hole Burning*

Persistent spectral hole burning occurs when a permanent change in the absorption line is photoinduced by a narrowband laser tuned to a particular frequency within the inhomogeneous absorption line (see fig. 3.8). This effectively burns a hole in the inhomogeneous

**Figure 4.3:** Principle mechanism of spectral hole-burning for information storage in an inhomogeneously broadened medium (a). Due to a permanent photochemical process, certain frequencies may be depopulated, giving a "hole" in the frequency representation (b). The presence of a hole corresponds to a logic 1, the absence of a hole to 0, resulting in a bitlike storage medium over the inhomogeneous linewidth of the medium (c).

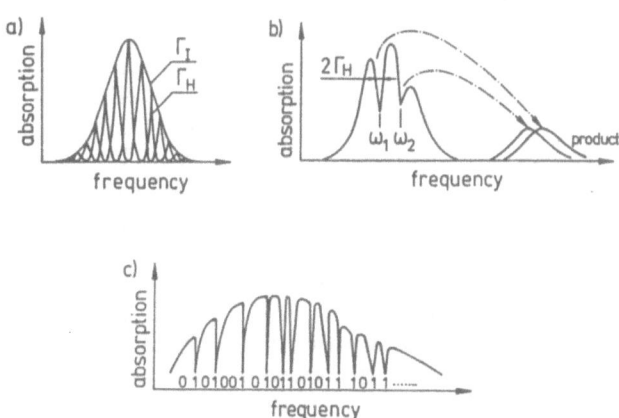

line profile. When the presence or absence of a hole at a certain frequency corresponds to a 1 or 0, the frequency domain can be used for information storage. This situation is depicted in fig. refpic:holeburning. The storage density of the method is defined by the ratio of the inhomogeneous to the homogeneous linewidth. This ratio can be as high as $10^4$, depending on the specific properties of the active center and host used, and is highly temperature-dependent, because phonon broadening (e.g. vibrations of the host lattice) causes the homogeneous linewidth to increase with temperature. Therefore, to achieve high storage densities in persistent spectral hole burning, it is necessary to write and read at very low temperatures. As a consequence, for most experiments, samples are immersed in a bath of liquid helium.

In spectral hole burning, holograms are recorded in the frequency domain by alteration of the optical properties of different subsets of molecules, and thus, in principle, the diffraction efficiency of one hologram is not affected by the presence of other holograms [81] — an effect that is not possible for photorefractive recording, in which the diffraction efficiency of an individual hologram scales as inversely proportional to the number of recorded holograms (see section 4.3). Therefore, recording in spectral hole burning materials does not need special recording schemes as they will be discussed in section 4.4 for photorefractive volume hologram recording. In contrast to photorefractive recording, there is no cross-talk noise resulting from violation of the Bragg-condition (see section 4.3), but cross talk may arise when holograms are recorded at high density because of the diffraction from holograms stored at nearby frequencies.

B. Kohler et al. have realized the storage of 2000 holograms in a photochemical hole-burning system [82], allowing for $10^5$ bits per memory access operation or image (see section 4.4). Moreover, the combination of hole-burning system storage with multiplexing techniques as angular and wavelength multiplexing allows the enhancement of that storage capacity [83] (see section 4.4). In section 7.4, a method to use the frequency as an additional optically parallel accessible degree of freedom to incorporate an error-corrective associative memory using persistent spectral hole burning is presented [84].

*Photon Echo*

As well as persistent spectral hole burning, photon echo generation has a processing speed that is $10^5$ times faster (nanosecond region) than two- and four-wave mixing effects in oxyde-type photorefractive crystals and therefore is comparable to the speed of refractive index grating formation in photorefractive semiconductors [85, 86, 87]. This fact is due to the excitation of a resonant transition that provides the forming of transient population gratings in the crystals rather than inducing transportation of charges.

Thus it is possible to store temporally and spatially encoded laser pulses — the first one being a long train of data pulses [88, 89], the second one being a series of multiple two-dimensional images [16] — in a single focal volume by focusing them into the crystal at the transition wavelength. The spectral interference pattern generated by the data and the write pulses is then recorded on the spectrally selective medium by modification of the materials absorption profile. This process is also known as spectral holography. In other words, the image processor in stimulated photon echo devices is the transition in a molecule. By utilizing the frequency dimension at each focal volume of the input laser, one can increase the storage density by many orders of magnitude over the conventional planar holographic storage technique. It is possible to store up to $1.6 \cdot 10^3$ bits in a single spot of 240 $\mu$m diameter [89].

In general, if there are many data pulses to be stored, it can be shown that the population modulation pattern is proportional to the Fourier transform of the entire temporal sequence of data and write pulses, allowing thus for multiple data pulses to be independently stored and recalled. In the limit of small pulse areas, we can obtain an echo profile by performing a temporal correlation between the write and read data pulse. Thus, the stored information can be retrieved by a read pulse into an echo image that carries spatial convolution and correlation of the input images [90]. If the write and the read pulses are both amplitude and phase modulated such that their temporal correlation resembles as closely as theoretically possible a single brief pulse to ensure high fidelity of the retrieved data [91], X.A. Shen and R. Kachru [92] achieved the storage of a 42 $\mu$m long stream containing 420 bits of data in a 40 MHz spectral channel in a $Eu^{3+}$:$Y_2SiO_2$ crystal. The holographic geometry of backward stimulated echo also permits high-speed optical processing of temporal and spatial data and — if one uses three excitation pulses as in photorefractive materials — phase conjugation of the incoming images [16, 17, 19].

In practice, photon echo experiments suffer from relatively short storage times in the range of normally seconds to hours and low diffraction efficiencies of less than 1% [93], making high density optical image storage difficult to be realized in these devices. Moreover, as in photorefractive volume holographic storage, the stored data are affected by many readouts. In photorefractive gratings, a new read or write cycle partially erases previously written information, whereas in photon echo storage, it is the resolution of the stored information that is severely degraded. However, recently, a repeatedly recall of the same stored data as many as eight times was realized in [92], without severe noise. Consequently, both methods require adapted data refreshment techniques to restore the data in the storage medium. Techniques of refreshment for photorefractive volume holographic storage devices will be presented in section 6.4. For photon echo devices, several methods are in discussion to refresh and enhance the efficiency by external-cavity amplifiers and enable to realize adaptive memories using phase-conjugate resonators [93].

*Combination of Photon Echo and Persistent Spectral Hole Burning*

In 1991, M. Mitsunaga [89] presented a novel type of frequency selective optical memory that writes and reads the data in both the time domain and the frequency domain applying a combination of time domain photon echo effects and frequency domain spectral hole burning (see fig. 4.4). The storage capacity of that device is largely enhanced compared to the single techniques, because with the hole burning memory, $N$ bits of information are stored as the distribution of holes, while with photon echo memory, the Fourier transform of $N$-bit temporal data is impressed into the inhomogeneous spectrum.

Both methods can be combined by dividing the inhomogeneous line into $M$ equally spaced slots. When a laser is tuned to a given slot, photon echo storage is performed that should have $N/M$ bits of time-domain storage. The excitation pulse width has to be long enough, so that its spectral width falls within the slot. Thus, by sacrificing the time resolution of the photon echo experiment, one can reduce the spectral width of the population grating impressed in the medium. Retrieval of the time domain data at a given address is achieved by simply tuning the read pulse to the corresponding frequency. In this way, a hybrid memory has $M$ randomly accessible addresses, and each address has $N/M$ bits of information.

*Two-Photon Processes*

Unlike other storage schemes such as the photorefractive effect and photon echo, two-photon effects provide a means of storing data into separate locations — which may represent single bits in a digital storage or interconnection device — throughout the entire volume without affecting the neighbouring bit locations. This is comparable to the effect of persistent spectral hole burning. In addition, the two-photon process — which may be realized in organic polymer and photochromic materials — has the benefits of high sensitivity and high speed. Two-photon absorption has also been introduced in section 3.2 and refers to the excitation of a molecule to an electronic state of higher energy by the simultaneous absorption of two photons. The first photon excites the molecule to a virtual state, while the second photon further excites the molecule to a real excited state. Since the intermediate state is a virtual state, the two-photon process is different from a biphoton process where a real intermediate state is present. The wavelengths of the two beams are such that, although neither beam is absorbed individually, the combination of the two wavelengths is in resonance with a molecular transition. Therefore, both beams must temporally and spatially overlap to allow for two-photon absoption. Since the two-photon process is localized to the small region of overlap, all points in the volume can be addressed individually. Depending on the wavelength of the two beams, which are incident on the material, the addressed location can be written or read. In addition, since two-photon absorption is based commonly on molecule transitions, the material is theoretically able to operate in the picosecond regime. Finally, the small size of the molecule and low cross talk between neighbouring bits should theoretically allow to reach the optical diffraction limit of about 1 $\mu$m.

Experimentally, a storage capacity about $10^6$ bits per memory access operation using Nd:YAG pumped solid state lasers as the optical power supply can be reached for such a single unit [94, 95]. For this purpose, several storage layers are realized in a cubic photochromic material, with each layer composed of a large two-dimensional array of interconnection points (data bits for digital storage). At the same time, the data are sent

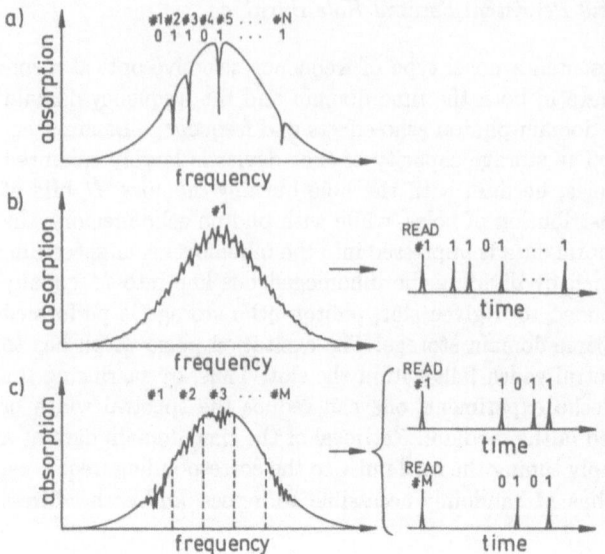

**Figure 4.4:** Inhomogeneous absorption spectra and readout processes in (a) hole-burning memory, (b) photon-echo memory and (c) hybrid memory. In (a), $N$ randomly accessible addresses that have one bit of information ar available, in (b) only one address is accessible that has $N$ bits of information. In (c), $M$ adresses are randomly accessible that contain $N/M$ bits of information (after [89]).

into the memory on the information beam. Due to the two-photon nature of the material, the data can only be stored in the selected layer. Thus, using about 100 planes of digital images, interconnection or memory devices with capacities upt to 1 Tbit/cm$^3$ combined with an enormous throughput of more than 100 Gbit/s can be achieved [95]. As in photon echo storage techniques, the short storage time and the erasure due to addressing of the system does not allow a nonvolatile storage, thus making it difficult to realize long-term memory devices. However, very recently, nonvolatile holographic storage with two-photon, two-step recording was demonstrated by S. Bai and K. Kachru [96], using an energy level slightly below the conduction band with a long life time. The long life time allowed the use of cw lasers with moderate intensities at near infrared wavelengths. In this case, the photons from the read beam have insufficient energy to promote electrons to the conduction band, causing only very little erasure. If the gating beam is supplementary made to be incoherent during readout, no erasure effects will appear [96]

### 4.1.3 Nonlinear Interconnection or Storage Methods — a Comparison

To summarize, there are several interesting and promising interconnection and storage techniques that utilize the parallelism of optics and — at least some of them — the volume nature of the materials to achieve high storage capacities. Some of the techniques mentioned above have short response times in the range of nanoseconds, but lack the possibility of high diffraction efficiencies or long-time data storage, as e.g. photon echo. Others as spectral hole burning allow for fixed data storage, but are not able to store analog data. The most critical points of the nonlinear three-wave mixing methods as photon echo or persistent spectral hole burning are their high energy requirement to create the effects, their low diffraction efficiency, their temperature stability requirements in the range of a few Kelvin and — as the most important point — the lack of possibilities of data adaptation in the case of hole burning.

Photorefractive volume holographic data storage is at the same time able to realize reconfigurable, adaptive gratings and to achieve high data storage capacities. Its big advantage is the possibility to write gratings that allow for high diffraction efficiencies, amplification and amplified phase conjugation of the stored signals. Although many photorefractive materials as semiconductors allow for quick data recording, materials that have the largest storage efficiency as oxyde-type photorefractive crystals lack high-speed recording of data. However, recent developments in crystal preparation and dotation let me hope that in the next future faster oxyde-type photorefractive materials will be available [97, 98]. These are the reasons why I estimate photorefractive data storage as the most promising of these volume storage techniques.

Consequently, my main emphasis for the following sections on interconnection and storage devices will be on photorefractive volume holograms. I will present possible implementation techniques for point- and page oriented interconnection storage in the next section in detail. Depending on the storage capacity, resolution and environmental conditions of the system in view, I will introduce several promising methods of superimposing large amounts of interconnections or pages with two-dimensional interconnection structures (images) in a single volume holographic medium.

## 4.2 Holographic Interconnection and Storage Methods

The precedent section revealed that incoherent and planar interconnection devices as well as coherent volume holographic media are suited for optical realizations of simple neural networks. However, if one wants to implement perceptron-like neural nets that require bipolar weight adjustment, incoherent processing becomes difficult. Although there are several possibilities to allow for bipolar incoherent weight processing by dividing the weights into subgroups that are treated differently, coherent optical techniques allow an easy realization of neuron excitation and inhibition by using amplitude and phase of the wave. Finally, multilayer neural nets require the processing of large data amounts through a multitude of connected layers. This can be accomplished in a very easy way in coherent devices, where multipass configurations may be used in single devices in order to simulate multiple layers.

Among the existing coherent interconnection devices, adaptivity is the second important selecion criterion. This can only be realized using erasable storage devices, as they are available by volume holograms using storage via the photorefractive effect. Their variability in storage time and efficiency because of the variety of materials that are available up to now makes them extremely well-suited for information processing and storage purposes. Consequently, the following section will treat mostly coherent realizations of the optical interconnection devices, treating incoherent implementations only aside.

### 4.2.1 Analogy Between Holography and Neural Interconnections

The principle recording and readout mechanism for volume holographic media is analogous to the one for planar holograms (see 4.2). You will find that the procedure is very similar to the process of two-beam coupling described in chapter 3.4 and the following ones. Indeed, the storage of images in volume holograms has the same origin as that interaction and it is the reason why two-wave mixing in photorefractive volume materials is often named *dynamic holography*.

How can these storage principles now be exploited for optical neural network algorithms? To get an insight in the analogy of holography and neural processing, let us refer to fig. 4.5. It is a schematic diagram showing the various components of a pair of neurons and the way in which they are connected in comparison to an optical analogue that uses the concepts of volume holography. This optical analogy was first introduced by D. Psaltis and coworkers in 1990 [99]. As you still know from the explanations in chapter 2, each neuron receives inputs through the synapses on its dendrites, it processes these inputs in a certain way and then broadcasts the result on the axon where it is picked up by the dendrites of other neurons. In the holographic analogue, the output of each neuron is a light beam and the activity of the neurons is encoded in the amplitude or the intensity of the optical signals. The input of each neuron is a light detector which senses the amount of light that is directed towards it. A holographic grating is placed in the path of the output beam of neuron B, which diffracts the incident light. The direction of the diffracted beam is determined by the period and the orientation of the grating. Thus, light from neuron B may illuminate the detector of neuron A, generating a signal in A as a result of the activity in B. We may say, that the hologram connects the two neurons, it is an *interconnection grating*, the strength of which can be modified by adjusting the modulation depth of the holographic grating. There are even some direct analogies between the compact structure of a neuron and its optical simulation using volume holography. The output light beam plays the role of the axon, broadcasting the signal from each neuron. The holographic grating represents the synapse and the optical pathways along which light is transferred from the hologram to the detector area are analogous to the dendrites. The device consisting of the light source, the detector and circuits that process the detected signal is reminiscent of the soma of the neuron. In real neurons, some processing tasks can take place on the branches of the dendritic tree, but in the optical simulation, all integration and computation tasks are concentrated on this integrated unit which I refer to from now on as the *optical neuron*. Having now established the basic structure of the optical analogue of a single neuron, the next issue is the way in which we may connect many of these units to form optical neural networks. The requirement for such a multiple image or pixel storage device is at first an easy storage mechanism. As a second criterion, it should offer the possibility of an independent, noise and cross-talk free recall of every arbitrary image or page with information units on it stored in the medium. You may imagine these multiple neuron structures as being equivalent to a multiple grating storage mechanism in optics or, what is the same, a multiple page or image storage system. Each grating between two pixels then is the interconnection for a single neuron and the totality of the gratings gives the structure of the input information — e.g. the image.

### 4.2.2 Spatial or Local Multiplexing of Interconnections

Suppose that a third neuron is to be connected to neuron A as well as neuron B. This can be done optically in two or even three ways which are illustrated in fig. 4.6. First, a second holographic grating can be imposed upon the same crystal, but this time it is located on a different place inside the crystal in such a way, that this second grating is tuned via a different spatial orientation and period to redirect light from this third neuron onto the

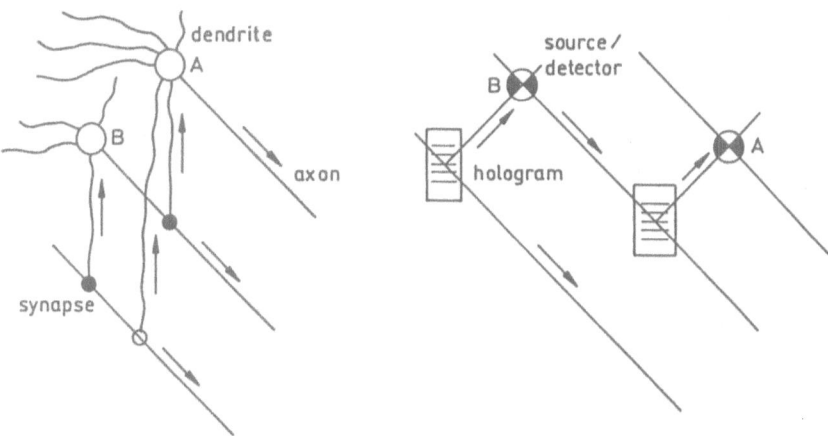

**Figure 4.5:** Comparison of the concepts of a biological neural network (left) with its optical analogue (right).

same input (or detector) of neuron A. In that version (see fig. 4.6, middle), each grating has a region of its own in the crystal which is not occupied by other gratings. This method holds as well for plane as for volume holograms and has many applications in all those cases where the information should be directed from one neuronal grating to another — as e.g. in multisignal switches in optical information processing units or in neural network inerconnection systems that have been trained before operation (prescribed learning) and only need a fixed interconnection device. However, planar holograms are — beneath their sensitivity to misalignments and the fact that they fail to realize reconfigurable interconnection devices — not able to store a large number of interconnections using that local storage principle because of the limited size of the two-dimensional hologram. Therefore, volume holograms have a larger potential for realizations of all-optical, parallel neural network interconnections and that is the reason why I will focus in the following my attention on realizations using volume holograms as point-to-point interconnection devices.

The most promising ideas using that optical interconnection concept in photorefractive volume materials will be discussed in detail in the next sections. However, I want to mention already at this point that the main disadvantage of it is the fact that the storage capacity is again limited by the available size of the hologram or crystal plane perpendicular to the incident beam. Because the region necessary for one interaction process is fixed by the beam waist and the resolution of the crystal — which in turn is limited by the number of available donors in the material and the registration mechanism — only a limited number of independent interconnections is possible for a given storage medium size. Even if one profits of the three-dimensional volume nature of photorefractive crystals and stores several planes of interconnections in the crystal, that limitation still exists. Because photorefractive crystals as e.g. $LiNbO_3$ are actually easily available at a size about 1 cm$^3$, and because the diameter of a focused beam pair will be in the range of 10 $\mu$m, about $10^9$ interconnections may be approximately realized with that technique. Even when larger crystal are available — a size of about $10 \times 10 \times 1$cm$^3$ may be reasonably achieved in the next years — the capacity of interconnections will rise only

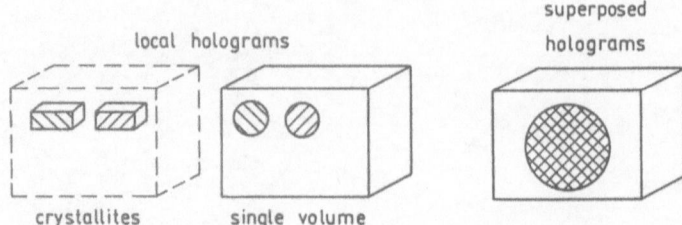

**Figure 4.6:** Different storage mechanisms for planar or volume holograms: local storage in crystallite arrays (left) or in volume materials (middle) have to be compared with multiplexing techniques where all gratings share the same volume taking advantage of the selectivity of the Bragg condition (right).

by a factor of 25. In the following sections, I will discuss these aspects of storage capacity and its limitations in more detail and will compare the different storage methods and their abilities.

An alternative method to connect a larger number of optical neurons in parallel is to arrange several crystallites in a grid with each of it carrying the information of a single image or interference grating (see fig. 4.6, left). However, in that case, the accurate arrangement of equal crystals is required to obtain equal results in each unit concerning the intensity inside the crystal (same absorption and scattering effects for all crystallites), the strength of the refractive index modulation (same amount of impurities or donors for all crystallites) and the value of the coupling coefficients (same geometries and intensity relationship of the interacting beams). Therefore, the technical meaning of that method is only of secondary importance.

### 4.2.3  Distributed Multiplexing of Interconnections

The second important storage principle is to superimpose the second grating with the first in the same volume of the holographic medium (see fig. 4.6, right). This second grating then has to be distinguished from the first one by another degree of freedom of the system in such a way that the two gratings do not interact with or disturb each other. These parameters may be the polarization, the angle of incidence, the wavelength or even the phase of the reference wave. In that case, several thousand to ten thousand images or pages with digital data may be stored or interconnected in a single spatial region, thus allowing for a much higher storage capacity as spatial multiplexing. The most important techniques of distributed interconnection multiplexing will be discussed in detail in section 4.4.7.

The underlying principle of all these methods exhibits a second difference between the real, biological neuron and its holographic counterpart. The synapses are not localized at the intersection between axon and dendrite, but are implemented instead in a distributed manner, each grating sharing the entire volume of the registration medium with all the others. Moreover, the distributed storage principle reveals the fundamental difference for registration of planar and volume holograms, because this method is only useful in volume materials. To understand this fact, let us return to the principle mechanisms of recording and readout of data in planar and volume holograms. As sketched in fig. 4.7

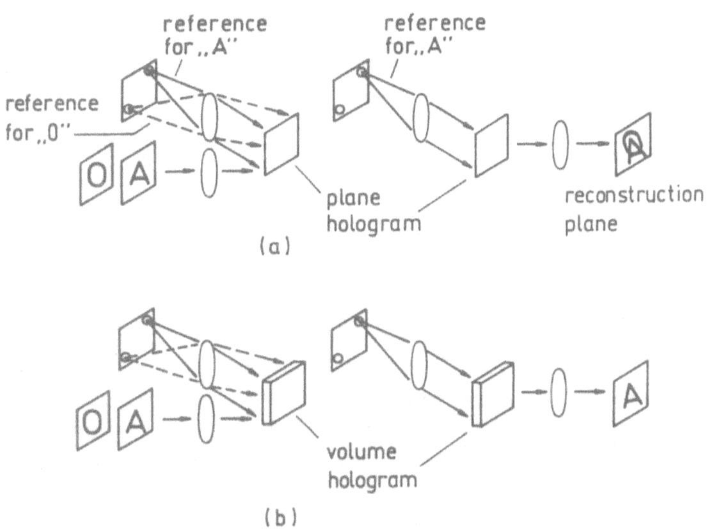

**Figure      4.7:**
Schematic represen-
tation of the differ-
ence between holo-
graphic storage in
planar (a) and vol-
ume media (b) us-
ing the selectivity
of the Bragg condi-
tion.

(a, left side), a reference wave and a plane, information carrying wave are simultaneously registrated or "written" on the holographic plate. For reconstruction, the reference is simply addressed on the plate and the image stored with that reference is reconstructed in the plane defined by the optical lens configuration of the writing process. What will happen if we try to write more than one image in a distributed manner in a single holographic plate? At a first glance, it should be possible to enregister a lot of images on one holographic plate if we choose one of the waves — for reasons of an easy experimental setup for information storage retrieval this would be in most cases the reference wave — to be different for each storage process. "Different" means, that one parameter of the reference wave is changed for every exposure — one can choose e.g. the wavelength or the k-vector (incident beam angle) of the incoming wave as well as the transverse phase or the polarization state. Unfortunately, a holographic plate is not able to discriminate these images in the reconstruction process. If you want to selectively read out one image from a two-dimensional storage device, there is no possibility of a cross-talk or disturbance-free recall. All images stored will be partially read out by the reference wave belonging to a single image — and they will be superimposed in such a way that they disturb each other as shown in fig. 4.7 (a, right side).

This behaviour changes if we profit from the third dimension available in volume holograms (fig. 4.7, b). In that case, the refractive index grating which was created by the two interacting writing beams, defines a set of planes at which during readout the incoming reference is partially diffracted and reflected. To recall the image at the output of the crystal, the components being reflected and diffracted at different parts have to interfere constructively to reinforce each other. This means that the path difference for rays from adjacent planes must be an integer multiple of the wavelength or fulfil the *Bragg condition*

$$2d \sin \vartheta = m \cdot \lambda \qquad (4.1)$$

This condition selects a certain angle of incidence $\vartheta$ for a given wavelength $\lambda$ of the reference wave and a crystal grating spacing or the distance $d$ between two adjacent planes which allows solely for reconstruction of the image. It is named after W.L. Bragg

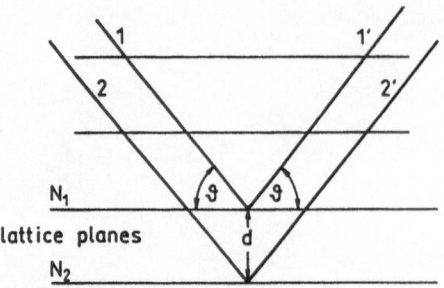

**Figure 4.8:** Bragg condition for constructive interference of beams reflected from a thick volume hologram at an angle $\vartheta$.

(1890 - 1971) who found it in 1913 for the case of x-ray reflection in crystallography. In other words, the selectivity of the Bragg condition in volume holograms supplies the possibility to discriminate images which are recorded in a distributed manner in the same crystal, each of it sharing the entire volume of the holographic medium with all others. Consequently, the parameters of the Bragg condition are the ones that can be varied in order to achieve gratings that can be read out independently from each other. Beneath the angle (*angular multiplexing*) and the wavelength (*wavelength multiplexing*), which are obviously present in the Bragg-condition, this can also be the phase of the reference beam (*phase encoding or phase-coded multiplexing*).

The advantages of the distributed holographic synapses are high storage density and ease of device fabrication. The number of distributed synapses that can be packed into a hologram of volume $V$ is proportional to $V/\lambda^3$ [100], where $\lambda$ is the wavelength of the light. This corresponds to $10^{12}$ synapses per cm$^2$ for a typical operating wavelength of $\lambda \approx 1$ $\mu$m. There are other factors, primarily related to the physical properties of the material used, that can prevent us from realizing this upper limit — and I will discuss them in the next section — but that number is not imaginable to be realized with allocated storage centers.

The biggest disadvantage of these distributed connections compared to the localized storage technique is the reduced control of individual synapses or grating strengths. On the one hand, each individual synapse is stored in an arrangement of interconnections for one registration, a so-called page. This page will be adressed in most cases with one parameter. On the other hand, the adjustment of one synapse may inadvertently affect other synapses as well — e.g. the registration of a new grating in a multiple storage process always partially erases previously written ones. Therefore, the interconnection weights corresponding to the weakened grating are lowered in an uncontrolled way. Accomodating this limitation is perhaps the most crucial research issue in this field. I will discuss the possibilities to handle this limitation further below. To overcome these problems, several attempts have been made to combine the two methods, resulting in *spatioangular* or *shift multiplexing*. Spatioangular multiplexing is a hybrid of both angular and spatial multiplexing, in which holograms are formed in spatially overlapping, but not completely superimposed regions of the crystal and are distinguished from one another by use of angularly multiplexed reference beams (see section 4.4.7).

### 4.2.4  Realization of Bipolar Interconnections

One of the biggest advantages of *coherent* optical volume holographic interconnections is their ability to realize bipolar optical interconnections that allow for excitation and inhibition of the interconnection weight strength. This feature is essential for realizations of perceptron-like neural networks and therefore for the important class of multilayer neural networks with error-driven backpropagation algorithms. There are several methods to implement bipolar weighting in holographic interconnections. I will explain their features and limitations and introduce shortly several concepts to realize bipolar interconnections even in incoherent systems.

*Inhibitory Weights by Incoherent or Coherent Erasure*

D. Psaltis et al. [101] used an incoherent erasure process to achieve subtractive weights. In their system, the photorefractive crystal was placed at the image plane of a spatial light modulator. A movable piezoelectric mirror was used to provide either a coherent reference or an incoherent one. For additive changes to the hologram, a coherent reference was provided so that the hologram was strengthened, but for subtractive changes, the reference beam was made to be incoherent so that nonuniform incoherent erasure of the gratings stored in the interconnection gratings resulted. However, this method does not fully take advantage of the phase sensitive nature of holography.

To realize even the subtractive weights in a coherent way, one can profit from the fact, that selective erasure of a grating can be realized when the grating is rewritten with a $\pi$ phase shift relative to the original grating. Already in 1965, D. Gabor et al. demonstrated that the superposition of two $\pi$-phase-shifted hologams can be used to perform coherent subtraction (or addition) of two-dimensional transparencies [102]. The double-exposure technique is a two-stage process in which a phase shift is induced on one of the control beams between recordings. The conventional ways of inducing this phase shift are either to electrooptically phase modulate the beam or to reflect it off a piezoelectrically driven mirror. This last technique has been applied by J.P. Huignard and coworkers to the partial erasure of information contained in a transparency in the presence of other holograms that are angularly multiplexed outside the Bragg selectivity bandwidth of each other [103]. Finally, in 1989, A. Marrakchi has extended this principle of coherent erasure by double exposure of phase-shifted holograms to the case of elementary gratings that implement real-time optical interconnections in photorefractive crystals [104]. The technique yields a continuously graded interconnection strength between two spatially separated planes, e.g. the input and output planes in single-layer networks. When several holographic gratings are recorded to simulate matrix-matrix or vector-matrix interconnections for fan-in or fan-out purposes, it is possible to modify and control one interconnection weight without significantly affecting the other gratings. Combined with the large storage capacity available in volume holographic recording, this double-exposure technique is suitable for the optical implementation of learning neural networks with continuously variable weights. In practial realizations, phase shifting devices are required that shift the phase exactly to the values of 0 and $\pi$. I will describe such a binary phase modulator when describing phase encoding as a means to superimpose pages in photorefractive storage media (see section 4.4).

*Interferometric and Liquid Crystal Phase Retardation for Coherent Bipolar Weights*

If such an accurate phase shifting device is not available, a phase control system based on interferometric phase detection can be used. The configuration described in [105] is known as a double Mach-Zehnder interferometer because both input ports are used. The beam from a first laser source is split into two paths by a dielectric beam splitter to intersect within the holographic medium to create a grating. Another laser source is positioned so that the directions of the beams transmitted and reflected by the beam splitter traverse precisely the same paths as those resulting from the first source. Stokes principle governs the relationship between the amplitude reflection and transmission coefficients seen by the first source ($r$ and $t$, respectively) and those seen by the second source ($r'$ and $t'$). If we compare now the two resulting transmissions seen by both sources, we find that

$$t = t' \qquad \text{and} \qquad rt^* + r't^* = 0 \tag{4.2}$$

where $^*$ denotes the complex conjugate of the coefficients. Because the amplitude of the hologram due to source 1 is proportional to $rt^*$ and that due to source 2 is proportional to $r't^*$, the two gratings are mutually $\pi$ out of phase —provided that the beams are aligned accurately. Such a phase shift of $\pi$ may also be used for parallel coherent subtraction of images and will be further discussed in section 6.1.

Coherent binary weight values can also be realized using this principle and the phase retardation properties in liquid crystal light valves (LCLV). Using phase-only light modulators [106] , this phase retardation can be adjusted continuously up to $2\pi$ by varying the biasing condition. Therefore, the phase modulation capability of the LCLV can be used to control the phase retardation of the reference beam, and a bipolar weight hologram can be effectively recorded. During the recordings, a positive excitatory neural weight hologram can be recorded first with the reference beam phase retardation set to 0°. Then a negative inhibitory neural weight hologram can be recorded by offsetting the phase retardation of the reference beam by 180°. The excitatorily and inhibitorily weighted holograms have to be superimposed spatially to avoid interference effects.

*Polarization Encoding for Bipolar Weights*

Finally, it is possible to use the two orthogonal states of polarization (e.g. vertically and horizontally linear polarized states) to realize the two channels of positive and negative interconnection weight values [107, 108]. The function of that method can be easily seen by the following example: If one wants to realize a zero result of the input activation value by a weight, the linearly polarized input light (e.g. vertically polarized) is rotated by 45° and then enters the interconnection device. Assume that this one is a birefringent device (e.g. storage crystals for volume holographic interconnections or liquid crystal devices for spatial light modulator interconnections). After passage through the interconnection device, the beam is incident upon a polarizing beam splitter, where the input is split evenly between the vertical and horizontal detection channels. Now, a second rotation may recombine the two contributions. If the rotation is chosen in such a way that both contributions may interfere destructively, a zero contribution results. This principle may also be implemented using electronic subtraction of the two different polarization contributions. If the desired output activation value is +1, the incident optical polarization is unchanged as it is transmitted through the interconnection device and is totally reflected by the polarizing beam splitter to the vertical output channel. If the desired output is -1, the incident light has its polarization rotated by 90°, and it is transmitted by the polarizing beam splitter to the horizontal output channel. Generally, output activation values between plus and minus one require partial rotation of the incident optical polarization.

A full algorithm of the outer product operation followed by a bipolar weighting of the resultant elements can be realized using this polarization encoding technique. Especially liquid crystal devices, as liquid crystal televisions, are well suited for that encoding technique, because the birefringence effect of the liquid crystals in dependance on the applied voltage can be used to realize the different positive and negative changes in the interconnection weights [109]. Thus, the spatial capacity of the interconnection device can be increased by a factor of four over implementations using other techniques.

Although these methods are best suited for easy realization of bipolar optical neurons, a problem arises when one wants to detect the output in such a coherent system, since the detector senses power or the square of the amplitude of the output light. For this reason, the output from a single detector can not be negative. Even worse, all realizations that somewhere in the network use an incoherent optical operation inherently lack the feature of realizing bipolar weighting, because here only the intensity is used to carry the information. However, if one wants to realize bipolar weighting in these neural net implementations, however, several tricks can be applied to achieve artificially a bipolar operation.

*Dual Channel Realization of Excitatory and Inhibitory Weights*

The most simple way is to divide the set of interconnections in two subsets — *dual-rail or spatial encoding*. One is used for learning of the excitatory weights, the other one for the inhibitory weight changes. After having finished the learning cycle, both sets are electronically subtracted weight per weight, thus realizing finally the combination of excitation and inhibition [70]. Another way is to construct two optical neural setups so that one is used for learning with the original patterns and the other with the complementary patterns. This is equivalent to realizing a subset with negative weight values. The outputs for the two detectors are then added when using the system for classification. Both methods have in common that they reduce the interconnection storage capacity by a factor of 2, which might be a dramatic and untolerable value for large neural networks.

These electrooptic methods of providing two different subsets of interconnection weights, one that is responsible for realizing excitatory weights and one for realizing inhibitory ones, which are finally subtracted, is roughly equivalent to the optical model of D. Psaltis, where subtraction is achieved by incoherent erasure of the interconnection gratings. These dual channel methods are critical in preserving the phase of the process. However, they may also be used for incoherent interconnection systems allowing to realize inhibitory as well as excitatory weights in incoherent configurations.

K. Wagner and coworkers proposed already in 1987 a method to obtain an incoherent inhibitory interconnection that is based on the idea of using an inverted sigmoidal transfer characteristic for the negative arm of the bipolar neuron [110]. This type of negative differential reflected intensity versus incident intensity characteristic can be obtained over a limited operation regime from a properly detuned nonlinear etalon. In this limited regime, the inverting sigmoidal characteristic can be written as one minus the standard sigmoidal response of a saturating neuron. We will discuss the principle application of such a bipolar weighting when presenting self-organizing networks in chapter 10.

All the advantages of volume holograms shown above have led to the preference of volume holograms in all complex optical neural net applications. Therefore, the methods for multiple grating storage and neural network interconnections are restricted in most cases to volume holograms. Naturally, some of these principles may also be applied to planar holograms, but the main emphasis of our further investigations is on volume holograms. However, planar holograms have found other areas of application instead. They are used as stabilizers in large optical feedback circuits and as fixed holograms for image reinforcements. If we combine the results of that section with the ones of the precedent chapter, photorefractive volume crystals become obviously the material of choice for all storage tasks in optical neural networks. It is their capability of performing read, write and erasure tasks with a high nonlinear response in connection with their adaptive storage capacity that makes them the most popular material for these applications for several years.

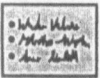 **Summary**

Neural processing and the principles of holographic storage have a lot of conceptual ideas in common. The optical neuron may best be realized using a volume hologram:

- The holographic grating plays the role of the synapse or interconnection device, the strength of which can be modified by adjusting its modulation depth.
- The input of each neuron is a light source or detector, whereas the output of each neuron is a light beam. It is analogous to the axon, broadcasting the signal from each neuron. The activity of the neuron is encoded in the amplitude or the intensity of the optical signal.
- The optical pathways along which light is transferred from the hologram to the detector area are analogous to the dendrites.

To store the gratings in the volume holographic material, plane and volume holograms may be used. In comparison, the latter one revealed to be best suited for optical storage and interconnection purposes in neural networks, because of several reasons:

- The storage capacity is enhanced because of the possibility of using the third dimension.
- Read-out of different images can be performed without cross-talk noise.
- There are several methods to exploit the third dimension:
  - Each hologram has its own location. All gratings form a grid of different holograms in the volume.
  - Holograms are superimposed with a distributed location of the gratings. Here, the selectivity of the Bragg condition allows for the superposition of several images or independent information by simply changing one parameter in the registration process — wavelength, incident beam angle or relative phases of the reference wave.
- Bipolar weight interconnections can be realized using
  - coherent, selective erasure due to $\pi$-phase-shifted gratings or interferometric interaction
  - incoherent optical erasure due to defined, incoherent illumination.
  - polarization encoding of excitatory and inhibitory weights in both orthogonal polarization directions

– division of the interconnection set into two different subsets that are taught with the original and their inverse patterns, which are subsequently added or with two equivalent pattern sets that are subsequently subtracted electronically.

## 4.3    Local Volume Holographic Interconnections

In that section, the possibility of arranging different holographic gratings in a volume by locally distributing them in the volume will be discussed in detail. Depending on the page distribution and the capacity of the interconnection volume, the volume hologram may connect an input vector of data to an output vector, an input matrix to an output vector or finally an input matrix to an output matrix with the same dimension. All these principles may be performed using photorefractive two- or four-wave mixing as well as self-pumped phase conjugation. I will show a couple of examples that have been realized using these methods and implementing different interconnection schemes. The case of self-pumped phase conjugation is somewhat special among these interconnection devices, because it distributes its local interconnection over a continuum of spatially and angularly distributed gratings, thus being a mixture between local and distributed interconnections. At the end of this section, this mechanism will be explained in detail, because it may overcome the capacity limitations that are limiting local matrix-to-matrix interconnections.

### 4.3.1    Two-port Operators: Vector-Vector-Interconnections

In two-wave mixing applications, point-to-point interconnections have been named *two-port operators* by D.Z. Anderson [111] who first developed that idea in photorefractive volume holograms.

Such a two-port device, which is shown in fig. 4.9, is simply one with two vector inputs and two vector outputs, which uses only two dimensions of the volume. Therefore, it can be regarded as a vector-vector multiplication or interconnection unit. When the medium in fig. 4.9 is exposed to the two optical vector fields, a matrix of gratings develops in it, which scatters every ray of a beam travelling in one direction into the other direction, which can be — as e.g. in fig. 4.9 — perpendicular to the first one. Each grating in the matrix is thus a history-dependent function of the outer product of the input fields of the photorefractive medium, provided that the total fraction of energy diffracted by each grating is small. The equations for each grating are exactly the ones for two-wave mixing presented in section 3.4.3. It is in principle possible to program the grating strength at each point to tailor the connection matrix in a specific manner. The programmed gratings as a whole behave as a two-port operator.

D.Z. Anderson and coworkers used the operation depicted schematically in fig. 4.9 to program the operation in photorefractive iron-doped LiNbO$_3$. The apparatus used to program the photorefractive two-port as a projection operator is shown in fig. 4.10. The argon-ion laser is spatially filtered to provide a plane wave to the spatial light modulator. The latter is constructed from a modified liquid crystal television and is converted into a phase modulating device (see section 4.4.4). The beamlets are configured in that scheme by that liquid crystal television screen, allowing for vectors with $N = 90$ input pixels.

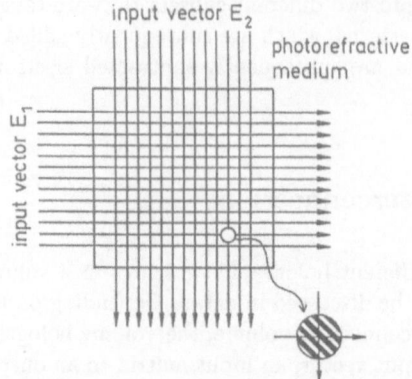

**Figure 4.9:** Photorefractive two-port operator: the input fields are depicted as vectors of rays. A grating at each ray crossing causes light of a ray travelling in one direction to be scattered into the other direction (after [111]).

The modulator is fitted with a mask to the geometry of the vectors and is driven by a microcomputer. The two binary vectors have identical arms. They are arbitrarily labelled 1 and 2. In arm 2, an attenuator is inserted to adjust the intensity of the light and a piezo-mirror to adjust the phase. To store several vectors, sets of orthonormal vectors have to be chosen. If appropriate orthogonal vectors are chosen for the input ports, up to seven vectors have been demonstrated to be storable in that device. Primary limitations of this device arise — as in most of the optical networks described in the next chapter — due to the fluctuations and the finite resolution of the spatial light modulators used as the input devices, giving rise to cross talk for the vectors to be stored.

If one wants to implement such a two-port operator in neural network applications, the question is, which neural network model of those described in chapter 2 is the one appropriate or analogous to two-port operation? As I have stated already above, two-port operation is the summation of vectors realizing the outer product rule in the interconnection regions. Therefore, the two-port operation simulates interconnections as they are required in neural nets where the outer product rule plays an important role. Two-port operators may also be implemented in networks with feedback or backpropagation learning, where the summation of the feedback signal with the original input to the previous input requires a two-port device. Here, two input ports and one output port are used, the extra output being free or not accessible.

An example for the application of two-port operators in optical neural network simulations is a novelty filter as described e.g. by T. Kohonen (see further reading of chapter 2). Its function is interesting especially to process and reduce data from moving scenes and will be described in more detail in section 6.2. Using feedback and the addition of a nonlinearity at the output of one of the ports, an outer product memory may be constructed which corresponds to a real-time version of an optical ring resonator memory (see section 6.5 for further explanations).

Naturally, all vector-vector interconnections can be modified to realize fan-in interconnections that interconnect several channels at the input to a single one at the output. Problems that may arise when combining the beamlets at the output are due to interference effects. The elimination of such effects is possible with the use of arrays of incoherent sources for readout, temporal sequencing, random phase modulation of the readout beams or allocated volume configurations. The performance of these different techniques and their consequences for the realizations of fan-in devices have been dis-

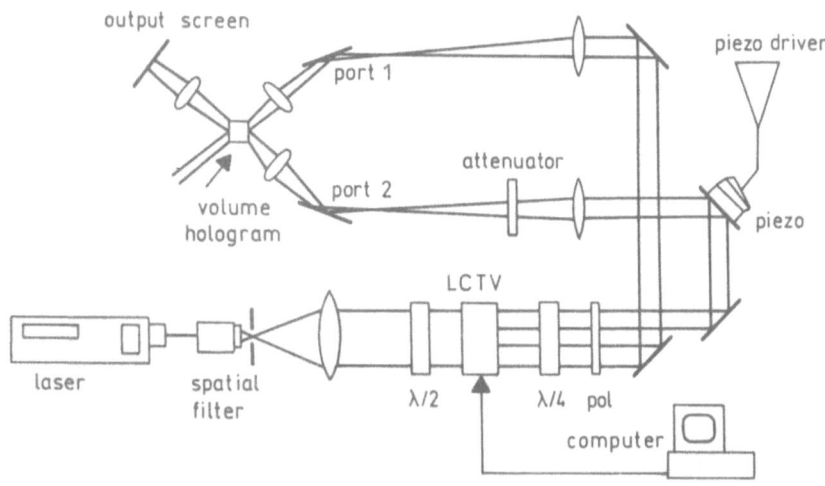

**Figure 4.10:** Optical apparatus for programming a vector-vector projection operator. The inputs to the photorefractive crystal are arbitrarily labelled as port 1 and port 2. One-dimensional information imposed on the beams by the LCTV is imaged into the photorefractive medium. A piezo-mounted mirror and an attenuator allow one beam to be multiplied by a complex constant (after [111]).

cussed in detail by A. Marrakchi and coworkers [112]. For those that are interested in detail in that question, I refer to this excellent publication and the references therein.

### 4.3.2 Matrix-Matrix-Interconnections

The interconnection of two-dimensional matrices is no longer possible using devices that are themselves only two-dimensional. As we have seen in the previous sections, the interconnection medium has to be able to support a number of interconnection nodes that correspond to the product of the two ports to be interconnected. Consequently, for matrix-matrix interconnections at least a three-dimensional interconnection medium is required, leading us immediately to volume holographic interconnections. The inherent three-dimensional storage properties of volume holograms are only completely exploited when full matrix-matrix-interconnections are realized, which can be performed again using photorefractive two- or four-wave mixing configurations.

*Matrix-Matrix Interconnections Using Photorefractive Two-Wave Mixing*

The most basic and principle configuration that uses photorefractive volume holography in order to realize a matrix-matrix interconnection is shown in fig. 4.11. It is based on storing a volume phase grating for each pair of input-output points to be interconnected. The light emanating from the input plane is collimated by a lens, and the diffracted light from the volume hologram is connected to the output plane by focusing the diffracted plane wave with a lens. From this basic principle, a variety of configurations using two-wave

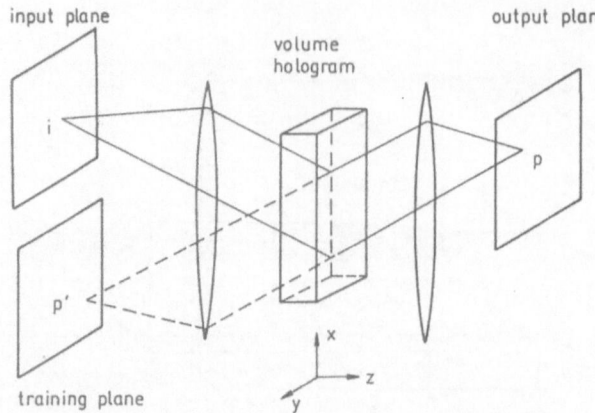

input plane

volume
hologram

output plane

i

p

p'

x

z

training plane

y

**Figure 4.11:** Scheme of the volume holographic matrix-matrix interconnection, given by the interaction of an input pixel $i$ and an output pixel $p$: a grating is stored by the interference of two Fourier-transformed beams of the sources $i$ and $p'$ in the volume hologram. Pixel $p'$ is the inverted image of pixel $p$. During readout, the stored grating interconnects the pixels $i$ and $p$.

wave mixing to realize the interconnection weights have been derived. A straight forward configuration is shown in fig. 4.12.

The matrices to be multiplied are given by the transversal distributions $a_{mn}$ and $b_{kl}$. While all the light sources shown in the figure are of the same nominal wavelength, each source differs from its neighbouring source by a frequency $\delta\omega$ that satisfies $\delta\omega \gg 1/\tau$, where $\tau$ is the photorefractive response time. If each source frequency is denoted $\omega_m$, the optical amplitude distribution immediately following the spatial light modulator that contains the matrix $a_{mn}$ is given by

$$E_{mn} = a_{mn} \cdot \exp(i\omega_m t). \tag{4.3}$$

The same source array is rotated by 90° and is directed through the second spatial light modulator to yield $b_{kl} \exp(i\omega_k t)$. By imaging each distribution through a slit, we obtain

$$A_n(t) = \sum_m a_{mn} \cdot \exp(i\omega_m t), \qquad B_l(t) = \sum_m b_{kl} \cdot \exp(i\omega_k t) \tag{4.4}$$

at the crystal plane.

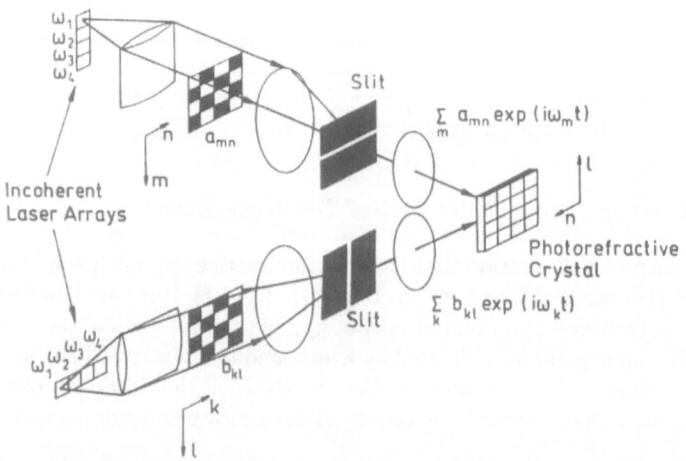

$\omega_1$
$\omega_2$
$\omega_3$
$\omega_4$

n

$a_{mn}$

m

Slit

$\sum_m a_{mn} \exp(i\omega_m t)$

l

Incoherent
Laser Arrays

l

n

Photorefractive
Crystal

$\omega_1 \omega_2 \omega_3 \omega_4$

$b_{kl}$

Slit

$\sum_k b_{kl} \exp(i\omega_k t)$

k

l

**Figure 4.12:** Matrix-matrix multiplication based on photorefractive two-beam coupling (after [113]).

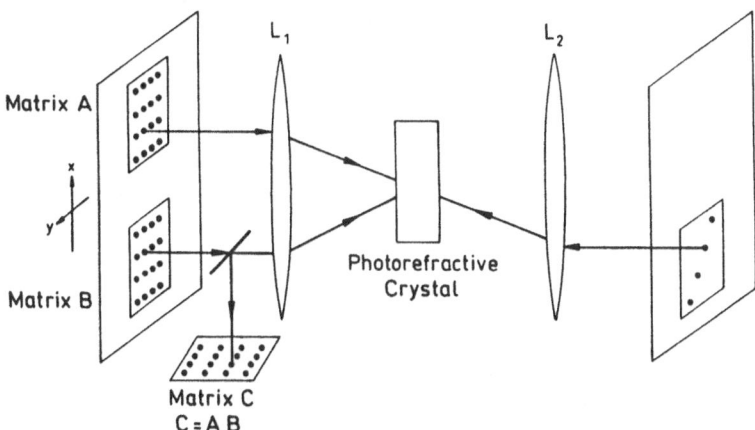

**Figure 4.13:** Matrix-matrix multiplication based on grating degeneracy in photorefractive two-beam coupling (after [114]).

Due to the refractive index modulation by the photorefractive crystal which includes a time-averaging operation with the integration time $\tau$, the steady-state amplitude of the grating formed in the crystal is given by [113]

$$\Delta n \propto \frac{\sum_m b_{ml} * a_{mn}}{\sum_m |a_{mn}|^2 + |b_{ml}|^2}. \tag{4.5}$$

For input matrices with $\sum_m |a_{mn}|^2 + sum_k |b_{kl}|^2 \approx$ constant, the right-hand side of eq. (4.5) is the desired matrix-matrix product. In general, deviations are obtained due to the normalizing effect of the denominator. The matrix-matrix-product can then be accessed holographically with a plane-wave reading beam to be detected or further processed coherently.

Another tricky way to realize matrix-matrix multiplication using two-beam coupling exploits grating degeneracy in photorefractive media in conjunction with an incoherent laser array [114]. Specifically, multiplications are implemented by photoinduced index gratings whose amplitudes are determined by the interference between coherent beams, while summations are implemented by grating degeneracy. This degeneracy is based on the Bragg condition in thick volume gratings and will be explained in detail in section 4.3. Here, it is sufficient to know that normally, the interconnection of two two-dimensional matrices requires a four-dimensional interconnection volume. Because volume holograms are only three- dimensional, consequently several points in the planes to be interconnected will be connected with the same interconnection grating. The result is a grating degeneracy, preventing several points to be interconnected independently from others.

Fig. 4.13 shows the schematic diagram that describes the principle of operation of the matrix-matrix multiplication. Both matrices $A$ and $B$ (both $N \times N$) are placed at the front focal plane of a lens. At the rear focal plane, a photorefractive volume holographic storage medium is inserted to record the multiplication of the two matrices. The recorded information is read out be a set of reading beams, which consists of $N$ diagonally aligned point sources placed at the front focal plane of a second lens, which has its second focal plane also at the location of the storage medium. Thus, the recording beams are propagating from left to right, while the reading beams are propagating from right to left. The diffracted readout beams are directed by a beam splitter to the output plane,

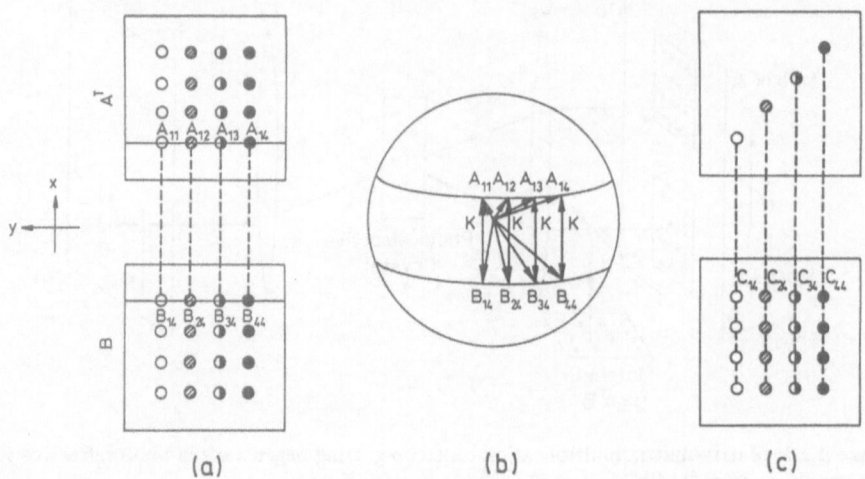

**Figure 4.14:** Arrangement for elements during recording and readout. (a) At the input plane, pixels with the same filling are coherent, and those with different fillings are mutually incoherent. (b) Degenerate gratings in the momentum space. (c) Readout configuration: diffraction occurs among pixels with the same shading (after [114]).

which is again located at the focal plane of the first lens. The arrangement of the matrix elements is shown in fig. 4.14.

To realize matrix-matrix multiplication, the illumination of the matrices $A$ and $B$ is chosen so that all pixels within each line along the $x$ direction are mutually coherent while pixels with different $y$ values at the input plane are mutually incoherent. During recording, gratings are formed according to $A_{ij}B_{jk}^*$, where * denotes the complex conjugate of the wave amplitude. Within these $N^3$ gratings formed, only $N^2$ have different wave vectors. Thus, the index grating is given for weak gratings and neglecting all gratings with small grating wave vectors by the sum of four terms [114]

$$\Delta_n \propto \frac{1}{I_0}\text{Re}[(A_{11}B_{14}^* + A_{12}B_{24}^* + A_{13}B_{34}^* + A_{14}B_{44}^*) \cdot \exp(-i\vec{K}\vec{r})] \qquad (4.6)$$

where $I_0$ is the total averaged intensity and $\vec{K}$ is the common grating wave vector. When this grating is read out, the diffracted beam amplitude is proportional to the grating amplitude, which is exactly the desired matrix element

$$C = \sum_{j=1}^{4} A_{1j}B_{j4}^*. \qquad (4.7)$$

During readout, all elements of $C$ can be read out in parallel by using $N$ point sources that are diagonally aligned at the reading plane (see fig. 4.14c). Each of the reading points reads out $N$ nondegenerate gratings, giving $N$ diffracted points in a line along the x direction at the output plane.

**Figure 4.15:** Wave-vector diagram of anisotropic self-diffraction in BaTiO$_3$. The writing beams are both in extraordinary polarization, the self-diffracted beams are in ordinary polarization, which is orthogonal to the optical c-axis (after [115]).

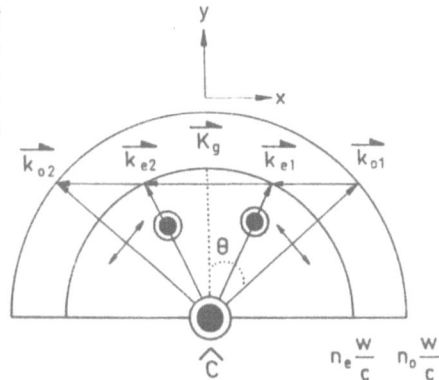

Experimentally, C. Gu and coworkers realized a system that was able to process the multiplication of two $4 \times 4$ binary (0,1) matrices, giving the correct analog multiplication result. The capacity of such a matrix multiplication system is determined by the thickness of the crystal, the wavelength of the laser, the spot size of a pixel and the numerical aperture of the lenses. If proper alignment is done, such an optical system may handle matrices with up to $1000 \times 1000$ elements in a parallel and therefore quick way.

To show the variety of possibilities to realize matrix-matrix multiplication using two-beam coupling in photorefractive crystals, a third example should be mentioned. It is based on anisotropic self-diffraction in photorefractive BaTiO$_3$. The input matrices are carried by two incident beams with special Bragg-matched incident angles. The output matrices are produced by anisotropic self-diffraction with the polarization orthogonal to those of the incident matrices. By thresholding the output, this architecture is particularly suitable for optical logic interconnections and optical switches.

Anisotroptic self-diffraction takes place in crystals, in which the matrix of effective electrooptic coefficients has zero diagonal elements, thus inhibiting coupling between two beams having the same polarization. However, coupling between different polarizations becomes possible because of nondiagonal tensor elements. Thus, when the incident beams are e.g. extraordinarily polarized, the diffracted beams may become ordinarily polarized (see fig. 4.15). To exploit this fact for matrix-matrix multiplication, only two incident beams are necessary, having for the case of a $2 \times 2$ interconnect two layers of interaction

**Figure 4.16:** Schematic diagram of matrix-matrix multiplication using anisotropic self-diffraction in photorefractive volume holograms (after [115].

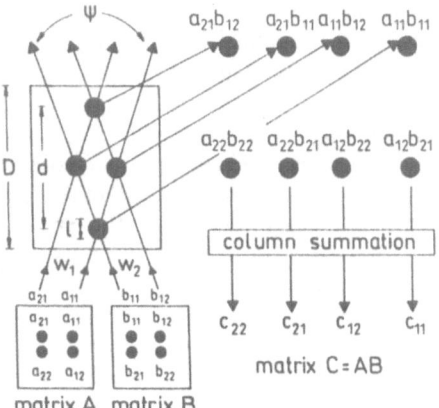

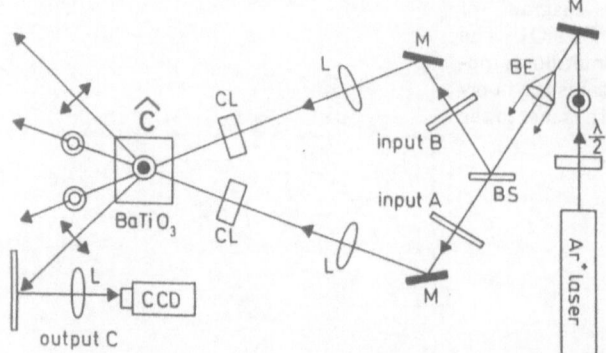

**Figure 4.17:** Experimental setup to realize matrix-matrix multiplication using anisotropic self-diffraction in photorefractive volume holograms (after [115]).

region in the vertical direction (see fig. 4.16). Each layer is confined in a horizontal plane and consists of the interaction units of the corresponding row of the first input matrix with the same row of the second input matrix. Consequently, for an $N \times N$ input matrix, there are $N$ interaction layers inside the crystal and each layer has $N^2$ interaction spots. In total, we achieve then the required $N^3$ interaction regions inside the crystal. The right-hand part of fig. 4.16 shows the second step of the matrix-matrix multiplication. Summation of the diffracted beams along the vertical column finally produces the element of the output matrix.

The experimental realization of that method is shown in fig. 4.17. C.C. Sun and coworkers [115] used an Ar-ion laser and a BaTiO$_3$ crystal as the interconnection medium. They realized a $4 \times 4$ matrix operator giving the results shown in fig. 4.18. Because in photorefractive beam coupling the diffracted intensity is a nonlinear function of the modulation depth of the input signals, this method is not appropriate for ordinary analog matrix-matrix multiplications. Instead, by thresholding the output, the system is suitable for optical interconnections and switches which do not require analog summation of the weights.

One of the most important suggestions of to improve two-beam matrix-matrix-interconnections has been presented in 1990 by J.E. Ford and coworkers [116]. He transmitted the input of an optical interconnection of two-dimensional arrays through a phase-coded mask and then correlated it with a control image constructed from the interconnection weights. The control image is a sum of phase-coded tensor components and can be manipulated in a variety of ways and still perform the required interconnection. In particular, the phase-code density is a free parameter that can be used to trade off between the output signal-to-noise ratio and the control image space- bandwidth product. If the correlation is made by using a dynamic medium such as a photorefractive crystal, fast reconfiguration is possible either by switching to a new control image or by changing the input phase code to access one of several holographically prestored interconnections.

The conceptual construction of the algorithm in shown in fig. 4.19 in three steps. First (fig. 4.19a), the correlation of the input array with the control image produces the output at sites imbedded in a field of noise. Using a supplementary phase coding of the input and the control images (fig. 4.19b) cuts that background noise relative to the signal. The final control image is produced in the third step (fig. 4.19c) by compression (overlapping) of the encoded interconnection, reducing thus the control space-bandwidth product at the cost of superimposing noise on the signal sites. In combination with changing the

**Figure 4.18:** Inputs and experimental results of matrix-matrix multiplication using anisotropic self-diffraction. The slanted outputs of (b) and (c) where caused by the inclination of the output plate (from [115]).

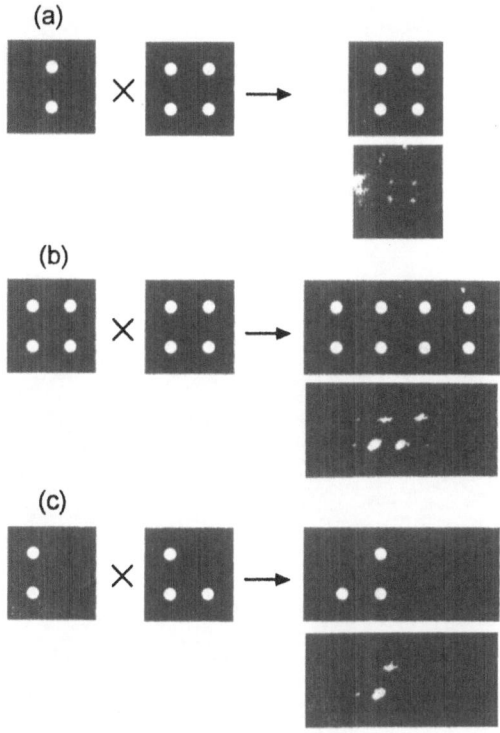

wavelength for every interconnection schedule to be stored, J.E. Ford and coworkers realized broadcast as well as random interconnections for three different wavelengths in a single crystal.

This method significantly enhances the signal-to-noise ratio and showed impressively, how the phase information inherent to photorefractive volume interconnections can be exploited to enhance the performance of these devices. We will discuss the role of the phase in interconnection and storage devices in more detail in the following section.

*Matrix-Matrix Interconnections Using Photorefractive Four-Wave Mixing*

Beneath two-wave mixing, especially four-wave mixing (see section 3.5) — the nonlinear wave coupling process in which three input waves mix in a photorefractive volume medium to yield a fourth wave that is the phase-conjugate replica of one of the incoming waves — is fairly well-suited for the multiplication of high density information bearing signals, because it combines the features of reliable grating storage (high grating modulation depth), high connection rates (high coupling coefficients) and cross-talk-free interaction (aberration correction).

To illustrate the features of matrix-to-marix interconnections in the case of four-wave mixing, let us have a detailed look at a configuration using photorefractive volume storage devices suggested by P. Yeh et al. [118]. For convenience, I assume a matrix multiplication operation between two $N \times N$ matrices that can be written as

$$C = A \cdot B \qquad \text{with} \quad C_{ij} = \sum_k A_{ik} B_{kj}. \tag{4.8}$$

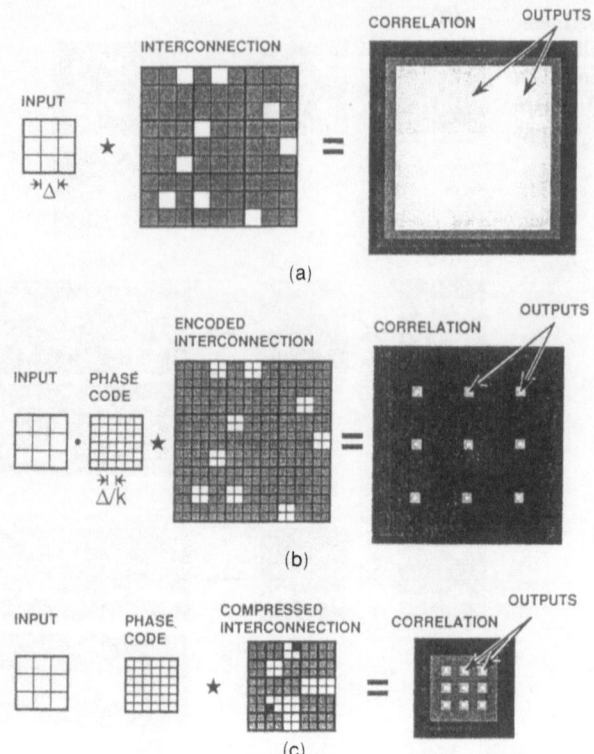

(a)

(b)

(c)

**Figure 4.19:** Conceptual construction of the matrix-matrix multiplier algorithm using phase-coded optical correlation. (a) Correlation of the input array with the control image produces the output at sites imbedded in a field of noise. (b) Phase coding the input and the control images cuts background noise relative to the signal. (c) The final control image is produced by compression (overlapping) of the encoded interconnection, reducing the control space-bandwidth product at the cost of superimposing noise on the signal sites (from [117]).

Eq. (4.8) shows, that matrix multiplication consists of two main operations in analogy to vector-vector multiplication: a parallel multiplication and a summation. In the four-wave mixing configuration shown in fig. 4.20, beams 1 and 2 contain the information about two matrices $A(x,z)$ and $B(z,y)$, respectively. Each beam consists of a matrix of beamlets. In the nonlinear medium, the two matrix-carrying beams form an interference pattern and thus a volume grating is formed. The grating contains all information of the products of the elements of the two matrices and can be written as

$$\Delta n = n_2 \cdot A(x,z) \cdot B^*(z,y)e^{i\vec{K}\vec{r}} + c.c. \tag{4.9}$$

where $\vec{K}$ is the grating vector formed by the two interfering beams $A$ and $B$. This volume grating of the refractive index can be read out by a third beam, which may simply be a plane wave. The diffracted output beam consists of the contributions from each point of the grating integrated along the beam path and can be written as

$$C(x,y) \sim \int A(x,z) \cdot B^*(z,y)\, dz \tag{4.10}$$

with the integration carried out along the beam path. This integration completes the operation of matrix multiplication. The information about the product of these two matrices is now impressed on the transverse spatial distribution of the diffracted beam.

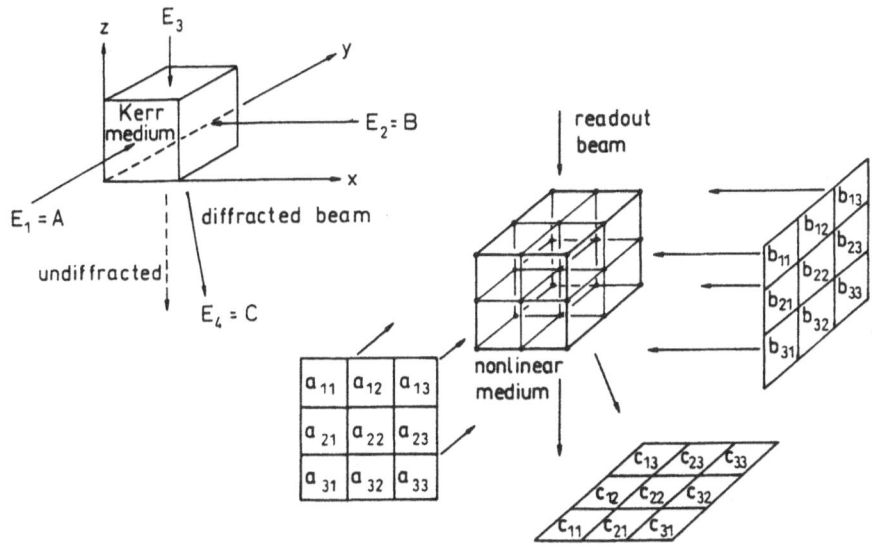

**Figure 4.20:** Schematic diagram illustrating the basic idea of optical matrix-matrix multiplication using nonlinear four-wave mixing (after [118]).

However, in practice, the dimensions of the matrices will be limited by cross talk between various channels due to diffraction when the Bragg condition is violated and due to imperfections in the optical components used in the experimental setup as e.g. lens aberrations, absorption and scattering. Owing to the phase matching requirement that the readout process imposes upon the operation, the readout beams have to be incident along a direction that satisfies the Bragg diffraction condition which has been set by the writing beams in order to achieve high efficiency.

There are several other configurations to achieve matrix-matrix multiplications, the best known up to now among them being colour multiplexing and spatial convolution. In the first one — named *colour matrix multiplication* — the basic idea is to decompose the problem into matrix-vector multiplications and carry out all of these operations simultaneously in parallel, but using a different wavelength for each matrix-vector pair. After multiplication, all of the colour components are then combined by a prism into a single row which has to be expanded again by anamorphic optics to match the mask representing the elements of the second matrix. After proper element-by-element multiplication and summation into the column, the resulting multicolour output is demultiplexed into different colour components that together represent the final product. The whole process is schematically shown in fig. 4.21. I will come back to that means of utilizing the frequency as a parameter to encode information in the context of image storage methods in the next chapter.

In the second case of matrix-matrix multiplication using spatial convolution by means of four-wave mixing, the nonlinear crystal is located at the common Fourier plane of the input matrix masks rather than at the common image plane as in the examples discussed up to now. J.O. White and A. Yariv [119] have demonstrated that spatial convolution and correlation of two (two-dimensional) patterns can be achieved in real-time using four-wave mixing in the common Fourier plane of the input patterns and recording the phase-conjugate output at the corresponding object plane. The encoding scheme for

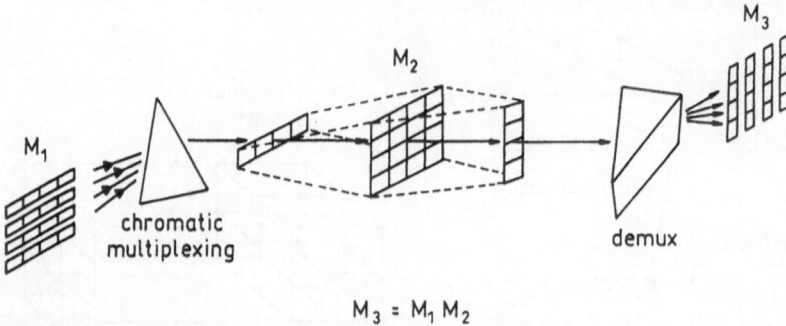

$$M_3 = M_1 M_2$$

**Figure 4.21:** Schematic diagram illustrating the basic idea of optical matrix-matrix multiplication using colour multiplexing (after [118]).

that case is different from those used in the approaches discussed earlier. Here, matrix-matrix multiplication in full parallelism is achieved by space-multiplexing by means of spatial convolution using degenerate four-wave mixing. Up to now, a large number of investigations have been performed to improve the performance of these matrix-matrix interconnection schemes. Finally, four-wave mixing can be also exploited to realize non-linear computational operations, as e.g.quadratic interconnections.

To realize equations of the form

$$y = \sum_{i=1}^{N} \sum_{j=1}^{N} a_{ij} x_i x_j = (\vec{x})^T \vec{A} \vec{x} \tag{4.11}$$

where the $a_{ij}$ are the elements of the interconnection matrix and the $x_i$ are the neuron inputs, the input vector is fanned into two $N \times N$ matrices (one for $\vec{x}$ and one for $\vec{x}^T$). These matrices are imposed upon the two pump beams of the photorefractive four-wave mixing process. The $A$ matrix interferes with the corresponding element in the $\vec{x}$ matrix. The $\vec{x}^T$ matrix is imaged into the crystal so that each element is exactly counterpropagating with respect to the elements in the $\vec{x}$ matrix, The output phase-conjugate matrix will then contain elements that are the coherent product ofthe elements of the three input matrices, giving the components of the sum in the equation above. the double sum in eq. (4.11) can subsequently be performed with a lens, where the detected power is the output of the quadratic operation.

G.N. Henderson and coworkers [120] realized such a system, using a $4 \times 4$ binary weight matrix that was interconnected in photorefractive $BaTiO_3$ as the four-wave mixing device. By extension of that idea, higher-order polynomial processors can also be developed. With the supplementary development of coding schemes for multipolar, analog and complex numbers, this processor can be used to implement quadratic and higher-order neural networks for solving a variety of signal- and image processing problems. I will discuss implementation of higher-order neural networks in more detail in section 9.2.

### 4.3.3 Matrix-Vector Interconnections

Now that we have investigated vector-vector operations (two-port operators) and matrix-matrix operations, it is nearby to look at how vector-matrix operations can be realized using volume holographic operations. Again, the number of interconnections that are

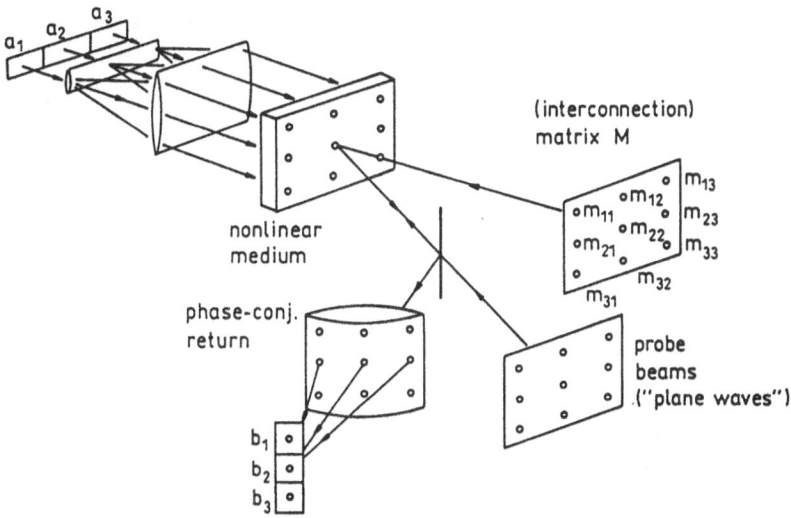

**Figure 4.22:** Schematic diagram illustrating the basic idea of optical matrix-vector multiplication using four-wave mixing (after [118]).

required for these connections suggest to use interconnection media that are based on volume interconnectivity. The most important schemes suitable for matrix-vector multiplication were proposed by D. Psaltis and N. Farhat [121] as well as P. Yeh et al. [118, 122] and rely on performing a matrix-matrix multiplication by expansion of the vector into a matrix. The systems differ only in the implementation of the adaptive interconnection matrix. Whereas in the first realization by D. Psaltis and N. Farhat in 1985 still a conventional optical interconnection matrix was proposed, later on realizations using photorefractive feed-forward two-wave mixing volume holographic configurations have been presented. In a lot of successfully implemented systems — as the system of P. Yeh et al. — four-wave mixing is the building mechanism of the interconnection weights (fig. 4.22). As an example for the system using four-wave mixing interconnection weights, consider a discrete case in which the multiplication of an $N$ element vector and an $N \times N$ matrix has to be performed. The vector is fanned out into $N$ rows of identical vectors, creating thus a second matrix. These $N \times N$ beamlets are directed to a nonlinear medium as well as the matrix, which also contains $N \times N$ beamlets. Then, the procedure is equivalent to the one applied for matrix-matrix multiplication: The beamlets interfere in such a way that each beam of the matrix is counterpropagating relative to the corresponding beam of the vector. Thus, there are $N \times N$ spatially separated regions, pumped by a pair of counterpropagating beams. Now, $N \times N$ probe beams (plane waves) are directed onto the medium in such a way that each probe beam can propagate through an intersection region. As a result of four-wave mixing in every of these regions, each probe beam generates a phase-conjugate beam that is proportional to $M(i,j) \cdot a(j)$, where $a(j)$ is the $j$th element of the vector $\vec{a}$ and $M(i,j)$ is the matrix element denoted with $(i,j)$. By using a cylindrical lens, we obtain a summation over $j$, resulting in

$$b(i) \sim \sum M(i,j) \cdot a(j) \tag{4.12}$$

The results of these operations are shown in fig. 4.23.

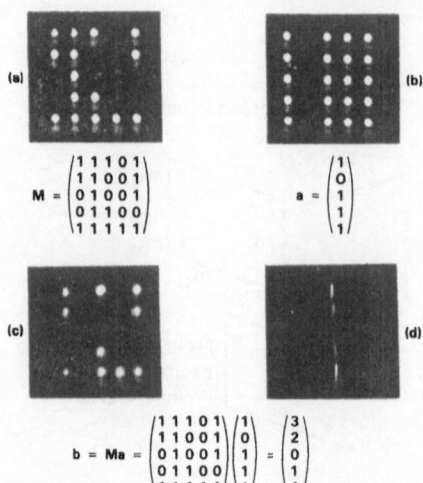

$$M = \begin{pmatrix} 1 & 1 & 1 & 0 & 1 \\ 1 & 1 & 0 & 0 & 1 \\ 0 & 1 & 0 & 0 & 1 \\ 0 & 1 & 1 & 0 & 0 \\ 1 & 1 & 1 & 1 & 1 \end{pmatrix}$$

$$a = \begin{pmatrix} 1 \\ 0 \\ 1 \\ 1 \\ 1 \end{pmatrix}$$

**Figure 4.23:** Photographs showing the matrix and the vectors: (a) input matrix, (b) input vector, expanded into a matrix with five identical rows, (c) resultant phase-conjugate reflection matrix, (d) resultant vector of a matrix-vector multiplication after summation of (c) by a cylindrical lens (from [118]).

$$b = Ma = \begin{pmatrix} 1 & 1 & 1 & 0 & 1 \\ 1 & 1 & 0 & 0 & 1 \\ 0 & 1 & 0 & 0 & 1 \\ 0 & 1 & 1 & 0 & 0 \\ 1 & 1 & 1 & 1 & 1 \end{pmatrix} \begin{pmatrix} 1 \\ 0 \\ 1 \\ 1 \\ 1 \end{pmatrix} = \begin{pmatrix} 3 \\ 2 \\ 0 \\ 1 \\ 4 \end{pmatrix}$$

Such a scheme for matrix-vector multiplication can also be used for matrix-matrix multiplication by decomposing a matrix into column vectors and then multiplying the matrix with each of the column vectors. As a result of the nature of the four-wave mixing process in the medium, this matrix-vector multiplier operates on the field amplitudes and thus can be used to handle matrices and vectors with complex elements — or in other words may also be suited to be implemented in neural nets that need bipolar weights. This is possible if the device is handled as a completely coherent one, having the phase of each beamlet maintained uniformly over the transverse dimension of the beamlet and in the summation process. If these phases are not uniform over the beamlets, the final step of summation becomes incoherent. Consequently, such a matrix-vector multiplication operates then on the intensities and thus may handle only positive numbers.

P. Yeh and coworkers demonstrated these operations with a series of specific examples using various architectures on the basis of four-wave mixing and phase conjugation. To imprint the matrices onto the two counterpropagating pump waves, masks are used on the basis of amplitude spatial light modulators which are configured in such a way to suppress cross talk between the interfering waves due to Bragg degeneracy. They demonstrated successfully that both the pixel-by-pixel product and the summation can be done inside the nonlinear medium to achieve the desired matrix-vector product.

In all the matrix-vector or matrix-matrix multiplication schemes described above, simultaneous alignment of the pixels of the matrices or the matrix and the vector is a major task, particularly for a large number of pixels. Misalignment of the pixels leads to severe errors. For a given size of the matrix mask, the density of elements increases as the dimension of the matrix increases. Thus, the requirement on the alignment becomes even more strict. In practice, the critical alignment required is likely to impose a limit on the matrix dimension (length of a row $N$), which is in the order of 100 or less, depending on the specific architecture.

Another scheme for optical matrix-vector multiplications, the principle of which may also be used for implementations of matrix-matrix operations, uses a phase conjugator in conjunction with a spatial light modulator to eliminate that pixel-by-pixel alignment requirement at the cost of some reduction in parallelism ($N^2/2$ instead of $N^2$ intercon-

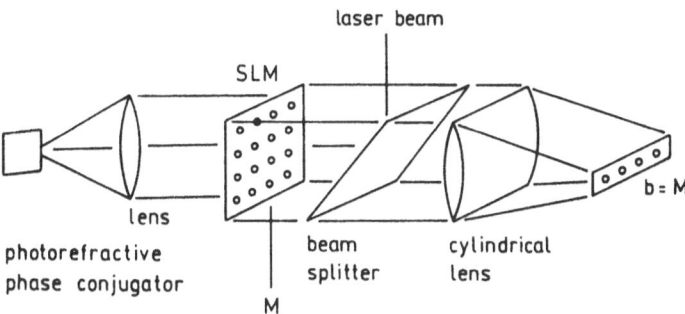

**Figure 4.24:** Schematic diagram illustrating the concept of matrix-vector multiplication using a photorefractive phase conjugator in conjunction with a spatial light modulator (after [118]).

nections). It has been reported by A.E. Chiou et al. [118] and its basic concept is depicted in fig. 4.24. Here, phase aberrations due to imperfections in the optics are self-corrected by the phase conjugation process. Its principle of operation is as follows: The spatial light modulator (SLM) first impresses the matrix information onto the input laser beam. This beam is then incident into a self-pumped phase-conjugate mirror, which stores the matrix information after a finite grating formation time. When the matrix information is removed from the SLM and all of the pixels are turned into a maximum transmission condition, the phase-conjugate beam that contains the reconstruction of the matrix information exists for a finite duration, which depends on the strength of the input (read) beam. During this time, if the next frame of the SLM carries the vector information, parallel multiplication is performed as the phase-conjugate beam propagates back through the SLM. Here, the vector is represented as a two-dimensional array of $N$ identical column vectors, where $N$ is the dimension of the vector. A cylindrical lens in the output port performs the summation. The dark storage time during which the matrix information can be retrieved is determined by the photorefractive material and the pumping configuration. It ranges from seconds to microseconds. In this architecture, the same input beam is used for alternately writing and reading the hologram.

Up to now, we imagined a multiplication of vectors and matrices of illuminated input masks. However, if we imagine the input vector representing the signals being realized by an array of lasers and the output vector being an array of detectors, we achieve a proper optical interconection system between laser and detector arrays. In combination with a photorefractive interconnection medium and a spatial light modulator input device, it is possible to realize reconfigurable optical interconnects with very high efficiency [118, 123]. The potential of reconfiguration is based on the formation of dynamic holograms in the crystal, allowing for nonreciprocal energy transfer in two-wave mixing in order to perform certain interconnection schemes. In fig. 4.25a, a simple one-dimensional case is shown to explain the concept. The input beam is split by a beamsplitter into the reference and the signal, which is expanded by using a cylindrical lens before entering the SLM. In this example, the input beam is to be connected to detectors b and d as prescribed by the SLM. Due to beam coupling in the crystal, this interconnection is enhanced, resulting in an optical interconnection with a very high energy efficiency. Fig. 4.25b shows the same principle for laser and detector arrays. In this example, laser 1 is to be connected with detectors b and c, laser 2 with detectors a and d, laser 3 with detectors c and d, and laser 4 with detectors a and c. In general, the two-wave mixing described above may be viewed as a real-time holography in which the recording and readout occur simultaneously inside

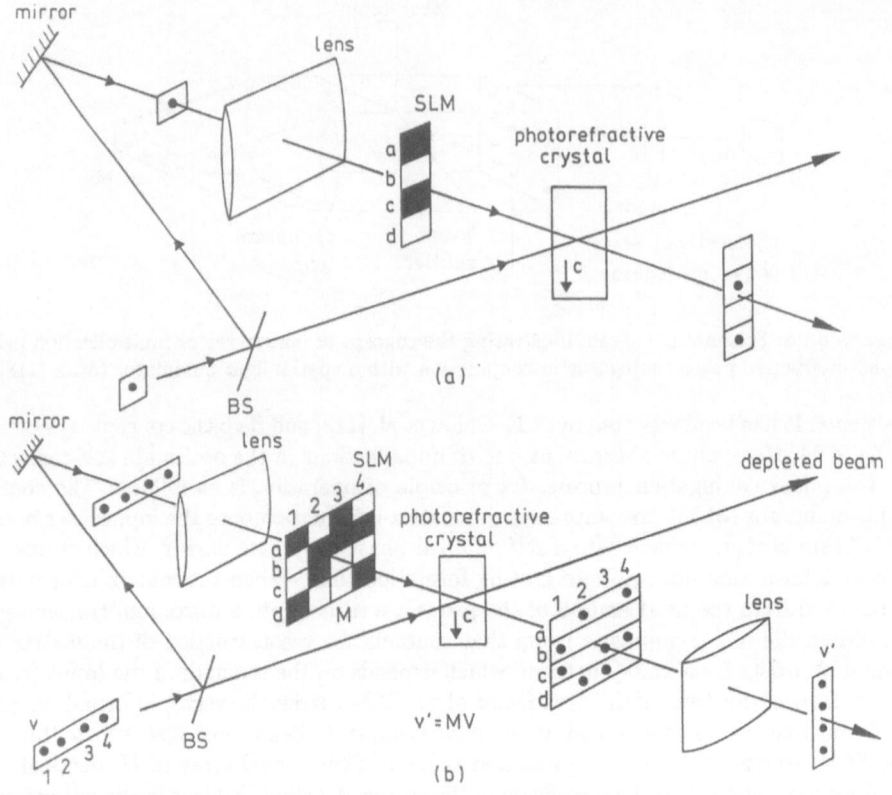

**Figure 4.25:** Schematic drawings of optical interconnections using dynamic photorefractive holograms with $N = 4$, for the case of $1 \times N$ (a) and $N \times N$ interconnections (b) (after [118]).

the photorefractive crystal. The beamsplitter and the spatial light modulator are used to generate the reference and object beams to record a volume hologram that represents the interconnection pattern as prescribed by the SLM. The energy coupling involved in two-wave mixing ensures a high diffraction efficiency during readout, allowing to realize crossbar switches very easily [122].

An interesting example for reconfigurable crossbar switching has been realized using wavelength tuning by S. Wu and coworkers [124]. The reconfigurability is given here by wavelength multiplexing different interconnection schemes in one volume of the storage crystal. That technique of page multiplexing will be dsecribed in detail in section 4.4. Here, it is sufficient to know that by changing the angle of incidence of one of the writing beams or by changing their wavelength, distinguishable interconnection patterns can be stored and recalled without cross talk. A possible configuration of such a reconfigurable crossbar structure is shown in fig. 4.26. Multiple reflection gratings are written in one piece of a LiNbO$_3$ crystal. The writing light should be slightly different from the reading light, e.g. a He-Ne laser ($\lambda = 633$ nm) can be used for writing and an array of laser diodes ($\lambda = 680$ nm) for reading. The wavelength difference between writing and reading may be compensated by modifying the writing angle according to the dispersion relation. The writing procedure is performed at a fixed writing angle and multiple exposures are carried out by rotating the crystal with an angular separation equal to the required

**Figure 4.26:** Realization of a reconfigurable optical crossbar switch (DA, LA are laser diode array and line detector array, respectively) (after [124]).

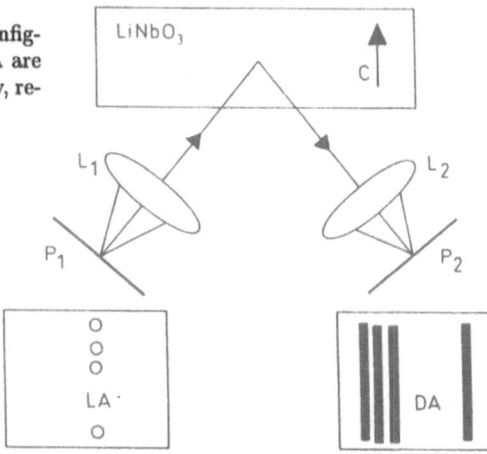

channel separation. In order to achieve equal diffraction efficiencies for each channel, a recording procedure as explained in sections 4.4 and 4.4 should be used. Moreover, the gratings can be fixed after writing in order to make them insensitive to the readout processes. A linear laser diode array is arranged along the nondispersion direction at the input plane. Since it is in the nondispersion direction, there would be no wavelength tunable range limitation for the separations between laser diodes, provided the total extent of the array is within the paraxial approximation. A one-dimensional line detector array is placed at the output plane with a separation of the line detectors equal to twice the channel separation. Different laser diodes can use the same set of gratings in the crystal. In the nondispersion direction, the light beams are reflected according to the law of reflection. By independently controlling the wavelength of the laser diodes, a reconfigurable electrooptical crossbar architecture can thus be realized. The principles of vector-matrix multiplication can also be used to form holographic interconnections in photorefractive waveguides, because an input vector corresponding to the incident field may be mapped to an output vector corresponding to the diffracted field via a matrix represented by the interconnecting hologram [125]. Volume holograms in waveguides offer a straightforward means of interfacing dynamically reconfigurable interconnections with integrated optoelectronic devices. Applications of thick holograms in waveguides have included grating couplers and distributed feedback lasers. The potential for information storage in integrated holograms is defined by the number of degrees of freedom which can be stored in a planar waveguide hologram. It scales with the area of the hologram divided by the square of the guided wavelength [125].

Many structures and implementations of these optical interconnections are well known since a couple of years and are described in detail in excellent early papers as e.g. [57, 112, 126, 127, 128, 129, 130]. Here, I had to pick out some of the most interesting examples of promising and forthcoming interaction schemes that may be implemented in all-optical or electrooptical networks in the near future. Because of the variety of examples, the choice is somewhat arbitrary. Therefore, the reader may refer to the sources cited above for a more complete overview over the implementations that are currently possible.

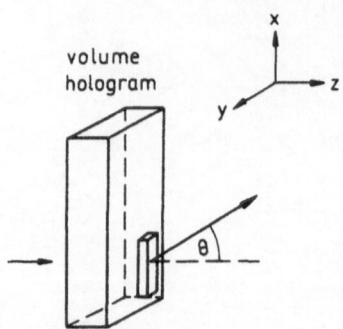

**Figure 4.27:** Definition of the resolution cell in a volume holographic medium.

### 4.3.4  Limitations of Pixel-to-Pixel Interconnections

At the end of this short survey on possible interconnection schemes using mainly volume holographic methods and before looking at the potentialities of nonlocal, distributed interconnection schemes, let us have a look at the principle limitations of these devices. These limitations set the frame in which interconnection schemes as well as image storage schemes may be applied in real neural net architectures, because they define the maximum storage capacity, the resolution of the system and the performance quality. For that purpose, refer to the exemplar scheme shown in fig. 4.11, where each resolution element or pixel at the first neural plane is estimated to be a location for a neuron. The light from a pixel is collimated and diffracted by a holographic grating. The diffracted light is then focused by a lens onto a pixel at the output neural plane.

Let us now consider connections between multiple neurons at the input and the output planes. A key criterion for the usefulness of these interconnections is their space-bandwidth capability. This criterion determines, to a large extent, the computational performance of an optical interconnection device. For an input containing $N^2$ distinct resolution cells (pixels or neurons), the required number of degrees of freedom is $N^4$ in the space variant, volume interconnection case, whereas it is only $N^2$ in the space invariant case. In other words, if we have $N^2$ pixels in the input plane, that is $N \cdot N$ neurons, and we want to interconnect independently each of the input neurons to all the output neurons, we would need $N^4$ interconnections, which implies that $N^4$ weights or gratings must be formed and properly stored in the hologram. As an example, for $4 \cdot 4 = 16$ pixels at the input plane, $N^4 = 256$ gratings have to be stored independently in the crystal. This leads to the question of the maximum number of gratings that can be distinguished in a hologram of volume $V$. An estimation of that number defined by the fundamental limits of diffraction was first given by van Heerden in 1963 [100], if one considers a cubic crystal with the volume $V = a^3$. A wave incident on the surface $A = a^2$ of the crystal produces a diffraction pattern with the solid angle $\phi \propto \lambda^2/a^2$ as shown in fig. 4.27. Thus, $a^2/\lambda^2$ independent directions can be distinguished in the crystal volume. In the same way, the number of independent wave vectors depends on the thickness of the volume medium. Thus, the global number of independent storage units in a crystal is given by

$$N \approx \frac{a^3}{\lambda^3} = \frac{V}{\lambda^3} \tag{4.13}$$

Assuming that the resolution of the holographic medium in any direction is $\delta > \lambda$, the total number of resolution cells is

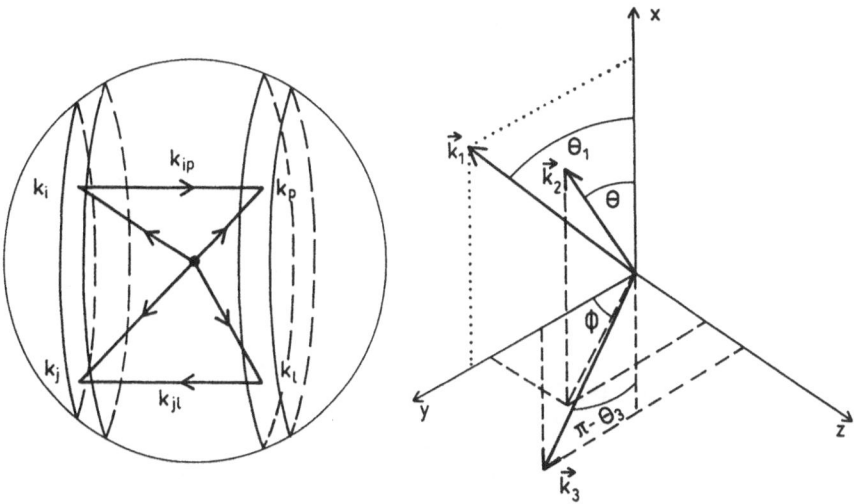

**Figure 4.28:** $\vec{k}$-space diagram illustrating the degeneracy of the gratings that connect point $(i,p)$ and $(j,l)$ (left) and wave vector diagram to calculate the phase detuning due to cross talk of the first order (right).

$$N \leq \frac{V}{\delta^3} \tag{4.14}$$

If this relationship is violated, an attempt to establish the interconnection between one pair of input and output neurons automatically specifies the interconnection for other pairs as well.

### Limitations due to Deviations from the Bragg Condition

Because the smallest resolution unit in a crystal is defined by the Bragg condition of the crystal (see eq. (4.1) in section 4.2), we can explain this situation also in the representation of the $k$-vectors of the interacting waves: If the divergence in $k$ between two neurons of the input plane, $\Delta k$, is smaller than the one defined by the Bragg condition, the reconstruction of these two pixels will fall onto the same area of the output plane. This effect is called *cross talk* and leads to noise disturbances in the reconstruction of the stored weights. In $k$-space, the Bragg condition defines thus two strips with the width $\Delta\Theta = \lambda/L$, where $L$ is the thickness of the crystal. The two strips are parallel circles on the wave normal sphere. The planes in which the strips lie are perpendicular to the grating vector $\vec{k}_{jl}$. If one chooses for a fixed input pair $(i,p)$ a second input pixel $j$ outside the strip (see fig. 4.28), no cross-talk noise will appear at the output plane in point $p$. . If pixel $j$ is in contrast situated inside the strip defined by $i$ and $p$, cross-talk noise will appear for all planes, except if the output pixel is not located at the corresponding strip in the output plane. If these criteria can be met for all input and output pixels, then this cross talk, called *first-order cross talk* will be completely eliminated. The required width of the strip in fig. 4.28 is determined by several factors, including diffraction due to the transverse aperture of the hologram and the angular sensitivity of diffraction from a thick gratings as it is given by the Bragg condition. The result of this cross-talk noise can be seen impressively when we attempt to interconnect two complete two-dimensional grids in

 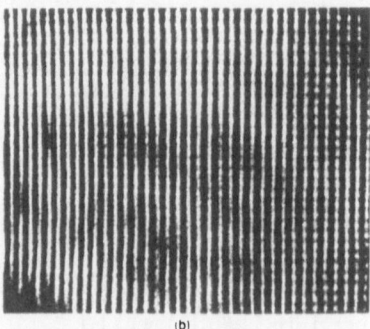

(a)                                                    (b)

**Figure 4.29:** Experimental demonstration of Bragg degeneracy using infrared hologram recorded in c-cut BaTiO$_3$. (a) object, (b) reconstructed hologram of the object. Vertical smearing is symptomatic of Bragg degeneracy (from [131]).

the volume hologram. In fig.4.29, the object plane consisted of a periodic rectangular grid pattern, and the reference plane was a uniformly filled rectangle. The interference pattern of these two planes was written in a photorefractive BaTiO$_3$ crystal in a configuration that has low two-wave mixing gain and thus low beam fanning. The result during readout with the reference plane is a vertical smearing of the image. To avoid this first-order cross-talk noise, we can either increase the volume of the crystal or decrease the density at which the neural planes are populated — the so called *fractal grid* arrangement of input and output points. Using the second strategy, we have to bear in mind that, for $N^2$ neurons at the input and the output plane to be interconnected, $N^4$ interconnections are necessary. If we imagine a grid of $S^2$ points, at which the $N$ neurons can be distributed, the number of neurons has to be $N^4 \leq S^3$ or, if $\sqrt{N} \times \sqrt{N} = N$ pixels in the input and output plane have to be interconnected, the number of nodes or neurons has to be $N^2 \leq S^{3/2}$ (see eq. (4.14)).

The most simple solution is to fill one plane (e.g. the input object plane in fig. 4.11) with the maximum number of pixels ($N^2$) and reduce the other plane to a single line ($N$). This configuration is the one chosen in all page multiplexing devices and the one that is common, when a simple, maximum storage and free-of-cross talk configuration is desired. It was investigated in detail in [132, 133].

In this case, the maximum number of pixels in the input plane still holds the condition mentioned above: $N \cdot N^2 = N^3$, or for the case of $N \times \sqrt{N}$ pixels: $N^{1/2} \cdot N^{2/2} = N^{3/2}$. Other, similar configurations are shown in fig. 4.30 for an example of patterns that samples $N^{3/2}$ out of $N$ for $N = 8 \cdot 8 = 64$. We derive these patterns using a process of elimination. Each time we attempt to add a new neuron to the input (output) sampling grid, we have to check whether this new neuron is already connected to one of the neurons selected previously at the output (input) plane by an existing grating. If it is, we eliminate the position of the new neuron from the sampling grid. If it is not, we select this position for a new neuron, which implies that new gratings are established to connect this new neuron to all the neurons now selected at the output (input). In addition, we eliminate all the positions at the output (input) to which this new neuron is connected by existing gratings. If we systematically continue that iterative procedure, we arrive finally at the complete sampling grid.

J.R. Wullert and Y. Lu [134] derived the conditions for such a grid configuration theoretically using the diameter of an input pixel, the pixel distance or pitch and the

**Figure 4.30:** Nondegenerate sampling configuration in the reference plane for holographic interconnection storage. The object plane is estimated to be completely filled with 8 x 8 information points.

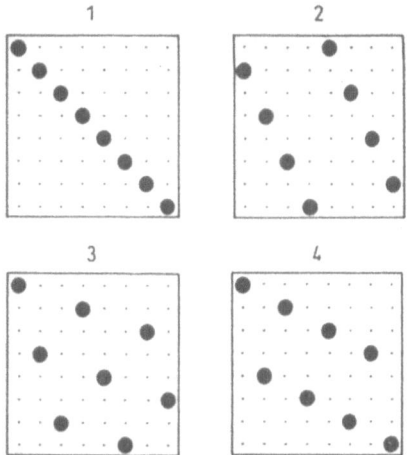

angle of the closest reference beam angle to the object beam. They could show that beneath the fundamental limitations arising due to the Bragg selectivity of the storage medium, the quality of the optical system severely influences the storage capacity. The key point of this derivation is that the optimum capacities possible with reasonably sized components, while they are not quite terabytes, are still desirable, but they can only be obtained with large, high-quality, fast lenses. Moreover, the results highlight a trade-off between holographic storage density and the number of bits accessed in parallel. Counterintuitively, they found that for a fixed degree of parallelism, higher capacities are possible with coarser pixel pitches. This is a result of the need for larger lenses when larger pixels are used. Rather than increasing the capacity, thus higher input-output resolution increases the parallelism of the data access, improving the data rate.

The situation becomes slightly different, when the size of the object and the reference plane becomes remarkably different, especially when the object plane is smaller than the reference plane. Because cross talk arising from reference pixels, that are out of the size of the object plane does not fall onto the area of reconstruction of the object points, it is allowed to add several rows of reference points in order to enhance the capacity. Possible situations are shown in fig. 4.31. The upper configuration of 4.31 is the one normally used for the case of storage of completely filled object planes, having here a rectangular size that is a fourth in length compared to the reference plane. In holographic image storage, this method, implemented by the authors in [135] and by F. Mok and coworkers [77], is known as *fractal-space multiplexing.* [136]

There are also other sources of cross-talk due to Bragg-mismatch, that arise when the first-order cross talk is removed completely by an arrangement of the input-output configuration as described above. Second-order cross-talk effect result from light waves that are first diffracted by a grating from an input wave at pixel $i$ and then rediffracted by a second grating and are directed to the output pixel $p$. Therefore, two gratings are involved in the noise generation process. All the second-order light waves resulting from diffraction by two intralayer gratings (gratings between pixels in one layer) or two interlayer gratings (gratings between pixels from different layers) are negligible because in the geometry in fig. 4.28 they are phase mismatched and thus they do not contribute to second-order cross-talk effects. Therefore, the principal source of second-order cross talk is diffraction from the interlayer gratings followed by rediffraction from the intralayer gratings. However, this contribution to cross talk results in a signal-to-noise ratio that is

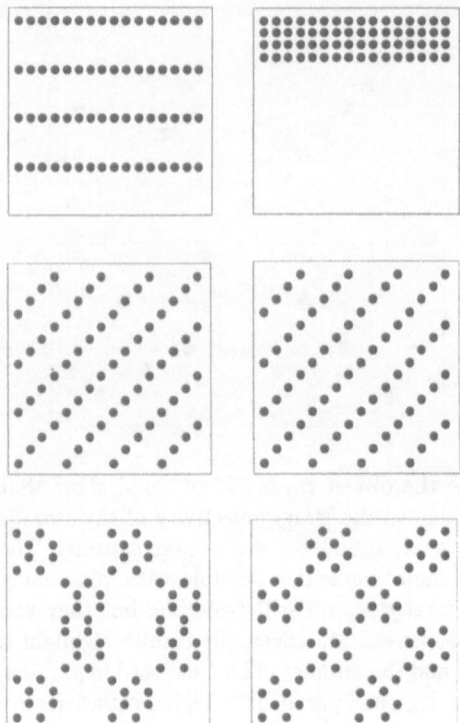

**Figure 4.31:** Nondegenerate sampling grids for holographic interconnection storage. Each pair of sampling grids is such that the holographic grating that interconnects any point on the input grid with any point on the output grid is unique to that input-output-pair (after [99]).

much lower than the one for first-order cross talk. A detailed description of second-order cross talk can be found in [137], as well as criteria for their suppression and for optimal implementation of volume holographic interconnections with reduced first and second-order cross talk. Third-order cross talk arises when light originating form the $i$th pixel is diffracted by three separate interlayer gratings and is ultimately directed at the output pixel $p$. However, because the signal to noise ratio here is proportional to $M^4/3$ (M = number of gratings to be superimposed) [133], third-order cross talk is not expected to be a serious concern for large networks.

Up to now, I have discussed the limitations arising by the conditions imposed by Bragg's law in volume holograms onto the registration and readout process. A detailed analysis of these $N$-to-$N$ volume holographic neural interconnections and the restrictions imposed by the Bragg condition can be found for the special application of neural network interconnections in [99, 138].

*Limiations due to the Nature of the Photorefractive Effect*

If we use bulk photorefractive materials as storage or interconnection devices, another limit arises which is based on the nature of the photorefractive effect. The crucial question is here, how many gratings can be created by the free charge carriers with noticeable diffraction efficiency. A limitation of the storage capacity arises in that case, because the reservoir of charge carriers in the bulk material is limited or, in other words, there is an exposure saturationfor photorefractive media. Therefore, the whole charge carrier density created by all interconnection gratings written in the crystal has to be smaller than the acceptor density at each point of the interaction volume. If we assume a maximum diffraction efficiency of $\eta_0$ if only one grating is written into the volume, the efficiency $\eta_1$ drops for $N$ gratings proportional to the square of the number of gratings:

$$\eta_1 = \frac{\eta_0}{N^2} \tag{4.15}$$

Because the original diffraction efficiency can be approximated utilizing the theory of coupled waves by H. Kogelnik [30] to be

$$\eta_0 \approx \left( \frac{\pi L \Delta n}{\lambda \cos \Theta} \right)^2 \tag{4.16}$$

where $\Theta$ is the angle between the writing beams in the material, we get the number of holograms that can be written independently in a volume to be equal to

$$N = \sqrt{\frac{\eta_0}{\eta_1}} \approx \frac{\Delta n_{max}}{\Delta n_{min}} \tag{4.17}$$

where we used the notation $\Delta n_{max}$ for the maximum index modulation that can be achieved for a single hologram registration process (equal to $\Delta n_0$). $\Delta n_{min}$ stands for the minimum detectable index modulation (equal to $\Delta n_1$).

With these expressions, we can estimate that for typical values of $\Delta n \approx 3 \cdot 10^{-5}$, $\lambda \approx 0.5 \cdot 10^{-6}$ m, $l \approx 0.5 \cdot 10^{-2}$ m we obtain a one-grating efficiency of $\eta_0 \approx 0.89$. If we assume furthermore that with standard detection devices, efficiencies up to $10^{-8}$ can be easily measured, we arrive at a storage capacity in the range of $10^5$ interconnections or gratings. What we learn from eq. (4.15) is that the storage capacity is strongly dependent on the diffraction efficiency for a single hologram and the sensitivity of the detectors used to obtain the smallest diffraction efficiency. These two values limit the range of $\eta$ in the two directions and is therefore called the *dynamic range of the diffraction efficiency* too. Therefore, the dynamic range of the crystal is one of the most important parameters and finally the limitation of the storage capacity, much more than the limitations arising from the Bragg condition. A datailed analysis of this limitation in comparison to other storage materials as photochromic materials can be found in [139].

Finally, the effect of cross-talk noise which results from the nonuniform amplification that is present in photorefractive two-wave mixing in certain materials — mainly those that display diffusion-dominated charge transport as $BaTiO_3$, — should be shortly considered (see also section 3.4 for details on photorefractive two-beam coupling). When the pump and the probe beams are Fourier transformed and interact in such a crysal, the incident pump-to-probe intensity ratio is a function of the position. Therefore, the two-wave mixing gain that is determined by the incident intensity ratio is also position dependent. Imagine the orientation of the crystal in such a way that the probe beam will be amplified. When this amplified probe beam is is Fourier tranformed back again, the image is degraded, resulting in cross-talk noise in the photorefractive interconnection. In addition, the finite contrast ratio of the spatial light modulator that may be used to specify the interconnection pattern affects the performance of the interconnection and may enhance this cross-talk noise. A way to reduce these effects is to move the crystal away slightly from the exact Fourier plane, or, equivalently, adding a random phase to each input channel. I will discuss these possibilities in more detail in section 4.4 and show experimental realizations using both suggestions. A detailed analysis of this cross talk due to beam coupling amplification can be found in [140].

### 4.3.5  Interconnections Using Self-Pumped Phase Conjugation

All previous optical holographic implementations of interconnection models have used a single grating in a photorefractive crystal to store the connection weight between two neurons. This approach relies on the Bragg-condition to avoid cross talk between different neurons. However, as we have seen in the previous section, even in that case cross talk may still occur if too many interconnections are formed in the storage volume, because of the angular degeneracy of the Bragg-condition. I have discussed in the previous section some solutions to that cross talk problem by arranging the neurons in special patterns in the input and output planes. Unfortunately, these methods result in subsampling and incomplete utilization of the field of the incoming wave.

This disadvantage may be suspended by using a different storage principle as it takes place e.g. in self-pumped or mutually pumped phase-conjugate mirrors. If we switch back to the principles of self-pumped phase conjugation (see section 3.5.2), there is a fundamental difference to two- or four-wave mixing. Here, light from each pixel or neuron of the input plane writes volume gratings in the crystal by interfering with light from all other pixels in the output plane, forming a continuum of gratings within the crystal by wave mixing effects such as beam fanning and two-beam interaction. Thus, the production of the phase-conjugate of the input beam results in a system that is both massively parallel and globally interconnected. The Bragg-degeneracy is broken in that case by distributing each interconnection weight among a continuum of cascased, both angularly and spatially distributed gratings. Since the Bragg-condition must be satisfied at many spatially separated gratings, cross talk between neurons due to the conical Bragg-degeneracy associated with a single grating is greatly reduced, making subsampling of the input or output planes no longer necessary. As shown in the Ewald sphere construction of fig. 4.32, two gratings in series are sufficient to break the Bragg-degeneracy since only one set of beams can simultaneously satisfy both Bragg cones associated with the two gratings. Thus, by forcing each read beam to diffract from a series of gratings, cross talk due to Bragg-degeneracy can be eliminated.

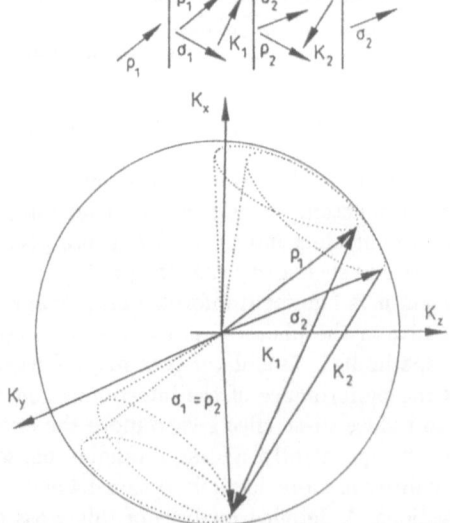

**Figure 4.32:** Ewald sphere momentum-space diagram for Bragg matching to two gratings in series. Only a single input-output wave-vector pair can lie on the two Bragg-cones and satisfy the Bragg-condtition at both gratings simultaneously (after [141]).

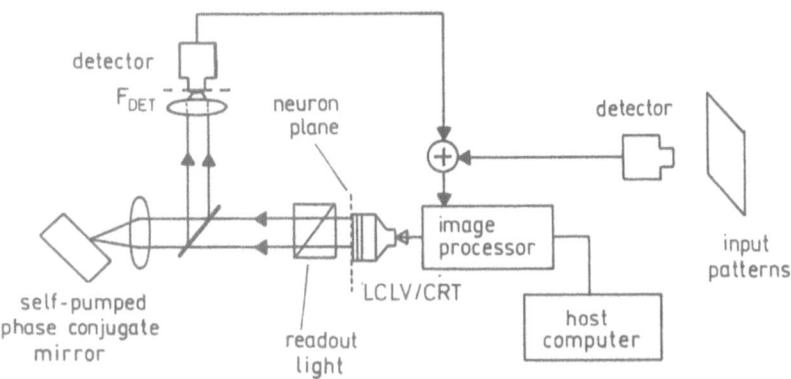

**Figure 4.33:** Configuration for interconnecting optical neurons with a self-pumped phase-conjugate mirror (after [141]).).

The only limiting condition that this method imposes on the storage medium is connected to the second limitation discussed above: The photorefractive crystal has to have sufficient capacity to store the additional gratings required for that distributed storage procedure. Provided that the storage capacity of the crystal is not exceeded, $N^2$ fully interconnected neurons can be implemented without subsampling, representing an increase by factors of $\sqrt{N}$ and $N$ in the number of neurons and weights, respectively, over the single grating per weight approach of two- and four-wave mixing photorefractive interconnections.

Based on this approach, Y. Owechko and B. Soffer [141] presented two configurations for interconnecting optical neurons using these *multiepoch grating storage* devices. The architecture of one device — with one input and a self-pumped phase conjugation process – is shown in fig. 4.33. The connection of the LCLV/CRT input device and the CCD camera detector to a host computer allows the combination of the device with electronic computing in order to realize outer product learning and error correction capabilities when the system is set up in larger neural networks.

In the same way, optical interconnections based on continuous distribution of angularly and spatially distributed volume holographic gratings can be formed in mutually pumped phase-conjugate mirrors. In these devices, two separate input beams (which may be mutually incoherent) are incident upon two opposite faces of a high-gain photorefractive crystal such as $BaTiO_3$, producing two phase-conjugate output beams. Light from each pixel or neuron in the two input planes writes volume gratings in the crystal by interfering with light from all the other pixels in that input plane, forming a continuum of gratings within the crystal by dynamic two-wave mixing effects such as beam fanning.

The production of the phase-conjugate of each input beam results in a system that is both massively parallel and globally interconnected. Outer-product learning using a bipolar error signal can be accomplished by changing the relative phase of the individual neurons and selectively erasing portions of the photorefractive gratings. While the self-pumped phase-conjugate mirror also takes advantage of distributed photorefractive holograms in order to reduce cross talk, a mutually pumped phase-conjugate mirror has

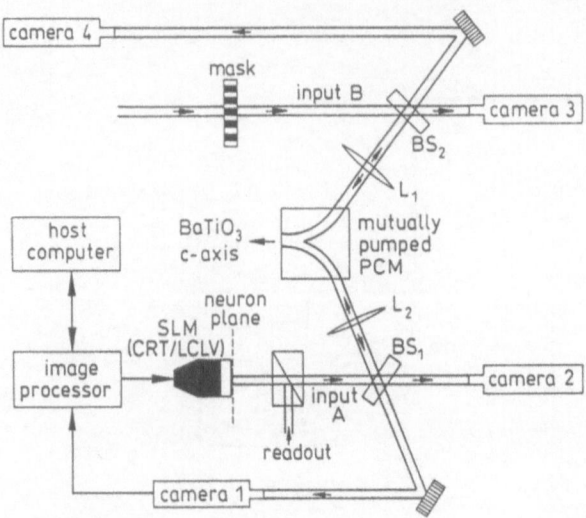

**Figure 4.34:** Architecture for an optoelectronic interconnection network using a mutually pumped phase-conjugate mirror (after [143]).

two independently controllable input beams, each of which forms the conjugate of the other beam. This results in no connections between neurons in the same input beam. The configuration has advantages when implementing neural networks in which intralayer connectivity is not desired. It also effectively doubles the number of neurons in the input planes by permitting the use of a second input spatial light modulator. An easy-to-understand overview over the various advantages of mutually-pumped phase conjugators for optical interconnection devices are described in [142].

The architecture used to experimentally demonstrate that second device, the mutually pumped phase-conjugate mirror interconnection method, is shown in fig. 4.34. The input neural plane is presented by a LCLV as in the self-pumped device, the output planes representing the phase-conjugate returns of the input beams are detected by a CCD-camera to allow the system to be set up in neural networks. The video signal from camera 1 can be processed electronically further on a point-by-point basis to implement programmable nonlinear thresholding functions. This signal can then be displayed on the LCLV in order to adjust weights by using outer-product learning. In implementing neural network models such as backpropagation, an error signal would be formed electronically and displayed on the LCLV to adjust the weights between neurons. With that system, Y. Owechko and coworkers realized cross-talk suppression between adjacent pixels and heteroassociation [143] as shown in fig. 4.35. Similarly, beam fanning as a non-conjugating method that uses the same multigrating principle to avoid Bragg degeneracy but does not suffer from hologram sharing, can be considered as a means to store distributed interconnection gratings. As shown in fig. 4.36, the fanned light can be used as a reference beam when it is interfered with a second, unfanned object beam to form a holographic connection that suffers neither from Bragg degeneracy (because multiple-cascaded gratings store each interconnection) nor from hologram sharing (because the conjugate beam is not used for readout). An unfanned object beam is used because object-beam fanning would degrade the quality of the reconstructed object image. During recording, the object beams form a set of signal gratings in which the connection weights are stored. Both the signal and fanning gratings are multiplexed angularly and spatially. The initial fanning of the reference beam needs only be performed once at the beginning of the learning session. Subsequent

**Figure 4.35:** Demonstration of
heteroassociation and global interconnectivity. (a) input beam
B consisting of a 700 pixel array,
(b) the corresponding phase conjugate of beam A for the input
beam B (heteroassociation). (c)
input beam B consisting of only a
subset of 35 pixels, reading out in
(d) the phase conjugate of beam
A as a whole (after [143]).

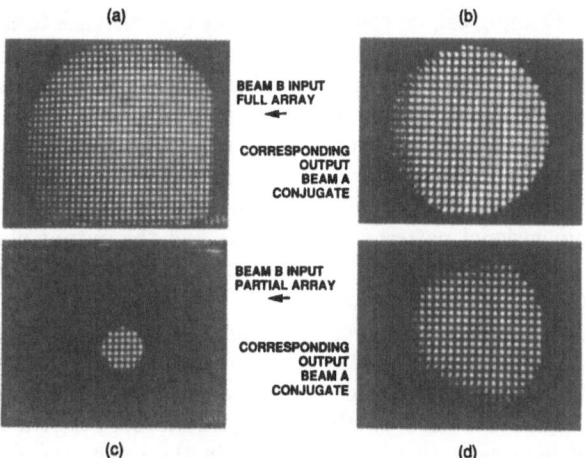

signal gratings are formed with fast two-wave mixing time constants rather than with
the slow fanning time constant. Upon readout each reference wave must match the Bragg
condition at a multitude of fanning gratings, which breaks the Bragg degeneracy. In addition, the individual components or beamlets, reconstructed by the signal grating are
all in phase and sum coherently; hence the aggregate diffraction efficiency can be high
even though the diffraction efficiency of any individual grating is small. Moreover, the
low coupling geometry of the individual signal gratings greatly reduces distortions owing
to beam coupling. Crosstalk tests performed by Y. Owecko [131] showed that the Bragg-
degeneracy advantages of self-pumped phase conjugation configurations are maintained,
but hologram sharing is eliminated.

**Figure 4.36:** Schematic illustration of
weighted interconnection formed using multiple grating holography. (a) Reference
beam fanning, (b) Recording, (c) readout
(after [131]).

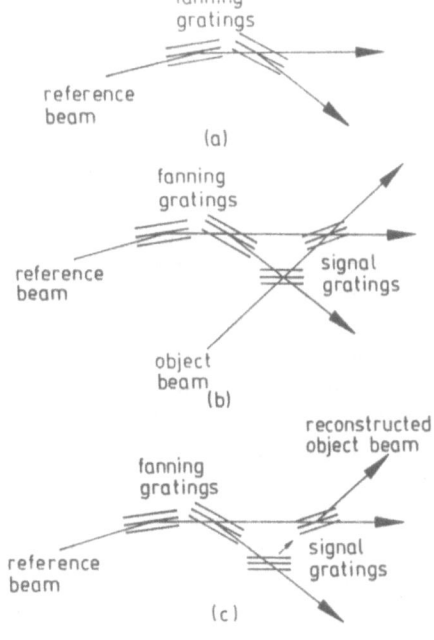

Fanning can be controlled so that the reference beam fans, and the object beam does not, by adjusting the orientation of the beams relative to the crystal. The fanning process generates high gain-length product fanning gratings that divide each reference beam into a set of beams at different orientations and locations.

To summarize, distributing each connection weight among a number of cascaded gratings that vary in position and orientation, has several advantages: First, Bragg-degeneracy is eliminated, which permits the placement of pixels in arbitrary two-dimensional patterns in the input and the output planes, leading to a new flexibility in holographic recording. Thus, the utilization of spatial light modulators and detectors is maximized because every pixel can be used without fear of cross talk. Second, distortions owing to beam coupling are reduced greatly, resulting in undistorted weight vectors and good quality reconstructed images. Finally, experimental measurements by Y. Owechko [131] indicate that grating erasure is apparently reduced. This might be due to the variation in spatial grating locations, I suggest. This in turn increases the number of holograms that can be stored in the same crystal for a given signal-to-noise ratio. The concepts of distributed grating interconnections will be further discussed in section 8.2 and section 9.1 when they are used to implement one- and multilayer perceptron algorithms.

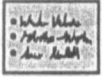 **Summary**

Optical interconnections may be realized using one- or two-dimensional arrays of sources as fibers, lasers or spatial light modulators that are connected via a nonlinear interconnection medium, e.g. a volume hologram to an array of detectors. Modes of interconnections are

- broadcasting
- point-to-point interconnections
- crossbar switching

Among nonlinear point-to-point interconnections using photorefractive volume holograms, there are several possibilities to connect the input to the ouput:

- Two-port operators can be built using two-beam coupling configurations. They perform the outer product rule in the interconnection regions by creating a refractive index grating that is proportional to the product of the incoming beam amplitudes. Therefore, two-port operators simulate Hopfield- or Boltzmann neural net architectures where the outer product plays an important role.
- Matrix-matrix interconnections can be performed in a similar way. The two matrices that interfere in the photorefractive medium form a volume grating, which contains all information about the products of the elements of the two matrices.
- To avoid cross talk, matrix-matrix multiplication can also be performed using colour multiplexing, where the problem is decomposed into matrix-vector multiplications with different wavelengths for each matrix-vector pair. After multiplication, all of the colour components are then recombined by a prism.
- Matrix-matrix multiplication may also be performed using spatial convolution in the Fourier-plane by means of degenerate four-wave mixing.

- Vector-matrix operations can be realized by extending the vector into a matrix with equal rows and then performing one of the matrix-matrix algorithms.

Limitations of these schemes arise due to

- the global number of independent storage cells that can be realized in a photore-fractive crystal,
- the Bragg-selectivity in the volume holographic medium,
- the limited number of charge carriers that are available to realize the creation of the volume holographic grating or the dynamic range of the crystal's diffraction efficiency and
- eventually due to the recording mechanism of the refractive index gratings.

These limitations require subsampling of the input and/or the output planes and reduce the storage capacity severely. Beneath these local interconnection schemes, self-pumped phase conjugation allows for the realization of interconnections with the following features:

- They are realized in a continuum of spatially and angularly distributed gratings.
- Since the Bragg-condition must be satisfied at many spatially separated gratings, cross talk between neurons due to Bragg-degeneracy is greatly reduced.
- Consequently, subsampling of the input and output planes is no longer necessary.

## 4.4  Distributed Volume Holographic Interconnections

Up to now, I have discussed local storage principles or *allocated volume multiplexing techniques* for single point-to-point interconnections, which are useful for all kinds of bit-information processing tasks.

However, what will happen instead if we want to store and interconnect a whole image or a fixed data page of points in the volume hologram? In that case, the requirements, the Bragg condition imposes upon the number of indepenent gratings in the volume can not be accounted for by restricting the interconnection topology of the object or image plane to relatively sparse fractal grids. Thus, we have to look for other methods which enable the storage of the whole set of input pixels representing the complete information page without losing information in that input plane. The most simple method has already been sketched in the previous chapter: It consists of reducing the output grid on the reference recording grid adequately to reduce cross talk, resulting e.g. in a single line of output points in the reference plane.

However, in this case, only one page can be stored in the whole storage area. Although it is possible to allocate several of these volume grating regions one beneath each other in a crystal volume, the restricted material dimensions allow in that case only for a limited number of pages to be stored. Therefore, it is more convenient to let the volume be shared by a number of superimposed holograms. Then, all gratings occupy the same location in the crystal volume. Although in this case of *shared volume multiplexing*, a single interconnection weight in one page can no longer be controlled independently, the whole page can be distinguished from each other by varying an external parameter. This multiplexing parameter ensures individual addressing of a single information page without cross talk and thus enables to access in parallel pages with a large number of information pixels. Because every optical system supplies only two spatial dimensions perpendicular to the light propagation direction, it is necessary to exploit the storage capacity inherent to that method with a supplementary, independent parameter. This multiplexing parameter enables to encode the data on every page of the information in order to be addressed during readout without cross talk.

The most simple technique is *polarization multiplexing*, where the two gratings that are to be superimposed have perpendicular polarization states. The disadvantage of that method is, that it enables only two different states to be encoded and, what is an even more serious restriction, does not account for the different coupling coefficients of photorefractive crystalline materials in the different polarization directions.

Another multiplexing technique, that becomes more and more attractive because of the actual development of broadband laser sources, is *wavelength multiplexing*. In that case, the holographic gratings are recorded with different wavelengths. If the wavelength interval between adjacent pairs of gratings is larger than a value about $1/100 \, \lambda$, these gratings are completely independent from each other. The only technical disadvantage of the system is that for a large number of gratings to be superimposed, the laser source as well as the storage crystal have to be tunable over a range of several ten to several hundred nanometers — a fact that can actually not be realized technically. However, the combination of wavelength multiplexing with other techniques to multiplex gratings in volume holography seems to be very attractive and allows to enhance the storage capacity in a single location.

The most important and most widely spread multiplexing technique up to now is *angular multiplexing*. In that technique, one or both of the interfering beams are varied in their incident angle for different pages to be stored. Each of the holograms superposed will selectively answer during reconstruction, depending on how exact the angle of recording is reproduced. There are a couple of methods to technically adjust the angle of incidence, which will be presented in the next section.

Finally, the method of *phase-coded multiplexing* can be derived from angular multiplexing. In this case, the phase of the reference wave is modulated in its plane perpendicular to the propagation direction. For that purpose, the reference wave is divided into a subset of $N$ fields, which are modulated independently from each other in phase. The set of adjustable phases then represents the address of a certain page. If the relative phases of the references are selected in such a way that all unwanted contributions interfere destructively during readout, a reconstruction without cross talk is possible. Because the angular separation of the single phase references is chosen to be larger than the one prescribed by the Bragg condition, this method has the same capacity as angular multiplexing, but does not require moving parts that change the angle of the incident waves or energy consuming modulation devices. I will describe this method in detail in one of the following sections.

All these shared volume multiplexing techniques have the disadvantage that each writing of a new grating on the same location as the precedent ones tends to erase these gratings written previously. Therefore, special dynamic recording algorithms have to be applied in order to guarantee that at the end of the recording process each grating has the same grating strength. Actually, two techniques are used to avoid asymmetric grating strengths due to erasure effects during writing. The first one, *sequential recording*, includes the exponential response time of the material into the recording procedure, thus giving a detailed dynamic writing schedule for every page to be recorded. The second one, *incremental recording*, writes every page for a very short time and cycles through all images until the desired grating strength is achieved. I will discuss these two principles in the following section in combination with angular multiplexing.

The big advantage of allocated volume holographic storage is the overwhelming storage capacity that can be achieved in this way in a single interaction region. Storage densities up to terabytes of interconnection points can be achieved in volumes smaller than 1 cm$^3$. Combined with transfer rates larger than 1 Gigabit per second, and random access times less than 100 $\mu$seconds, these devices are extremely attractive for analog as well as digital data storage.

Before I will describe the technical realization of these multiplexing techniques in detail, I want to introduce a general mathematical formalism for Bragg-selective multiplexing that allows to describe angular, phase-coded and wavelength multiplexing in the same notation and shows their differences clearly. Moreover, a mathematical description of the different writing techniques will be given.

### 4.4.1 Mathematical Formulation of Bragg-Sensitive Multiplexing

 Due to the Bragg selectivity of volume holographic media, superimposed holograms have to be separated angularly or in wavelength by an amount that is larger than the value defined by the Bragg condition. The larger the "distance" from that lowest lewel or the more the reference beams are sparsely selected (wavelength or angles are separated by an amount greater than the wavelength or angular selectivity of the medium), the more cross talk is reduced. However, at the same time, the storage capacity of the system decreases. Therefore, most of the implementations based on Bragg selective holographic storage try to exploit the minimum angular or wavelength separation, Bragg condition still allows. In the following derivation, I will show how the Bragg condition defines rules how to modulate the reference beam in order to achieve independent reference waves for all images or pages to be stored. During the recording procedure, the image beam, whose complex amplitude

$$A_m(\vec{r}) = A_m(\vec{r}) \cdot e^{i\vec{k}_m \vec{r}} + cc. \tag{4.18}$$

carries the information of the $m$th image in the amplitude $A_m(\vec{r})$ and simultaneously interferes with all $M$ reference beams inside the photorefractive medium. Each of these references is encoded with the multiplexing parameter carrying the complex spatial amplitude

$$\mathcal{R}_m(\vec{r}) = R_m \sum_{l=1}^{M} \sum_{k=1}^{M} a_{klm} \cdot \exp\left(2\pi i \frac{\cos \alpha_k}{\lambda_l} \vec{e}_{kl} \vec{r}\right). \tag{4.19}$$

Here, the set of adjustable parameters given by the wave vector $|\vec{k}_{kl}| = (\cos \alpha_k / \lambda_l)r$, $k, l = 1, \ldots, M$ am the complex factor $a_{klm}$ for reference beam number $k$ during the writing or reading process of the image number $m$ represent the address of the $m$th image. $\alpha_k$ is the angle between the wave vector of the reference beam and the crystal axis and $\lambda_l$ is the wavelength of the reference beam.

In order to take full advantage of the Bragg selectivity in thick recording media, the wave vectors $\vec{k}_{kl}$ have to be chosen in such a way that the gratings written by different reference beams with one signal wave are independent of each other. In other words, no reference beam should fulfil the Bragg condition of gratings that have been created by the interference of other reference waves with one of the signals: $\vec{k}_{kl} - \vec{k}_m \neq \vec{k}_{k'l'} - \vec{k}_{m'}$, for all $k, k', l, l' \in 1, \ldots, M$ $(k \neq k', l \neq l')$ and arbitrary $\vec{k}_{m'}$ with $|k_{m'}| = 2\pi / \lambda_{l'}$.

After successful recording of all $M$ images using angular, wavelength or phase-coded multiplexing, the factors $a_{klp}$ with fixed $p$ represent the address of the image number $p$. That address is the essential code to reconstruct every image independently from all others without cross talk for all three multiplexing techniques. The special conditions on $a_{klp}$ for each multiplexing technique will be derived in the following.

First, let us have a look at how photorefractive materials come into play during that recording process. The modulation of the refractive index in the volume holographic medium can be described for the case of a fixed angular configuration of the information bearing signal ($\cos(\alpha_m) = \cos(\alpha_s) = $ const. for $m = 1, \ldots M$) and the same wavelength for a reference and its appropriate signal beam by

$$\Delta n(\vec{r}) = \sum_{m=1}^{M} \sum_{l=1}^{M} \sum_{k=1}^{M} \Delta \bar{n}^{klm}(\vec{r}) \cdot a_{klm}^{*} \cdot \exp\left\{ 2\pi i \left( \frac{\cos \alpha_s}{\lambda_l} - \frac{\cos \alpha_k}{\lambda_l} \right) \vec{e}_{klm} \vec{r} \right\}. \quad (4.20)$$

The local modulation strength $\Delta \bar{n}^{klm}(\vec{r})$ of the gratings contains the amplitude and phase information of the respective image and can be written as:

$$\Delta \bar{n}^{klm}(\vec{r}) = g_{kl} \cdot A_m(\vec{r}). \quad (4.21)$$

Here, the factor $g_{kl}$ depends on the material and the experimental geometry. Moreover, it depends on the wave vector of the grating $\vec{K} = \vec{k}_s - \vec{k}_{kl}$, because it is proportional to the product of the effective electrooptic coefficient and the space charge field normalized by the local amplitudes of the interfering waves:

$$g_{kl}(\vec{K}) \propto r_{\text{eff}}(\vec{K}) \cdot \left| \frac{E^{\text{sc}}(\vec{r})}{A_m(\vec{r}) R_m} \right| (\vec{K}). \quad (4.22)$$

In the case of an ideal system, where the effective electrooptic coefficient $r_{\text{eff}}$, the space charge field $E_{\text{sc}}$ and the amplitude of the reference beams are independent of the special set of reference beams we consider and where the interaction among the reference beams can be neglected, the factor $g_{kl}$ can be substituted by a global sensitivity $g$:

$$\Delta \bar{n}^{klm}(\vec{r}) = g \cdot A_m(\vec{r}). \quad (4.23)$$

With eq. (4.23), the pattern of index modulation in the crystal (see eq. (4.20)) becomes

$$\Delta n(\vec{r}) = g \cdot \sum_{m=1}^{M} A_m(\vec{r}) \sum_{k=1}^{M} a_{klm}^{*} \cdot \exp\left\{ 2\pi i \left( \frac{\cos \alpha_s}{\lambda_l} - \frac{\cos \alpha_k}{\lambda_l} \right) \vec{e}_{kls} \vec{r} \right\}. \quad (4.24)$$

For readout of the image number $p$, the reference wave $\mathcal{R}_p$ with the parameters of encoding $k', l'$ has to be used. Taking the Bragg selectivity into account, the amplitude of the diffracted wave becomes

$$A_p^{\text{diff}} = R_p \cdot g \sum_{m=1}^{M} A_m \left( 2\pi i \frac{\cos \alpha_s}{\lambda_l} \vec{e}_{kls} \vec{r} \right) \sum_{k=1}^{M} \sum_{l=1}^{M} a_{klm}^{*} a_{klp} \quad (4.25)$$

Consequently, to achieve selective readout of one of the images without cross talk, we have to address the system with codes fulfilling the condition

$$\sum_{k=1}^{M} \sum_{l=1}^{M} a_{klm}^{*} a_{klp} = \begin{cases} 0 & \text{for} \quad m \neq p \\ M & \text{for} \quad m = p \end{cases}. \quad (4.26)$$

Considering $a_{jkm}$ as the elements of a tensor $A$, eq. (4.26) requires the unitarity of $A$: $X_{ijk} \cdot X_{ijk}^* = I$, where $I$ is the unitary tensor. Eq. (4.26) is a generalized condition to read out an image stored in a volume hologram without cross talk. Depending on how we chose the factors $a_{klm}$, we may easily formulate the conditions that have to be fulfilled in order to realize wavelength, angular or phase-coded multiplexing. We will derive these special conditions in sections 4.4.3 - 4.4.5, when discussing these techniques in detail.

### 4.4.2   Dynamic Recording Procedures

If we want to store several pages (or images) with equal weight strength (which is equivalent to the diffraction efficiency in image storage language) using angular multiplexing as well as every other allocated volume multiplexing technique, we have to take into account that in a dynamic, reconfigurable recording medium, each new registration tends to erase previously written ones. Therefore, the registration time for each image has to be adapted in such a way that all holograms have the same diffraction efficiency at the end of the recording cycle. In the following, I will describe the two most promising techniques to achieve multiple hologram storage with equal diffraction efficiencies. For our photorefractive neural net implementations, this hologram decay due to illumination implies a weight decay that limits the number of cycles a training algorithm can run on an optical system, because earlier exposures are erased as the training progresses. In section 6.4, I will show how dynamic copying or resonant refreshment can overcome this problem by storing the strength of the hologram through feedback. Another way of bypassing the weight decay problem is to use local algorithms, since they do not require long training sequences. In this case the large storage capacity of three-dimensional holograms can be used to synthesize the large networks that are required. Algorithms that implement hybrid techniques using features of local algorithms in which each hidden unit is trained separately and the training method is not iterative, whereas the representations of that result are distributed, will be presented e.g. in section 9.

*Sequential Recording*

The traditional method to account for that problem is the *sequential recording scheme*. In that method, the first hologram is written up to saturation of all space charge carriers among the acceptors. Then, the registration time for the second hologram is chosen in such a way that both holograms will have the same diffraction efficiency at the end of the procedure. The third hologram is then again recorded for a duration that is adapted in such a way that all three holograms will have the same efficiency and so on (see fig. 4.37). This method was first described in 1989 by A.C. Strasser et al. [144], but seems to be the method of choice since the first multiple image holographic storage experiments by e.g. L. d'Auria, J.P. Huignard et al. [145] or H. Kurz [75]. In a more general way, the recording scheme for $M$ images or pages leads to an equilibrium condition for the recording and readout phases as indicated in fig. 4.38, which postulates that the increase in the refractive index modulation $\Delta n$ during the time interval $\Delta t$ has to be equal to the decrease in $\Delta n$ during $(M - 1) \cdot \Delta t$.

This is equivalent to the mathematical notation

$$\frac{d\Delta n_w}{dt} = -(M - 1)\frac{d\Delta n_e}{dt} \qquad (4.27)$$

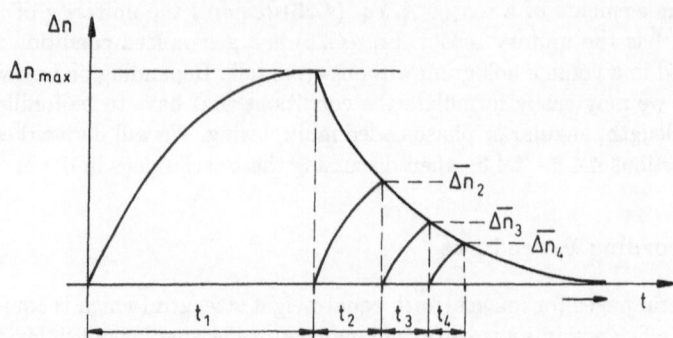

Figure 4.37: The principle of sequential recording in volume holographic media.

with the writing and erasure refractive index modulations $\Delta n_w$ and $\Delta n_e$, respectively. Note that here $M$ is the notation for the number of pages or images, each containing $N \times N$ pixels or interconnection points.

The number of holograms which can be registrated with that procedure can be derived using eq. (4.27) for the procedure of sequential recording (see [146]) and is given by

$$M \approx \frac{\Delta n_{\max}}{\Delta n_{\min}}, \qquad (4.28)$$

using $\Delta n_{\max}$ and $\Delta n_{min}$ as the notation for the maximum refractive index modulation that can be achieved and for the minimum index modulation that can be detected. Because of $\Delta n \propto \eta^2$, this is proportional to

$$M \propto \sqrt{\frac{\eta_{\max}}{\eta_{\min}}}. \qquad (4.29)$$

This is exactly the equation I have already derived in chapter 4.3 (see eq.(4.17)), showing that both effects, the photorefractive dynamics as well as the recording procedure, impose the most severe limitations upon the storage capacity in photorefractive volume holograms. In more general words, the exposure time of the $m$th hologram is expontially decreased in the superposition by an amount that corresponds to the square root of the increase in the diffraction efficiency $\eta$. For the writing time of the $m$th hologram $t_m$, this gives [144]

$$t_m = t_{\max} \cdot \exp\left\{ -\frac{(M-1) \cdot t_w}{2\tau_e} \right\}, \qquad (4.30)$$

provided that the writing time for the first hologram is given by $t_w$ and the erasure

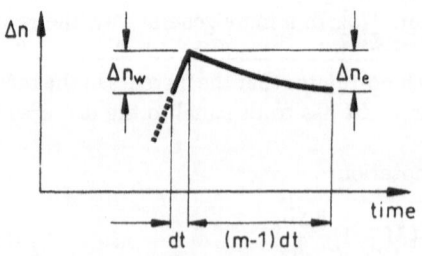

Figure 4.38: Equilibrium condition for identical refractive index modulation for all gratings written into a single volume holographic interaction region.

time constant by $\tau_e$. In experimental observations, a preparation light pulse enhances significantly the diffraction efficiency that can be achieved during registration using sequential recording. A sequential exposure schedule that takes into account these two time constants can be realized by a process with does multiple cycles of exposures and is described by [147]

$$t_{m-1} = -\tau_w \cdot ln\left\{1 - \left(1 - e^{\frac{t_m}{\tau_w}}\right) \cdot e^{\frac{-t_{m+1}}{\tau_s}} \cdot \left(\frac{r_1 + r_2 \cdot e^{S_{m+1}\left(\frac{1}{\tau_s} - \frac{1}{\tau_l}\right)}}{r_1 + r_2 \cdot e^{S_m\left(\frac{1}{\tau_s} - \frac{1}{\tau_l}\right)}}\right)\right\}, \qquad (4.31)$$

where $S_m = \sum_{i=m}^{M} t_i$, $t_m$ is the exposure time for the $m$th hologram, $\tau_w$ is the writing time constant, $\tau_s, \tau_l$ are the short and the long time constant, respectively, $r_1$ and $r_2$ are the weight values for the contribution of the short or the long time constant, respectively, and $M$ is the total number of exposures in each cycle.

Almost all the inplementations described in section 4.4 using angular reference wave multiplexing, use that technique. Although it requires the exact knowledge of the crystal parameters in order to determine the maximum refractive index modulation that can be achieved, up to 10.000 holograms [80]have been written using that algorithm with a resulting diffraction efficiency that is rather homogeneous, with deviations from the mean value in general less than 15%. A simple method to obtain maximized photorefractive holographic storage exploiting sequential recording has been given by E.S. Maniloff and K.M. Johnson [148].

For the special case of photorefractive interconnections or page-oriented storage, where spatial light modulators are used as the input devices, A. Goldstein and coworkers realized a modification of this scheduled recording procedure. Instead of changing only the exposure time for each data page, they iteratively increased the spatial-light modulator transfer function gain simultaneously with the exposure time. In this way, they could show with computer simulations that all implementations for optical neural networks that are based on weight writing in photorefractive volume holograms — as e.g. backpropagation algorithms, Widrow-Hoff and perceptron learning rules — show improved learning performance compared with results for networks trained without scheduling [149].

Sequential recording can also be implemented to realize beam fanning-based photorefractive image storage using angularly multiplexed fanned reference waves. The principle of that method has been described in section 4.3 and is based on the fact that fanned light can be used as a reference beam when it is interfered with a second, unfanned object beam. The advantage of that approach is, that recording of holograms in a fanning crystal of BaTiO$_3$ does not show Bragg degeneracy, hologram sharing cross talk or distortions owing to beam coupling. Even when using a reduced reference beam, e.g. 50% of the original reference, the reconstructed image is virtually identical to the original.

Using an orthogonal grid pattern to generate multiple reference waves, 30 images with $3 \cdot 10^4$ pixels or bits per image could be superimposed using that multiepoch recording of random reference patterns [131]. An interesting result is obtained concerning the mean hologram diffraction efficiency relative to the number of holograms. If multiple holograms are recorded coherently in a single exposure, then we would expect the diffraction efficiency per hologram to decrease as $1/M$, where $M$ is the number of holograms (compare to the calculations in sections 4.3 and 4.4). If the $M$ holograms are recorded in $M$ time-sequenced exposures, then we would expect the diffraction efficiency to decrease as $1/M^2$ owing to grating erasure. For the case of time-sequenced recording using fanned references, the data fall between the $1/M$ and the $1/M^2$ curves. After Y. Owechko [131], this

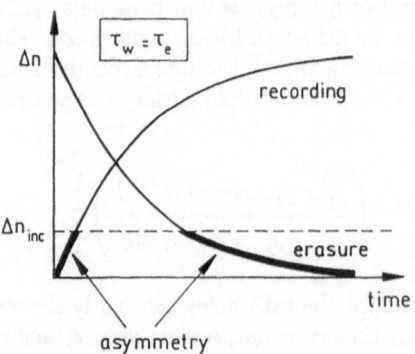

**Figure 4.39:** Asymmetry in writing and erasure time that appears when utilizing the incremental recording procedure in volume holographic grating recording.

may be caused by the spatially multiplexed nature of the fanning gratings and the associated spreading and filamentation of the reference beam. Light from a particular reference pixel is directed by the fanning grating to the signal gratings that connect it with object pixels. Reference and signal gratings that are not associated with this reference pixel may have less spatial overlap; therefore, each grating tends to be erased only by light beams that are connected by this grating. The gratings are therefore erased less than we would expect from the above analysis, which assumes complete overlap between all of the beams. This improved efficiency performance may be an additional positive feature of the fanned reference recording technique, making it well-suited for implementations with a high degree of interconnections, as multilayer perceptrons with backpropagation features. Therefore, I will discuss the realizations of backpropagation neural networks using cascaded photorefractive grating holography by beam fanning or self-pumped phase conjugation later on in more detail.

*Incremental Recording*

An alternative method of hologram recording was introduced in the early nineties by several authors, using an incremental step algorithm. Here, I will present the procedure of Y. Taketomi et al., which has already shown to have a good performance in photorefractive multiple image recording using different recording principles as angular or phase-coded multiplexing. In that approach, each of the $M$ holograms is recorded with a series of incremental exposures, each of it being short compared to the material's response time. Because for low index modulation (as they exist at the beginning of a writing process), the slope of the writing curve is much steeper than that of the erasing one (see fig. 4.39), some or at least the first of the holograms written with the first increments remain after all $N-1$ others are incremented. During recording, each image-reference pair is sequentially displayed repetitively cycling through all $M$ images. Thus, the incremental recording technique allocates a short recording time increment for each grating and repeats the recording cycle. The hologram gratings gradually increase in refractive index modulation (or diffraction efficiency) as each cycle is completed as shown in fig. 4.40. The recording process reaches a saturation, when the growth rate equals the erasure rate. This leads to the same equilibrium condition as in the case of sequential recording, giving the same estimation value for the number of holograms that can be achieved for a specific, fixed index modulation. Moreover, both approaches achieve exactly the same index modulation in exactly the same time [151].

**Figure 4.40:** The principle of incremental recording in volume holographic media (after [150]).

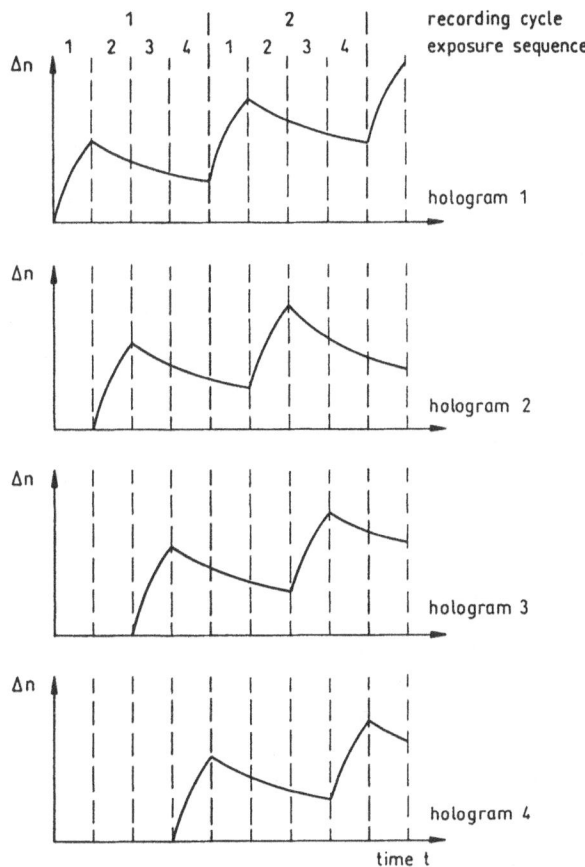

The advantage of that method is, that is does not depend on precise knowledge of the material characteristics as in the case of sequential recording, where the recording time values for each cycle have to be evaluated before recording the sequence of images. In particular, the material's response time and the maximum index modulation that can be obtained are not required to be known exactly in incremental recording. Moreover, incremental recording is advantageous when recording analog grey-scale images. Because the refractive index modulation for each grey-scale image increases gradually with repeated recording cycles, it is expected that the incremental recording will provide uniformly multiplexed gratings for the grey-scale images. That result was found experimentally by H. Sasaki and coworkers [152]. In contrast, scheduled recording fails to provide uniform multiplexing results independently of the grey scale of the writing beam intensity and its order in the exposure sequence. However, the difficulty in implementing the incremental recording approach is that the fringe pattern of the recording beams has to be reproduced exactly in every cycle. Even a difference of more than a fraction of a fringe will prevent the recorded increments from reinforcing each other, disrupting the recording process. In the same manner, the total recording time and the fluctuations of the diffraction efficiency during the recording cycle are affected by the grey scale. However, in real implementations,technical limitations arise due to the repeated shutter operations that are necessary in each cycle. They introduce temporal delays in the recording procedure, causing incre-

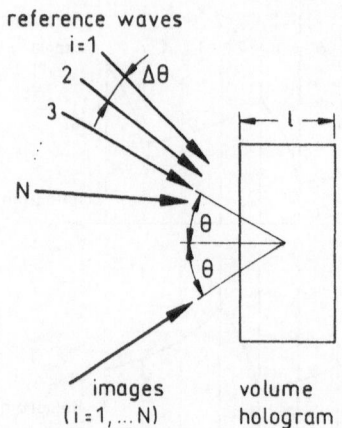

**Figure 4.41:** Principle configuration to realize angular multiplexing in volume photorefractive media.

mental recording to last longer in real implementations than sequential recording for the same conditions.

### 4.4.3 Angular Multiplexing

Angular multiplexing, which has already been described in this book in sections 3.4.3 and 3.4.4 in the context of multiple beam storage in photorefractive media, is shown again in fig. 4.41 for the case of image recording. The principle of that method is to change for every page of $N \times N$ pixels to be interconnected the orientation of the input-output $k$-vector or the written interference grating relative to the orientation of the crystal. This can be accomplished by rotating either the crystal or the incoming reference wave by an angle that is larger than the minimum angle defined by the Bragg condition .

Among the multiple possibilities to realize that situation technically, methods that change the reference angle mechanically (via piezo-driven mirrors or acousto-optic deflectors) or rotate the crystal in order to change the grating vector angle relative to the crystal orientation, require an exact positioning in order to write each grating exactly into the same interaction region into the holographic medium. For the case of utilizing acousto-optic devices, one has to take into account that frequency shifts appear during modulation, which need to be compensated in order to reproduce the same frequency in the object and the reference arm of the interconnection device. These compensation and adjustment techniques make the method slow compared to modulation techniques using display technology. Liquid crystal spatial light modulators can be introduced into the reference arm in order to define the different directions of reference beam incidence in combination with a focusing Fourier lens. Because in most cases, this is realized by enlarging the beam into a plane wave that transmits the modulator and by blocking most of the modulator fields in order to realize the different angular positions, this method is strongly energy consuming.

From a mathematical point of view, angular multiplexing can be derived from eq. (4.26) using the formalism described in section 4.4.1 when the factors $a_{klm}$ are chosen to be delta-functions.

Then, eq. (4.26) reduces for the case of equal wavelengths ($l = $ const.) to

$$\sum_{k=1} M \, \delta_{km}\delta_{kp} = \left\{ \begin{array}{ll} 0 & \text{for} \quad m \neq p \\ M & \text{for} \quad m = p \end{array} \right. \tag{4.32}$$

If angular multiplexing is used as the storage principle of choice, both methods of recording dynamics as described in the previous sections may be implemented, but bear both the possibility to fail or to give rise to noise. For the case of sequential recording, the angular multiplexing procedure must ensure the same coupling constants for each incident angle to preserve the same material response time. Assuming a storage capacity of about 10.000 images for angular multiplexing with a minimum difference angle of about 0.01 degrees, the angular range that is covered gives a remarkable dependence of the coupling constants on the angle (see sections 3.4 and 3.4) For incremental recording, in contrast, the crystal must be rotated to within a few milliradians of its position in the previous cycle for each incremental recording step — an accuracy difficult to achieve. In both cases, a fixed spatial amplitude modulation in the reference arm can provide a plane, at which angularly multiplexed reference beamlets can be generated in a definite way. This may help to overcome the problems mentioned above, but dramatically decreases the intensity available in the crystal volume and during the registration process.

Up to now, angular multiplexing is the most widely used multiplexing technique. Using electrooptic or acoustooptic modulators to change the reference beam angle, already in 1973, L. d Auria and J.P. Huignard realized a system that was capable of storing 100 images in a single location of a LiNbO$_3$ crystal [145, 153, 154, 155]. Although their approach was not pursued in the following years because of the lack of input and detection devices that match the capabilities of the holographic memories, their pioneer system showed already the potentials of volume holographic storage. The most impressing demonstration in these years was the recording of 100 pages of $10^6$ bits contents each, realized by storing binary data masks and telephone directory pages, done by H. Kurz [75].

At the beginning of the nineteers, a renaissance of volume holographic data storage enhanced the storage capacity dramatically. In 1991, F. Mok and coworkers realized the storage of 500 analog high-resolution pages [77], again realized in photorefractive LiNbO$_3$. These images were stored in the image plane, using mechanical mirror rotation to fix the different angular positions and a liquid crystal TV to introduce the data pages into the system. In 1993, F. Mok realized the storage of 5000 holograms using the same system, but an acousto-optic reference beam deflection device. The ultimate number of pages stored in a single interaction region was done in 1995 by D. Psaltis and G.W. Burr [80]. Having a page size of about 320 x 220 bits, and storing up to 10.000 holograms in a single interaction region, this gives a storage capacity of 88 Megabytes per single interaction region. By finally realizing up to 16 of these interaction regions in a single LiNbO$_3$ crystal, D. Psaltis and G.W. Burr realized up to 160.000 data pages or 1.4 Gbyte in their memory.

The storage of digital data pages in such a memory using angular multiplexing was also realized at the beginning of the nineteers. The most fascinating realization done by L. Hesselink and coworkers showed in 1994 [156, 157], that the storage of digital data pages having up to 320 x 220 bits may result in the storage of whole video films, allowing to record and display more than 300 data pages in a single region. By combination of several of these regions in a single crystal volume, about 1200 pages can be stored in the memory, allowing to store video sequences up to several minutes. In these digital realizations, attention has been paid to the need for supplementary means to reduce the bit-error-transfer-rate of data by error correction procedures. In fact, without these correction codes, bit-error-rates of about $10^{-4}$ to $10^{-7}$ can be achieved, which is evidently far away from the bit-error-rates of $10^{-12}$ - $10^{-14}$ as they are required and achieved by other digital data storage systems. Although the use of such error correction coding is often perceived to cost user capacity owing to the storage of redundant data for the purpose of correcting errors, the use of a suitable code can decrease the minimum signal-to-noise ratio required for reliable operation, thus allowing to increase the user capacity finally. Here, I do not want to go into the details of digital error correction. Those who are interested in the information theoretic concepts of error correction in volume holographic memories, might refer e.g. to [158, 159, 160, 161] and the citations therein for further reading.

These digital memories allow for digital neural network concepts. In contrast to analog realizations, that have to cope with nonuniformities of optical components, noise and a limited dynamic range, which tend to limit the scalability of these systems, they allow to threshold and restore binary logic levels, leading to robust system operation, superior scalability and better throughput. However, for that purpose, both, the interconnection weight images and the input data images have to be decomposed into the sequences of bit planes. Thus, fan-in and fan-out optical interconnection systems have been realized, combined with memory updating (error-backpropagation) by selective erasure [162].

Improvements of angular multiplexing concepts concerning the compactness and robustness of the systems for industrial applications have been realized in the last two years by several groups. At IBM, M.P. Bernal and coworkers realized a demonstrator system that is capable of realizing different multiplexing techniques and to investigate the bit transfer function of the system [78]. At Rockwell, I. McMichael and coworkers demonstrated in their compact system based on acoustooptic angular multiplexing the storage of a 30Kbyte file containing compressed color images, what is equivalent to the storage of 20.000 holograms in 20 locations in the memory crystal [79]. In order to optimize systems that suffer from crystal inhomogeneities as striations or scattering noise, A. Aharoni and coworkers showed that readout of an angularly multiplexed memory with the phase-conjugate of the reference beam, resulting in the reconstruction of the phase-conjugate of the stored page, eliminates cross talk and distortions [163]. An actual overview over the most promising realizations of volume holographic and photorefractive page-oriented memory realizations can be found in the May Issue of Applied Optics in 1996.

However, all these realizations still are dealing with the difficulties of angular multiplexing that arise due to the mechanical moving parts of the system or those components that introduce absorption losses of frequency shifts into the system. A way out of these difficulties was proposed in 1990 by the author and her coworkers: Instead of an amplitude modulation, why not taking advantage of the information processing capabilities that ly enclosed in the phase as a wave parameter? Therefore, a spatial phase modulation instead of an amplitude device was used to provide a phase-coded set of reference beams instead of amplitude modulated ones in the recording phase of multiple image storage. The principle of independent recall of the images relies on the effect of destructive interference of undesired contributions to the image in recall. Thus, fast switching and a high degree of reproducibility are ensured during that registration and recall process. The registration method was called *phase encoding* or *phase-coded multiplexing* and has become a promising and effective registration method since its first presentation at the Topical Meeting on Photorefractive Materials, Effects and Devices in 1991 in Boston / USA. Up to now, phase encoding has revealed to be able to store analog and digital data pages [146, 135, 164] and allows simultaneously for image summation, subtraction und inversion [135, 165, 166]. Moreover, phase encoding can be used for image encryption [167] and to realize pseudo-deep holograms [168].

### 4.4.4  Phase-Coded Multiplexing

Since a couple of years, the phase has been discovered as a means of impressing information on different storage media. Although direct phase addressing as it is operating in phase encoding had taken until 1990 to be elaborated, there have been several interesting storage ideas in the area of phase modulation for image and information storage before.

D.Z. Anderson showed in 1987 [111], that the phase information itself can be stored by superposition of reference and image waves having exactly the same phase modulation over the whole transversal plane. Their experimental realization of an optical interconnection device based on phase modulated beams has been discussed in section 4.3.1.

In order to obtain a better signal-to-noise ratio in angular multiplexing, J.E. Ford et al. [116] proposed in 1990 to impress a supplementary, homogeneous and binary phase modulation on a set of angularly multiplexed reference waves. The combination with wavelength multiplexing lead to the successful realization of optical matrix-matrix interconnections with reconfigurable interconnection types and has been shown in section 4.3.2 (see fig. 4.19).

The idea of direct and pure phase encoding goes back to 1977, when T.F. Krile et al. [169] multiplexed several holograms onto a single place of the recording medium by using a random phase-only diffuser mask in the reference arm. This mask was changed in position definitively for each exposure in the recording sequence, thus providing each of the references and thus each of the exposures with a unique phase pattern which is characteristic for the appropriate image. Consequently, any one of the holograms multiplexed can be reconstructed by illuminating the storage medium with the corresponding arbitrarily phase encoded reference beam. The aim of T.F. Krile and coworkers was to enhance the number of stored images in a planar holographic medium. Thus, the advantages of utilizing the Bragg selectivity in volume holographic storage devices has not been exploited in that case. This missing step, the implementation of an angularly multiplexed system whose cross talk is reduced by applying a supplementary random phase mask to the reference beam was done in 1993 by H. Lee and S.K. Jin [170]. In their method, interconnections are made by a pair of random patterns and plane waves that interfere in the material and generate the corresponding volume hologram. Interconnections are made by reading the volume hologram with the same random patterns used to write the hologram. To ensure independent interconnections, the random patterns have to be mutually uncorrelated.

The major difficulty of all these random-phase approaches for multiplex holography is the fact that a reference beam usually reconstructs not only the wave fronts from its corresponding hologram, but also scrambled wavefronts (most often noise) from one or more of the other holograms. This difficulty is usually obvious as *cross talk* during recall of a single image — spurious parts of another image pattern may appear and disturb the reconstruction of the desired image. Reconstruction is best and cross talk is reduced when each pairwise cross-correlation function of arbitrary image pairs is made as small as possible. For the case of random phase encoding, cross talk becomes significant when the number of stored holograms grows larger, because random phase values then will be probable to duplicate at least in a certain area of the transverse encoding plane, giving phase codes that are no longer mutually uncorrelated.

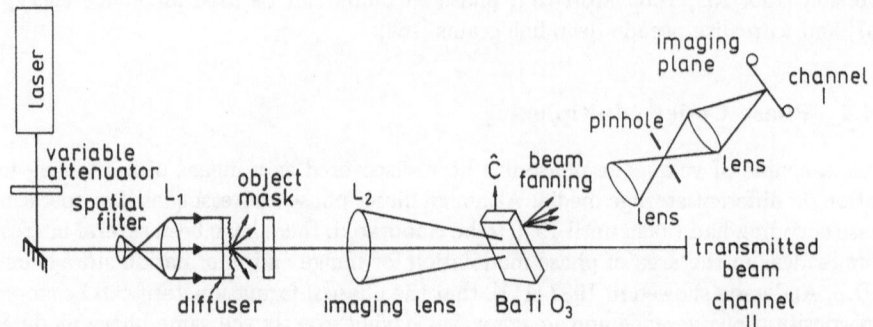

**Figure 4.42:** Experimental arrangement of a photorefractive storage system based on the random-phase speckle coding technique in a beam fanning geometry (after [171]).

A third random phase storage-retrieval system was introduced in 1996 by K. Kamra and coworkers [171]. Its multiplexed address is based on speckle random patterns as a carrier on which the page information is impressed together with the beam fanning concept. Each page is therefore recorded in the same volume, coded with a different random phase pattern by positional coding of a diffuser plate. For decodification, the volume holographic optical processor can be accessed by giving the correct code, that is the stored image registered in the photorefractive crystal is retrieved by placing the diffuser at the same position, as was employed for recording. Again, the system is very sensitive to the code — the position of the diffuser — as in the case of interconnections based on beam fanning or self-pumped phase conjugation (see section 4.3), resulting on the one hand in a significant reduction in cross talk, but on the other hand again in a strong sensitivity in positioning together with a new augmentation of cross talk when realizing a large number of stored pages.

With the same geometry, a dual-channel optical processor can be realized. Such a processor simultaneously displays the object as well as the contrast-reversed object respectively at the filtered beam fanning port and the transmitted signal. This effect finds applications in image algebra, and logical NOT operations. The most interesting point of that system, which is shown in fig. 4.42, is that instead of mitigating the degrading influence of speckles and beam fanning, the system advantageously employs these two, normally bothersome features themselves in order to realize a coherent optical memory system.

It was until 1990, when the author and her coworkers realized the first deterministic and binary phase encoding method for reference waves in volume holographic storage. A slightly different phase encoding storage method was proposed a short time later by Y. Taketomi and the group of S.H. Lee of the University of California, USA [172]. In our phase-coded multiplexing, each of the $M$ images to be registrated is stored with $N$ pure and deterministic phase-coded reference beams, generated by a phase-only spatial light modulator from a plane wave. The set of adjustable phases of these reference beams represents the address of the stored image and is orthogonal to all other phase addresses. "Orthogonal" in that case means that the other phase codes are linearly independent from the first one, if one writes down the different phase values in a vector representation. Phase vectors that are multiples of each other may not realize two independent phase codes. It is that orthogonality that causes complete extinction to the first order of any unwanted images because phase codes being orthogonal ensure destructive interference of all these contributions. This fact give also rise to the name *orthogonal phase encoding* of that multiplexing technique.

To take full use of the Bragg selectivity in thick recording media, the angular sparsing between adjacent reference beams is chosen to be large enough to satisfy the Bragg condition in all directions. The result is, that all reference beams are allowed to overlap with the image in the recording volume and interact independently with it.

As an advantage of that method, we obtain a light-efficient recording procedure without alignment problems during writing and readout cycles. Moreover, orthogonal phase-code addressing provides easy, reliable and immediate retrieval of the images. Especially, theoretically no cross talk and therefore noiseless reconstruction appears because the reconstructions of the undesired images interfere destructively at the output to produce zero intensity.

### Mathematics of Phase-Coded Multiplexing

 How can phase encoding be described with the formalism introduced in section 4.4? A scheme of the recording arrangement is shown in fig. 4.43. During the recording procedure, the image beam, whose complex amplitude

$$\mathcal{A}_m(\vec{r}) = A_m(\vec{r}) \cdot e^{i\vec{k}_m \vec{r}} + cc. \tag{4.33}$$

carries the information of the $m$th image in the amplitude $A_m(\vec{r})$, simultaneously interferes with all $M$ reference beams inside the photorefractive medium. Each of these references is encoded with a phase modulation carrying the complex spatial amplitude

$$\mathcal{R}_k(\vec{r}) = R_k \cdot e^{i\varphi_k^m} e^{i\vec{k}_k \vec{r}} + cc. \tag{4.34}$$

thus resulting in a phase set for a page to be stored of

$$\mathcal{R}_m(\vec{r}) = R_m \sum_{k=1}^{M} a_{km} e^{i\varphi_k^m} e^{i\vec{k}_k \vec{r}} + cc. \tag{4.35}$$

This equation is equivalent to eq. (4.19), now adapted to the case of phase encoding. Here, the set of adjustable phases $(\varphi_1^m, \varphi_2^m, \cdots, \varphi_M^m)$ represents the address of the $m$th image. In the readout-cycle, the holograms are recalled with these $N$ reference beams and the corresponding phase addresses $(\varphi_1^m, ..., \varphi_M^m)$. The change in the diffracted amplitude $\mathcal{R}_m = R_m \cdot e^{i\vec{k}_m \vec{r}} + cc.$ can be expressed via a first derivative equation with respect to the space coordinate $z$ along the $k$-vector using the coupled-wave analysis and the material equations. In the case of an ideal system, where the effective electrooptic coefficient $r_{\text{eff}}$, the space charge field $E_{\text{sc}}$ and the amplitude of the reference beam $n$, $P_n$ are independent of the special set of reference beams we consider and where the interaction among the reference beams can be neglected, the change in the diffracted amplitude with respect to the $z$-direction is given for reconstruction of image $p$ by:

$$\frac{\partial R}{\partial z} = i \cdot e^{i\psi} \cdot \frac{\pi}{\lambda} \cdot \sum_{m=1}^{M} \sum_{k=1}^{M} \Delta n_k P_k \frac{A_m P_k^*}{(I_m + I_R)} \cdot e^{i(\varphi_k^p - \varphi_k^m)} \tag{4.36}$$

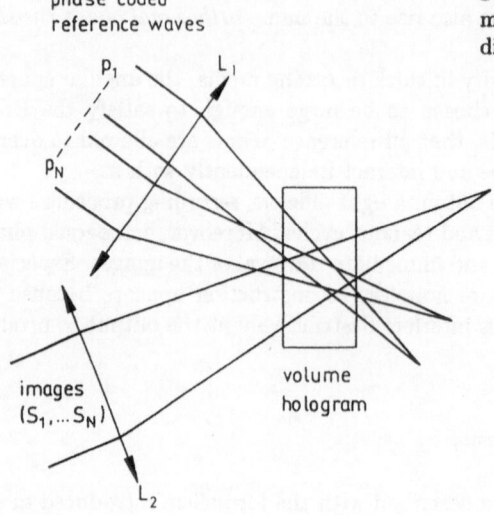

phase coded
reference waves

**Figure 4.43:** Schematic of phase-coded multiplexing in photorefractive volume media.

images
$(S_1, \ldots S_N)$

volume
hologram

where $\psi$ is the phase of the image beam, $I_m = A_m A_m^*$ is the intensity of the $m$th image beam, and $I_R = \sum_{k=1}^{M} P_k P_k^*$ is the total intensity of one set of $N$ reference pixels.

Taking advantage of the Bragg condition, the reconstruction wave $A^{\text{diff}}$ is then given as formulated in eq. (4.25)

$$A_p^{\text{diff}} = R_p \cdot g \sum_{m=1}^{M} A_m \left( 2\pi i \frac{\cos \alpha_s}{\lambda} \vec{e}_{ks} \vec{r} \right) \sum_{k=1}^{M} a_{km}^* a_{kp} \tag{4.37}$$

In order to retrieve the $p$th image ($R_p \equiv A_p$) without cross-talk noise, we have to address the system with phase codes fulfilling the condition:

$$\sum_{k=1}^{M} a_{km}^* a_{kp} = \sum_{k=1}^{M} e^{i(\varphi_K^p - \varphi_K^m)} = \begin{cases} 0 & \text{for } p \neq m \\ M & \text{for } p = m \end{cases} \tag{4.38}$$

Thus, the general formalism of multiplexing can be adapted to phase encoding by setting $| a_{ij} | = 1$. This equation can therefore be solved using matrix algebra methods, because the problem is formally equivalent to the case of finding matrices $X_{ij}$ that fulfil the condition:

$$(X_{ij}) \cdot (X_{ji})^* = I \tag{4.39}$$

where $I$ is the unity matrix. The solution of that equation is an irreducible basic matrix, where all column vectors of $X_{ij}$ are orthogonal. This was exactly the reason why this algorithm was called *orthogonal phase encoding*.

Consequently, we have to choose orthogonal phase codes to retrieve an image without cross talk. This orthogonality can be reached with a variety of phase codes, being fractions of $\pi$. However, from a technical point of view, binary phase codes using the phase values 0 and $\pi$ instead of phase codes that are based on multiples of e.g. $\pi/2$ or $\pi/3$ are easier to realize and to adjust with phase-only modulators that are currently available. The construction of orthogonal phase codes may be obvious for a small number of phase codes, but when $N$ grows larger, the Walsh-Hadamard transform (see e.g. [173]) has

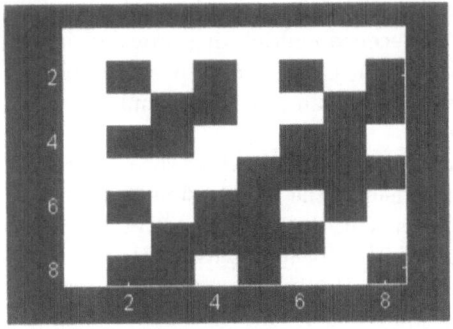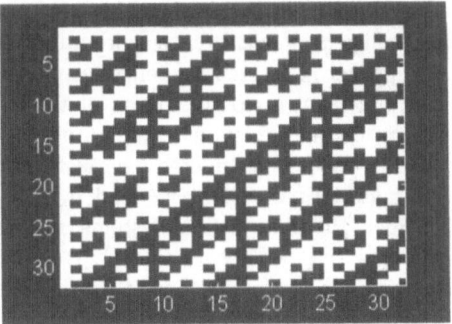

**Figure 4.44:** Examples of Walsh-Hadamard codes that can be used for binary phase encoding. $N = 8$ (left), $N = 32$ (right).

to be used to construct the orthogonal set of phase addresses properly. In general, the construction rule for a binary phase encoding matrix $H_N$ from the first Walsh-Hadamard matrix

$$H_1 = \begin{pmatrix} 1 & 1 \\ 1 & -1 \end{pmatrix} \tag{4.40}$$

is given by

$$H_N = \begin{pmatrix} H_{N-1} & H_{N-1} \\ H_{N-1} & -H_{N-1} \end{pmatrix} \tag{4.41}$$

In that case, the binary phase matrix to realize the storage of $N = 8$ independent images as an example is given by

$$H_8 = \begin{pmatrix} 1 & 1 & 1 & 1 & 1 & 1 & 1 & 1 \\ 1 & -1 & 1 & -1 & 1 & -1 & 1 & -1 \\ 1 & 1 & -1 & -1 & 1 & 1 & -1 & -1 \\ 1 & -1 & -1 & 1 & 1 & -1 & -1 & 1 \\ 1 & 1 & 1 & 1 & -1 & -1 & -1 & -1 \\ 1 & -1 & 1 & -1 & -1 & 1 & -1 & 1 \\ 1 & 1 & -1 & -1 & -1 & -1 & 1 & 1 \\ 1 & -1 & -1 & 1 & -1 & 1 & 1 & -1 \end{pmatrix} \tag{4.42}$$

In fig. 4.44, the binary Hadamard codes for $N = 8$, and $N = 32$ are shown.

Using this reconstruction rule,$N$ has at a first glance to be a power of 2, seeming to restrict the number of phase codes to that rule. However, in order to accomodate the phase codes to the available number of phase fields or pixels in a phase-only spatial light modulator and to achieve the maximum possible storage capacity of a phase-code multiplexed holographic memory, it is possible to generate phase codes whose length (or order) is not a power of two. For that purpose, various methods of combining different types of Hadamard matrices are possible, allowing thus to construct matrices up to the order $N = 4m$, with $m$ being an odd integer. A complete list of these codes for all orders up to $N = 4m < 400$ is given in [174]. Because there is no upper limit of $N$ from an algebraical point of view, limitations are only imposed by the resolution, the Bragg selectivity and the dynamic index modulation range of the photorefractive material as discussed in the previous sections. Thus, the limitations of our method are the same which appear using the usual angular multiplexing technique. However, there are several advantages in implementing phase-only addressing. A pure phase address is naturally light-efficient. What is even more important, phase-coded multiplexing provides simpler and quicker generated reference beam patterns. If some cross-talk noise is tolerable, we can implement a closely spaced phase addressing which can provide a simple reference generation mechanism with high storage capacity.

Up to now, we considered a perfect system (eqs. (4.36), (4.38)), where the photorefractive material parameters $r_{eff}$ and $E_{sc}$ are independent of the set of reference beams we consider. However, in real systems, the nonuniformity in the photorefractive diffraction efficiency ($r_{eff,i} \cdot E_{sc,i} \neq r_{eff,j} \cdot E_{sc,j}$) and in the reference beam intensity ($|P_i|^2 \neq |P_j|^2$) as well as the imperfections of the phase modulator ($\varphi_i^m = \varphi_{0,i}^m \pm \epsilon$) may give rise to a certain amount of cross-correlation noise [175]. Moreover, for a high number of closely spaced phase addresses, the Bragg degeneracy perpendicular to the crystal axis and the propagation direction may contribute to an increased noise level, thus indicating that an arrangement of the phase codes that is asymmetric relative to the direction of Bragg degeneracy is helpful, e.g. by aligning the different phase codes in one or only a couple of adequately separated lines.

*Image Storage Using Phase-Coded Multiplexing*

Experimentally, our group has achieved in 1993 a storage capacity of 64 analog images with that method using a nematic liquid crystal phase-only modulator to create the phase references [135] and a computer-generated Fourier hologram to separate them.

In order to realize phase-only modulation, the LC-molecules in the phase-modulating element have a parallel planar orientation (refer to section 4.1.1). The application of an AC voltage in the range of 2 - 5 V to the cell leads to a tilt of the rod-shaped molecules. Due to the optical anisotropy of the liquid crystal molecules, the effective refractive index can be controlled in this way without changing the polarization and thus the amplitude transmittance of the device. The structure of the ITO electrodes of the LC display has been designed for the experimental conditions in order to realize 64 independent phase codes in a relatively small modulation area. Therefore, the active areas are aligned in two lines of 32 phase fields. Each phase field can be actively addressed by an AC voltage, thus avoiding instabilities of the phase modulation due to fluctuations which occur in the case of electronic multiplexing. With the geometrical arrangement of the reference beams in two lines, a simple configuration of fractal multiplexing (see section 4.3.4) has been realized. The separation of the two lines is realized in such a way that cross talk reconstructions do not overlap with the image reconstruction and thus can be neglected.

The experimemtal setup is sketched in fig. 4.45. We used an argon ion laser at an output wavelength of 514 nm. The reference beam system is produced by a computer generated focus array hologram, which creates the 2 x 32 spot pattern in its Fourier plane, where the phase modulator is located. By the help of several additional lenses, the reference beams overlap with the signal beam in the crystal volume. In order to achieve dynamic characteristics of the crystal that are approximately equal for all beam components of the hologram, a narrow spatial frequency spectrum is realized by a long focal length of the focusing lens — a necessary condition for a phase-coded system when it is realized as in our case using a photorefractive storage medium that displays energy coupling too, as $BaTiO_3$. Moreover, the crystal was oriented in such a way that energy coupling did not contribute to noise in the image reconstruction plane. Diapositives and amplitude modulator images of letters, images and spot patterns were used as images.

**Figure 4.45:** Experimental realization of phase-coded multiplexing using a phase-only liquid crystal device with a capacity of 64 images. BS = beam splitter, M = mirror, L = lenses, ASLM = amplitude spatial light modulator, PSLM = phase modulator, CGH = computer generated focus array hologram, CCD = camera.

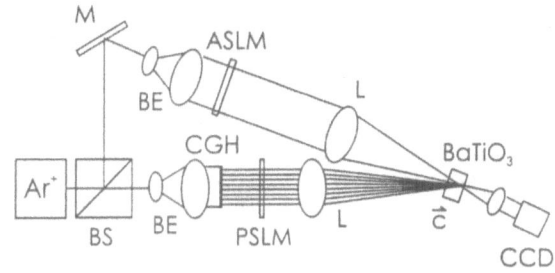

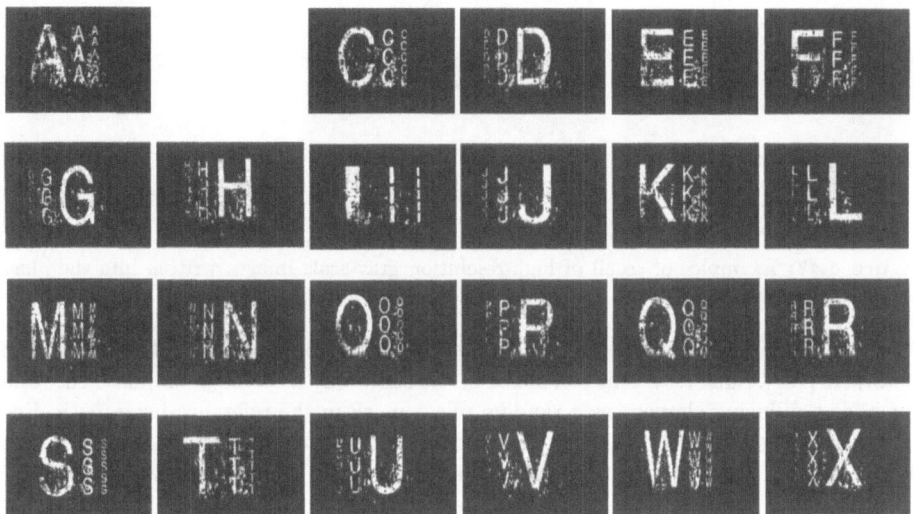

**Figure 4.46:** Examples of recall of 64 images in the phase-coded memory described in fig. 4.45.

64 images were written subsequently using the sequential recording technique from 5 minutes for the first to 2 seconds for the last image. The intensity of the signal beam and the reference system was approximately $10\text{mW/cm}^2$. Fig. 4.46 shows the experimental configuration and a selection of the 64 images, recalled using the corresponding phase address. Cross talk is not visible in the reconstructions. Fig. 4.47 shows several examples of high-resolution gray-scale images, that have been written by a spatial amplitude modulator into a $LiNbO_3$ memory crystal.

Digital data storage is also possible with about 3000 data points per page, showing that the main limitation for digital data storage is not given by the method itself, but by the quality of the storage crystal used. Some improvement in crystal quality is still necessary to achieve a sufficiently low error rate required by digital data storage. Similar encouraging results, showing that phase encoding is capable of storing a large number of data pages, have meanwhile been obtained by several other groups [164, 176]. Actually, investigations are in progress to enhance that capacity up to the number that is realized using angular multiplexing [177, 178].

The combination of orthogonal phase-coded multiplexing with a supplementary ran-

**Figure 4.47:** Examples of recall of high-resolution gray-scale images written into the phase-coded memory described in fig. 4.45 by an amplitude modulator.

dom phase key has been realized as an extension of phase encoding by J.F. Heanue and coworkers [167]. Such a system is realized in the same way as orthogonal phase encoding, but adds a diffusing element as a random phase mask in the reference-beam path. Each hologram is then recorded with its phase code and an additional random phase mask that is realized by a certain position of the phase mask. Thus, the low cross-talk performance of orthogonal phase encoding is retained, but data are stored in a secure way, because a given hologram cannot be read out without knowledge of the corresponding mask. However, again as for random phase multiplexing, cross talk becomes important as the number of holograms to be stored rises.

Another extension of phase-code multiplexing is based on the theory of superposition of two mutually co-orthonormal functions. In a more optical language, those superimposed holograms are recorded by the pairs of object and reference waves, whose complex amplitudes are proportional to two arbitrary complete sets of discrete, orthonormal functions as they are represented e.g. by the unity matrix or the Walsh-Hadamard matrices — one for the object waves and another for the reference waves. Such a hologram will contain only the cross-modulation gratings produced by the interference of the reference waves with the object waves, and not the intermodulation gratings produced by the interference of reference or object waves with themselves. These intermodulation gratings are eliminated by integration of their contributions of all the superimposed holograms, because the sum of the amplitudes of these gratings from different superimposed holograms is equal to zero [179]. Although these gratings may only be recorded in ideal media with linear response to the dielectric permeability and independent hologram writing, these volume holograms of coorthogonal waves might have important applications.

What has this theoretical proposition in common with phase-coded multiplexing? Because binary phase-coded references based on the Walsh-Hadamard matrices represent a set of orthonormal functions, they may be used to realize these volume holograms together with another set of orthonormal functions, that should play the role of the object wave set. Such a second orthonormal set can be found using Kronecker $\delta$-functions. In other words, the object waves have to be object point sources, represented by the unity matrix. If these two matrices or the phase-coded references together with the object point sourcres are recorded into the crystal volume, switching of optical channels can be easily realized. Moreover, the gratings can be used to expand each incoming light distribution into a set of Walsh functions, thus allowing to optically realize an expansion similar to the Fourier transform that can be done by far-field diffraction by a normal biconvex optical lens. Although the experimental confirmation of these algorithms is still in a premature state, optical switching and expansions of object points into Walsh functions have been shown by V.V. Orlov and A.R. Bulygin [179], demonstrating the potential of their application of binary Walsh phase-codes to optical interconnections.

In a similar way, phase-sensitive holographic gratings can be recorded with different phases with angularly multiplexed reference waves. When the relative phases of the signal gratings to be stored give an orthogonal set of phases, again orthogonal photorefractive grating storage is possible, this time achieved between different orthogonal signals and multiplexed reference beams. During readout, this configuration allows to realize constructive ord destructive interference between different signals, allowing to produce maximum and near-zero diffraction, respectively [180].

Although phase encoding — which can also be considered as a modification of angular multiplexing which is based on the same principles — seems to be the most advantageous method of storage and interconnection in volume holographic storage applications, a third promising method deserves to be mentioned in that context: The idea of changing the wavelength as an encoding parameter of volume holographic storage. In that case, successive holograms may be recorded and reconstructed using different wavelengths of the writing and readout beam. Consequently, this method is known as *wavelength encoding* or *wavelength multiplexing* in literature.

### 4.4.5  Wavelength Multiplexing

Wavelength multiplexing uses the second parameter in the Bragg condition (see section 4.3.4) to be modified in order to realize independent gratings in a volume storage material. This can be done by changing the wavelength of every grating to be stored, thus changing the wavelengths of the object and reference beam together for each writing pair. Another possibility is to use different wavelength for writing and reading, thus allowing nondestructive interconnection at the expense of adapting the angular configuration in order to match the Bragg condition. In the following, I will present several interesting implementations for different materials (photorefractives and hole burning materials) and applications (interconnections as well as page-oriented memories).

In 1993, T. Weverka, K. Wagner and M. Saffman presented a compact method of writing holographic gratings in spectral hole burning materials at separate wavelengths [83]. In that method, the gratings are completely independent if they are separated by a sufficient wavelength interval. The ingenious idea of their compact method is to implement a combination of both, angular and wavelength multiplexing. While a first line of interconnections is written with angularly multiplexed waves, supplementary lines are written with different wavelengths. While the Bragg selectivity allows distinct pixels in the same row to couple through separate gratings, there is Bragg degeneracy along the column direction if the spacing is chosen to be too small and allows that cross-talk

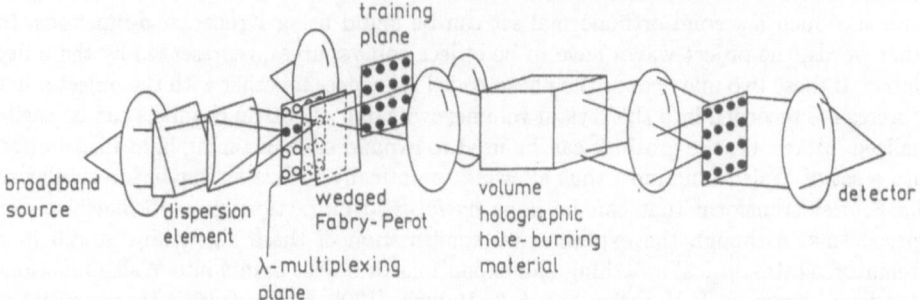

**Figure 4.48:** Wavelength-multiplexed volume holographic interconnect for storage in a volume holographic hole-burning material (after [83]).

contributions interfere with the image reconstruction (compare to section 4.3). This can be eliminated with the method of fractal space multiplexing (see section 4.1) or by using a distinct wavelength for each row in the direction of Bragg degeneracy.

Thus, $N$ wavelengths can provide an additional factor of $N$ interconnections in order to come nearer to the case of full interconnection between $N^2$ inputs and $N^2$ outputs with $N^4$ degrees of freedom. The spectrally specified gratings necessary to realize this can be written using persistent spectral hole buring (see section 4.1). By cooling a suitable inhomogeneously broadened medium to cryogenic temperatures, the homogeneous linewidth becomes narrow. This enables a large number of independent holograms to be wavelength-multiplexed (see section 4.1 for further details on spectral hole burning).

The multiplex factor is limited by the ratio of the inhomogeneous to homogeneous linewidth of the relevant electronic transition. Up to now, materials have been developed which have the potential to store up to 10.000 wavelengths-multiplexed holograms [181]. However, holographic detection of the spectral hole implies that both, absorption and refractive index changes contribute to the grating strength, which results in a broadened effective linewidth. This will reduce the number of gratings that can be independently stored to the order of 1000, which is in the same range as the limitations that are present in angular and phase-coded multiplexing. Fig. 4.48 shows the schematic arrangement of an experimental setup using this type of combined wavelength- and angular multiplexing to realize a volume holographic interconnection system.

The second important class of implementations of wavelength multiplexing is based on recording holograms at different wavelengths in photorefractive volume media. Again, as for the case of hole burning materials, wavelength multiplexing comes first into ones mind as a possibility to enhance the capacity of other multiplexing techniques by adapting it for a series of different wavelengths. S. Campbell and coworkers [183, 182] realized in this way a memory that uses the conventional setup for 90°- angular multiplexing, combined with the possibility to use up to five wavelengths. By enlarging the angular separation in such a way, that the wavelength selectivity can be included into the Bragg selectivity of the material, $5 \times 400$ images could be stored in photorefractive LiNbO$_3$. The cross-talk-free, very good quality images during recall are presented in fig. 4.49.

In 1989, R. McRuer and coworkers presented an interconnection system that interconnects a set of input signal sources to a set of output signal detectors using multiplexed volume gratings written in photorefractive BGO [184]. Transmission gratings that are created when both writing beams are incident on the registration medium from the same

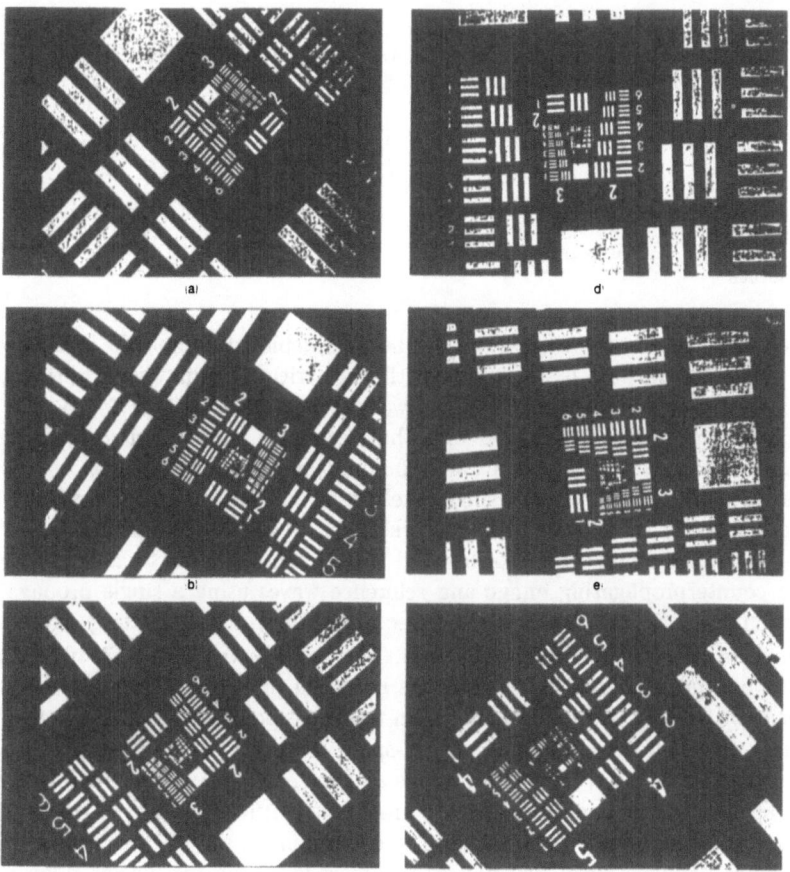

**Figure 4.49:** Experimental results from the storage of 2.000 pages (5 wavelengths times 400 angles) using sparse-wavelength angular multiplexed storage in photorefractive $LiNbO_3$. (a) $\lambda_1 = 457.9$ nm, $\theta_1 = 43.5°$, (b) $\lambda_2 = 476.5$ nm, $\theta_{100} = 44.25°$, (c) $\lambda_3 = 488.0$ nm, $\theta_{200} = 45.0°$, (d) $\lambda_4 = 501.7$ nm, $\theta_{300} = 45.75°$, (e) $\lambda_5 = 514.5$ nm, $\theta_{400} = 46.5°$. (f) shows a magnification of (c), giving the resolution of about 138 pixels per mm (group six, element 1) (from [182]).

side, are written or erased using planar wave fronts of wavelength $\lambda_\omega$ to which the crystal is sensitive. To provide nondestructive readout of the interconnection gratings, the signal beams have a longer wavelength $\lambda_s$. Therefore, the wavelength of the system does not have to be tunable. Moreover, the two-wavelength dynamic optical interconnect forms an optical crossbar in which the order of complexity is reduced from $N^2$ to $N$ when compared to that of a general two-wavelength $N \times N$ holographic switch. Complexity is defined here as the number of writing beams required to generate the $N^2$ interconnection nodes.

In section 4.3, I have shown that in general, to diffract each of $N$ signal inputs to $M$ signal outputs, the number of gratings needed equals the product $N \cdot M$. The generation of this number of independent gratings requires $2N \cdot M$ writing beams if the recording geometry demands two distinct beams per grating. However, by restricting the writing beams to propagate on the surface of a cone, the resulting set of gratings can be strictly Bragg matched by longer-wavelength signal beams propagating on a similar conical sur-

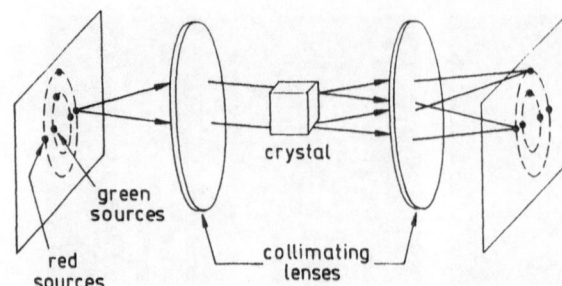

**Figure 4.50:** Two-wavelength conical volume interconnect system based on multiplexed volume gratings stored in a photoefractive crystal (after [184]).

face with a different cone angle. This situation is depicted in fig. 4.50. Here, each signal beam (input or output) is associated with one writing beam, so the total number of writing beams needed for an $N \times M$ network is reduced from $2N \cdot M$ to only $N + M$. R. McRuer and coworkers realized with that technique a $2 \times 3$ switch by using a BGO crystal with 514 nm writing beams and 633 nm signal (readout) beams.

The most attractive system using wavelength multiplexing up to now has been introduced recently by A. Yariv and coworkers. It is based on recording holograms not in the traditional configuration of forward two-wave interaction (transmission grating), but on counterpropagating image and reference waves using a single propagation axis (reflection grating). In that case, the $k$-vectors of the recorded gratings are distributed in a uniform way throughout the $\vec{K}$-space. This fact allows a reduction in the cross talk which is normally inherent to all methods based on the Bragg selectivity if the spacing between two grating points is smaller than the corresponding Bragg angle. One method to obtain this uniform distribution in $\vec{K}$-space is shown in fig. 4.51. Here, holograms are recorded by counterpropagating image and reference waves. All the reference waves propagate along the same axis, which we may label with $z$, but each of it with a slightly different wavelength. The $\vec{K}$-surface of a single picture is shown in fig. 4.51a. Fig. 4.51b shows what happens when two images are recorded using two beams differing in frequency by $\Delta k = \Delta\omega/2\pi$. The increase in the information content of a given image causes a mostly lateral spread of the places, allowing for a more uniform coverage of $\vec{K}$-space. The advantage of that configuration is, that the mismatch parameter $\Delta k$ that determines the relative intensity of the cross-talk noise, is nearly constant in magnitude so that the level of cross talk is essentially independent of the information content (or spatial detail or resolution) of the images [185]. This indicates the potential for high data storage capacity of that storage scheme. Because the curves in fig. 4.51 remind of an orthogonal configuration, the scheme is also called orthogonal data storage by some authors [185].

An experimental realization of that idea has been presented in 1993 by S. Yin and coworkers [186]. In their experimental setup using photorefractive $LiNbO_3$ as a storage medium, they implemented a tunable laser diode at $\lambda = 670$ nm with a tuning range of about 12 nm. In the 90° reflection geometry (c-axis in beam direction), they were able to realize wavelength steps of $\Delta\lambda = 0.1$nm, resulting in the storage of 26 images in that memory (see fig. 4.52).

However, even this promising application shows that there is a severe technical limit in wavelength multiplexing due to the limited wavelength tuning range of the laser, that prevents to store as many holograms in one location as in angular or phase-coded multiplexing. Imagine e.g. the attempt to store 5.000 holograms in one location, as has been realized in angular multiplexing. Even with a wavelength separation of $\Delta\lambda = 0.01$nm,

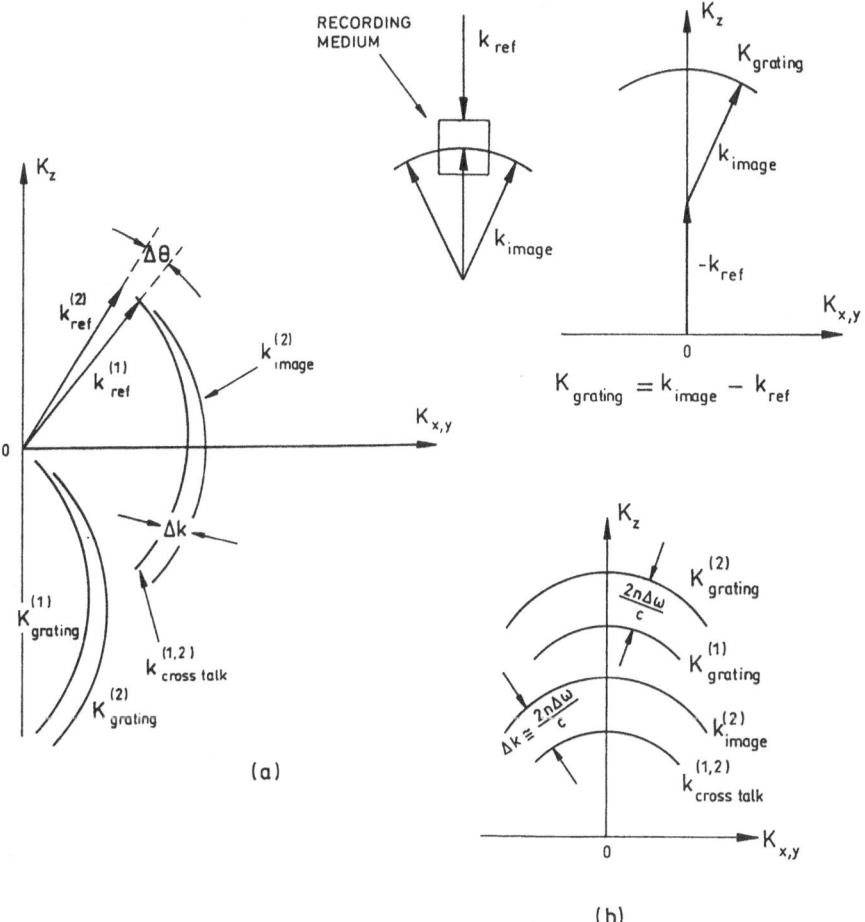

**Figure 4.51:** $\vec{K}$-space diagrams for volume holography that uses angular multiplexing in the transmission geometry (a) and orthogonal wavelength multiplexed data storage in the reflection grating configuration: single hologram (above), multiple hologram storage (below), (after [185]).

which is still a factor of 10 below the current experimental realizations, we would need a wavelength tuning range of 50 nm, combined with the ability of the material to exhibit the same sensitivity over that wavelength range — data that are actually difficult to achieve.

Consequently, techniques are needed to increase the potential capacity of wavelength multiplexing by combinations with other techniques. In *coded-wavelength multiplexing*, introduced by M.C. Bashaw et al. [187], a discrete angular sprectrum of plane waves is used for this purpose at each frequency, and assigns a complex amplitude to each constituent plane wave. Moreover, the amplitude codes are chosen so that they are mutually orthonormal, as is the case for phase encoding or angular multiplexing at one wavelength. Coded-wavelength multiplexing utilizes the properties of Bragg mismatch in a counterpropagating geometry to increase the number of pages. In an ideal Fourier holographic arrangement, points in the image plane correspond to plane waves in the

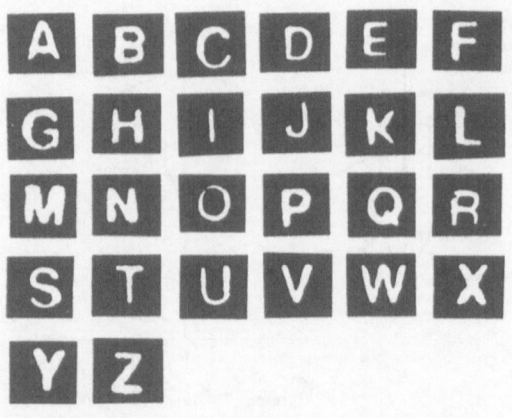

**Figure 4.52:** Reconstructed holographic images using wavelength multiplexing in the reflection grating geometry and a specially doped Ce:Fe:LiNbO$_3$ crystal (from [186]).

holographic medium. In the counterpropagating arrangement, any small change in angle of the reference beam results in a close match to the Bragg condition (fig. 4.53a). Cross talk generated by these degenerate reconstructions remains Bragg-matched, is therefore not attenuated and propagates at angles different from the desired reconstruction, corresponding to different spatial locations in the image plane (fig. 4.53b). If the signal domain is bounded by limiting the observation plane to a given region of the image plane, cross talk between two images can be avoided by writing the first holographic page at one reference beam angle and choosing the second reference angle so that reconstruction of the first image does not overlap the second image. This principle has already been used for fractal space multiplexing (see section 4.3). For different wavelengths, in contrast, the wave vectors of the signal and the reference waves change without incurring Bragg degeneracy, and Bragg mismatch significantly attenuates cross talk in the ordinary manner. Fig. 4.53c shows that appropriately selected signal and reference wave vectors result in a unique grating for each pair of wave vectors.

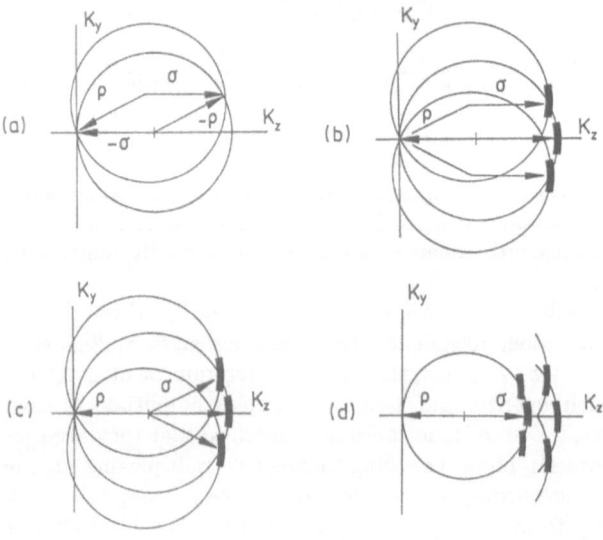

**Figure 4.53:** 2D reciprocal space diagrams showing $K = \sigma - \rho$. Each circle shows the location of all grating vectors that may be written by a particular reference wave vector $\rho$. $\sigma$ is the signal wave vector. (a) Wave vector pairs $(\sigma, \rho)$ are degenerate to wave vector pairs $(-\sigma, -\rho)$. (b) Storing of multiple signals (grating spectra do not overlap). (c) Reconstruction of degenerate signals: a reference wave vector is almost Bragg-matched to a number of signal spectra. (d) Angular-encoded wavelength multiplexing: grating vector spectra (shaded strips) are unique for each stored grating (after [187]).

**Figure 4.54:** Reconstruction of a test target that is stored in different orientations at four angle-wavelength combinations in order to realize encoded wavelength multiplexing. (a) $(\lambda_1, \theta_1)$, (b) $(\lambda_1, \theta_2)$, (c) $(\lambda_2, \theta_1)$, (d) $(\lambda_2, \theta_2)$ (from [187]).

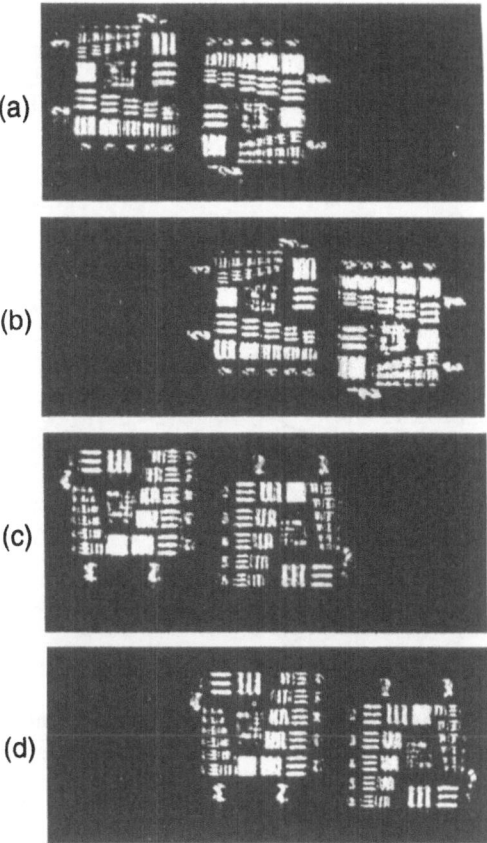

Coded wavelength multiplexing generalizes this procedure by assigning orthonormal codes to constituent plane waves determined by reference wave vector pairs chosen to avoid degeneracy. The increase in the total number of pages is proportional to the number of codes at each wavelength.

M.C. Bashaw and coworkers realized a storage configuration using a tunable dye laser and photorefractive $LiNbO_3$ as the storage device, that was able to store four images with plane reference waves drawn from a domain of two angles $(\theta_1, \theta_2)$ and two wavelengths $(\lambda_1, \lambda_2)$. The field of view of interest is the central portion of each image in the reconstruction (see fig. 4.54), which could be isolated by limiting the detector to the desired portion of the image plane. An image stored at $(\theta_1, \lambda_1)$ is reconstructed with the same reference wave in the central portion of fig. 4.54a. An image stored at $(\theta_2, \lambda_1)$ generates significant optical cross talk, which has the same strength as the reconstruction of the first image. The optical cross talk appears to the left of the image of interest, and therefore does not contribute to system cross talk. Figs. 4.54b-c show reconstructions at other angle-wavelength combinations, the cross talk appearing in each case to one side of the window of interest.

In a slight variation of that principle, it is possible to generate a memory that stores different pages at one wavelength and each of it at different angles. Fig. 4.55 shows, that the reconstruction of four images with counterpropagating wave vectors — note that this corresponds to angular multiplexing in the reflection grating configuration — generates a

**Figure 4.55:** Reconstruction of a test target that is stored at four different angles with the same wavelength in reflection geometry (from [187]).

mosaic composed of each image in a different location. This allows to build up a page as a mosaic of different smaller pages, giving this technique also the name *mosaic multiplexing*. In another variation of that principle, a digital wavelength-multiplexed holographic data storage system was built, able to store 60 Kbyte digital image file in a single hologram stack in a lithium niobate crystal. A raw bit-error rate of $10^{-4}$ was achieved [188].

### 4.4.6 Electric Field Multiplexing

Finally, *electric field multiplexing* in noncentrosymmetric media is possible by tuning the material parameters of the recording medium such as refractive index or lattice parameters while keeping the external parameters (wavelength and angles) fixed. Experimentally, this can be done by applying an external dc electric field that alters the index of refraction through the electrooptic effect, effectively changing the recording and reconstruction wavelength in the storage medium. Thus, electric-field multiplexing is closely related to wavelength multiplexing. A. Kewitsch [189] et al. demonstrated this concept in a preliminary experiment by electrically multiplexing two volume holograms in a strontium barium niobate crystal (SBN).

A mechanism similar to electric field multiplexing can be realized when using the voltage-controlled photorefractive effect in paraelectric crystals, called *electroholography*. Up to now, we discussed the photorefractive effect as a means to enable the recording of optical information in a crystal by spatially modulating its index of refraction in response to the light energy that it absorbs. Charge carriers are photoionized from their traps, transported, and eventually retrapped, forming a space charge field spatially correlated with the exciting illumination. This space-charge field induces a modulation in the index of refraction through the electrooptic effect. However, in the paraelectric phase, one can control the efficiency of the effect by applying an external electric field to the crystal

**Figure 4.56:** Schematic diagrams of one neuron with its holographic synaptic connections (a) and of a two-dimensional array of electroholographic neurons (b) (after [190]).

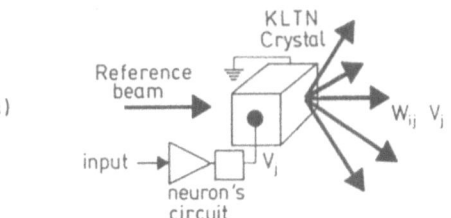

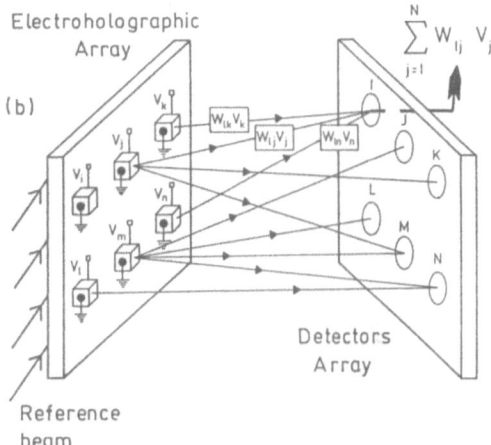

during the reconstructing stage. In general, the diffraction efficiency is proportional to the local photoinduced changes in the index of refraction ($\delta(\Delta n)$). In the paraelectric phase, the electrooptically induced modulation of the index of refraction depends quadratically on the polarization, as $\Delta n(x) \propto P^2$, where $P$ is the low-frequency polarization. When a space-charge field $E_{sc}(x)$ is formed in the crystal, the modulation that it induces in the index of refraction and that contributes constructively to the diffraction is given by [190]

$$\delta(\Delta n) = n_0^3 g \epsilon^2 E_0 E_{sc}(x), \qquad (4.43)$$

where $n_0$ is the refractive index and $g$ is the effective quadratic electrooptic coefficient. In eq. (4.43), it is assumed that the working temperature is approximately 5 - 15°C above the phase transition. In this temperature range, the dielectric constant $\epsilon$ is very large, and the electric field $E$ is given by $E = E_{sc}(x) + E_0$, where $E_0$, the applied electric field, is below the saturation level, so that $P = \epsilon E$. At this temperature range, the switching frequency is limited to approximately 1 MHz by the critical slowing down of the dielectric response in the vicinity of the phase transition. Thus, it can be seen that the information-carrying space-charge field contributes to the diffraction efficiency only in the presence of an external electric field. An attractive crystal for applications that exploit the voltage-controlled photorefractive effect is potassium tantalate niobate (KLTN) [191], because it requires only moderate fields of 0 - 2.5 kV/cm for controlling the diffraction efficiency between 2% and 75%, while having at the same time a phase transition at 15°, that allows to have the working point at room temperature.

The essence of electroholography is to use the capability provided by the voltage-controlled photorefractive effect for controlling light beams by electronic circuits. Thus, in a neural network based on electroholography, the electronic circuit emulates the body of the neuron as shown in fig. 4.56a. In this way, an electroholographic array can be used

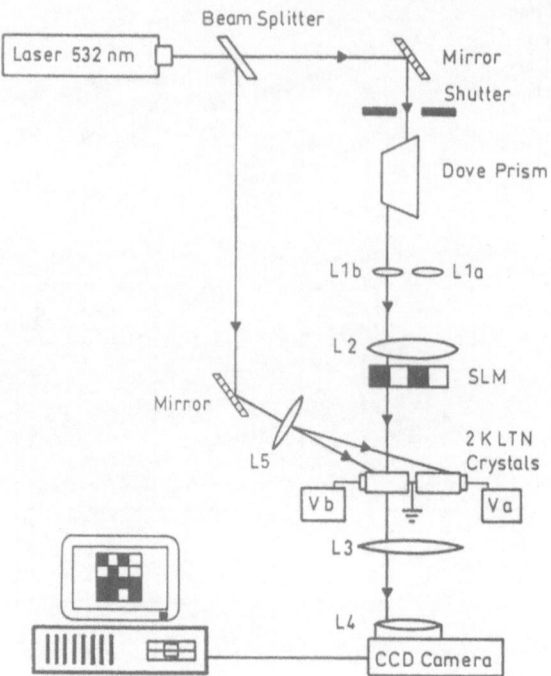

**Figure 4.57:** Experimental setup for recording holograms on two small crystals based on electroholographic storage (after [190]).

to realize matrix-matrix interconnection schemes (see fig. 4.56b), which needs only one reference beam. This permits to realize parallel reconstruction. A.J. Agranat and coworkers tested their electroholography concept on a system containing two small electroholographic crystals [190]. With the experimental setup shown in fig. 4.57, two holograms (the letters H und U) were written into the system. During reconstruction, the reference beam is incident upon both crystal "pixels", and the stored holograms are reconstructed according to the voltages applied during recording. A striking phenomenon of this sytem is, that by changing the voltage during readout, the ratio of reconstruction of both holograms can be varied, giving the potential to superimpose in an analog way differently weighted contributions of both holograms.

## 4.4.7 Combined Multiplexing Techniques

All three Bragg-selective principles discussed in that chapter in order to enhance the number of interconnections which might be stored simultaneously and independently in a volume hologram location have about the same storage capacity. Thus, combinations of these methods may also be realized without changing the storage capacity. One solution, which was already proposed in a special application by K. Weverka et al. in the precedent section (see section 4.4.5), is to combine angular and wavelength multiplexing. The other possibility is the inverse way of that method, which means encoding wavelength-multiplexed holograms with a discrete sprectrum of plane waves. This technique, called *coded-wavelength multiplexing*, was also presented at the end of the last section. A comparable solution concerning phase-coded multiplexing, which has been suggested by the

**Figure 4.58:** Purely spatially multiplexed holograms (a) and spatioangu-
larly multiplexed holograms (b) (after
[194]).

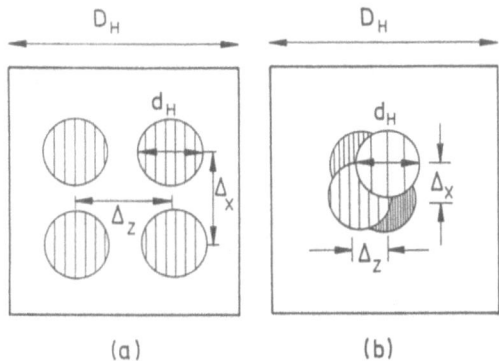

(a)          (b)

author and her coworkers [135, 192], is to combine wavelength and phase-coded multiplexing in order to preserve the advantages of both methods.

Moreover, there are several solutions to combine spatial and angular multiplexing in
order to enhance the storage capacity of the whole volume of the storage material. The
most simple method is to realize angular multiplexing in different spatial locations. Using
this technique, G.W. Burr and coworkers realized in 1995 a storage capacity of 4.000
pages, distributed on 8 locations, each having a storage density of 500 holograms [193].
Other interesting combination methods are *fractal space multiplexing*, which I mentioned
already in section 4.3.4, *spatioangular multiplexing*, *shift multiplexing* and *peristrophic
multiplexing*.

*Spatioangular Multiplexing*

Spatioangular multiplexing is a hybrid of both angular and spatial or local multiplexing. The aim of that combination, realized e.g. by S. Tao and coworkers [194, 195], is
to combine the advantages of both, the high storage density of images that is possible
by using angularly multiplexed volume holograms, and the low cross talk possible by
using spatially multiplexed Fourier-transform holograms. The storage principle of this
technique compared to pure angular multiplexing is illustrated in fig. 4.58. If spatially
multiplexed holographic spots of diameter $d_H$ are separated by distances of $\Delta_x$ and $\Delta_z$,
they will not overlap if $\Delta_x, \Delta_z > d_H$, but this limits the number of holograms that can be
recorded (fig. 4.58a). If the separation is diminuished, the holograms overlap (fig. 4.58b).
Consequently, when one hologram should be replayed, the reference beam replays several images at once, resulting in cross talk. In the spatioangular multiplexing recording
scheme, Fourier-transform holograms are formed in spatially overlapping regions of the
crystal and are distinguished from one another by use of variously angled reference beams.
The spatioangular technique introduces a slight spatial separation $\Delta_x, \Delta_z \ll d_H$ between
adjacent recordings, together with a slight change in the reference angle $\delta\theta$ ($\delta\theta >$ the
angular width of the holograms to avoid cross talk). For purely spatial multiplexing, the
number of bits that can be stored in a unit area has been estimated to be dependent on
the filling factor $F = \sqrt{M_s} \cdot d_H/D_H$, where $M_s$ is the number of multiplexed holograms
and $D_H$ is the lateral dimension of the crystal approximated by a square. Then, the
number of bits is $S = F^2\alpha^2/4\lambda^2$ where $\alpha$ is the angle of the images from a fixed point of
the memory plane [196]. By partially overlapping holograms with the use of spatioangular multiplexing, one can increase the filling factor $F$ substantially, thus increasing the

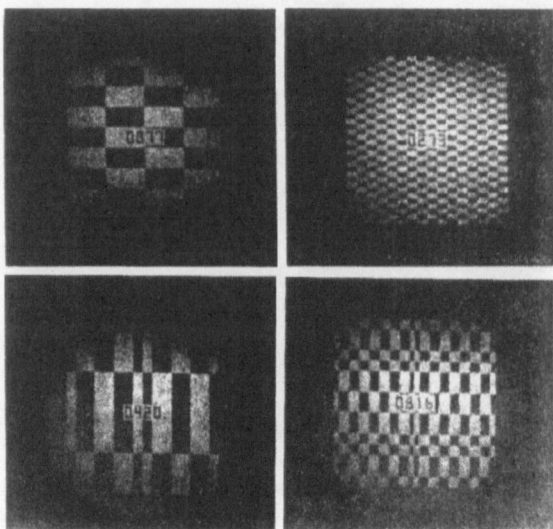

**Figure 4.59:** Four images recalled from a spatioangular-multiplexed photorefractive memory with a storage capacity of 750 images in photorefractive LiNbO$_3$ (from [194]).

storage capacity. Moreover, after S. Tao et al., the number of holograms overlapping a given hologram is reduced, giving a reduced $M$ and thus higher diffraction efficiencies per hologram. The authors realized experimentally a storage density of 750 holograms with high resolution and an averaged diffraction efficiency of about 0.5% ±0.2%. In fig. 4.59, four replayed images are shown, indicating that no cross talk is obvious. As for angular, phase-coded or wavelength multiplexing, a special exposure procedure to achieve uniform diffraction efficiencies is necessary for the novel spatioangular multiplexing technique. However, in contrast to these methods, spatioangular multiplexing exhibits an erasure effect that depends not only on the erasure time but also on the overlapping area (strictly volume) of two holograms. If the erasure is assumed to be proportional to the overlapping area, an erasure weight array can be calculated, which is used to determine an equivalent erasure time for each recording. Then the recording procedure can be performed using these erasure times in analogy to sequential hologram recording (see 4.4). Moreover, in spatioangular recording it is difficult to find methods to refresh the stored data due to partial erasure during readout of a hologram.

Compared to pure angular or pure spatial multiplexing, spatioangular multiplexing effectively increases on the one hand the storage capacity by more fully using the crystal volume for the interaction of the writing beams. Moreover, the system allows for parallel recall and parallel correlation of all stored images without the need to use multiple readout wavelengths. On the other hand, a complicated recording procedure to guarantee uniformity of diffraction and the problem how to refresh partially erased hologram gratings are limitations of that method that have not yet been overcome.

*Shift Multiplexing*

Shift multiplexing is a derivation from spatioangular multiplexing in a way that is adapted to the use for recording on optical disks. Moreover, a reference beam is used consisting of a spectrum of plane waves. In this context, shift multiplexing is similar to phase encoding, where a spectrum of different phases is used for encoding the reference beams. Multiplexing is achieved by shifting the recording medium with respect to the signal and

**Figure 4.60:** Geometry for shift multiplexing in the Fourier plane. $F$, $F_r$ are the focal lengths of the Fourier transform lenses (after [197]).

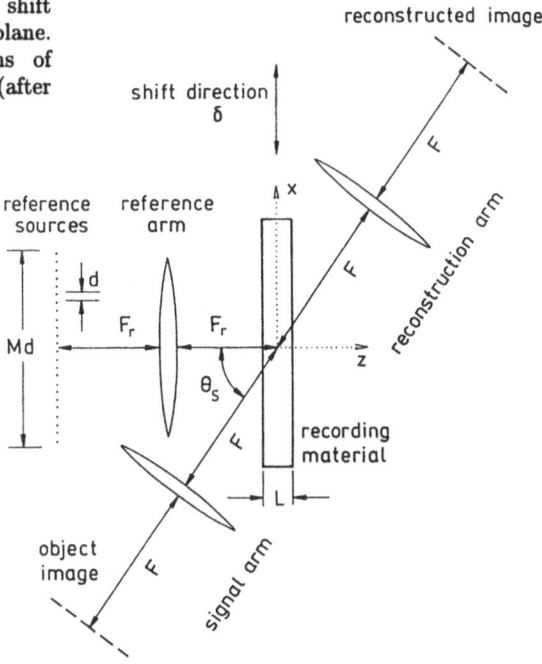

reference beams. Alternatively, the two beams can be translated in tandem with respect to the stationary medium. The idea of shift multiplexing is shown in fig. 4.60 for the case of storing Fourier transform holograms. The reference originates from an array of $N$ point sources located at the front focal plane of a Fourier lens and centered around the optical axis $z$. The lens transforms the field into a fan of $N$ plane waves. The angular separation is uniform, given by $\Delta\theta \approx d/F_r$, where $d$ is the distance between successive point sources and $F_r$ is the focal length. Thus, the angle of incidence of the $m$th component is [197]

$$\Delta\theta_n \approx \left(n - \frac{N-1}{2}\right) \cdot \delta\theta, \quad n = 0, \ldots, N-1. \tag{4.44}$$

The angle of incidence of the central component of the signal with respect to the $z$ axis is denoted by $\theta_S$.

As in phase-coded multiplexing, the reference wave for a single image or page consists itself of $N$ components, in the case here of $N$ plane waves. Thus, we can think of the recording as consisting of $N$ separate holograms recorded simultaneously. On reconstruction, each plane wave in the reference fan reads out not only the hologram that it recorded but also all the holograms recorded by the other plane waves of the reference fan. These reconstructions — called ghosts — produce images that are shifted with respect to the primary reconstruction as a result of the change in the readout anlge relative to the recording angle. The ghosts are Bragg-mismatched by an amount that is roughly proportional to the angular separation between the plane-wave component that originally recorded the hologram and the component that is reconstructing it. For the hologram recorded between the central signal component and the $n = 0$th reference component the amount of Bragg mismatch is given by $\Delta k_z = 2\pi l \tan\theta_S \cdot \Delta\theta/\lambda$, when read out by the $\pm l$th reference component. The same relationship holds approximately for the other holograms. Because the diffraction efficiency of these Bragg mismatched holograms is proportional to [197]

$$\eta(\Delta k) = \text{sinc}^2\left(\frac{\Delta k_z L}{2\pi}\right),\tag{4.45}$$

where $\text{sinc}(x) = \sin(\pi x)/\pi x$, and $L$ is the thickness of the recording medium. Independent readout of the stored pages without cross talk requires that the angular separation $\Delta\theta$ between the reference components has to be chosen in such a way that the sinc function of eq. (4.45) vanishes. Thus, the ghosts will be eliminated, leaving a clean reconstruction. This requirement can be formulated as [197]

$$\Delta\theta \approx \frac{\lambda}{L\tan\theta_S}.\tag{4.46}$$

Moreover, shift multiplexing includes a shift $\delta_1 = \lambda/M \cdot \Delta\theta$ for each hologram with respect to each other during recording, defined by the zeros of the intensities of the diffracted field as a function of the shift.

Shift multiplexing is particularly well suited for the implementation of holographic three-dimensional disks. One can readily implement a three-dimensional disk with this method by simply using the disk rotation (which is already part of the system intended to permit accessing of information in different locations on the disk surface) to implement the shift. In systems that are based on holographic disks as storage materials, the storage density $\mathcal{D}$ per unit area is limited by the thickness-dependent angular selectivity of the medium. If $N_P$ is the number of pixels per page, $b_P$ is the pixel size and $M$ the number of pages to be stored, this density is approximately given by [197]

$$\mathcal{D} \approx \frac{M\cos\theta_S}{b_p^2(1 + M\delta_1 + \cos\theta_s/N_p b_p)}.\tag{4.47}$$

This allows to store about $\mathcal{D} \approx 21$ bits/$\mu$m for $M = 100$, $N_p = 1000$ and $b_p = 2$ $\mu$m and with $L = 0.1$ mm, $\theta_S = 30°$, $\Delta\theta = 0.5°$. Using spherical instead of plane reference waves, the minimum shift between two pages rises slightly, but allows to implement the focusing lenses of disk systems as natural multiplexing devices. There are two major drawbacks of shift multiplexing I want to mention here. First, shift multiplexing requires accurate alignment of the reference wave separation and the shift between different writing positions. Therefore, small deviations in one of these parameters causes immediately cross talk, giving only low signal-to-noise ratios in real experiments. Second, when no fixing is used after each exposure, the decay of stored holograms during writing a new one has to be taken into account as for all other multiplexing techniques. However, because the holograms are only partially overlapping, the exposure schedule can not be realized as easily as for sequential or incremental recording (see sections 4.4 and 4.4). Consequently, even when applying a tricky exposure schedule accounting for the overlap ratios of the different waves as in [198], different diffraction efficiencies for different pages and consequently again, contributions of cross talk do arise during recall of the images. However, G. Barbastathis and coworkers [198] realized a memory that was able to store 600 analog, but binary data pages.

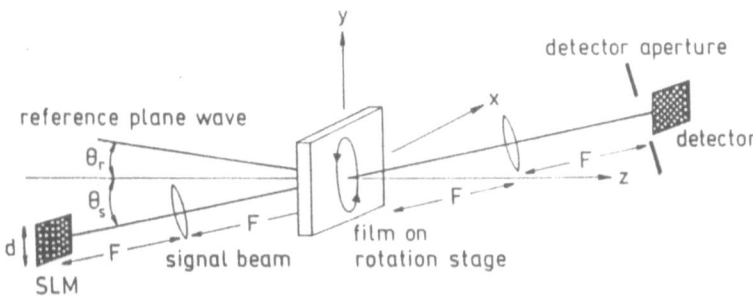

**Figure 4.61:** Principle of peristrophic multiplexing. SLM = spatial light modulator (after [199]).

*Peristrophic Multiplexing*

In peristrophic multiplexing [199], the material or, equivalently, the recording beams are rotated with the axis of rotation being perpendicular to the recording medium's surface (see fig. 4.61). The rotation does two things. It shifts the reconstructed image away from the detector, permitting a new hologram to be stored and viewed without interference, and it can also cause the stored hologram to become non-Bragg-matched. In addition, peristrophic multiplexing can be combined with other multiplexing techniques such as angular and wavelength multiplexing to increase the storage density and with spatial multiplexing to increase the storage capacity of the system. K. Curtis and coworkers realized such a combination of peristrophic and angular multiplexing in photopolymer films, rotating the film about 3° for each peristrophic and for each angular position and using a sequential recording algorithm. They achieved the storage of 295 holograms with an average diffraction efficiency of about $4 \cdot 10^{-6}$ [199]. Again, the problems of that combinational multiplexing system lie in its sensitivity to cross talk and in the inhomogeneous diffraction efficiency of different pages because of the complicated interaction of recording and erasure of previously written pages that takes place with different speeds at different locations of one page.

Although experimental realizations of all the combinations of different multiplexing schemes described above have not yet shown a technical or storage capacity advantage over the use of a single method, the field of image storage is a rather exciting, expanding one at the moment, attracting the interest of researchers of a lot of scientific areas. Actual new results in that area are given in the Mai issue 1997 of Applied Optics — a special issue dedicated to optical memories. Therefore, the next future will surely bring new results from all these methods and find ways to enhance the storage capacity further.

## 4.4.8  Limitations of Distributed Storage Techniques

Because distributed holograms are written with the same circumstances as local holograms, of course the same limitations as discussed in section 4.3.4 are valid for distributed, multiplexed storage of data pages. These cross-talk contributions are called interpage cross-talk moise in the case of page-oriented storage, because they appear among different pixels at one page. They have been investigated e.g. in detail in [200], concerning their influence on the storage capacity. However, the most severe limitation in that case arises due to cross talk between stored pages during readout, an effect that is not considered in local storage concepts. Here, I do not want to go into the details of cross-talk noise for the different multiplexing schemes. However, to compare their potentials and their limitations, several important results of cross-talk noise will be discussed in the following section.

In general, the limits of three-dimensional storage capacity have different origins. First, for recording of each image or page, scattering noise and material inhomogeneities will result in a reduced capacity of the page. Second, the number of holograms is affected by the multiplexing technique and configuration chosen. Moreover, there is a capacity dependence on the wavelength or geometry used for the storage configuration as well as a strong dependence on the effects of lens or other optics abberations in the whole setup. To account for all these noise sources in a common formulation is rather difficult, because every setup and every storage technique changes the ratio of these contributions significantly. However, a general result of these overall limit investigations shows a trade-off between holographic storage density (given by the number of pages to be stored) and the number of bits that can be accessed in one page in parallel [134].

The main reason for all cross-talk effects due to holographic page multiplexing is the Bragg degeneracy — the fact that the Bragg condition does not impose a strict, step-like condition on the reconstruction, but represents a smooth dependence on the angular deviation. Therefore, although Bragg selectivity ensures that the image pattern associated with the particular reference point that is used for readout is reconstructed with the highest efficiency, other stored images are also reconstructed with lower efficiency and with distortions. The sum of such Bragg mismatched components in the readout presents a form of cross-talk noise and limits the storage capacity of the system. When the reference points are distributed two-dimensionally at the reference plane in a statistical way, the cross-talk limited storage capacity is due to the degeneracy of the Bragg condition in the direction perpendicular to the grating vectors given by [201]

$$M_{max} = \frac{nL}{4\lambda \, \mathrm{SNR}_R^2},$$ (4.48)

where $L$ is the material thickness, and $\mathrm{SNR}_R$ is the signal-to-noise ratio required to reach this storage capacity. However, as we have seen in section 4.3, cross talk is significantly reduced when the reference points are restricted to the direction of the grating vectors or are arranged in grids that obey the distribution of this direction. In all other cases, a major portion of the cross-talk noise terms are due to Bragg matched readout of these degenerate gratings. For the case that the reference points are limited to the direction of the grating vectors, cross-talk noise is a function of the position at the output plane for a given readout beam. Fig. 4.62 shows the relative cross-talk noise — noise/signal — as a function of the hologram number for the case of angularly multiplexed images, in comparison with wavelengths multiplexed ones.

The resulting storage capacity is then given by [202]

$$M_{max} \approx \frac{2nLf}{\lambda \, d \, \mathrm{SNR}_{re}}$$ (4.49)

where $f$ is the focal length when writing Fourier-holograms, $d$ is the linear dimension of the output plane (object image), and $\mathrm{SNR}_{re}$ is the required signal-to-noise ratio in terms of intensity. The differences between E.G. Rambergs result (eq. (4.48) and eq. (4.49)) is first, that $\mathrm{SNR}_R^2 = \mathrm{SNR}_{re}/2$.

A more fundamental difference is, that E.G. Rambergs result represents a statistical average for the reconstruction fidelity, not taking into consideration that, for statistically

**Figure 4.62:** Signal-to-noise ratio as a function of the hologram number for angular and wavelength multiplexing.

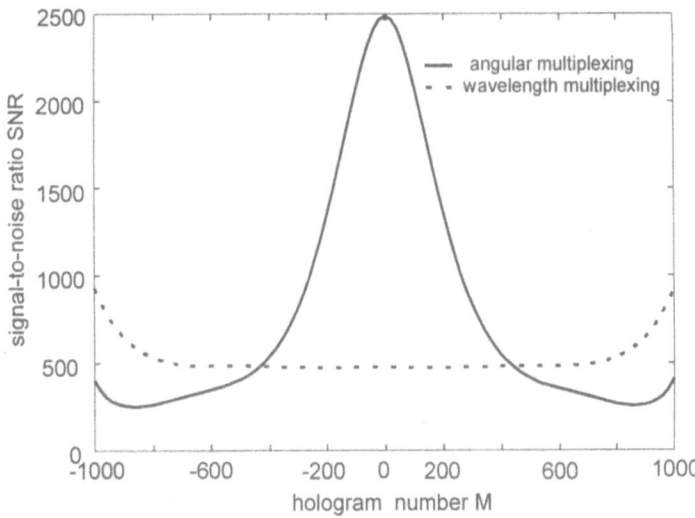

distributed reference patterns, the SNR itself is a randomly varying value, giving for some images SNR that might be less than unity and for others very high SNR values. In practice, it is more desirable to have regularly arranged reference points and similar SNR values for all reconstructions. This is included in the result above (4.49), which is a deterministic one. For that case, when arranging the reference points in a single direction and thus degeneracy noise is completely eliminated, the noise level can even be zero in the center of the output images. Thus, eq. (4.49) represents an upper limit for the SNR of the worst hologram, while most holograms will have a much better SNR than this one.

In most of the storage devices presented in the precedent section, the registration configuration was a 90° one, for which the Bragg selectivity is greatest. This optimal operating point was identified by E.N. Leith et al. [203], but was not considered feasible for the thin holograms they investigated at that time.

In this case, the limit on the storage capacity can be calculated in the Fourier regime by using the coupled-wave theory, including dispersion of the medium in its grating vector response [204] for the reconstruction of output $p$ as

$$M = \frac{2nL}{\lambda \, \text{SNR}} \cdot \frac{\beta}{K_{py,\max}}, \tag{4.50}$$

where the signal bandwidth is given by $K_{py}/\beta$ ($K_{py}$ being the deviation from the zero position in the output plane and $\beta$ being the same for the z-direction). Thus, with increasing signal bandwidth, the SNR increases too (see fig. 4.62).

The same derivation can be done for phase-encoded multiplexing, resulting in [204]

$$M \approx 2 \cdot \frac{2nL}{\lambda \, SNR} \cdot \frac{\beta}{K_{py,\max}}. \tag{4.51}$$

This result is of the same order of magnitude as the result for angular multiplexing. Reduction by approximately one half occurs because the cross talk is the page-average, rather than the worst-page, as it has been calculated for angular multiplexing.

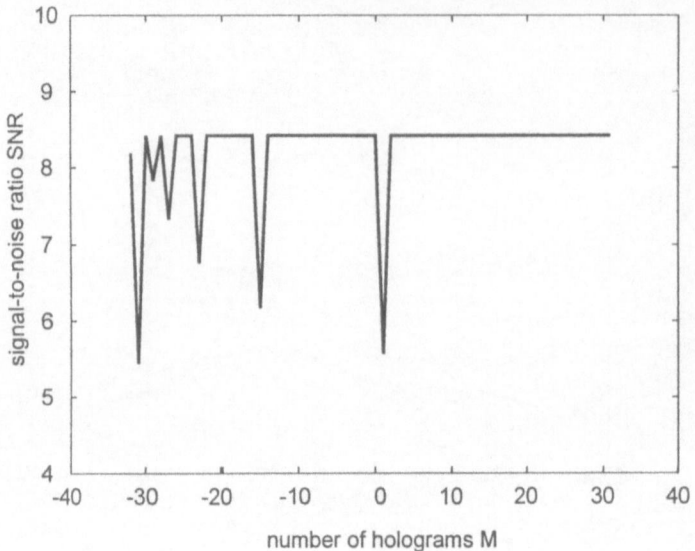

**Figure 4.63:** Signal-to-noise ratio versus hologram code number for N = 64 holograms stored with phase-coded multiplexing after [205].

In phase-coded multiplexing, each hologram can have almost the same good SNR, if one bad code is excluded [205]. In fig. 4.63, the SNR is shown versus the number of holograms, indicating that several single codes contribute most to the averaged SNR. When these codes are left out and the resulting SNR is compared to the one of angular multiplexing, phase-code multiplexing exhibits a significantly better SNR than angular and wavelength multiplexing. This situation is depicted in fig. 4.64.

For wavelength multiplexing finally, the same derivation gives for the case of reflection grating recording [206]

$$SNR \approx \frac{4f^2}{x_2^2 + y_2^2},\tag{4.52}$$

where $X_2$ and $y_2$ are the positions of the output plane. The signal-to-noise ratio as a function of the hologram number is shown in fig. 4.62 for wavelength multiplexing in comparison with angular multiplexing. Here, the higher the hologram number, the higher the signal-to-noise ratio grows, compared to angular multiplexing, where SNR is best in the center of the hologram numbers and decreases quickly with rising hologram number.

In comparison, the worst-case SNR for phase-coded multiplexing decreases very quickly with rising number of holograms, when the code with the worst SNR is left in the encoding process. Taking this code word out and recording $M-1$ holograms with $M$ plane-wave references results in a worst-case SNR that is approximately three orders of magnitude better than that of angular multiplexing. In contrast, if we take the SNR for angular multiplexing at the maximum position $Y_2$, averaged over all holograms, we obtain an average SNR that is still more than two orders of magnitude less than that for phase-coded multiplexing for large $M$. Therefore, well-chosen phase codes suppress, even on an average, the cross talk. For wavelength multiplexing, the worst-case SNR for 500 holograms, with all other parameters the same as for angular and phase-coded multiplexing, is $1.4 \times 10^3$ compared with $10^4$ for angular multiplexing and $10^7$ for phase-coded multiplexing.

Although the above analysis does not take into account the influence of the writing procedures or the reduction of the dynamic range by noise gratings, it gives an insight

**Figure 4.64:** Signal-to-noise ratio versus the total number of holograms for phase-coded, wavelength, and angular multiplexing after [205].

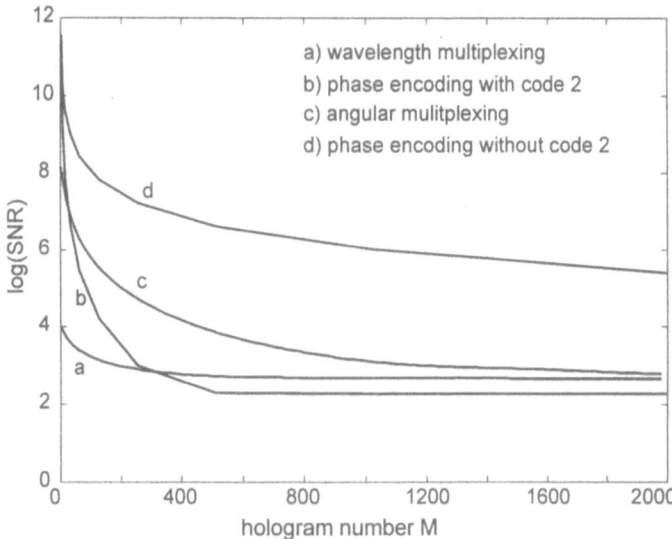

a) wavelength multiplexing
b) phase encoding with code 2
c) angular mulitplexing
d) phase encoding without code 2

in the upper limit of the potentialities of the different multiplexing techniques, allowing to judge them for different applications in storage or interconnection schemes.

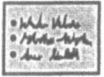 # Summary

For storing and interconnecting pages in volume holographic storage media, distributed storage is the appropriate means. Here, all gratings are superimposed and share the same volume. To enable independent, individual addressing of arbitrary pages, several different recording techniques, all based on the selectivity of the Bragg condition, can be applied.

- In angular multiplexing, it is the the reference beam angle that is changed for each recording process.
- In phase-coded multiplexing, each page is stored with $N$ purely and deterministically phase encoded reference beams. The set of adjustable phases of these references represents the address of the stored page and is orthogonal with all other phase addresses. It is that orthogonality or linear independence of the phase codes, that causes complete extinction to the first order of any unwanted pages because phase codes being orthogonal ensure destructive interference of all these contributions.
- In wavelength multiplexing, a tunable laser source is used to superimpose gratings at different wavelengths. If the reflection geometry is used to write gratings, cross talk can be reduced.
- Wavelength multiplexing can be combined with the other two techniques by using one wavelength to write several gratings that can be distinguished by their angle or phase code. The next wavelength is then still available to form further independent gratings.

- Spatioangular multiplexing is a hybrid of both angular and spatial multiplexing which introduces a slight spatial separation between adjacent recordings, together with a slight change in the reference beam angle.
- Peristrophic multiplexing is based on rotating the material or the recording beams, shift multiplexing on changing the interaction position in the material.
- Electric field multiplexing can be realized in noncentrosymmetric media by tuning the material parameters (e.g. via an external dc field), while keeping the external parameters fixed.

All those techniques give rise to slight cross talk due to the Bragg selectivity during readout, which depends strongly on the writing configuration and the technique used. When the worst code, giving rise to the most cross talk, is omitted, phase-coded multiplexing exhibits a signal-to-noise ratio of $10^7$, which is up to four orders of magnitude higher than other multiplexing techniques. Because in all these techniques, each new registration tends to erase previously written ones, adapted write cycles have to be implemented to ensure that all holograms to be stored have the same diffraction efficiency at the end of the recording cycle and are stored in such a way that the whole dynamic range of the material is adequately exploited.

- In sequential recording, different writing times are applied for each image. The first image is written until saturation, whereas the writing time of the second is chosen in such a way that both gratings have the same diffraction efficiency. The third and further images are adapted in the same way.
- Because sequential recording requires the knowledge of the crystal writing and erasing characteristics, incremental recording is based on recording the holograms with a series of incremental exposures each of it being short compared to the material's response time. The procedure is cycled until saturation is reached.

## 4.5   Interconnection Schemes Using Three-Wave Interaction

Although photorefractive volume holograms are the most promising and technically most developed candidates for high capacity interconnections, some other nonlinear optical storage and interconnection concepts show interesting and prospective features. Among them, the third-order nonlinear optical effects of photon echo and spectral hole burning allow for coherent interconnections or page-oriented image storage (see section 3.3.5). Therefore, I want to present at the end of that chapter some exemplary realizations with these methods showing the actual state of the art in that area.

### 4.5.1   Interconnections in Photon Echo Devices

Using three-wave interaction in photon echo devices, a phase-conjugate replica of the data image may be stored in the probe. A typical experimental setup to realize image storage or optical interconnections for such a three-pulse backward-echo experiment is sketched in fig. 4.65. In this geometry, the input laser pulses are spatially separated, permitting the spatial modulation of individual pulses for the purpose of image processing or neural network applications. In addition, this geometry permits the use of optical fibers for image transmission. The experiments performed by X.A. Shen and R. Kachru were realized on the $^3p_0$ - $^3H_4$ transition in a $Pr^{3+}$:$LaF_3$ crystal which was mounted

**Figure 4.65:** (a) Schematic illustration of the experimental apparatus for image storage and image processing in time-domain optical photon echo memories. BS's: beam splitters, BE's: beam expanders, M's: mirrors, DL: dye laser, OD's: optical delay lines, CCD: intensifying CCD-camera, SF: spatial filter. The port labelled "optional" is used to demonstrate phase conjugation. (b) Temporal sequence of the input and the echo pulses (after [85]).

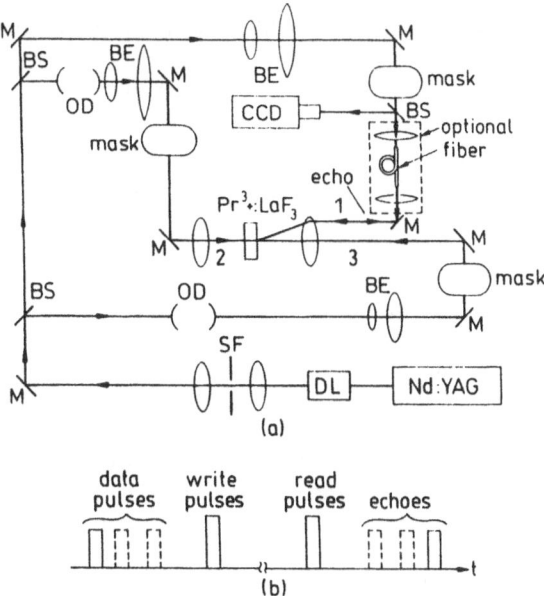

in a flowing helium vapour cryostat at about 6 K. This material has a short storage time of about 100 ms and permits to run experiments at 10 Hz. A Nd:YAG-pumped, pulsed dye laser was used to resonantly excite the $Pr^{3+}$:$LaF_3$ crystal at 477.8 nm. The laser pulses (5 ns long and 10 GHz linewidth) were first split into three parts by beam splitters, two of which were optically delayed. A spatial filter was installed before the beam splitters to improve the spatial profile of the laser pulses. The individual laser pulses were expanded and collimated with beam expanders to illuminate transmission masks, and the images generated by the masks were then focused onto the sample in a nearly coaxial configuration. The masks were placed at the front focal plane of the transform lenses, while the sample was placed at the back focal plane, ensuring thus the exact Fourier transform of the input images at the sample. The echo emitted in the direction opposite to that of the data pulses owing to the phase-matching requirement, was detected by a gated intensified CCD camera.

X.A. Chen and coworkers could realize the storage and recall of a standard U.S. Air force pattern as an image. In order to realize optical point-to-point interconnections, masks can be introduced in the incident waves. The write and read pulses were plane waves to simulate delta function masks. Resolutions up to 3.17 linepairs/mm could be resolved, primarily limited by the recording system and the fact of storing a Fourier-transformed image in a small sample area, thus cutting off the high-frequency components. The use of the first pulse as a data pulse in this counterpropagating geometry ensures high-quality echo images, because the recalled image is a phase-conjugate replica of the data image and any distortions caused by optics are removed upon retrieval.

### 4.5.2 Interconnections in Spectral Hole Burning Devices

Coherent matrix-matrix or matrix-vector interconnection schemes may also be implemented optically using persistent spectral hole burning (see section 4.1 for a detailed explanation of the effect). O. Ollikanen et al. [84] used a persistent spectral hole burning material comprising two different organic impurity molecules. The partially overlapping inhomogeneous absorption bands of the two impurities give in the 615 to 623 nm wavelength region a nearly "flat" absorption frequency interval of width about 150 cm$^{-1}$, where the transmission of the sample before the write-in exposures varies between 1% and 2%. The sample is positioned inside an optical cryostat and immersed in liquid helium. At 2 K, the decay time of the written-in holes in this particular material was slower than 10 hours. As a laser source, a picosecond Rhodamine 6G or DCM dye laser synchronously pumped by a mode-locked Argon ion laser is used. The dye laser can be tuned in about 40 steps within the 150 cm$^{-1}$ spectral interval. The quantum yield of the hole-burning process is about 1%. An average exposure energy of 1 to 10 mJ is required per cm$^2$ of the sample area and per cm$^{-1}$ of the spectral interval, in order to increase the transmission of the sample at the illumination wavelength by a factor of 2. Depending on the linewidth of the dye laser, the exposure time per spectral hole or hologram lasted from ten to several hundreds of seconds — which is comparable to the writing time using photorefractive volume holography. The read out is performed by collecting the transmitted or the diffracted light with a lens and by focusing it onto the detector, a conventional photomultiplier or a streak camera. Read out can be accomplished in a fraction of a second, but longer detection accumulation times (averaging) yield better signal-to-noise ratio.

In their experiment, A. Ollikanen and coworkers [84] wrote the memory matrix by tuning the laser sequentially to 12 different burn-in frequencies. At each frequency, the sample is exposed by directing the laser beam through a mask slide consisting of 12 horizontal stripes. Each stripe projects upon a corresponding section of the persistent spectral hole burning plate. If the calculated value of the memory matrix element $T_{ij}$ is zero, then the $j$th stripe of the mask used for exposure at the $i$th frequency is opaque and no hole is burnt at the corresponding spatial and frequency element of the probe plate. Only if the memory matrix element $T_{kl}$ is not zero (e.g. equal to 1 for a binary system), the corresponding $k$th stripe of the mask for frequency $l$ is transparent, and a hole is burnt in. Once the memory is stored, it can be interrogated by an arbitrary 12-bit probe vector. Probe vectors are coded as masks with corresponding transparent and opaque stripes in a way similar to that described above. A parallel beam of light comprising all the 12 frequencies is directed first through the probe mask and then through the sample so that each stripe of the mask projects at the corresponding section of the probe plate. Each of the 12 spatial elements of the input vector is connected to each of the 12 frequency-domain output vector elements via the 144 connections materialized by the persistent spectral hole burning plate. Consequently, the spatially integrated transmission of the probe mask along with the probe plate, measured at the 12 pre-fixed frequencies, implements the scalar vector-matrix product operation.

As an extension of that experiment, it is possible to use the time delay as a supplementary variable, allowing to store in the memory even temporally changing laser pulses. Another extension is to use the second spatial coordinate to realize three-dimensional memory devices to interconnect images. To materalize the interconnection matrix, three physical coordinates are used — two orthogonal spatial coordinates and the frequency — where each variable has 32 different values. The scheme of the experimental arrangement is presented in fig. 4.66. The expanded dye laser beam is divided into two parts by a beam splitter. At the persistent spectral hole burning plate, the two beams (object and reference beam) cross at an angle of 10°. The path of the object beam contains a holder with an interchangeable mask. The laser beam projects the mask upon the sample so that

**Figure 4.66:** Scheme of an interconnection matrix realized with spectral hole burning. The memorized information consists of 32 basic vectors. 32 different masks, each consisting of 32 ×32 pixels, are used to write a $32 \times 32 \times 32$-element memory matrix. Four different 32 × 32 masks are probed as the input, each containing four error bits (as compared to the closest matching basic vector) (after [84]).

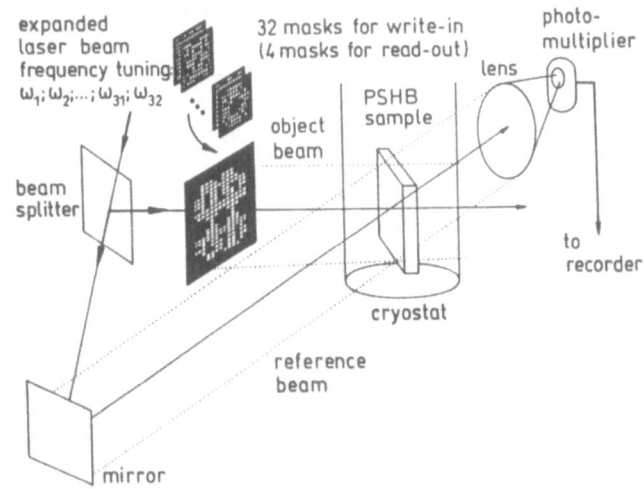

every spatial element of the mask coincides with a corresponding spatial element of the storage plate. The reference beam possesses a plane wavefront and uniformly illuminates the whole storage area of the probe plate. Holograms are stored at those spatial elements and frequencies for which the value of the corresponding memory matrix element is not zero. Only at these positions is the write-in matrix transparent to the illuminating object laser beam; at all other positions the mask is opaque and no holograms are recorded. The exposure procedure is carried out for each of the 32 frequencies by tuning the dye laser sequentially to 32 fixed frequency positions defined by an intracavity etalon. The memory has been interrogated with four different probe vectors, each of which resembles one of the four stored vectors but contains four error bits. During read out, the object beam is attenuated to avoid erasing of the memory by further hole burning. The reference beam is blocked. To represent the probe vector, a 32 × 32-element mask with transparent and opaque elements, that correspond to the self-product of the probe vector, is prepared. It is placed at the position where the memory write-in matrix has been located. At the output of the memory, a focusing lens collects the light diffracted from different spatial elements of the persistent spectral hole burning hologam onto the cathode of a photomultiplier. Although the system is still in a premature state, A. Ollikanen et al. arrived at recognizing three out of the four probe vectors to be tested correctly by the memory — the sequence of bits obtained after thresholding of the diffraction signal reproduces a corresponding stored vector that is the closest match to the probe vector. Thus, a simple associative memory (the features of these systems will be described in detail in section 7) has been realized using the technique of persistent spectral hole burning. Moreover, the setup allows in principle for parallel read out detection of frequency-domain bits by using a broadband illumination and a monochromator at the output.

### 4.5.3  Distributed Storage in Spectral Hole Burning

Spectral hole burning was also used to realize page-oriented memories. In contrast to photorefractive volume holographic memories, the diffraction efficiency of one hologram is not affected by the presence of subsequent ones. However, the source of cross talk in these memories is the diffraction from holograms stored at nearby frequencies. B. Kohler et al. [82] realized an experiment in which 2000 image holograms were stored by use of a swept recording technique that substantially reduces cross talk and is expected to be of importance for maximizing the storage capacity of spectral hole burning memories. A hologram recorded in a spectral hole burning material is characterized by an absorption as well as a refractive index grating. The finite spectral extend of each of these gratings leads to coherent interference effects when multiple holograms are stored at nearby frequencies. The electric-field amplitude of the light diffracted from the absorption grating decays asymptotically as $1/(\omega - \omega_0)^2$, where $\omega$ is the readout frequency for a hologram recorded at $\omega_0$. The light diffracted from the refractive index gratings, however, exhibits an asymptotic $1/(\omega - \omega_0)$ dependence, which is the principle cause of interference between spectrally nearby holograms. If $N$ holograms are recorded at equally-spaced frequency intervals $\Delta\omega$, the diffracted signal at the center frequency of any of the holograms will contain contributions from the refractive index and absorption gratings of the other $N-1$ spectral holes. The cross talk will be most severe for holes at the low- or high-frequency end of the region where the holes are stored. Since the total diffraction is given by a coherent sum of the light waves diffracted from the absorption and refractive index gratings from each hologram, B. Kohler et al. [82] were able to reduce cross talk by controlling the hologram phase, which is just the spatial position of the interference pattern incident upon the sample. If recorded with a phase difference of $\pi$ between pairs of holograms, the background could be reduced when multiple holograms were stored. This technique is most successful when all holograms have the same amplitude and is thus less applicable to the storage of gray-scaled images.

A more general scheme for avoiding cross talk in frequency-selective materials is to control the phase within each individual hologram. This can be done by sweeping the recording frequency while simltaneously altering the phase of the hologram, creating a so-called *swept hologram*. A swept hologram consists of a superposition of holograms recorded at slightly different frequencies with a phase that chanes linearly with frequency. As a result, the intensity distribution of the light in the different diffraction orders from a swept hologram is strongly asymmetric, whereas the diffraction efficiency for a large number of swept holograms is homogeneously distributed and exhibits a remarkable reduce in cross talk. The experimental setup to store 2000 swept holograms in such a spectral hole burning memory is shown in fig. 4.67. The sample is kept at 1.2 K in a liquid-Helium kryostat. A single-frequency tunable dye laser is used for frequency-selective storage. During recording, the spatial phase of the holograms can be altered by use of the piezoelectric transducer (PZT). Under computer control, an image displayed on the LCD is recorded. During readout, this beam is blocked, and the image is reconstructed by diffraction from the reference beam and detected by a CCD camera.

### 4.5.4 Distributed Storage by Frequency- and Time-Domain Multiplexing

Finally, I want to mention the possibility to realize optical memories by the combination of the different methods described up to now. One possibility is the combination of frequency- and time-domain multiplexing. As I have already shown in section 4.1.2 (see fig. 4.4), the inhomogeneous line can be divided into $M$ equally spaced slots, each representing a location of a spectral hole. When a laser is tuned to a given slot, photon-echo storage can be performed that could have $N/M$ bis of time-domain data. The only condition for the excitation pulse width here is that is has to be long, so that its spectral width

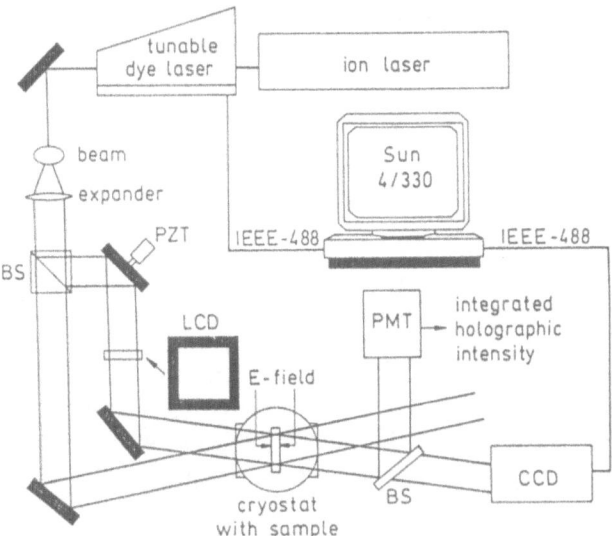

**Figure 4.67:** Experimental setup for holographic image storage using spectral hole burning. (PZT = piezoelectric transducer, BS = beam splitters, E-field = dc-electric field, PMT = photomultiplier tube, after [82]).

falls within the slot. In other words, by sacrificing the time resolution of the photon-echo experiment, one can reduce the spectral width of the population grating impressed in the medium. Retrieval of the time-domain data at a given address is achieved by simply tuning the read pulse to the corresponding frequency. In this way, the hybrid memory has $M$ randomly accessible addresses and each address has $N/M$ bits of information. The first demonstration of such a memory was presented in 1991 by M. Mitsunaga et al. [89], who realized a 1.6 kbit memory, with 16 bits in the time domain and 103 addresses in the frequency domain in an $Eu^{3+}:Y^2SiO_5$ crystal.

Moreover, it is possible to combine Bragg-angle multiplexing and spectral-hole multiplexing in a spectral hole burning material. The result is a complete four-dimensional memory, having a cross-talk limited storage density in each spectral hole that is the same as that achieved with conventional holographic materials. In addition, the total storage density is increased by a factor equal to the number of spectral holes compared to the storage capacity of pure angular multiplexing [207].

As an application of hole-burning memories, H. Sasaki and coworkers [208] realized recently a novel technique for directed pattern recognition based on frequency and spatial domain holography using a cryogenic $Eu^{3+}:Y_2SiO_3$ crystal. The fast holeburning speed of the crystal allows a moving image to be contineously recorded in it by a frequency-scanned laser. Using Fourier transform holography permitted direct pattern recognition of a 25s movie stored in the crystal to be realized. The correlation spot continuously followed the moving object with the same time resolution as the reproduced movie, which was in the range of 10 ms.

Photon echo as well as persistent spectral hole burning techniques suffer at the present state from limitations as low resolution, low diffraction efficiencies, stray light noise, very short storage times in the range of hardly microseconds, mostly even nanoseconds and large, sensitive and high-power requiring experimental apparatus. However, they open new horizons for optical coherent high-density data storage, because they are theoretically able to increase the number of interconnections by several orders of magnitude compared to photorefractive volume storage devices. Therefore, they may be comparable to volume holographic devices in the near future, provided that the possibility to operate the system at lower energy levels and at reasonable temperatures becomes available.

 ## Summary

Using three-wave interaction techniques as spectral hole burning and photon echo, pages (images) as well as interconnection schemes can be stored. In the first case, two-dimensional distributions of information are introduced into the beam path of the signal wave, in the second case, masks that define the input pixel configurations can be used. To enable independent, individual addressing of arbitrary images, different probe vectors have to be used, storing or recalling the data in the time or in the frequency domain.

- For photon echo implementations, the signal is encoded as the first data pulse, while the reference follows at different temporal delays for different signals to be stored (temporal domain encoding). Using e a phase-conjugate geometry to store the images, it is possible to compensate for distortions of the optics during recall.
- In spectral hole burning, interconnection memories can be realized by tuning the laser source to different burn-in frequencies for each page to be stored (frequency domain encoding). The memory can be addressed by the appropriate probe vector frequency in order to recall the image or data stored. Using a broadband source comprising all reference frequencies for recall, all pages can be recalled at once.
- The time delay can be used as a supplementary variable in hole burning, allowing to store even temporally changing laser pulses.
- The second spatial coordinate can be used to realize three-dimensional memory devices to interconnect images. Three physical coordinates are used — two orthogonal spatial coordinates and the frequency.

 ## Further Reading

1. P. Yeh *Photorefractive nonlinear optics and optical computing*, Opt. Eng. 28 (1989), p. 328-343.
   A concise description of the possibilities to realize optical computing algorithms using photorefractive materials.
2. R.J. Collier, C.B. Burckhardt, L.H. Hin, *Optical holography*, Academic Press, New York (1971).
   Still a concise introduction to holography, including volume holographic registration and their restrictions.
3. L. Solymar, D.J. Cooke, *Volume holography and volume gratings*, Academic Press, London (1981).
   A very profound derivation of volume grating recording and reconstruction in different materials as well on coupled wave theory.

4. F.T.S. Yu and S. Jutamulia, Eds., *Optical Storage and Retrieval*, Marcel Dekker, Inc., New York, 1996.
   This novel collection of application of optics to storage and neural networks contains, among others, an interesting chapter on *Optical Neural Network Architectures*, by T. Lu and D. Mintzer.

5. S. Bains, *Holographic storage: From pixel to bits*, OE Reports, Mai 1995,p.2.
   A short introduction into holographic data storage and its application to digital data.

6. W.E. Moerner, ed, *Persistent spectral hole burning: science and applications*, Vol. 44 of Topics in Current Physics, Springer Verlag, Berlin (1988). A well-written introduction into the principles of hole burning and an overview over possible applications.

7. E. Manykin and V. Samartsev, *Optical Echo Spectroscopy*, Nauka, Moskow, 1984.
   A more theoretic work on photon echo and its applications to optical processing and spectroscopy.

# 5 Nonlinear Thresholding

Now that we have got an insight into the possibilities to realize all-optical interconnection devices for different storage and interconnection purposes, we have to focus our attention on the second necessary key component of a neural structure: The nonlinear decision or thresholding element. The threshold unit is the key element of a neural net, because its slope decides, whether the net is able to solve nonlinear decision problems. Together with the interconnection unit, it forms the heart of any neural net and provides the means to realize simple optical neural networks, e.g. with associative memory features. Therefore, both components together are often called the optical neuron itself. If we are able to construct these two key compents of neural net devices, we are ready to construct already a variety of simple optical neural networks as they will be presented in the third part of that book.

The nonlinear neuron is a device whose transmission shows a saturation behaviour dependent on the weighted sum of all activation inputs to the neuron. This weighted sum may be performed using the elements described in the previous chapter. The operation of the neuron consists now in thresholding or nonlinearly processing this weighted sum. The optical neuron may operate slowly compared to modern digital circuitry due to the distributed and parallel nature of the information processing and can have any one of a variety of saturation properties I have already described in section 2.1 (see fig. 2.3) including hard limiting, sigmoidal, or other nonlinear responses.

This essential component of even the most simple neural net can be realized optically with a broad variety of nonlinear elements, among them again photorefractive nonlinearities as well as liquid-crystal or semiconductor bistable electronic devices, organic materials as bacteriorhodopsine, Fabry-Perot resonators filled with an optically nonlinear material, microchannel spatial light modulators or the Stark effect in multiple quantum well structures. Here, no requirements on the capacity of the device medium are imposed on the component of choice. In contrast, the limiting conditions are given by the demand of a high switching speed, which is most important for a proper and quick decision element.

In that chapter, I want to discuss some of the recently developed possibilities and concepts to realize the types of threshold functions mentioned already in section 2.1 with different nonlinear materials, their advantages and inconveniences. It is not my aim to describe all existing methods and devices, but to give a basic overview over those concepts that fit in best with the interconnection and storage concepts I have discussed in the previous chapter.

Although coherent optical threshold elements bear a lot of advantages such as the possibility to process amplitude and phase of a beam and to take advantage of the inherent parallelism of optics, technological constraints make it actually difficult to realize pure optical switches whose properties satisfy all the requirements of all-optical or even coherent processing. At the moment, semiconductor devices seem to be the most promising structures of realizing threshold in optical or electrooptic systems. Therefore, I will describe a selection of electrooptic and all-optical devices for the case of thresholding elements, that may be able to be adapted in pure or almost pure optical neural net realizations.

The nonlinearity in a neural net has to weight the inputs which have been summed up in the interconnection device (compare to fig. 2.2). It is defined by its threshold or offset $\Theta$ (see also section 2.1). Three important types of nonlinearities have already been shown in fig. 2.3. These are *hard limiters* or *Heaviside-functions* (fig. 2.3a), which divide the decision region by a step function and which may be as well unipolar (border values for infinity are (0,1)) or bipolar (-1,1), *threshold logic* (fig. 2.3c) operators which have a linear, but smoothly increasing decision boundary and are most often used as unipolar thresholders and finally *sigmoidal functions*. These latter ones may be represented by the unipolar Fermi function (fig. 2.3d) or the bipolar tangens hyperbolicus function (fig. 2.3b).

The most limiting disadvantage of the step- or Heaviside function is that only linear separable decision problems can be solved adequately. In section 2.3 I have shown that even a combination of several of these functions with different threshold levels is not able to solve the problem, because hard limiters reduce the information necessary to initialize changes which result in a solution that is closer to the input in the next iteration. Consequently, the solution of nonlinear decision problems is only possible by adjusting the step-like threshold function into a function with a smoothly changing region between both decision boundaries. Mathematically, this requirement implies that the function which is able to solve nonlinear decision problems has to be differentiable at every point and in an unambiguous way. Depending on the slope of the "indecisive" region, there are several functions available that are able to solve nonlinear decision problems. Most often, sigmoidal functions are used (see fig. 2.3). Depending on the system structure, they are defined by the border values $f(x \to -\infty) = 0$ and $f(x \to \infty) = 1$ for unipolar threshold systems or by $f(x \to -\infty) = -1$ and $f(x \to \infty) = 1$ for bipolar systems. In between these two extremal values, the sigmoid function is characterized by a continuously differentiable, monotonously increasing activity potential, which is given by the Fermi-function $f(x) = 1/1 + e^{-\beta x}$ for the case of unipolar systems, while the case of bipolar values is most often represented by the tangens hyperbolicus function $f(x) = \tanh(-\beta x)$. In both cases, the input-output characteristic depends on the sign of the incoming signal. In the case of a bipolar function, the answer of the system preserves the sign, while for the case of a unipolar threshold functions, all negative incoming signal values are reduced to 0 and positives are set to 1.

Let us now have a look at the different optical implementations and how they meet the conditions of the threshold functions described up to now. Because smooth threshold functions are required to solve nonlinearly separable decision problems, it will be that class of threshold functions, I will focus my attention on for optical realizations of neural networks. Most of the decision, which threshold functions fits best to a certain optical neural net system depends on the system configuration itself. If one is using pure intensity modulating devices in the neural net setup — which is the case if one is using e.g. optoelectronic light emission diodes or detector arrays or spatial light modulators to write the information into the system — naturally only the intensity part of the optical information is preserved within one cycle in the the optical neural net. In contrast, all phase information of the incoming amplitude is lost. Therefore, these systems are unipolar and the use of the Fermi function would be a natural choice for the threshold function. However, if a bipolar function should be realized in that case, it can be done artificially by introducing a second set of channels into the system. That set is constructed in parallel to the original set, but here all the information is carried that has the opposite sign of that in the first set of channels. Although this concept can be implemented in all unipolar systems without problems, one has to bear in mind that it reduces the storage capacity by 50%, because each information bit needs two channels to be treated correctly (see also section 4.2).

In order to preserve the bipolar input-output character that is offered by the structure of neural nets in a natural way, one may use phase-preserving thresholding elements as e.g. photorefractive or phase conjugating elements. Because the proper restauration of the optical phase in each cycle of the neural net requires coherent light operations, systems that implement bipolar operations are very sensitive to disturbances from outside, requiring therefore high stability conditions or correcting elements to avoid wrong decisions.

Because of these disadvantages of both implementations of neural nets, it is hard to judge which of them is the better one. Therefore, the decision for unipolar or bipolar thresholding will mainly be given by the conditions the system imposes on the realization. Therefore, the following sections will present some of the most promising realizations of different threshold functions and will describe for which systems and how they are suited for certain neural net applications. Because most of the thresholding operation principles become clearer if they are described together with the neural network architecture they have been designed for, several simple neural net configurations and associative memory setups and their appropriate threshold realization will be presented in that chapter. Among them, there will be as well unipolar as bipolar architectures.

## 5.1   Incoherent Thresholding for Unipolar Optoelectronic Neural Networks

The design of electronic nonlinear thresholding elements and their combination with optical interconnection devices offer the possibility of using on the one hand the strength of optics — linear transformations, massive interconnectivity and parallelism — and on the other hand the strength of electronics as point nonlinearities and programmability, e.g. of different thresholding functions. As an advantage, these systems are based on well-known techniques and are therefore easy to implement, reliable and adaptive to a lot of available electronic image processing systems. Therefore, optoelectronics has become for many scientists the most promising technology for the implementation of neurons and neuronal thresholding. However, these advantages can only be gained at the expense of the loss of phase information and coherence of the system, allowing only cascaded systems that are still utilizing the bottleneck of transformations from optics to electronics and back.

Among the various devices that can be used to realize optoelectronic thresholding nonlinearities, electrooptic liquid crystal light valves are widely spread. Novel, forthcoming devices have also been fabricated using the field shielding effect in semiconductors, programmable microchannel spatial light modulators, organic switching materials and multiple quantum well structures. Here, I want to focus my attention to some systems built up with these materials that are adapted to the interconnection devices presented in the precedent chapter and that may be implemented in larger optical systems or in more complex neural networks as error driven backpropagation systems.

### 5.1.1 Electrooptic Liquid Crystal Devices

Liquid crystal electrooptic switches are most often used as electrooptic threshold devices because of their ease of fabrication, reliability and their intrinsic bistable behaviour depending on the incoming intensity.

A simple optoelectronic threshold operation based on a liquid crystal television (LCTV) has been proposed by A. Bergeron and coworkers [209]. Their system allows off-line operation that does not break the beam path or requires a new laser source with the help of a tapped input beam. Thus, this device, that maintains translation invariance, can be put whereever a threshold operation is needed in an optical neural network or pattern recognition application. The optoelectronic implementation of the proposed optoelectronic thresholder is presented in fig. 5.1. An input signal is tapped by means of a beam splitter and a CCD-camera and then transferred to a controller. The controller acts in turn on the thresholder to modify the input according to the tapped data. Thus, the threshold operation is achieved by means of a feedback loop. The attenuator, which is a polarizer that changes the transmission intensity by means of its orientation, controls the threshold value. Finally, the threshold is applied with the help of a LCTV coupled with a polarizer. When a signal is injected upon the beam splitter, it is tapped, attenuated and transferred by imaging to the CCD camera. The camera signal is then transferred back to the LCTV. This operation modifies the transmittance of the LCTV by decreasing the transmission when the input signal is small. At points where the image exibits high signal values, the CCD camera saturates and opens the LCTV to let the incoming signal pass unchanged. As the camera detects the intensity, the decrase of the value of a point with a small value is proportional to the square of its value at the preceding iteration. Because the iteration process is continuous in time, after a few iterations values below the threshold will vanish, whereas values above the threshold will stay uncahnged expcept for the attenuation induced by the polarizer. This polarizer is required for the correct operation of the LCTV. The addition of a polarizer between the cubic beam splitter and the CCD camera allows acting on the detected intensity and controlling the threshold value of the system. Because the attenuator is located in the feedback loop, it does not affect the maximum amplitude transmitted by the LCTV, which corresponds to the output signal. When the critical aligment betwen the CCD camera and the LCTV is done properly, the system exhibits a variable sigmoid function that is able to threshold two-dimensional image distributions or interconnection patterns in a reliable way. One of the applications of that system is the winner-takes-all function, when an automatic maximum peak extraction function is added to the system. The winner-takes-all function then retains the input point of a scene showing the highest light intensity while eliminating every other point. When the beam splitter is replaced by a holographic sampling device, which retains only a small portion of the light to be tapped, and when an automatic control is introduced for the attenuator, which enables to find the position where only the intensity of the maximum value of the input scene is transmitted, a simple winner-takes-all system can be realized [209].

Although liquid crystal spatial light modulators are well-known and technologically well-developed devices, their operation characteristic may cause a problem in thresholding for backpropagation neural net implementations, that can significantly affect learning. The reason lies in the fact that gray scales (or weight values) in these devices are often discretized, as a result of the particular SLM device physics or as a result of the limited accuracy in the drive electronics. However, the performance of backpropagation algo-

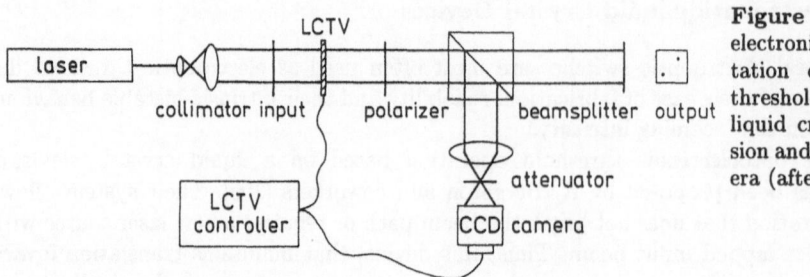

**Figure 5.1:** Opto-electronic implementation of a simple thresholder using a liquid crystal television and a CCD camera (after [209]).

rithms is extremely poor for less than approximately 1000 weight levels. In principle, to represent the gryy scale in the weights needed for learning, a smooth linear characteristic over a reasonable voltage range is desired. In practice, the smoothness of the transmission-voltage characteristic may be limited both by the physics underlying spatial light modulation operation and by the driving circuit parameters. A. Von Lehmen and coworkers [210] found, that by the addition of noise to the interconnection weights, a convergence of such a net can even be realized having only 16 grey levels, showing also the robustness of the backward error propagation algorithm to surprisingly large amounts of noise.

### 5.1.2   Semiconductor Devices

The neuronal functions described above can be realized electronically in simple transistor circuits [211]. The basis circuit for implementing a thresholding function with optical inputs and optical outputs using these devices is based on the following principle, introduced by S. Lin and coworkers. The gate voltage on the driving metal oxide semiconductor field effect transistor (MOSFET) is controlled by the input circuit, which consists of a photodetector acting as an optical input port and a biasing element, which can be either a photodetector or a transistor. The switching characteristics of the circuit are determined by the current-voltage characteristics of the photodetector and the biasing element. If one draws the characteristic curves of these two devices, the intersection point of both determines the node voltage. When the current in the photodetector exceeds the threshold current set by the biasing element, the gate voltage switches from ground to its maximum, which in turn switches on the driving transistor. This causes current to flow through a light emitting diode (LED) and the output light intensity to increase to its high value. Since the current through the LED is roughly proportional to the gate voltage, the nonlinear input-output relationship is determined almost exclusively by the input circuit. The sharpness of the threshold function is determined by the relative flatness of the current-voltage curves of the two devices in the saturation regime. The threshold becomes sharper as the output impedance of the devices in the saturation regime becomes larger. The leakage current through the gate of the driving MOSFET also affects the switching characteristics of the circuit. The extra current that is drawn through the gate needs to be supplied by extra photocurrent, which tends to increase the optical threshold level.

Since the LED's are incoherent light sources, only positive values can be implemented directly with these circuits, allowing only for unipolar realizations of neuron activation functions. If the weights have to be bipolar, there are two ways to represent bipolar signals in an incoherent optoelectronic system. The first method is to add a constant bias to all the bipolar weights before they are recorded in the optical system. In this case the input signal to each neuron is a positive quantity with the desired bipolar signal riding on the bias. In the system proposed by S. Lin et al. the control signal (either optical or electronic) on the biasing element adjusts the threshold of the circuit. The second method for representing bipolar signals is the conventional spatial encoding using two channels already mentioned above (see section 4.2) and consists of separating spatially the recorded positive and negative weights. The inner product between the input vector to the neuron and each of the two sets of weights is formed separately, and the results are subtracted electronically before thresholding. The circuits I have described here can be used in this mode if the biasing element is also a photodetector, identical to the input detector. Then, the positive signal is routed to the signal port, and the negative signal is routed to the biasing port. The gate voltage saturates as the difference of the positive and the negative signals gets large and positive (or negative). When both inputs are equal, the gate voltage is half of the saturation voltage. Therefore the circuit implements a sigmoidal function of the difference of the positive and negative input voltages, as desired. S. Lin and coworkers [211] realized three different monolithically integrated neuron circuits on Gallium Arsenide (GaAs) based on the LED as the optical ouput and a MOSFET as the amplifying transistor. The circuits differ in the type of photodetector used — bipolar phototransistors, meta-semiconductor-metal (MSM) detector and an optical field-effect transistor (FET).

Another possibility to realize all-optical semiconductor neurons for backpropagation or error driven learning networks is to use the nonlinear effects inherent to semiconductor devices. M. Ziari and coworkers [212] presented in 1990 the first attempt to use the field shielding effect in CdTe to realize an infrared optical nonlinear neuron. This semiconductor nonlinear effect being very similar to the photorefractive effect, is a charge transfer effect wherein an optical beam creates a photocharge which is then separated and alters the electric field pattern in the material. The change in the electric field through the linear electrooptic effect (see section 3.2) changes the indices of refraction and hence the birefringence. As shown in fig. 5.2, the incident beam creates a conduction band electron density from deep impurity levels. These electrons drift in the applied electric field into adjacent dark regions where they are trapped. The resultant negative charge density in the dark region and the compensating positive charge density in the illuminated regions create a space charge field which is opposite in sign relative to the applied electric field. The steady state is reached when the flow of electrons from the illuminated regions into the traps is balanced by the thermal reionization of the traps. Thus at low intensities little charge is trapped, and the field is seen by the beam is the applied field. At higher intensities, a larger space charge field builds up which again balances the flow of electrons into and out of the traps and the beam sees only a small remaining field. Therefore, through the electrooptic effect, the material's indices of refraction and hence the birefringence are a function of the beam intensity in a saturating manner. This effect is very similar to photorefraction but can be readily seen in materials where the electron trap density is too low for sizable photorefractive gratings to form.

Field shielding has been used to demonstrate infrared power-limiting, self-switching and all-optical switching. The model of the field shielding neuron, resulting in a characteristic sigmoidal function relative to the incoming intensities is shown in fig. 5.3. The neuron has a sensitivity to low power activation inputs and a saturation response for

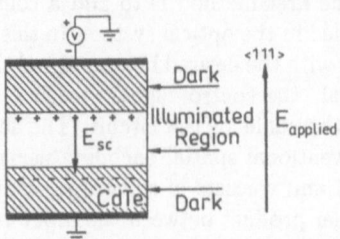

**Figure 5.2:** Field shielding electrooptic nonlinearity in CdTe where a uniform beam illuminates the middle region of the crystal and an opposing space charge field is created (after [212]).

inputs of about 1 mW/cm$^2$ at the 1.06 $\mu$m wavelength. This form of the neuron can be used at wavelengths from 0.9 to 1.4 $\mu$m with relatively low insertion losses ($\leq$ 1.0 dB).

The big advantage of the system is, that the response of the neuron can be altered in order to match the requirements of particular forms of neural networks. One alteration is to use the temporal response of the neuron in a synchronous network to achieve a softer saturation characteristic. The longer the response time of the neuron is (up to about 2 $\mu$s), the smoother the response sigmoid function will be. Moreover, various responses for a neuron inside an optical cavity have been simulated with that model, leading again to different steepnesses of the response curve depending on the power reflectivity of the mirrors in the cavity.

The most promising result however is the fact that even bidirectional neurons for multilayer networks may be constructed. This nonlinear neuron may thus be used as a hidden unit in a layered feed forward network which utilizes error backpropagation learning in analogy to the neuron presented in the previous section for coherent error backpropagation.

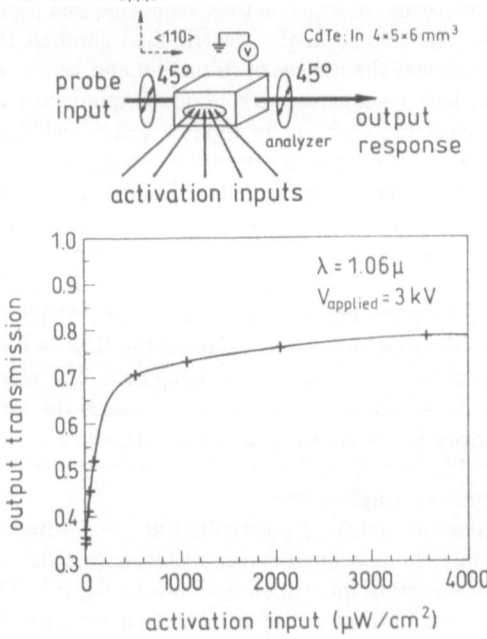

**Figure 5.3:** CdTe-based neuron where a low intensity polarized probe input is incident on the crystal. The activation inputs are applied to the crystal and alter the local electric field, which is sensed by the probe beam. The transmission of such a CdTe-based neuron versus the activation input intensity at 1.06 $\mu$m is shown below (after [212]).

**Figure 5.4:** Transmission of an etalon filled with an electrooptic CdTe crystal versus the normalized electric field in the < 111 > direction. The etalon mirrors have 70% reflectivity and the electrooptic material is assumed to be lossless. The inset shows the index ellipsoids formed by the two polarizations (after [212]).

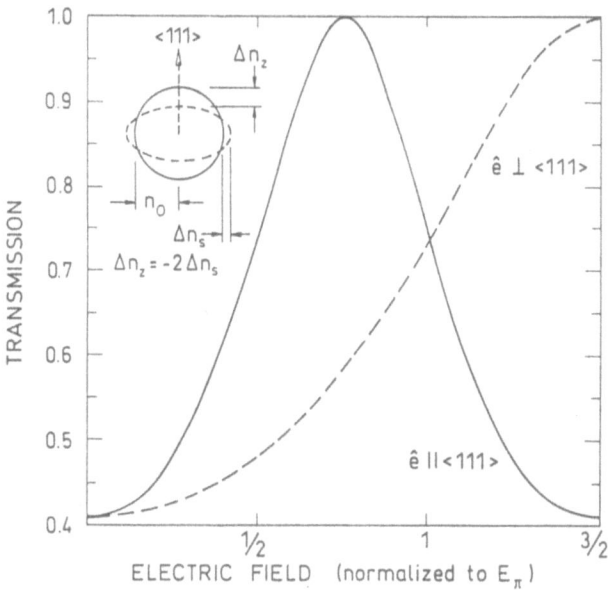

The field shielding nonlinearity in CdTe, when combined with a Fabry-Perot etalon and orthogonal polarizations for the backward and forward beams, can also be shown to give a good approximation to the desired bidirectional response. The neuron makes use of the difference in electrooptic index change for polarizations perpendicular and parallel to the applied electric field in a certain direction in crystals such as CdTe or GaAs. As shown in fig. 5.4, if the CdTe crystal is placed inside an etalon, the etalon will sweep through two cavity resonances in the parallel polarization for every single resonance in the perpendicular polarization as the electric field is changed. The figure suggests that the beam polarized perpendicular to the < 111 > direction could be used as the saturation forward signal, and the beam polarized along the < 111 > direction could be used as the backpropagating error signal. The resulting backpropagation learning neuron is shown in fig. 5.5. The backward beam is polarized along the < 111 > direction and the forward beam is polarized perpendicular to < 111 >. The activations or presynaptic inputs can be unpolarized and collinear or perpendicular to the other inputs as shown in the figure. A birefringent plate is placed in the etalon to provide a fixed shift between the resonant frequencies of the two polarizations. In this device design, the backpropagating error signal is the transmitted beam, and the forward thresholding beam is the reflected beam. Although other outputs can be used for the backward or forward outputs, this configuration gives the highest contrast ratio for the forward thresholding signal. At low activations, the forward beam is at resonance and the backward beam is at antiresonance. At high activations, the forward beam is at antiresonance and so is the backward beam again. At intermediate activations, it is the backward beam which is at resonance, exactly as the error backpropagation prescribes it.

The field shielding neuron can be used with excitatory and inhibitory inputs by using two activations with different positions relative to the sensing beam. This concept of splitting each neuron into two channels has already been suggested in 4.2. Although it reduces the capacity of the neuron by a factor of two, it is in certain cases easier to implement than a coherent neuron. For the infrared neuron of M. Ziari and coworkers,

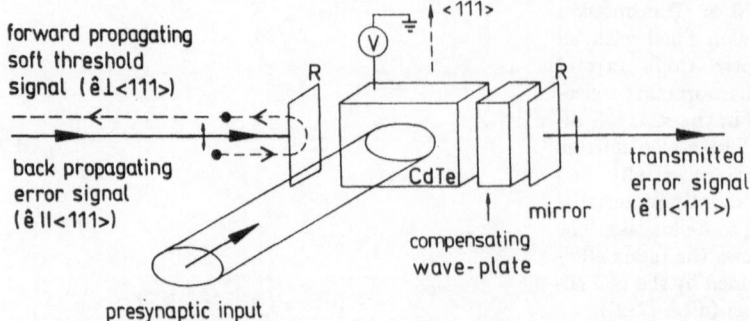

**Figure 5.5:** Bidirectional neuron for the hidden layers of a backpropagating error network. The reflection of a probe beam is used as the forward response and is polarized orthogonal to the backward response, which is the transmission of an error signal polarized parallel to the $< 111 >$ direction (after [212]).

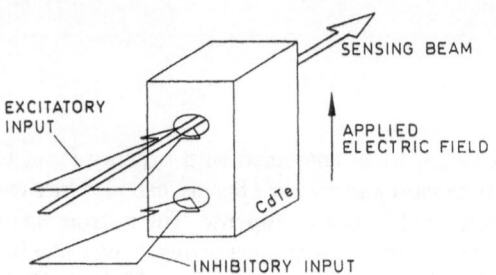

**Figure 5.6:** CdTe neuron with excitatory and inhibitory activation inputs. The excitatory input overlaps the weak sensing beam, and the inhibitory input can either be above or below the sensing beam (after [212]).

the activation input can be either collinear or at right angels to the sensing beam path, but the two beams must overlap. If the activation beam is above or below the sensing beam, the electric field seen by the sensing beam will increase instead of decrease as the activation beam intensity increases. This is because the voltage across the crystal is constant, and a decrease of the field in the region of the activation beam must result in an increase in the field in the remainder of the crystal. The activation beam, therefore, acts as an excitatory signal when it overlaps the sensing beam and as an inhibitory signal when it is above or below the sensing beam — without requiring any coherence of the excitatory and inhibitory beams. This concept is illustrated in fig. 5.6.

To summarize, the field shielding neuron demonstrated by M. Ziari and coworkers can be activated by several incoherent beams, has a wide infrared spectral range, can be used with microsecond pulses and its saturating response can be taylored by several methods. Moreover, its combination with a Fabry-Perot etalon makes it well-suited for error backpropagation networks. Because it is an effect that is present in a large range of materials, it should also be observable in several photorefractive materials such as GaAs, InP or BSO.

### 5.1.3  Spatial Light Modulators

The principle of using microchannel spatial light modulators (see section 4.1.1) is to image the beam distribution to be thresholded on the modulation side of the modulator and reflect it back, now weighted with the image matrix or vector which has been incident on

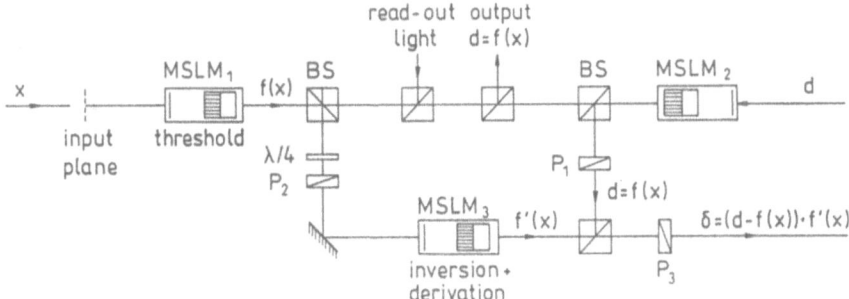

**Figure 5.7:** Experimental setup of an all-optical error signal generator using microchannel spatial light modulators (MSLM). The operations performed by MSLM1, MSLM2 and MSLM3 are soft thresholding of the net input, addressing the target and taking the derivative of the thresholding function, respectively. M indicates mirrors, H.M. half-mirrors (after [213]).

the modulator during the learning phase. This feature is based on the special capabilities of microchannel spatial light modulators (MSLM). They are able to internally sum and store control images incident on its detector side. This summation can be stored for periods up to months as an internal charge distribution inside the MSLM. The MSLM can also operate in a subtraction mode, where the control image incident on the detector is subtracted from the internally stored image. The MSLM output image at any time is the product of the stored image and the read image. Finally, the MSLM is also capable of operating in a thresholding mode, in which a thresholded version of the control image intensity is internally stored. It is that final feature that is the most important one for its implementation in neural network architectures. With that electrooptic thresholding device, a lot of optical implementations of associative and backpropagation networks have been proposed.

H. Yoshinaga and coworkers proposed and experimentally realized an all-optical error signal generation based on the generalized delta rule (see section 2.3) for backpropagation learning in optical multilayer networks [213, 214]. In that system, the MSLM device performs the key functional operations for the error signal, namely subtraction, soft thresholding and taking the derivative of the thresholding function. The experimental realization of that threshold device is shown in fig. 5.7. Soft thresholding by realizing the sigmoid function and subtraction are carried out optically by MSLM1 and MSLM2, respectively. The derivative of the thresholding function is obtained using a quarter-wave plate ($\lambda/4$) and MSLM3. Linearly polarized light from an Argon ion laser ($\lambda = 514$ nm) is used for readout. The target image addressed on MSLM2 and the reflected light from MLSM1 bearing the thresholded net input signal are read and then passed through a polarizer (POL1), which is used to convert the polarization modulated light into intensity modulated light. In this way the target and the thresholded net input signal are subtracted, yielding a positive value. The derivative of the net input is taken as follows: The reflected light from MSLM1 is phase shifted by a quarter-wave plate and then relayed to polarizer POL2, resulting in a negative derivative of the thresholding function for the net input. MSLM3 inverts the result into a positive derivative. By illuminating MSLM3 with the previously obtained subtraction signal and passing the reflected light from MSLM3 through polarizer POL3, multiplication of the subtraction signal and the positive derivative output signal are accomplished.

Up to now, similar systems and system improvements to that one accomplishing sub-

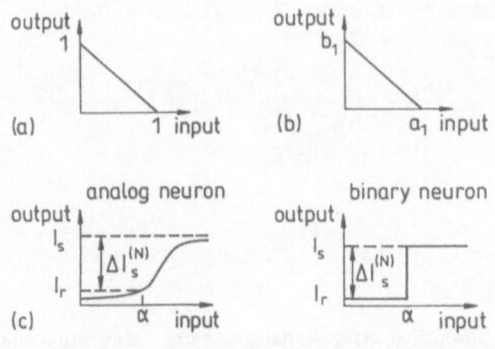

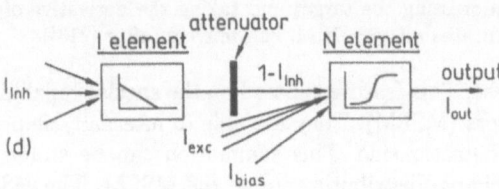

**Figure 5.8:** Illustration of an incoherent optical neuron model using spatial light modulators. (a) normalized inhibitory ($I$) element, (b) unnormalized $I$ element, (c) nonlinear ($N$) element, and (d) the incoherent optical neuron structure (after [215]).

traction and thresholding for error backpropagation networks using liquid crystal or microchannel spatial light modulators have been investigated by several authors. The system proposed by C. Wang and coworkers [215] is able to subtract inhibitory from excitatory neuron inputs by using two device responses. Their incoherent optical neuron is able to accomodate positive and negative weights, excitatory and inhibitory inputs and nonnegative neuron outputs and it can be used in a variety of neural network models. This incoherent optical neuron comprises two elements: an inhibitory ($I$) element and a nonlinear output ($N$) element. The inhibitory element provides an inversion of the sum of the inhibitory signals, the nonlinear element operates on the excitatory signals, the inhibitory element output, and an optical bias to produce the output. The inhibitory element is linear, while the nonlinear threshold of the neuron is provided entirely by the nonlinear output device. Fig. 5.8a-c show the characteristic curves of the $I$ and $N$ elements, respectively. The structure of the incoherent optical neuron model is illustrated in Fig. 5.8d. The output of the normalized I element is given by

$$I_{\text{out}}^{(I)} = 1 - I_{\text{inh}} \tag{5.1}$$

and of the $N$th element is given by

$$I_{\text{out}}^{(N)} = f(I_{\text{out}}^{(I)} + I_{\text{exc}} + I_{\text{bias}} - \alpha) \tag{5.2}$$

where $I_{\text{inh}}$ and $I_{\text{exc}}$ represent the total weighted inhibitory and excitatory neuron inputs, respectively. This include any lateral feedback signals as well as inputs from other layers. f() is the nonlinear output function of the neuron, $I_{\text{bias}}$ is the bias therm for the $N$th element, which can be varied to change the threshold, and $\alpha$ is the offset of the characteristic curve of the $N$th element. If $I_{\text{bias}}$ is chosen to be equal to $\alpha - 1$, the output of the $N$th element is given by

$$I_{\text{out}}^{(N)} = f(I_{\text{exc}} - I_{\text{inh}}) \tag{5.3}$$

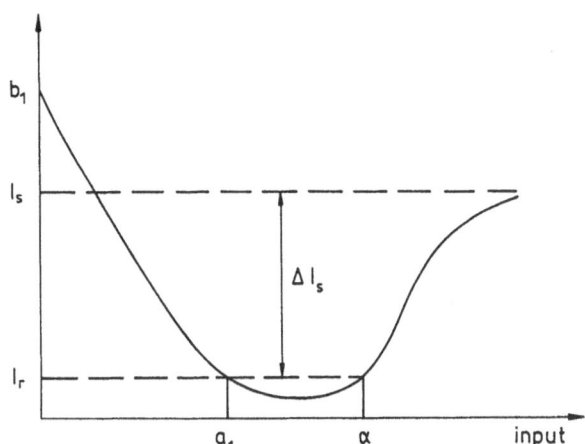

**Figure 5.9:** Characteristic response of a Hughes twisted-nematic liquid crystal light valve, a possible device for the homogeneous incoherent optical neuron model (after [215]).

which is exactly the desired subtraction.

In general, the $I$th element will not be normalized (see fig. 5.8b). In this case, the offset and slope of its repsonse can be adjusted using $I_{bias}$ and an attenuating element (neutral density filter), respectively, again enabling proper subtraction, provided that the unormalized $I$ element have a gain $\geq 1$. More quantitatively, the characteristic curve of the unnormalized $I$ and $N$ elements (see fig. 5.8b and c) can be modeled as

$$I_{out}^{(I)} = b_1 - \frac{b_1}{a_1}I_{inh} \quad \text{for} \quad 0 \leq I_{inh} \leq a_1$$
$$I_{out}^{(N)} = f(I_{in}^{(N)} - \alpha) \quad \text{for} \quad \alpha \leq I_{in} \tag{5.4}$$

It may be based on a twisted nematic liquid crystal spatial light modulator as well as on microchannel spatial light modulators provided that the input-output response of the device is of the form shown in fig. 5.9. These systems have two important advantages. One is the fact that cascades of these operations as they are necessary in multilayer optical neural networks can be easily built using MSLM's. A.D. Fisher and coworkers [216] even constructed a multimodular system of thresholding devices that is able to realize associative paradigms such as Hebbian Learning, the Widrow-Hoff learning algorithm and differential learning procedures (see section 2.2).

As a second advantage, the error signals are readily available to adaptive optical multilayer networks in which photorefractive crystals are used to determine the connection weights between processing units.

### 5.1.4  Organic Materials

Organic materials are one of the most forthcoming novel thresholding devices, because of their possibility of easy and low cost fabrication. Among them, bacteriorhodopsin (BR) has become very famous because of its application in storage as well as in bistable devices. BR is a pigmented protein which is found in the purple membrane of the bacterium *halobacterium halobium*. The purple membrane is the only crystalline membrane found in nature, and the BR molecule is the only protein found in the purple membrane. The principal role of BR is to operate as a proton pump activated by light in a wide range of wavelengths around 570 nm. As a result of photon excitation, the BR molecule undergoes cyclic molecular transformations which result in a proton being pumped from the inside to the outside of the bacterium. At least six of these transformations have been detected as color changes in the vitamin A related pigment, retinal, which is covalently linked to the BR protein.

For our purposes of optical thresholding, it is important to note that these transformations are also associated with the generation of charge transitions across the membrane. The kinetics of the optical and electric properties of BR can be significantly perturbed by treating the membranes with a pH = 10 borate buffer. Such a treatment stabilizes the intermediate state of the photocycle described above. With this simple methodology, completely erasable and oriented BR films can be obtained. D. Haronian and A. Lewis [217] have used these films to impress a variety of words and symbols with high resolution. This gives the possibility to use the BR medium as an electrooptic molecular medium with a memory that is erasable. In addition, the molecular characteristics of this medium can be selectively adressed with the appropriate choice of light pulses to stimulate either BR oder the intermediate state M of the cyclic molecular transformation. Depending on the resolution of the experimental devices, the optical and electric information can be stored in principle in a pixel composed of a single molecular unit or some multiple of this unit. Near room temperature, this allows allows for the storage of information for indefinitively long storage times. The authors [217] could realize with that material a read and write memory. For thresholding purposes, they constructed a special BR-cell, which allows to produce closed and open loop circuit voltages that are proportional to an off/on switching of the threshold function.

 # Summary

In optoelectronic thresholding devices, the combination of conventional electronics or optics with the adaptivity that is required for systems that are able to learn is only achieved at the expense of the loss of the phase information. These devices can be built using

- liquid crystal light valves that transfer the output to the electronic threshold device.
- liquid crystal electrooptic switches that are controlled by electronics to switch only the brightest beams, thus allowing them to reenter the network.
- semiconductor devices having a current-voltage characteristic that is proportional to the threshold function and that is driven by a photodetector.
- the field shielding effect in semiconductors — the change in the electric field and thus in turn in the index of refraction of the optical material through the linear electrooptic effect — can be used as the photorefractive effect to produce optical sigmoidal functions in the microsecond range. It can be combined with a Fabry-Perot etalon to give a bidirectional response for multilayer error backpropagation networks.
- microchannel spatial light modulators that can operate in a thresholding mode, in which a tresholded version of the control image intensity is internally stored.
- organic materials like bacteriorhodopsin that enable to realize hard-limiting or soft sigmoidal threshold functions.

All these threshold functions are unipolar implementations, requiring the dual channel or spatial encoding techniques described in section 4.2 to realize bipolar neurons.

## 5.2 Coherent Thresholding for Bipolar Optical Neural Networks

The advantages of all-optical concepts for thresholding devices have already been discussed in the previous section. The most important of them are on the one hand the possibility to omit all conversions of optics into electronics and back, which often cause a bottleneck in the beam processing operations. For example, for coherent systems that may realize bipolar neurons, the output signals remain completely in the optical domain, do not lose their feature to interfere and to transport information encoded on the phase of the wave and can be used readily for the next stage of processing in a cascaded system. On the other hand, all-optical methods use the inherent advantages of optics as parallelism and cascadability, allowing to perform thresholding in many input channels in parallel because of the intrinsically high bandwidth of coherent operations. Together, coherent and incoherent systems can perform thresholding on many input channels in parallel because of the intrinsically high bandwidth.

In the following paragraphs I will show that photorefractive input-output characteristics as they exist for two-wave mixing, four wave mixing or self-pumped phase conjugation are well-suited to implement sigmoidal functions that are able to threshold multilayer neural networks. Moreover, resonators containing a phase conjugating element and a saturable nonlinear device or resonators containing a laser amplification medium have been suggested to be used as sigmoidal thresholding elements. Bidirectional neurons for backpropagation having a forward sigmoidal nonlinearity and its derivative as the backward operation nonlinearity in contrast can be realized using nonlinear bistable resonators or Fabry-Perot etalons. Finally, computer generated holograms are a flexible means to realize variable thresholding filters. They are especially suited if one wants to realize several different filter functions in a single element without changing the experimental setup for every new thresholding function.

### 5.2.1 Photorefractive Two-Beam Coupling

If we want to realize the bipolar sigmoidal function shown in fig. 2.3c, we have to look for a coherent thresholding process that is able to preserve the phase of the optical beam and at the same time produces a saturation of the output intensity for a wide range of input intensities.

A concept using a photorefractive component has been proposed in 1988 by several authors in parallel, among them H.M. Stoll and L.S. Lee [218] and the group of G. Roosen [219]. In the following, I will discuss the principle function of that concept — amplification and binarization — for the beam intensities. Of course, a similar discussion can be conducted for the amplitudes. From eq. (3.43) derived in section 3.4.3 we already know that the transmitted beam amplified by the two-wave mixing process shown in fig. 3.11 has the input-output characteristic:

$$s_{1,l}^2 = s_{1,0}^2 \cdot \frac{1+r}{r + \exp(-2\gamma l)} \tag{5.5}$$

with the intensity signal-to-noise ratio $r = |s_{1,0}|^2/|p_0|^2$ , the interaction length $l$ in the crystal and the coupling coefficient $\gamma$.

For photorefractive materials that have a diffusion-dominated charge transport mechanism with a phase shift of $\Phi = \pi/2$ between the interference and the refractive index

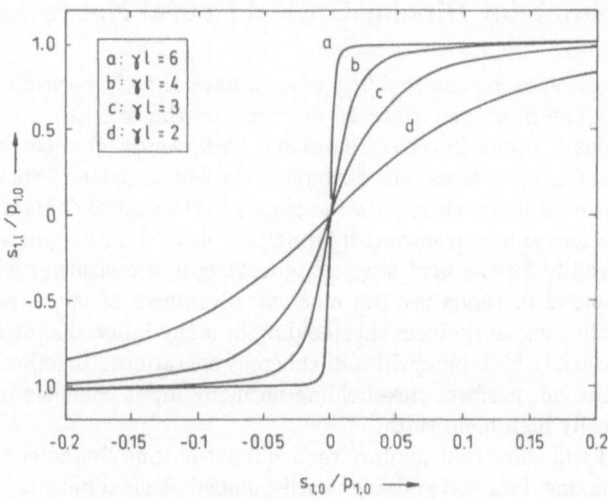

**Figure 5.10:** Numerical simulation of the sigmoidal threshold function using photorefractive two-beam coupling.

grating (see section 3.4), this transformation process preserves the phase of the input signal exactly. Let me denote the amplification factor as $G = exp(\gamma l)$. The behaviour of $s_{1,l}^2$ as a function of the input parameter $r$ can then be characterized by three extremal behaviours:

- For $r \ll 1$ and $Gr \ll 1$, the amplified intensity $I_1(l)$ is proportional to $I_1(0)$. The main slope, given by $G$, is thus large for large values of the gain-interaction length product. It is that region of gain saturation, where image amplification experiments are generally realized.

- When $r$ is increased (but $r \ll 1$), $Gr$ becomes much larger than unity, $I_1(l)$ remains quasi-constant and equal to the pump beam intensity: $I_1(l) = I_2(0)$, whatever the value of $I_1(0)$ is. In this region, $G$ varies as $1/r$. This situation that gives the level of the normalized output signal is achieved when e.g. $G \leq 3000$ and $r \approx 5\%$.

- If $r$ is about unity, the amplified output intensity is $I_1(l) = I_1(0) + I_2(0)$.

This short discussion shows, that a normalized level is obtained for a large number of input intensities. Moreover, this level can be modified by varying the pump beam intensity. The whole behaviour is shown in fig. 5.10 with the amplitudes rather than the intensities of the waves considered. To show how this photorefractive neuron response may be applied to neural network applications, let us have a look at an associative memory that uses Hebbian learning as introduced in section 2.2. In that case, the interconnectivity of neuron $n$ is given by the equation for the output $\mu_i$:

$$\mu_i = \frac{1}{N} \sum_{j=1}^{N} \sum_{m=1}^{M} x_i^m x_j^m x_j \tag{5.6}$$

where $N$ is the number of neurons and $x_i^m$ the $m$-th stored pattern, $x_i^m \cdot x_j^m$ is the interaction matrix $w_{ij}^m$ for the number $m = 1, 2, \ldots M$ of the stored information with the binary elements $x_i^m = \pm 1$. If we present a noisy pattern at the input of such a system with $p \cdot N$ error bits, $y_i$ has a Gaussian distribution with variance $\sqrt{M/N}$ and a mean value which increases from $(1 - 2p)x_i^m$ to $x_i^m$ — provided that all $x_i^m$ are independent, identically distributed variables with the mean value 0 and the variance 1 for a sufficiently large number $N$.

**Figure 5.11:** Experimental realization of the positive arm of a bipolar sigmoidal threshold function using two-beam coupling in photorefractive BaTiO$_3$.

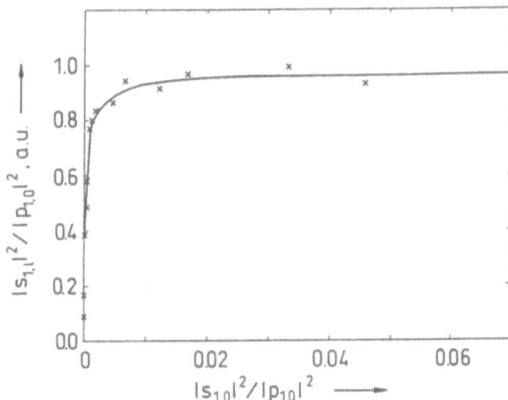

In this photorefractive realization of the optical threshold neuron, the synapse is equal to the output $s_{1,l}$ of the photorefractive neuron. In the case of binary $x_j$, the synapse output or the driving neuron input would be $s_{1,0}$ for the case of lossless optical interconnection. However, in reality, we have to take into account an intensity loss-factor $t = (1 - x^2)$ for each interconnection, leading to an output of the interconnection matrix or synapse of $t \cdot y_i$.

If we take into account the variance and the error bits too, the system can recognize an input successfully, if during the readout process the value of $t \cdot y_i$ remains in an interval that provides a constant or quasi-constant output. If $(s_{1,0}^2/s_{2,0}^2)_{\min} =: \sqrt{r_{\min}}$ and $(s_{1,0}^2/s_{2,0}^2)_{\max} =: \sqrt{r_{\max}}$ are the upper and lower borders of this region, we obtain the conditions:

$$(1 - 2p - \sqrt{M/N})\,t \;>\; \sqrt{r_{\min}} \tag{5.7}$$

$$(1 + \sqrt{M/N})\,t \;<\; \sqrt{r_{\max}} \tag{5.8}$$

The second condition is obviously always true, because the losses in the interconnection gratings are always high enough. The fact that the first one is fulfilled may be made plausible by the following estimation. The typical attraction basin area of a stable solution of a network is given by [219]

$$pN = N \left( \frac{1 - \sqrt{\dfrac{M}{M_{\max}}}}{2} \right) \tag{5.9}$$

with the maximum number of patterns which can be stored as stable states after Hopfield $M_{\max} = N/(2 \ln N)$ (see section 2.5). For large $N$, the first condition defines thus an upper limit for the losses, for which the net is still capable of producing a stable solution:

$$t^2 > \frac{N}{2M \ln N}\, r_{\min} \tag{5.10}$$

Experimentally, the sigmoid answer function has been realized with two-beam coupling by the author and her coworkers as shown in fig. 5.11. G. Roosen et al. [219] could show experimentally, that if a defined cross section is guaranteed for each channel, up to 625 neuron channels can be thresholded and processed in parallel.

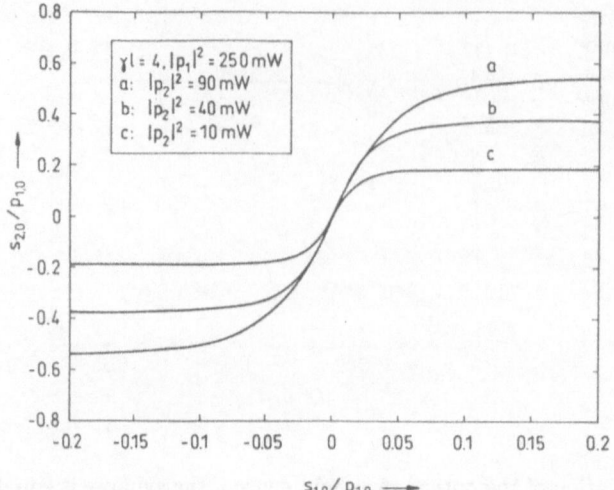

**Figure 5.12:** Numerical simulation of the sigmoidal threshold function using photorefractive four-wave mixing.

H.M. Stoll and L.S. Lee and coworkers [218] realized with that concept of the optical thresholding neuron element a complete continuous-time optical neural network capable of executing a broad class of energy-minimizing neural net algorithms. The network is based on a ring resonator, which contains a saturable two-beam amplifier, that acts as the optical thresholding neuron array element and two volume holograms as interconnection devices. This system will be described in section 11.2.

### 5.2.2 Photorefractive Phase Conjugation

Because the characteristic beam coupling functions are similar for two-wave mixing, four-wave mixing and phase conjugation, one would expect the same input-output characteristics for the case of a four-wave interaction. Indeed, as indicated in fig. 5.12, the sigmoidal function may also be simulated by four-wave mixing with simultaneous phase conjugation.

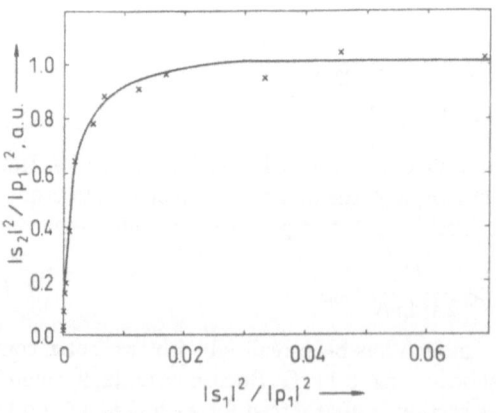

**Figure 5.13:** Experimental realization of the positive arm of a bipolar sigmoidal threshold function using four-wave mixing in photorefractive $BaTiO_3$.

**Figure 5.14:** Experimental realization of the positive arm of a bipolar sigmoidal threshold function using self-pumped phase conjugation in photorefractive BaTiO$_3$.

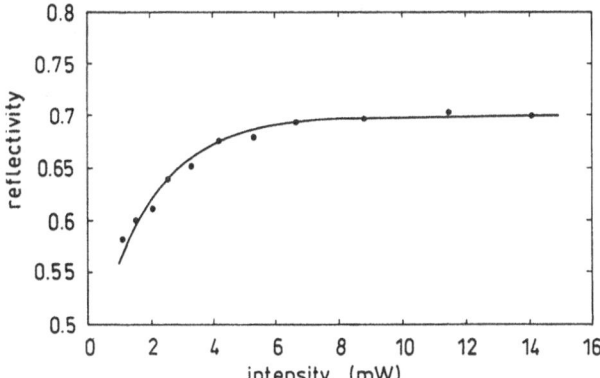

In experimental realizations (see fig. 5.13), this case is even more interesting than two-wave mixing, because the slope of the "undecisive" region of the sigmoidal curve may be varied by changing the intensity of the backward pump wave. This is a rather important fact for all feedback systems, because no new adjustment is necessary in that case to adapt the threshold for each neuron.

The disadvantage of four-wave mixing neural thresholding lies in its sensitivity to slight disturbances of the adjusted setup: Even small deviations in angle or frequency of one of the two pump waves destroys the exact phase conjugation of the incoming information bearing beam and creates response signals shifted in angle or frequency. This problem can be omitted by utilizing the process of self-pumped phase conjugation. In that case of beam coupling (see section 3.5), the threshold function may be realized as well bipolar (amplitude and phase thresholding) as unipolar (intensity thresholding). In the latter case, an operational threshold for small intensities can be realized, as shown in fig. 5.14. This allows to use the system as a coherent amplitude thresholding device or as an incoherent intensity thresholder, depending on the requirements of the neural net architecture. In order to enhance the reflectivity of the system, a semi-self-pumped system can be realized, where a supplementary reference wave supplies energy to the system. The response of the system in dependance on the input intensity ratio is the same as for ordinary self-pumped phase conjugation, but can be tuned by the reference beam [220].

Moreover, thresholding can be achieved in a semilinear self-pumped phase-conjugate mirror. This system is achieved when the backward reference wave is created by reflecting the input beam back into the crystal by an ordinary or a curved mirror (see section 3.5, fig. 3.17). Thus, external retroreflection occurs for the fanned beam, creating two four-wave mixing interaction centers as in self-pumped four-wave mixing. By introducing grating motion within the photorefractive crystal — e.g. by applying an external electric field — and using again the dependence of photorefractive response time on the intensity of the incident beam, a similar threshold dependence as for self-pumped phase-conjugation can be achieved. However, the threshold level is here a function of the grating velocity [221].

A similar system that supresses bistability effects has been realized by R. Yahalom and coworkers [222] by using a fiber-coupled-ring passive phase-conjugate mirror. Two mutually incoherent input beams are focused into a multimode graded-index fiber, each having a different angle. The mode-scrambled depolarized light diverging from the output end of the fibers passes through a polarizer and is then focused into a semilinear $BaTiO_3$ self-pumped phase-conjugate mirror. Again, the positive arm of a sigmoidal function can be realized with this configuration by changing the input beam intensities.

In summary, all photorefractive beam coupling processes are in principle suited to simulate the sigmoidal function required to realize optical thresholding devices. However, the technical demands of stability and easiness of adjustment of the devices for each neuron prevented most of the ideas presented in that area from a broader realization.

Some of the concepts in that area will be presented in the next section.

### 5.2.3   Nonlinear Resonators

Let us now have a look at how threshold functions may be realized in more complex neural net structures. For that purpose, I will give a short review of the most important features of these multiple layer networks before I will describe an example that is based on Fabry-Perot-resonators that are filled with a nonlinear material in order to create the responses necessary in these nets.

Multilayer neural networks have already been discussed in sections 2.3 and 2.6 and have been shown to be capable of performing complicated mappings from the input to the output plane by developing an internal representation and detecting features by a set of hidden neurons. Systems that implement error backpropagation allow for adaptive corrections of interconnection weights *during* the operation of the network. The generation of the error signal is thus a main task in neural network structures with backpropagation learning. In general, the error signal is the result of the comparison of the desired output and the actual output. If they are different, the error signal has to be formed in such a way that the interconnection will be modified in order to minimize the difference between the two outputs. In section 2.3.1, I have described how the threshold function must be modeled to realize a characteristic response that has a sigmoidal form in the forward direction and the shape of its derivative in the backward direction.

This learning procedure, defined as the adaptation of the interconnection weights, can be achieved through an iterative process called the generalized delta rule (see section 2.3). In detail, an input is initially introduced to the first layer, while the network's response at the final layer is compared to a desired or target value, and an error term is generated. The second phase of learning involves backpropagating the error through the same interconnection network and the hidden layers for the purpose of adjusting the weights according to the local value of the backpropagating error term. In a mathematical formulation, the weight change $\Delta w_{kj}$ has to be performed in such a way that the mean-square error energy function over the training set is gradually minimized

$$\Delta w_{kj} = \delta_k \cdot o_j. \tag{5.11}$$

Here, $\delta_k$ is the error signal of unit $k$ and $o_j$ is the output from unit $j$.

If we assume unit $k$ as the output layer of the network, its error signal for output units is given by

$$\delta_k = (t_k - o_k) f_k'(\mathrm{net}_k) \tag{5.12}$$

where the actual output $o_k$ is given by $o_k = f_k(\mathrm{net}_k)$, $t_k$ is the desired or target output and $\mathrm{net}_k$ is the net input to unit $k$. $f_k(\mathrm{net})$ is a logistic type of activation function for unit $k$ to execute a soft (in most cases a sigmoidal) thresholding and $f_k'(\mathrm{net})$ is its temporal derivative.

**Figure 5.15:** Schematic representation of the functional operations of eqs. (5.12), (5.13). $d$ is the desired, $y$ is the actual output of the net for a given input $x$, $f$ is the activation function, and $f'$ is its temporal derivative.

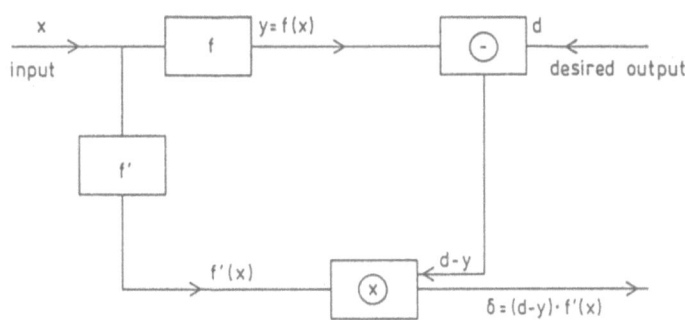

In analogy, the error signal for hidden units $j$ is determined in terms of $\delta_k$ as

$$\delta_j = \left( \sum_k w_{kj} \delta_k \right) \cdot f'_j(\text{net}_j) \tag{5.13}$$

Eqs. (5.12) and (5.13) is schematically represented in fig. 5.15. For simplicity, the error signal between the output and the hidden layers before the output based on eq. (5.13) is generated. By propagating the error signal back one layer, the same process can be repeated to generate the error signal for every layer — the extension is hence straightforward. This process is iteratively continued until all weights are adjusted so that the output matches the target value and the error term is reduced to zero. The reader should refer to the exhaustive explanations in section 2.3 for detailed explanations.

The physical implementation of such a network would require a bidirectional neuron with three important features. It should be activated by a forward activation input, it should have a differentiable nondecreasing forward response, such as the one presented in that chapter as e.g. the sigmoidal function, and finally a backward response that is the derivative of the forward one. This feature, already treated in section 2.3, can be intuitively understood by noting that the weights are changed according to the error signal. This signal is highly transmitted when the neuron is undecisive and is, therefore, uncommitted to being fully "on" or "off" in the forward direction. This situation is sketched in fig. 5.16. It is important to note for all optical implementations, that both

**Figure 5.16:** Bidirectional neuron for backpropagation, its forward mode saturating nonlinearity and its magnified derivative (after [223]).

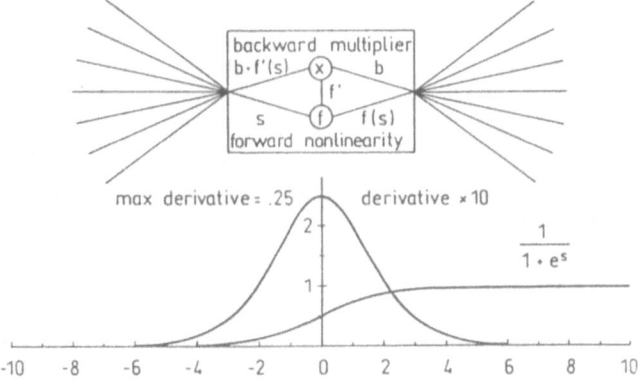

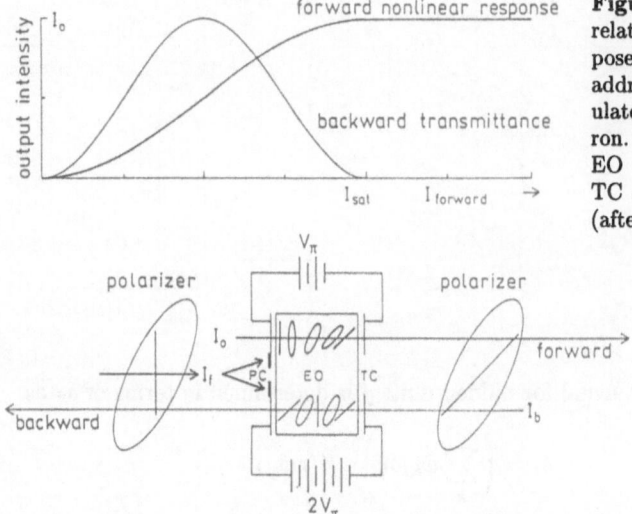

**Figure 5.17:** Input-output relations for a special purpose bidirectional optically addressed spatial light modulator backpropagation neuron. PC = photoconductor, EO = electrooptic material, TC = transparent conductor (after [223]).

the forward and the backward signals will travel through the same interconnection network, and their weight matrices should be the transpose of each other — an important simplification compared to electronic realizations of neural networks. The simplification of the implementation of complex interconnections would be further enhanced by a single bidirectional neuron device. Furthermore, it would be advantageous to use the same wavelength for both directions, in that the interconnection network may be wavelength sensitive in both its patterns and weights.

One possibility to achieve such a device is to taylor a special purpose, bidirectional detector-modulator pair array structure to generate the desired backpropagation neural response. However, in this case, one would have to use appropriate integrated electronic circuitry, thus losing the requirement of a coherent neural response. Although such a concept may be well-suited for nets with a small number of neurons, the individual neuron could become quite complicated with this conventional optoelectronic integrated circuit approach, requiring a larger electronic effort for high capacity networks.

A better solution for backpropagation neurons are therefore appropriately modified transmissive optical elements, as e.g. spatial light modulators. One possible structure of this type is illustrated in fig. 5.17. In this type of birefringent SLM, crossed polarizers are placed on either side of an electrooptic medium which is optically addressed by a photoconductor. A high voltage is applied across a transparent conductor in contact with the photoconductor on one side and a transparent conductor on the other. To use this type of electrooptic device as a backpropagation neuron, the induced birefringence must be doubled for the backward propagating wave to obtain a saturating forward nonlinearity while obtaining a derivative backward multiplication. This can be accomplished by a pair of photoconductors, both addressed by the same forward propagating beam, where one is used to modulate the forward propagating device which is biased with a certain voltage $V_\pi$, while the other is used to modulate an electrooptic device with a saturation voltage $V_{2\pi}$.

A short mathematical estimation using the forward beam intensity $I_f^m$, the backward beam intensity $I_b^m$ for the $m$-th iteration cycle and the saturation intensity $I_{sat}$ may illustrate the principle of that device: The forward propagating modulator is used to modulate a fixed intensity pump $I_0$ so that a single half-cycle of a saturating nonlinearity can be generated

$$I_f^m = \begin{cases} I_0 \sin^2\left(\frac{I_f^{m-1}}{I_{sat}}\right) & \text{for } I_f^{m-1} < I_{sat} \\ I_0 & \text{otherwise} \end{cases} \qquad (5.14)$$

The backward propagating modulation is used to multiply the backward propagating error signal by a function

$$I_b^{m-1} = \begin{cases} I_b^M \sin^2\left(\frac{2I_f^{m-1}}{I_{sat}}\right) & \text{for } I_f^{m-1} < I_{sat} \\ 0 & \text{otherwise} \end{cases} \qquad (5.15)$$

This last equation has exactly the form of the desired derivative multiplication.

However, several years ago some less complicated resonator structures have been found that accomplish these two functions required for backpropagation in a faster and more reliable way compared to the SLM technology. Therefore, our attention in the remainder of that section will focus on the performance of these promising devices to realize a coherent optical error backpropagation neuron. They are based on resonant structures involving nonlinear optical processes.

*Nonlinear Fabry-Perot Etalons*

The idea of D. Psaltis and K. Wagner from the California Institute of Technology [223] was to use nonlinear Fabry-Perot etalons for implementing these neurons, because they can perform nonlinear operations on arrays of coherent beams in a quite natural way, which allows the outputs to be used to record and modify interconnection holograms as they can e.g. be written in photorefractive volume materials. A soft thresholding operation can be performed on a forward propagating beam by decreasing the cavity detuning below the critical detuning needed for bistability. However, these optical neurons can not as easily implement the idealized derivative transmission required for backpropagation. But a similar peaked response can be obtained by operating a nonlinear etalon in the *probe mode* for the backward propagating error signal. In this mode, the Fabry-Perot resonance is scanned by the nonlinear dependence of the index on the intracavity intensity, which varies in response to the high power forward beam intensity. The weak backward propagating probe beam does not scan the cavity, but it is modulated by the current state of the cavity transmission function, which is the appropriate derivative type of response needed in the backward direction. The probe mode transmission is peaked at the resonance of the Fabry-Perot, which occurs when the sigmoid response to the forward beam reaches the upper level. The peak maximum is not exactly at the highest point of the slope of the forward beam's nonlinear sigmoid response, but since the forward and backward beams have different polarizations (or different wavelengths), the resonance function can be offset to achieve a properly positioned probe beam resonance peak.

In the polarization multiplexed case this shift can be induced by including a thin birefringent sheet in the cavity, or a tunable birefringence that can be caused by applying a static external electric field to the cavity. Moreover, the same effect can be achieved using two closely spaced cavities, both addressed by the same forward and backward propagating resolution spots. In this case one cavity is optimized to produce a sigmoid response of the forward propagating beam while blocking the backward propagating error signal, while the other cavity is resonant to the backward propagating beam.

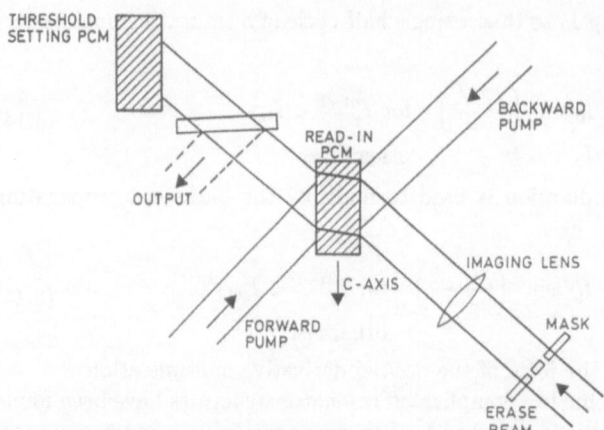

**Figure 5.18:** Architecture for an imaging threshold detector using a double phase conjugate resonator and an incoherent erase beam (after [224]).

Although all of these methods do not ensure a real gradient descent of the error backpropagating signal as it was presented for the backpropagating learning algorithm in section 2.3, D. Psaltis and coworkers expect that learning and convergence can be achieved in a multilayer optical network with the forward and backward response that can be obtained from these devices because of the robustness of this gradient descent learning procedure [223]. Several multilayer networks using error backpropagation based on this principle are shown in chapter 9, especially in section {sec:alloptmultipercep. .

*Phase-Conjugate Resonators*

A different approach to realize optical coherent thresholding by taking advantage of resonant nonlinear processes still goes back to photorefractive devices. In 1986, M. Klein and coworkers [225] demonstrated an imaging thresholding detector using a phase conjugate resonator provided by degenerated four-wave mixing. A semi-linear resonator, which is bounded by a conventional mirror at one side and by a phase conjugate mirror at the other side is the heart of the architecture. The device is based on spatially resolved grating erasure in the photorefractive medium from the backward side of the resonator phase conjugate mirror. In detail, the signal input to the device is in the form of an image bearing erase beam that is incident on the crystal, collinear with the resonator axis (refer to fig. 5.18). The erase beam optics are chosen to image the input spatial intensity pattern into the pumped volume of the crystal. It must be chosen to be noninterfering with the pump beams. This avoids the formation of undesired errorneous gratings which would degrade the device's performance. Such an erasure beam can be generated using a different wavelength for the pump beams, a different polarization or it could be made incoherent by increasing the path length difference between the pumps and the erase beam to a value greater than the coherence length of the laser. Finally, it could be generated by using an incoherent light source. The intense portions of the image then erase the photorefractive gratings and locally extinguish the resonator output; oscillations build up, however, at locations in the image where the intensity is too weak to erase the grating. The resultant resonator pattern at the phase conjugate mirror is then a highly nonlinear, inverted-contrast image of the input. The threshold itself can be modified in this setup by insertion of loss in the resonator or by changing the reflectivity of the phase conju-

**Figure 5.19:** Transfer curve for degenerate four-wave mixing and phase conjugate resonator based devices utilizing an erase beam (after [224]).

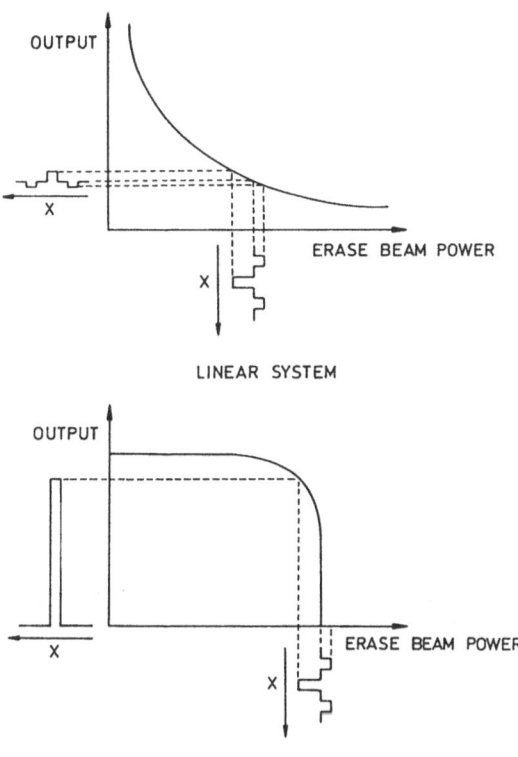

gate mirror itself. Moreover, the system can be improved in its performance by using a saturable absorber in the resonator to obtain real bistable operations depending on the input conditions [226]. The authors could show experimentally the enhancement of one or more elements in an array of picture elements with a high resolution. Thus the system is able to detect, whether one or more elements of a given image have an intensity above or below a set threshold value, even in the presence of a high level of background noise. Moreover, it is able to enhance the signal-to-noise ratio of the bright spots relative to the background.

A second improvement of the system is depicted in fig. 5.18. In this device, a second phase conjugate mirror is used instead of the ordinary mirror. The transfer curves for the degenerate four-wave mixing signal versus the input erasure beam intensity is shown in the upper portion of fig. 5.19. It is important to note, that for any operating point the output signal is linearly related to the input. The lower curve corresponds to the output of the phase conjugate resonator based device versus the erase beam power. If the input operating point is chosen to be near the threshold level, notice the existence of a nonlinear threshold operation. Those portions of the erase beam higher in intensity than the threshold produce no output, while those portions below the level produce a signal. Another advantage of the phase conjugate resonator based threshold device is that the output signal will oscillate at a power level close to the maximum output power level. This improves the device's signal-to-noise ratio for pixels in the "on" state compared to those in the "off" state.

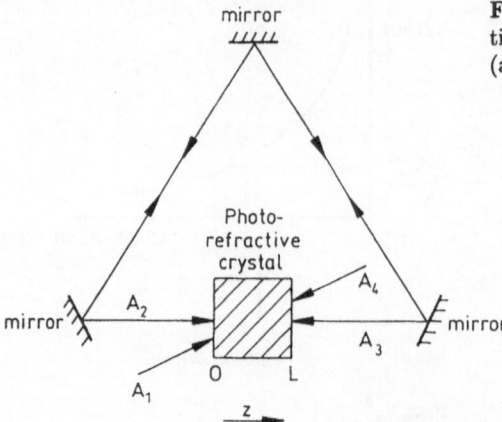

**Figure 5.20:** Schematic of a bidirectional photorefractive ring resonator (after [227]).

### Self-Oscillations in Optical Phase-Conjugate Resonators

A similar approach to nonlinear resonator thresholding of M. Klein and coworkers was proposed by C. Gu, P. Yeh and coworkers [227] of the University of California. They studied the properties of self-oscillation optical bidirectional resonators using nonlinear photorefractive four-wave mixing for parallel multiple channel optical thresholding. Beneath thresholding, they also achieved *maximum operation*, an operation defined as the one for finding the maximum value among more than one input variables. This is the function that is desirable in the implementation of neural networks of the winner-take-all type. . The principle of their method is to exploit the starting point of self-oscillations as a criterion for enhancement or positive thresholding.

Imagine a photorefractive bidirectional ring resonator pumped by two counterpropagating beams as shown in fig. 5.20. Self-oscillation in the ring rsonator occurs if the two counterpropagating beams $A_2$ and $A_3$ are generated without an incident probe beam ($A_2(z = 0) = A_{20} = 0$). The response of such a ring resonator is shown in fig. 5.21. Because the oscillating beams form a feedback loop with the ring resonator, a coupling constant *Gammal*, where $l$ is the interaction region of the beams inside the crystal, is required for establishing steady-state oscillation in the ring resonator, that is smaller than the one necessary in four-wave mixing (see above). In the region near threshold,

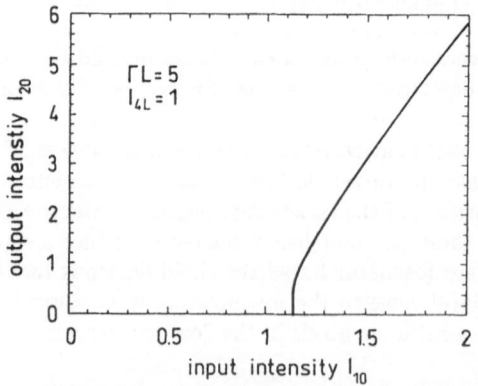

**Figure 5.21:** Output intensity $I_{20}$ as a function of the input intensity $I_{10}$ when the reference intensity is fixed as $I_{4l} = 1$ and the coupling constant is $\Gamma l = 5$ ($l$ = length of the interaction region inside the photorefractive crystal (after [227]).

**Figure 5.22:** The linear stability analysis shows that there may be an instability near threshold. Within the unstable region, cavity outputs change between zero and nonzero oscillation intensities and are sensitive to any pertubations (after [227]).

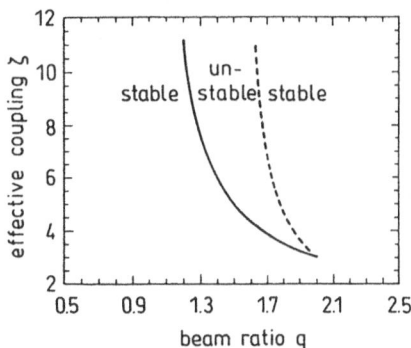

instabilities may occur in the resonant system. Fig. 5.22 shows the result of the linear stability analysis ($\zeta = \Gamma l/(1 - R)$ is the effective coupling constant, $R$ is the product of the mirror reflectivities and $q = I^1(0)/I_4(l)$ is the ratio between the two pump beams). When $\zeta > 1$, and $q > q_{th} = (\zeta + 1)/(\zeta - 1)$, the resonator will have nonzero oscillation intensities. When $\zeta < 3$, the ring resonator exhibits a stable steady state that depends on the beam ratio $q$. Finally, when $\zeta > 3$, a region near $q_{th}$ in which both trivial and nontrivital solutions are unstable always exists. When $q$ is beyond this region, stable oscillation can be maintained again. Within the unstable region, cavity outputs change between zero and nonzero oscilltion intensities and are sensitive to any pertubations. However, since the transitions from a stable state into instability occur by means of a saddle bifurcation, no self-pulsation or other instabilities are expected. This effect can be exploited to use the system as a thresholder and maximum operator. When an array of input beams with different intensities (represented in the figure by different line types) interacts at different locations inside the crystal with a respective array of reference beams with equal intensities, maximum operations can be implemented by adjusting the intensity of the reference beams.

As the respective pairs of input and reference beams interact within the crystal, optical thresholding can be performed in parallel. In addition, by adjusting the intensity of the reference beams, it is possible to identify the beam with maximum intensity. This is done by increasing (or decreasing) the intensity of the reference beams. Now let us implement the device in the feedback system described above in such a way, that the signal and its phase conjugate replay oscillate, whereas the pump waves supply the energy to enable self-oscillations. When the reference beam intensity is decreased, oscillation occurs when the intensity of the brightest input beam reaches the regime that allows self-oscillation. At this point, the brightest beam is selected and located. With this technique,

**Figure 5.23:** Principle of parallel thresholding and maximum operation (after [227]).

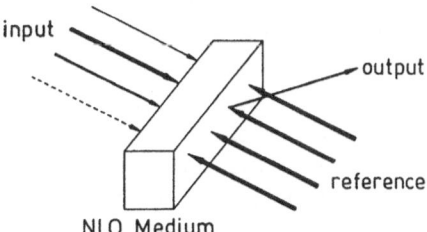

the comparison is done in parallel and the maximum can be found without measuring the intensities of the light beams electronically. This approach is extremely useful when the number of input beams becomes large. Moreover, similar properties also exist in other nonlinear media beside photorefractive crystals. Therefore, the principle is a basic one for all neural structures, where the maximum of several thousands or more input signals has to be selected. However, the system still has some problems, the most significant one being the instability near the threshold operation point. This fact makes the distinction between two signal beams with close intensity levels very difficult and sensitive to system faults.

 ## Summary

Coherent thresholding is attractive because of its capability to transport information encoded on the phase of the wave and the feature of coherent waves to interfere. Thus, bipolar neurons can be realized without doubling the number of interacting channels. The most important nonlinear threshold function, the sigmoid function, can be realized using

- photorefractive two-beam coupling in diffusion-dominated crystals in a saturable amplification mode.
- photorefractive phase conjugation. The slope of the "undecisive" region of the sigmoidal curve may be varied by changing the intensity of the backward pump wave.
- phase conjugate resonators, where thresholding takes place by spatially resolved grating erasure in the photorefractive medium from the backward side of the resonator phase conjugate mirror.
- self-oscillations in phase-conjugate resonators. The starting point of self-oscillations is exploited as a criterion for enhancement or positive thresholding.
- the laser, in a mode where the oscillation of one eigenmode inhibits the oscillation of all other states, creating a winner-takes-all system.
- computer generated holograms as variable filters. The nonlinearity in the filter plane is implemented using a nonlinearly transformed filter to store the associative images such that nonlinear correlations between the input image and the stored images are obtained.

Backpropagation networks require the following thresholding functions

- a differentiable, forward response function, such as the sigmoid function and
- a backward response that is the derivative of the forward one, having its maximum peak in the undecisive region with the steepest slope of the forward response.

They can be realized all-optically using

- birefringent spatial light modulators combined with a pair of photoconductors. Thus, the induced birefringence is doubled for the backward propagating wave to obtain a saturating forward nonlinearity and a derivative backward multiplication.
- nonlinear Fabry-Perot-etalons. A soft thresholding operation can be performed on a forward propagating beam by decreasing the cavity detuning below the critical detuning needed for bistability.

- In a similar manner, a nonlinear etalon can be operated in the probe mode for the backward propagating error signal. In this mode, the Fabry-Perot resonance is scanned by the nonlinear dependence of the index on the intracavity intensity, which varies in response to the high power forward beam intensity. The weak backward propagating probe beam does not scan the cavity, but it is modulated by the current state of the cavity transmission function, which is the appropriate derivative type of response needed in the backward direction. The probe mode transmission is peaked at the resonance of the Fabry-Perot, which occurs when the sigmoid response to the forward beam reaches the upper level.
- In many implementations, the peak maximum is not exactly at the highest point of the slope of the forward beam's nonlinear sigmoid response. However, the robustness of the gradient descent procedure in Boltzmann machines allows still convergence and learning.

 ## Further Reading

1. M. Wang, A. Freeman, *Neural Function*, Little, Brown, Boston (1987).
   This book describes in detail the different neuronal threshold functions, giving attention to problems as the realization of bipolar thresholding, exhibitory and inhibitory neuronal functions.
2. H. Haken, *Neural and Synergetic Computers*, Springer-Verlag, Berlin (1988).
   This introduction into synergetics contains several ideas how to use feedback systems as thresholders, exploiting self-organization.

# 6    Further Computing Elements

With the basic components described in the last chapters — interconnection and storage devices in chapter 4 and thresholding elements in chapter 5 — we are in principle able to build different simple neural network systems which would be able to perform associative tasks, feature extraction or solve nonlinear decision problems. However, for a large range of more complex networks as e.g. all multilayer networks using error backpropagation, it is convenient to have recursion to a number of computational functions that enable to detect and process differences between the desired and the actual output or that visualize temporal changes in the two-dimensional representation of the input.

Therefore, the present chapter gives an insight in the methods that are currenlty used to perform these computing tasks optically. Because these elements require a reliable compatibility with all the devices and elements discussed up to now in the network, I will concentrate the overview on those elements that can be connected easily with the devices described in the last two chapters. Naturally, as coherent devices have been favourited in that book, coherent optical and especially photorefractive devices represent the major contribution to that chapter.

First, the optical realization of algebraic operations necessary for image comparison — as addition, subtraction as well as temporal and spatial derivations — are presented.

Here, photorefractives revealed to be the most appropriate and successful device material for quick and high-resolution parallel optical calculations. The second important device for neural network architectures are filters, which enable to extract the important information out of the whole data set. Novelty filters e.g. extract what is novel or what is stationary in a two-dimensional scene. This element is helpful in networks, where changes of the desired output appear with short response times, but it is also an attractive device for several applications in data reduction for the processing of high-definition television images, for movement detectors and machine supervision in manufacturing environments. Among these devices, optical novelty filtering using photorefractive devices has become popular, because it uses the simple and reliable effect of beam coupling to ensure filtering.

Next, bistable elements, that are able to realize decisions between two states or realize a binary threshold behaviour are presented in a very short way, relating these elements to the requirements of optical neural networks. Bistable switching in all-optical neural nets may be best realized using resonant devices that allow for a well-defined switching characteristic. Therefore, I will discuss the effects of saturable absorbers on phase-conjugate resonators and the behaviour of photorefractive flip-flop memories in the present chapter. A very important category of devices for photorefractive memories are systems that are able to refresh the stored data, because every new readout or learning process tends to erase previously written information. Thus, methods to refresh the data optoelectronically or all-optically will be discussed in the fourth section of that chapter.

Finally, because every neural net is relied on the resonant feedback of the data for comparison, updating or processing, it is necessary to discuss the conditions and requirements, the different techniques of interconnection storage, thresholding and processing impose upon the feedback system. The problems arising for coherent optical feedback are much more severe than the ones arising for all-optical or electrooptic incoherent feedback, because one has to guarantee the stability and preservation of the phase as the optical information cycles through the net. Especially for the processing of two-dimensional information that is distributed transversally to the light propagation direction, one has to ensure high stability conditions. Fortunately, phase conjugation is a well-suited means to adapt larger nets to phase matching because of its ability of correcting aberrations and phase errors in the net. Therefore, I will discuss at the end of this chapter how the characteristics of oscillators and resonators using the elements described in the previous chapters have to be tailored in order to allow the realization of simple neural network configurations.

## 6.1 Optical Arithmetic and Logic Operations

The problem of how to arithmetically calculate in optics, how to add, subtract, multiply, differentiate or integrate has engaged scientists now for about fifty years. First, optoelectronic systems have been proposed, suffering from the already mentioned fact that electronic digital processing of images is slow because of its serial nature. Optical techniques in contrast offer the capability of parallel processing over the entire images. The parallel processing is versatile and inherently faster. The technique of optical image synthesis by the addition and subtraction of the complex amplitude of light was first described by D. Gabor et al. [102]. Its basic principle consists of spatially modulating the two images to be processed by periodic waves that are mutually shifted by a phase shift of $\pi$. In that way, the two images are consecutively recorded on a holographic plate and may then be read out subsequently of the composite hologram.

The attractiveness of that area of optical realization of arithmetic and logic functions results from its various fields of applications, among them quality control in manufacturing environment, vibration analysis, earth resonance studies, meteorology, growth studies, pattern recognition and bandwidth compression in communication systems. Most of these applications involve maintaining the changes of a certain spatial information or image at two instants or states. If the comparison is a temporal one, e.g. comparing two different states of the same object, the characteristic time interval between the two instants may range from a millisecond for vibrational analysis to several months for e.g. earth resonance studies. Here, time discriminating devices as they are represented by novelty or monotony filters are best suited for optical implementations.

### 6.1.1 Addition, Subtraction Using Phase-Conjugate Interferometry

The classical comparison method to fulfil tasks in image or object state comparison is interferometry, where two images are coherently overlapped at the output. For most cases, Mach-Zehnder or Michelson configurations are chosen, as illustrated in fig. 6.1. On the image planes, where images of the two objects overlap, image subtraction (addition) takes place by destructive (constructive) interference, provided the phases of the two wavefronts are 180° out of phase (in phase) throughout the whole overlapping region. If one can introduce and maintain a phase shift of 180° between the wavefronts of the two pictures at the output port, parallel subtraction can be achieved. In practice, the overlapped image intensity often drifts between the extrema because of the phase fluctuation due to ambient air currents and/or thermal drift of the interferometer body. Thus the setup is very

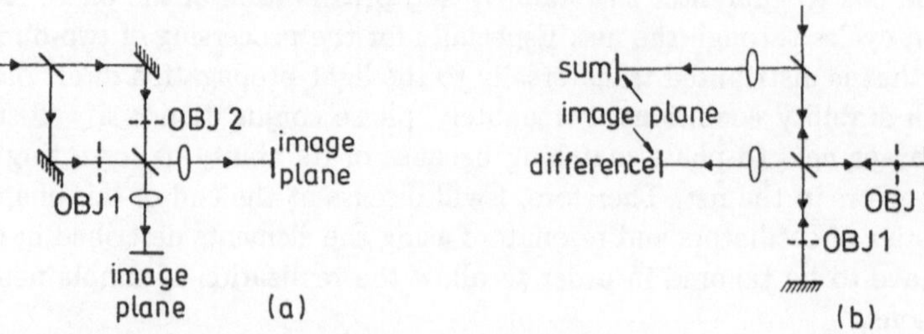

**Figure 6.1:** Basic idea of coherent image subtraction and addition using (a) Mach-Zehnder and (b) Michelson interferometers. (after [228]).

sensitive to environmental conditions. In addition, accurate subtraction requires critical alignment of the interferometer, which is extremely difficult to maintain in practice.

A way out of this class of problems in optical image subtraction is to use phase-conjugate Michelson interferometry — first introduced independently by N.G. Basov [229] and F.A. Hopf [10], later further developped by M.D. Ewbank [230] and photore-fractive two-beam coupling. The phase-conjugate interferometer is a modified version of the conventional optical interferometer with one or more of its conventional mirrors be-ing replaced by phase-conjugate reflectors. Although this system ensures overall stability and no adjustment problems, it has not been thoroughly explored for practical applica-tions in industrial environment. The rather slow development in the past was probably due at least in part to the high optical intensity or long response time requirements, low efficiency, and relatively expensive and bulky hardware involved in implementing a phase-conjugate mirror. Recent advances in nonlinear optical materials in general and photorefractive materials in particular have alleviated many of these problems. Phase-conjugate mirrors (PCM's), based on self-pumped configurations as introduced in chap-ter 3.5, exhibit reflectivities as high as 70% [231] while requiring only a few tens of milliwatts of optical input power. A phase-conjugate Michelson interferometer, for in-stance, is particularly suitable for real-time image subtraction and detection of temporal evolution (or changes) of a scene. As will be apparent from the discussion below, the process of phase conjugation significantly improves the versatility and dynamic stability of the conventional interferometric techniques by eliminating the effective optical path difference between the two arms and reducing the sensitivity to beam path fluctuations and misalignment. These novel techniques incorporating phase-conjugation have added a new dimension to conventional coherent image processing. Applications such as visualiza-tion of vibration modes [232], edge enhancement and intensity contour generation [233], spatial convolution or correlation [119], and detection of defects in periodic patterns [234] have all been demonstrated in real time.

In the following sections, I will focus onto the basic principle of a phase-conjugate Michelson interferometer (PCMI) as it has been realized by the author and her coworkers [235]. As illustrated in fig. 6.2(a), both the reflective and transmissive components (of the incoming beam) that propagate from the beam splitter to the two phase-conjugate mirrors are retroreflected back in a "time reversal mode" and recombined at the beam splitter. The "time reversal" characteristic of the phase-conjugate beams ensures that the relative phase delay due to any optical path difference and the slow phase fluctuation resulting from air currents along the paths and/or thermal drifts are canceled after the

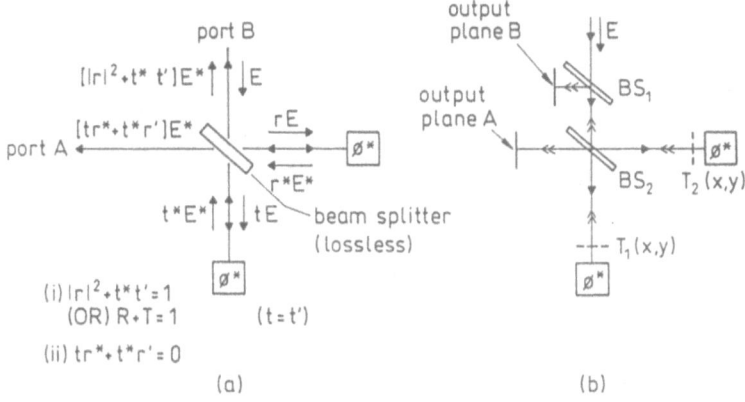

**Figure 6.2:** (a) Simple analysis of the Stokes equations for time reversal. (b) Basic principle of optical image subtracting using a phase-conjugate Michelson interferometer (after [228]).

round-trip passage from the beam splitter to the phase conjugator. What is probably as important as this is that the interferometer is self-aligning.

*Mathematics of Addition and Subtraction using Phase-Conjugate Interferometry*

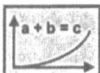  The phenomenon described above can be analyzed as follows. Let us introduce in both arms of the interferometer the amplitude transparencies $T_1$ resp. $T_2$. Then we obtain in the output plane of detection

$$I_{out} \propto T_1^2 + T_2^2 + 2T_1T_2 \cos \Delta\phi \qquad (6.1)$$

In that equation, $\Delta\phi$ represents the sum of all phase changes arising at the mirrors or at the phase-conjugating elements

$$\Delta\phi = \phi_r + \phi_r' - 2\phi_t + \phi_{pc_1} - \phi_{pc_2} \qquad (6.2)$$

with $r =|r| e^{i\phi_r}$, $r' =|r'| e^{i\phi_r'}$ and $t =|t| e^{i\phi_t}$, $t' =|t'| e^{i\phi_t'}$ being the reflection and transmission coefficients of the amplitudes in forward or backward direction of the beams and $\phi_{pc_i}$ being the phase change of the phase-conjugate mirrors. For that equation, we will see that a variety of logic or arithmetic operations may be realized by adjusting the phase parameter carefully. Assuming a loss-free beam-splitter ($\phi_r + \phi_r' - 2\phi_t = \pi$) and a perfect phase conjugation ($\phi_{pc_1} = \phi_{pc_2}$), we can realize *image subtraction*

$$I_{out} \propto (T_1 - T_2)^2. \qquad (6.3)$$

The case of $\Delta\phi = 0$ can be achieved by adding a supplementary mirror to the system, introducing one more phase shift of $\pi$ to the system. Then, we realize coherent *image addition*

$$I_{out} \propto (T_1 + T_2)^2. \qquad (6.4)$$

For $\Delta\phi = \pi/2$ we are able to construct an *OR-operation*

$$I_{out} \propto T_1^2 + T_2^2. \qquad (6.5)$$

If we substitute one of the images by a plane wave (e.g. $T_1 = 1$), we may even arrive at the *inversion* of an image (compare to eq. (6.3))

$$I_{out} \propto (1 - T_2)^2. \tag{6.6}$$

Finally, the combination of inversion and the OR-operation enables the realization of a NOR-function. Moreover, it opens the possibility of spatial differentiation. For this case, one has to place the image transparency before the first beam splitter into the course of the principal rays, instead of putting it into both arms of the interferometer. To understand the principle function of that case, consider the input field $E_{in} = |A_0|e^{i\phi_0}$ with the amplitude transmission $F(x,y) = |F_0(x,y)|e^{if(x,y)}$. The fields in the interferometer arms are then, neglecting all terms of propagation,

$$E_1 = |t||F_0||A_0|e^{i(\phi_0 + f + \phi_t)} \tag{6.7}$$

$$E_2 = |r||F_0||A_0|e^{i(\phi_0 + f + \phi_r)} \tag{6.8}$$

If we assume that both beams obtain the same reflectivity in the phase-conjugate photorefractive medium

$$R_{1,2} = |R|e^{i(\delta_{1,2} + \phi_{pm;1,2})} \tag{6.9}$$

with the phase shifts introduced by e.g. misalignments of the pump waves, $\delta_{1,2}$ and $\phi_{pm;1,2}$, we arrive at an expression for the conjugate field of

$$E_{3,4} \propto |F(x,y)||A_0|e^{-i(\phi_0 + f(x,y) + \Psi_{3,4})} \tag{6.10}$$

where $\Psi_{3,4}$ summarizes all phase terms of the equation above. If these waves are shifted relative to each other about $\Delta x$ resp. $\Delta y$ along the vertical or horizontal axis and if they interfere destructively in the output plane, we get the following intensity distribution at the output plane:

$$I_{out} \propto \{|F(x + \Delta x, y + \Delta y)||A_0(x + \Delta x, y + \Delta y)| \cdot \tag{6.11}$$
$$\cdot e^{-i(\phi_0(x + \Delta x, y + \Delta y) + f(x + \Delta x, y + \Delta y))} - |F(x,y)||A_0|e^{-i(\phi_0(x,y) + f(x,y))}\}$$

This expression is formally proportional to the total difference of the complex amplitude transmission $F^*(x,y)$. Thus, the *spatial differentiation* can be realized by

$$I_{out} \propto |dF^*(x,y)|^2 \tag{6.12}$$

It results in an enhancement of the edges of the image and can be used to visualize contours of the objects. The exactness of the differentiation in that system naturally depends on the resolution of it — the translations may be very small if the system has a high resolution.

*Experimental Addition and Subtraction Using Phase-Conjugate Interferometry*

To realize these operations experimentally, external as well as self-pumped phase conjugation may be used. The big disadvantage when externally-pumped phase conjugation is used is the high dependence of the precision of the phase conjugation on the adjustment of the pump waves, which is necessary to obtain exact results for all the operations mentioned above. The authors and her coworkers found that even small angular deviations in the range of milliradians of the counterpropagating pump waves are able to create supplementary phase terms [236] that can destroy the whole operation. Moreover, crosstalk effects and edge-enhancement of the conjugate signal may disturb the reconstruction of the signal wave and with it the exactness of the operation.

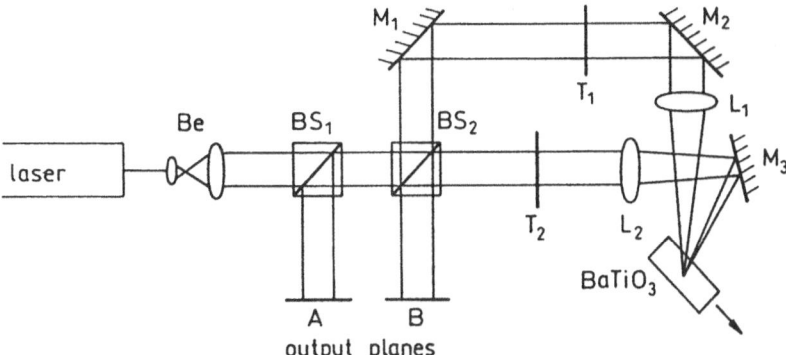

**Figure 6.3:** Scheme of the experimental setup to realize addition, subtraction and inversion using a single phase-conjugate mirror.

However, if one uses self-pumped phase conjugation, no supplementary degrees of freedom exist to adjust the pump waves which are built in a self-generating process in the crystal volume. Consequently, the sensibility of the system to adjustment is omitted, it is completely self-aligning. In their experimental realization, the author and her coworkers could show that the angular selectivity of the self-pumped phase conjugation process is sufficient to realize addition, subtraction and inversion of both interferometer arms in a single phase-conjugate $BaTiO_3$-mirror without polarization-decoupling. The experimental setup realized with an Ar-ion-laser ($\lambda = 514$ nm) as the light source (5 mW, $TEM_{00}$) is shown in fig. 6.3.

By choosing the incoming angles of both beams adequately, both incoming beams could be phase-conjugate with 65% reflectivity without any interaction or crosstalk. The input images and the resulting operations of addition and subtraction are shown in fig. 6.4. The resolution of the system is better than 200 lp/mm, while the response time depends on the total intensity incident on the crystal and ranges from some seconds for 50 mW up to several ten seconds for 10 $\mu$W.

As I have stated at the beginning of this section, all sorts of interferometers are in principle suited for the realization of the arithmetic operations as addition or subtraction using phase-conjugate interferometry, provided that constructive or destructive interference is operating and provided that the phases of both wavefronts are in or 180° out of phase. This promising principle has been exploited by a lot of different authors in the past years, using different nonlinear photorefractive phase-conjugate mirrors instead of conventional ones. The advantage of that configuration is, that the interferometer arms obtain formally exactly the same length. No complex adjustment algorithm is necessary to guarantee that feature. The first of these self-adjusting, dynamically stable and phase aberration-free interferometers has been realized by S.K. Kwong, G.A. Rakuljic and A. Yariv [237] in 1985. In 1986, P. Yeh and coworkers [238] constructed a similar setup using self-pumped phase conjugation as well as externally pumped phase conjugation via four-wave mixing. They could realize the functions of addition and subtraction as well as real-time contrast reversal or inversion of an image [239]. In 1988, A. Yariv's group improved their system by introducing orthogonally polarized beams [240], which was later on investigated in detail by H. Fujiware and coworkers [241], whereas N.A. Vainos and

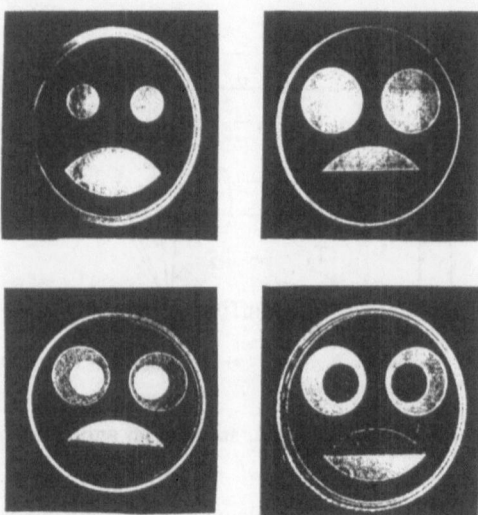

**Figure 6.4:** Upper row: image transparencies in both arms of the phase-conjugate Michelson interferometer; lower rows: addition via constructive interference (left), subtraction via destructive interference (right).

R.W. Eason [242, 243] realized such a device in photorefractive BSO with degenerate four-wave mixing and demonstrated the function of real-time image differentiation [244].

All these systems can also be realized in configurations using self-pumped phase conjugation. An example is the experimental demonstration using Cu:KNSBN, which enables to realize real-time image subtraction or differentiation [245]. A device offering the possibility to switch between addition and subtraction at one output channel was demonstrated by G.T. Hengst and W.B. Roh [246]. Finally, in 1991, H. Ogusu and coworkers from the University of Tokyo constructed an ingenious system with three incoming beams [247]. This three-beam-interferometer using self-pumped phase conjugation to return the beams enables to realize beneath addition (= OR-operation) and subtraction (= XOR-operation) the logic operations $\overline{XOR}$, NAND and $(T_1^2 + T_2^2)$.

This interferometric method of comparing images by addition, subtraction or inversion is best suited for all on-line comparisons of unknown images that are first presented to the system. However, if images have to be set in relation which are known to the system because they are stored e.g. in the system's memory, a second useful method based on phase-coded multiplexing to obtain addition, subtraction and inversion can be applied. I will present this novel method, first proposed by the author and her coworkers in 1993, in the following section.

### 6.1.2 Addition and Subtraction Using a Phase-Coded Memory

From 4.4 we know that multiple holograms may be stored in volume holographic media by changing the reference beam angle (angle or $\theta$-multiplexing), the recording wavelength (wavelength- or $\lambda$-multiplexing) or by phase encoding the reference beam (phase or $\Phi$-multiplexing). Among these methods, phase-coded holographic storage has revealed to have several advantageous features, resulting from the phase-only modulation of the reference beams, especially high energy efficiency and short readout times (see section 4.4.4). Here, I want to focus my attention on the fact that it is possible to realize the recall of linear combinations (e.g. additions and subtractions) as well as inversions of stored binary data pages using only a subset of reference beams, hence decreasing the selectivity of the phase address. The all-optical performance of these operations is based simply on the combination of an amplitude modulator with the reference beam phase modulator. Moreover, only one recall is necessary to produce the desired operation on the whole data page in real-time (processing time of the phase modulator), thus allowing for high-speed data page processing.

*Mathematics of Addition and Subtraction Based on Phase-Coded Multiplexing*

 In order to understand the principle of that method, let me first recall shortly the main principles of phase encoded multiplexing. It is based on the superposition of the image $A_m(\vec{r}) = A_m(\vec{r}) \cdot e^{i\vec{k}_m\vec{r}} + cc.$ to be stored with a set of reference beams having the amplitudes $\mathcal{R}_m(\vec{r}) = R \cdot \sum_k^N e^{i\vec{k}_k\vec{r}}e^{i\varphi_k^m}$, where $e^{i\varphi_k^m}$ is a definite additional phase term added on each reference wave. The resulting refractive index modulation in the crystal after the recording is thus given by

$$\Delta n(\vec{r}) = \sum_k^N \sum_m^N \Delta\bar{n}_k^m(\vec{r}) e^{i(\vec{k}_s - \vec{k}_k)\vec{r}} e^{-i\varphi_k^m} \qquad (6.13)$$

where $\Delta\bar{n}_k^m(\vec{r})$ is the modulation depth of the index grating between the $k^{th}$ reference beam and the $m^{th}$ signal beam.

During readout we obtain the diffracted signal $S_j^{\mathrm{diff}}$, provided that the Bragg condition is fulfilled for each reference beam. To reconstruct the $p^{th}$ image, we have to fulfil the condition

$$\sum_{k=1}^N \left( e^{(i(\varphi_k^j - \varphi_k^m))} \right) \equiv \sum_{k=1}^N \chi_{km}^* \chi_{kj} \overset{!}{=} \begin{cases} 0 & \text{for } m \neq \hat{j} \\ N & \text{for } m = \hat{j} \end{cases} \qquad (6.14)$$

The matrix $\chi_{km}$ represents the additional phase terms, $k$ the number of the reference beam while writing the $m^{th}$ image. Therefore, the phase codes of the images are given in that notation by the rows of a matrix. To obtain reconstruction without crosstalk of any arbitrarily addressed image, the matrix $\chi$ has to be unitary. If pure phase encoding should be used, all elements of $\chi$ have to have the absolute value one. Otherwise a combination of phase and amplitude multiplexing — which is exactly the same as angular multiplexing using an amplitude modulator instead of a phase modulator — is obtained. For the one-dimensional matrix as an example, this case is simply a description of pure angular multiplexing. Here, I will use binary phase codes (elements of $\chi$ are either 1 or -1 as in section 4.4), which correponds to a phase shift of either zero or $\pi$.

To obtain addition (subtraction) of several images, only a subset of the reference beams has to be used, having the same (different) additional reference beam phase factors. Eq. (6) then has to be changed during readout into

$$\sum_{k=1}^N \left( e^{(i(\varphi_k^j - \varphi_k^i))} \right) \equiv \sum_{k=1}^N \chi_{km}^* \chi_{kj}' \overset{!}{=} \begin{cases} 0 & \text{for } m \neq \hat{j}' \\ \frac{N}{m} & \text{for } m = \hat{j}' \end{cases} \qquad (6.15)$$

where $m$ is the number of images involved in one operation, $\hat{j}'$ may be any number of one of the images which shall be added or subtracted, and $\chi'$ is defined as a readout matrix that is different for every arithmetical operation

$$\chi'_{km} = \begin{cases} \chi_{km} & \text{for equal (different) phase factors in beam } m \\ 0 & \text{for all other cases} \end{cases} \tag{6.16}$$

Thus, $\chi'$ represents the operation of an additional amplitude modulator that realizes the addressing of the dynamic memory with only a subset of the original phase codes used for storage of the appropriate images. For the case of eight images stored in the phase encoded memory as an example, addition of image $A_1$ and image $A_5$ can be realized using

$$\chi'_{kj',\text{add}} = \begin{pmatrix} 1 & 1 & 1 & 1 & 0 & 0 & 0 & 0 \\ 1 & -1 & 1 & -1 & 0 & 0 & 0 & 0 \\ 1 & 1 & -1 & -1 & 0 & 0 & 0 & 0 \\ 1 & -1 & -1 & 1 & 0 & 0 & 0 & 0 \\ 1 & 1 & 1 & 1 & 0 & 0 & 0 & 0 \\ 1 & -1 & 1 & -1 & 0 & 0 & 0 & 0 \\ 1 & 1 & -1 & -1 & 0 & 0 & 0 & 0 \\ 1 & -1 & -1 & 1 & 0 & 0 & 0 & 0 \end{pmatrix} \tag{6.17}$$

Note that this matrix is also the correct one for realizing addition of image $A_2$ and image $A_6$, image $A_3$ and image $A_7$ or image $A_4$ and image $A_8$. As another example of addition, the addition of all eight images would require a matrix that suppresses all row except the first one.

For the case of image subtraction, an equivalent matrix as for the case of image addition can be defined, using the complementary phase codes which have to be blocked.

$$\chi'_{kj',\text{sub}} = \begin{pmatrix} 0 & 0 & 0 & 0 & 1 & 1 & 1 & 1 \\ 0 & 0 & 0 & 0 & 1 & -1 & 1 & -1 \\ 0 & 0 & 0 & 0 & 1 & 1 & -1 & -1 \\ 0 & 0 & 0 & 0 & 1 & -1 & -1 & 1 \\ 0 & 0 & 0 & 0 & -1 & -1 & -1 & -1 \\ 0 & 0 & 0 & 0 & -1 & 1 & -1 & 1 \\ 0 & 0 & 0 & 0 & -1 & -1 & 1 & 1 \\ 0 & 0 & 0 & 0 & -1 & 1 & 1 & -1 \end{pmatrix} \tag{6.18}$$

Again, a matrix that allows to subtract image $A_1$ and image $A_5$, is as well correct for subtraction of image $A_2$ and image $A_6$, image $A_3$ and image $A_7$ or image $A_4$ and image $A_8$.

### Experimental Realizations of Addition and Subtraction Using Phase-Coded Multiplexing

Our schematic setup is shown in fig. 6.5. An Argon ion laser (514 nm) is used to realize the signal and reference beams. In the reference arm, a binary phase modulator (see section 4.4) is combined with an amplitude spatial light modulator. Images are fed into the image arm using a ferroelectric liquid crystal display with $200 \times 200$ pixels resolution taken from a commercially available Sharp video projector. The storage crystal is a nominally undoped $BaTiO_3$-crystal. During recording, the reference arm amplitude modulator was set to maximum transmission and the images are stored using an incremental storing technique as described in section 4.4. All images are successively stored with time increments smaller than the response time of the recording medium. The cycles are then repeated until the desired diffraction efficiency is reached. The integral storage time

**Figure 6.5:** Scheme of the experimental setup to realize addition, subtraction and inversion of stored images using phase encoded hologram multiplexing. ASLM: amplitude spatial light modulator, PSLM: phase spatial light modulator, CHG: computer generated hologram.

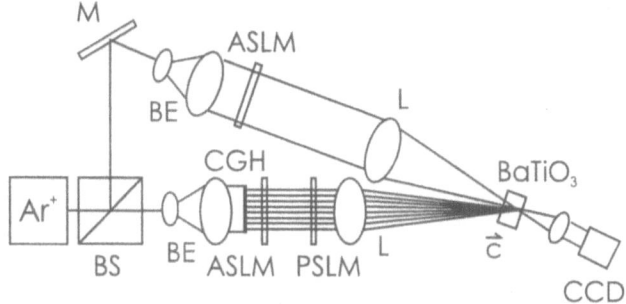

for each image in our experiment was about 10s. After storing the images, we obtained a diffraction efficiency of about 3 % for each image. Significant cross talk between images has not been detected. During readout, the amplitude spatial light modulator was used to select the appropriate subset of reference waves to realize the different operations. To demonstrate the realization of pairwise addition and subtraction, eight images have been stored, shown in fig. 6.6 (left) during recall using pure phase coded reference beams. Fig. 6.6 (right) shows several representations of combined recall that result in addition or subtraction. If these arithmetical operations are combined with a supplementary threshold device, it is possible to realize the logical operations OR for addition and XOR for subtraction. In a second experiment, we demonstrate the linear combination of three images. In fig. 6.7 (left), three stored images are shown during readout with pure phase coded reference beams. The linear combination of two (fig. 6.7, right, a,b) or three (fig. 6.7, right, c) of these images has been realized. In fig. 6.7c, the analog character of the operations is clearly visible. Moreover, it is possible to realize in a variation of the previous experiment the inversion of an image by subtraction of a stored plane wave image from the original image that has to be inverted, what is equivalent to a logical 'NOT' operation. This operation is shown in fig. 6.8. Thus, using an appropriate threshold device, the logical operations OR, XOR and NOT can be realized using the method of reduced reference beam recall in phase-coded multiplexing. Consequently, any logical operation can be performed by combining these basic operations. For example, a logical 'a AND b' can also be realized with NOT(NOT(a) or NOT(b)). This is a basic step to construct an optical arithmetic logical unit like the ones used in electronic computing by combining several of these devices, by feeding back the result into the memory, and by using the supplementary possibility of analog arithmetic operations. Recently, another demonstration has shown in a similar setup the possibility to realize pairwise addition and subtraction in a photorefractive $LiNbO_3$-memory that has stored three images recorded with sequential recording [165].

In summary, coherent image addition and subtraction can be performed in each optical system, where the superposition of images in phase or with a phase shift of $\pi$ is allowed. In holography, this realization is straight forward, because it can be done by simply reconstructing an image with a reference wave having a phase shift compared to the writing reference wave. Image two monochromatic light waves that form an interference pattern whereby the spatial position of the fringes on the sample depends on the relative phase of the interfering beams. A phase shift of one of the beams, e.g. the object beam, results in a corresponding phase shift of the interference pattern at the sample

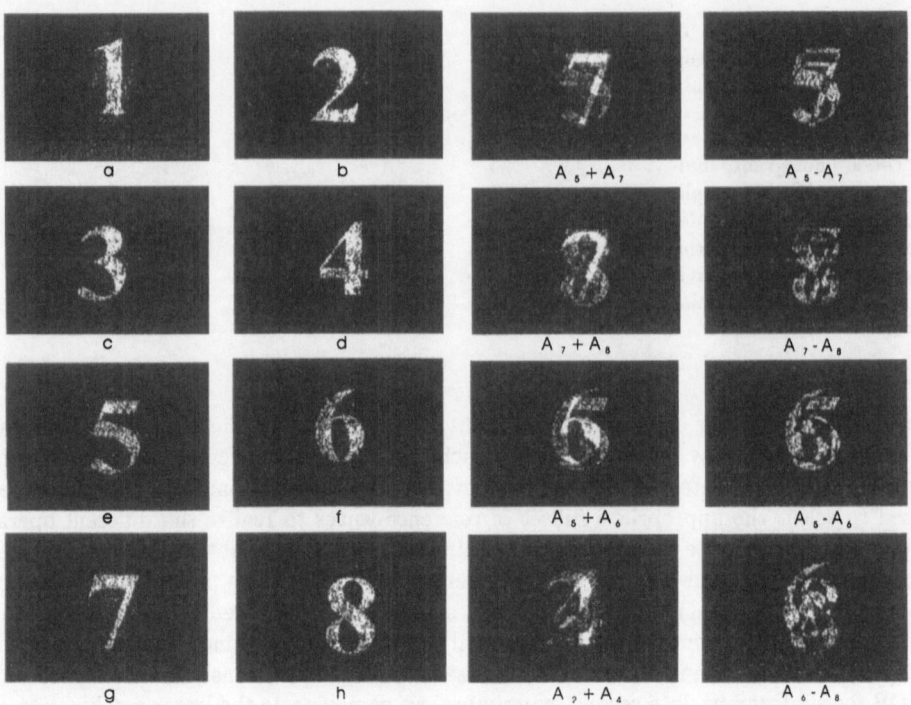

**Figure 6.6:** Eight images stored with binary phase codes during readout with the complete set of reference beams (left) and some examples of addition and subtraction of pairwise images: a) $|A_2 + A_4|$ b) $|A_5 + A_6|$, c) $|A_5 + A_7|$, d) $|A_7 + A_8|$, e) $|A_5 - A_7|$, f) $|A_5 - A_6|$, g) $|A_6 - A_8|$, h) $|A_7 - A_8|$.

described by the grating phase. The interference pattern may be stored by the interaction with a photosensitive material as for holographic plates, by the hole burning process as spatial modulations of the absorption coefficient and the refractive index or by the photorefractive effect as a pure refractive index modulation. When the two holograms have been successively stored, the complex wave amplitude diffracted from this resulting grating can be thought of as a coherent superposition of two holograms. The hologram efficiency is composed of the individual efficiencies and an interference term depending on the phase difference between the holograms. for phase difference 0, constructive interference is observed, and destructive interference is achieved for phase difference $\pi$.

In a similar way, spectral hole burning can also be exploited in order to realize optical arithmetics, a field that is known as *molecular computing*. In order to superimpose different holograms, the electric field can be used as a multiplexing parameter. An electric field changes the shape of the spectral hole burnt at zero field due to the shift of the transition frequencies caused by the interaction of the dipole moments and polarizabilities of the guest molecules with this externally applied field. Because the diffraction efficiency splits for higher field strengths, it is possible to write pairs of holograms in the frequency-electric field plane. When the phase can be modified supplementary to the electric field, e.g. by a moving piezo-driven mirror, image superpositions with different

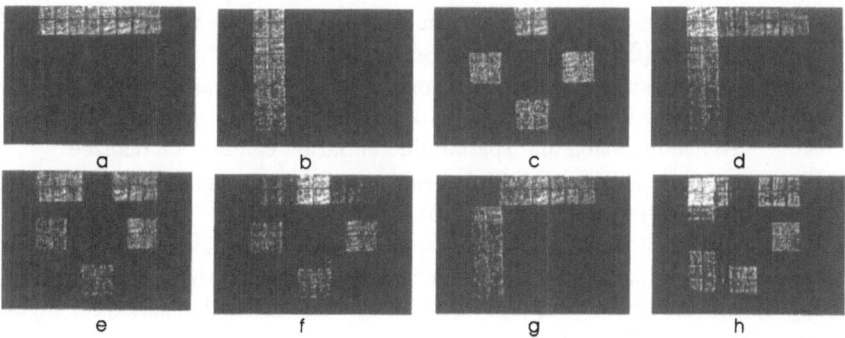

**Figure 6.7:** Three images stored in the phase coded memory, read out with the complete reference beam set (a,b,c) and addition and subtraction of the images by addressing the memory with subsets of the phase codes: d) $|A_1 + A_2|$, e) $|A_1 - A_3|$, f) $|A_1 + A_3|$, g) $|A_2 - A_1|$, h) $|A_1 + A_2 - A_3|$.

**Figure 6.8:** Inversion of an image by subtracting the image from a plane wave (upper image).

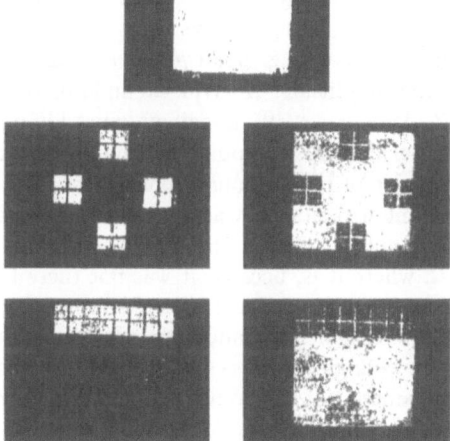

phases can be realized [248], resulting in addition and subtraction ("AND" or "XOR") operations of the images.

---

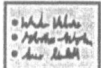 **Summary**

---

Optical techniques to add and subtract two-dimensional data are based on holographic comparison principles.

- The earliest method was to spatially modulate the two images to be processed by periodic waves that are mutually shifted by a phase shift of $\pi$. After successive recording on a holographic plate, the readout of the composite hologram results in subtraction of the data.
- In classical interferometers like Mach-Zehnder or Michelson interferometers, images

are added (subtracted) pairwise when introduced into the two interferometer arms by constructive (destructive) interference.

- In phase-conjugate interferometers, one or both mirros are substituted by phase conjugators, which enable compensation of phase distortions and aberrations. In these systems, inversion and spatial differentiation can also be realized.
- Phase-conjugate three- or multiple beam interferometers enable to realize addition and subtraction of more than two images at a time.
- Using volume holographic memories, phase-coded multiplexing can be exploited to realize addition, subtraction and inversion by addressing the memory with only an appropriate subset of the phase codes. Thus, this method is suited for high-speed data page memory processing.

## 6.2 Temporal Optical Operations

A temporal differentiation enables to look at what is new or changing in a scene compared to the inputs recent history. Therefore, these devices are often called novelty filters, too. Novelty filtering using an optical system is similar to temporal high-pass-filtering, where the final output sensor (e.g. our eyes or a CCD-camera) detects the optical intensity rather than the amplitude. A vivid example of the function of such a filter has been given by D.Z. Anderson [249]. Imagine a quiet lily pond viewed through a novelty filter. At first the filter displays the pond. In time, however, the filter adapts to its input and removes the image of the stationary pond from the output. On the other hand, a flying bug is a constant stream of newness, so the bug remains visible. The peculiar feature is that if a frog should leap from one lily pad to another, the frog would instantly appear in two places: where it is, because it was not there before and where it was, because its sudden absence is just as novel as its sudden presence. Such a device is important, when input data to a neural net contain too much information and need to be reduced. By only transferring the changing data — image two images to be compared as two moments of a scene — a big amount of data reduction and transformation of essential data is possible.

In contrast, when nearly everything in a scene is undergoing change, it may be desired to know what is static in the scene. This complement of a novelty filter is called a time integrating or *monotony filter*. The most celebrated use of the monotony filter is the detection of modes of vibrating objects — e.g. the eardrum. The application of holography to this purpose using holographic plates long preceded the invention of the optical novelty filter.

An optical version of both of these systems may be implemented by several different setups. The most widely spread possibilities of realizing novelty filters using photorefractive filter elements will be presented in that chapter. For more detailed explanations on the topic, refer to the excellent review of D.Z. Anderson and J. Feinberg [250].

### 6.2.1 The Ring Novelty Filter

The *ring novelty filter* was invented because of the observed behaviour of a ring resonator that used a photorefractive amplification medium or a phase conjugating medium in place of a conventional mirror. The output from this ring tends to zero at steady state, whereas a sudden change in the resonator length was observed to produce an intense output.

**Figure 6.9:** Ring resonator with a holographic mirror used as a novelty filter (after [250]).

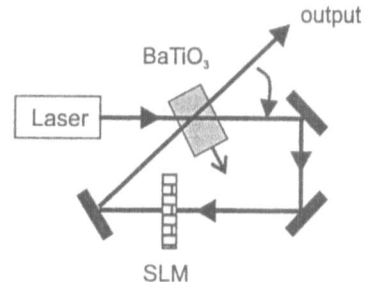

The ring resonator can be transformed into a novelty filter by placing a spatial light modulator inside the cavity as shown in fig. 6.9. The spatial light modulator is connected to a television camera, so that it encodes a live phase image onto the transmitted laser beam. The beam transmitted through the modulator is deflected by mirrors to reenter the crystal and intersect the original beam. These two beams create a volume phase hologram inside the photorefractive crystal, which in $BaTiO_3$ is spatially shifted from the intensity pattern by 90° (see section 3.4) and so couples the amplitudes of the two beams. The crystal's c-axis direction is chosen so that energy is coupled from the incident beam into the ring resonator cavity. Energy then builds up inside the resonator. Because the ring resonator can store energy, the intensity inside the ring can be greater than the incident intensity. D.Z. Anderson and coworkers could show, that the contrast of such a novelty filter is the same as for the two-beam coupling case [250]. In practice, the ring is a rather awkward geometry because the incident light must be expanded after traversing the photorefractive crystal to the dimensions of the LCTV and afterwards again be reduced to enter the crystal again.

### 6.2.2 Phase-Conjugating Interferometric Novelty Filter

A more practical geometry, which is analogous to the system described for the arithmetic addition and subtraction operations (see fig. 6.1 and section 6.1.1) and was demonstrated by D.Z. Anderson et al. in 1987 [249], uses four-wave mixing in an arrangement identical to that used to demonstrate optical phase conjugation (compare to section 3.5). As shown in fig. 6.10 and fig. 6.11, two configurations are possible in order to exploit phase conjugation for novelty filtering. In the first one (fig. 6.10), an incident plane wave is divided by a beamsplitter. One of the two resulting beams passes through an optical phase modulator that contains a picture of the live scene. Both beams are subsequently incident on the same phase-conjugate mirror. The conjugator generates the phase-conjugate beams, which propagate back towards the beam splitter. Under steady-state conditions, the information on the one phase-conjugate wave is exactly cancelled by its return trip through the phase modulator, so that it emerges as a plane wave. In steady state, the two phase-conjugated beams coherently recombine at the beam splitter so as to travel back onto the light source, with no light going into the "output port" of the beam splitter. This can be explained by drawing upon the analogy between phase conjugation and time reversal.Since no light came into the output port of the beam splitter, no light should leave this port. However, if the image on the modulator changes in a time faster than the response time of the phase conjugator, the beam emerging from the modulator will not be a plane wave. In particular, the wavefront will be altered for those locations in

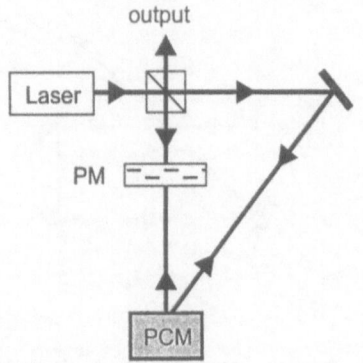

**Figure 6.10:** A phase conjugate novelty filter using a phase modulator in a self-aligning interferometer configuration (after [250]).

the modulator that are changing in time, and there will be an instantaneous signal at the output of the beam splitter. After the medium has had time to respond, the output fades again.

### 6.2.3  Novelty Filters Exploiting Polarization Encoding

Another implementation of the four-wave mixing novelty filter uses a polarization modulator instead of a phase modulator to impress a polarization-encoded image onto the beam (see fig. 6.11). The liquid crystal display from commercially available hand-held televisions is inherently a polarization modulator, because these televisions use two between-sheet polarizers in order to make the display modulate the intensity of the light. A polarized plane wave incident on the novelty filter setup is transmitted by the first polarizing beam splitter. The beam passes a liquid crystal display sandwiched between a pair of waveplates. The transmitted beam is then sent into a phase conjugator designed to accommodate any polarization. This is done by the following method: The beam is first divided by a second polarizing beam splitter into two linearly polarized components. Since a photorefractive phase-conjugator will only conjugate one component, a half-wave plate is used in one arm to rotate the polarization by 90°. The two beams are separately conjugated and then recombined at the polarizing beam splitter. The conjugate beam passes backwards through the modulator, and back to the first polarizing beam splitter. In steady state, the light returning to the first polarizing beam splitter will have its original polarization intact, and so the beam will be sent towards the laser. However, any sudden change in the image will alter the polarization of the beam returning to the first polarizing beam splitter, and so that part of the image will be deflected into the output port.

There are three important features of these systems that I want to emphasize. First, any change of the image will appear *instantly*, because the speed of the phase conjugator determines only the time required for an unchanging image to fade, but not for a change in the image to appear. Second, it is interesting to note that in both systems, while the output from the beam splitter or the polarizing beam splitter shows the novelty of the scene, the light that propagates back towards the laser carries the monotonous information. This information can be oberseved with an additional beam splitter placed directly in front of the plane wave source. Finally, the contrast of the four-wave mixing novelty filter types does not depend upon the coupling constant of the crystal alone.

**Figure 6.11:** A polarization novelty filter using a polarization modulator and two polarizing beam splitters (after [250]).

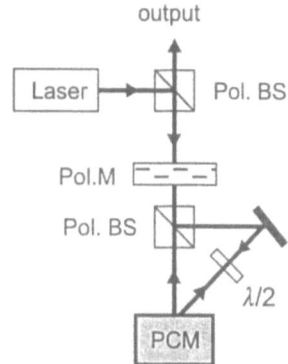

Two signals are subtracted at the (polarizing) beam splitter. If the phase conjugation is perfect, the contrast is infinite. The interferometer configurations are therefore well suited to perform novelty detection with any four-wave mixing media. In particular, one is not limited to photorefractive materials, with their large gain but slow response, and instead of them other nonlinear optical medium as sodium vapour or laser dyes may be used.

### 6.2.4 Two-Beam Coupling Novelty Filter

M. Cronin-Golomb and coworkers demonstrated an elegant and simple novelty filter that uses two-wave energy coupling [251]. Fig. 6.12, upper scetch, shows the image-bearing beam and a reference beam intersecting in a $BaTiO_3$ crystal. A similar configuration that enables to filter phase information is shown in the lower sketch of fig. 6.12 using photorefractive beam fanning as the novelty filtering process. Due to the 90° phase shift between the intensity interference pattern and the refractive index pattern in this crystal, in steady state the image-bearing beam transfers energy to the reference beam and emerges from the crystal in a severely depleted way. If the image-bearing beam suddenly changes, the energy coupling is momentarily defeated, and the transmitted image becomes more intense. A physical explanation for the two-wave mixing depletion of the image beam is destructive interference: the transmitted image beam and the deflected

**Figure 6.12:** A novelty filter that uses two-wave mixing (upper scetch) or beam fanning (lower scetch) to deplete energy from the image-bearing beam. SLM = Spatial light modulator, PM = phase modulator, c = c-axis of the photorefractive crystal.

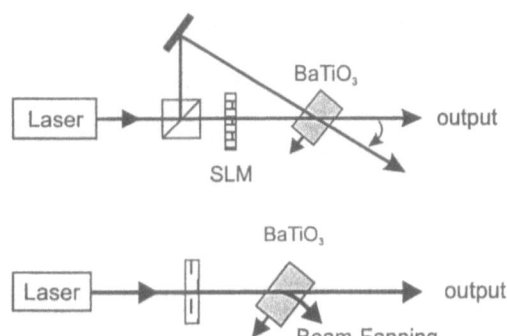

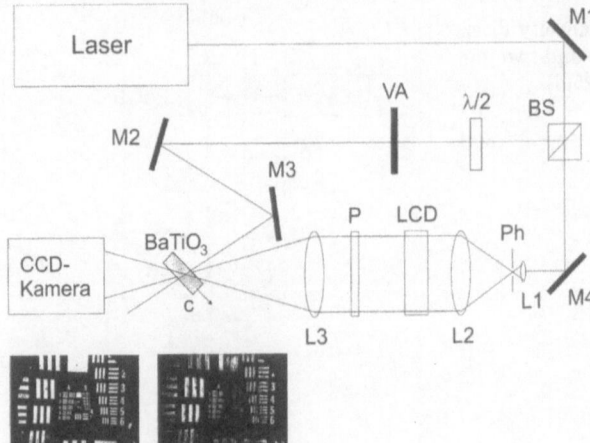

**Figure 6.13:** Experimental setup to realize two-beam coupling novelty filtering. The resolution of the system with and without beam coupling is shown in the inset. M = mirror, VA = variable absorber, BS = beam splitter, L = lenses, Ph = pinhole aperture, LCD = liquid crystal display, P = phase modulator.

reference beam destructively interfere on the output screen. The output screen then always contains two superimposed images: the real image transmitted through the crystal and the holographically-reconstructed image produced by the deflected reference beam. In steady state, these two images are of nearly equal amplitude and 180° out of phase with respect to each other, so that they tend to cancel. If one of the beams is altered, the destructive interference is destroyed and the screen becomes bright and displays the alternation. Note that in contrast to the four-wave mixing novelty filter, the image can be either phase or amplitude modulated or even both. However, any steady state polarization change in the image bearing beam will not be canceled in the output beam. The author and her coworkers have realized this configuration and could show that novelty filtering is possible with contrasts as high as 400:1 [252, 253], using photorefractive $BaTiO_3$ as the processing material (see fig. 6.13). An example of the performance of that system using an LCD display for data input is shown in fig. 6.14.

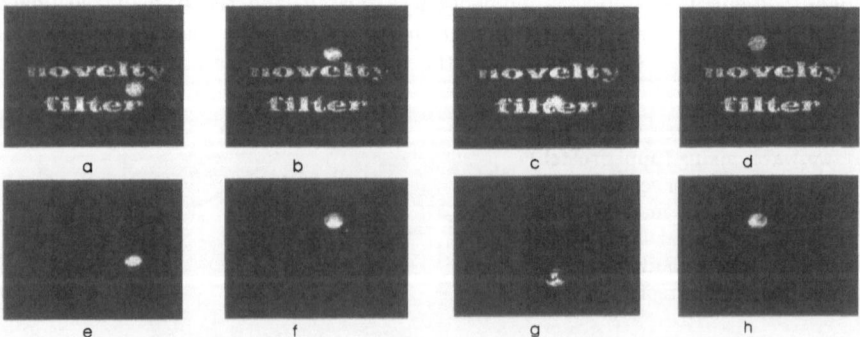

**Figure 6.14:** Filtering of a dynamic scene displayed on a liquid crystal display using a photorefractive two-beam coupling novelty filter. (a) - (d) non-filtered scene, (e) - (h) filtered scene.

**Figure 6.15:** Visualization of the gas flow out of a gas lighter (four different states) using a phase-sensitive photorefractive novelty filter as shown in fig.6.13.

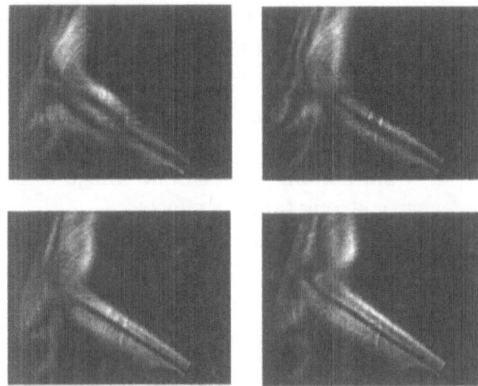

### 6.2.5  Novelty Filtering Using Photorefractive Beam Fanning

An even simpler version of a novelty filter that uses pure two-wave energy coupling was suggested by M. Cronin-Golomb[251], too and first demonstrated by J.E. Ford and coworkers [254]. It uses only one input beam into a BaTiO$_3$ crystal (fig. 6.12b). The device is similar to that one using two-beam interference, except that the pumping beam is replaced by amplified scattered light (beam fanning) from the single incident beam. Two-wave mixing with this scattered light inside the crystal depletes the stationary portions of the transmitted beam, so that only the changing portions of the scene are transmitted. Unfortunately, this device requires a large coupling gain, which is obtained by using a crystal of BaTiO$_3$ with only a few impurities as the one described by T. Tschudi and coworkers [255] or by using a crystal that is cut so that its input and exit faces are oriented at a 45° angle to the c-axis of the crystal, thereby taking maximum advantage of the large $r_{42}$ electrooptic Pockels coefficient of BaTiO$_3$. Depending on how the input image is focused into the crystal, this device is sensitive to changes in the amplitude, phase, wavelength or polarization of the incident beam. A phase change visualization is shown in fig. 6.15, where a gas stream out of a gas lighter is visualized.

The devices I have described are simple but effective all-optical processors that extract the temporal changes from a two-dimensional input. They perform easily high-pass, low-pass or bandpass filter characteristics. These are fundamental processing primitives that can be combined to perform complex operations. Thus these filters are especially suited for optical computers that can process an entire scene in parallel. Because they are simple devices, I believe that they may easily be developed further in order to realize important data compression operations in optical computing or neural network structures.

---

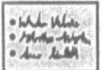 # Summary

---

A temporal differentiation enables to look at what is new or changing in a scene compared to the input's recent history. Devices that are capable of this operation are called novelty filters. They are similar to temporal high-pass filters. Novelty filters can be realized by

- pure two-wave-mixing, in a configuration that depletes the signal beam by the energy transfer. If the transmitted signal beam is observed, no signal is obtained for a stationary grating built up. However, all changes in a scene will not be present as diffraction gratings and are thus visible in the signal beam.

- a ring novelty filter, which is built up with a photorefractive two-beam coupling nonlinearity that couples energy from the incident beam into the ring resonator cavity.
- Beam-fanning in a simple image beam transmission in a photorefractive crystal with strong fanning effects.
- optical phase conjugation in an interferometric setup. Phase conjugation can be performed self- or externally pumped.
- polarization encoding in modulators that are included in four-wave mixing setups, because the photorefractive beam coupling effect is dependent on the polarization.

Many novelty filters have as well a monotony port, allowing to filter out stationary parts of the image. Applications of novelty filters are found in object inspection, state comparison as e.g. for a vibrating loudspeaker, movement detectors as e.g. for microscopic biologic data, data reduction in dynamic scene processing, or page subtraction or comparison.

## 6.3   Bistable Elements

Often, neural decisions have to be made which require a bistable operation path. Therefore, switches that enable to change between two states are required in these nets. Here, I want to take only a short glimpse on the field of bistable elements, choosing those from the vast field of devices that are well adapted to the nonlinear elements described up to now for our networks. Most often, liquid crystal bistable devices can be used to realize a bistable, pure optical switching behaviour. However, if a system is required that allows to preserve the coherent operation of the optical network simulation e.g. for bipolar networks, these systems need to be substituted by those that process the phase adequetely. It is that area that I want to discuss here in more detail, showing some exemplar systems that use resonators with photorefractive gain in order to realize bistable operations

### 6.3.1   Bistable Phase-Conjugate Resonator

M.B. Klein and coworkers [226] used a phase-conjugate resonator containing an intra-cavity saturable absorber to realize a bistable device. They showed that the resonator will be bistable if on the one hand the small-signal absorption of the saturable absorber exceeds the small-signal gain of the phase-conjugate mirror and on the other hand if the absorption saturates more readily than the gain (the reflectivity) of the phase-conjugate mirror.

A scetch of the setup is shown in figure 6.16. To control the bistabilty, an incoherent control beam was used that was coupled into the cavity. By adjustment of the control beam intensity as well as the intensity of the pump beams of the phase-conjugate mirror, the resonator could be tuned into bistable behaviour. In the absence of the control beam, no oscillations were possible in the resonator. When the control beam was turned on, the resonator switched to the "ON" state. The control beam could then be removed, and the cavity oscillation continued. In order to return to the "OFF" state, the resonator path was blocked.

**Figure 6.16:** Schematic diagram of
a linear phase conjugate resonator
containing an inctracavity saturable
absorber. Information is read into
the resonator by means of a separate
control beam incident upon the saturable absorber (after [226]).

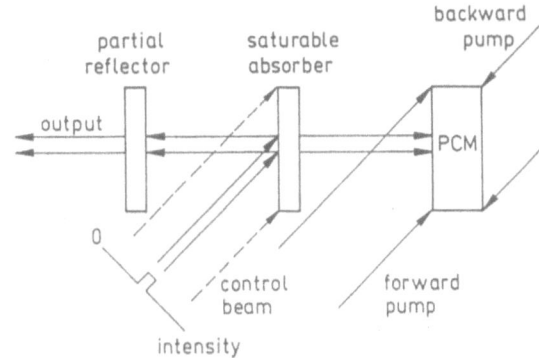

## 6.3.2 Photorefractive Flip-Flop

A quite different type of bistable element was presented in 1991 by D.Z. Anderson et al.
[256], a photorefractive flip-flop. Usually one thinks of the flip-flop as a memory and as
a binary counting unit. From another perspective, the flip-flop serves as the prototypical
example of a competitive system, in this case a bistable system that must settle in one
of two essentially identical states. Competitive interactions such as those that occur in a
flip-flop play a fundamental role in analog decision-making systems, particularly in analog
neural network processing. It is in that context that I want to present a photorefractive
version of this elementary processing system.

The basic principle of such a photorefractive flip-flop is shown schematically in fig. 6.17.
Two unidirectional ring resonators are pumped through two-beam coupling in a photorefractive crystal. For clarity, the two rings are drawn with separate photorefractive gain
and loss media. The competitive interaction between these two rings is achieved by additional photorefractive media. A fraction of the energy in each ring is taken out of the
resonator and brought to interaction with the other ring through two-beam coupling.
The photorefractive medium is oriented in such a way that each mode experiences a loss
that depends on the intensity of the other mode. When one mode is oscillating strongly,
it supplies a loss pump for the photorefractive two-beam coupling of the second mode.
This active loss, together with the passive losses of the cavity, exceeds the gain for the
second mode so that it can not oscillate. Thus one mode is on while the other one remains

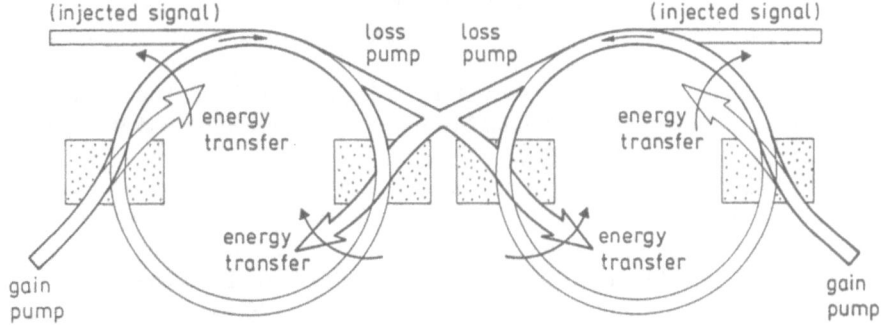

**Figure 6.17:** Conceptual scheme of the photorefractive flip-flop (after [256]).

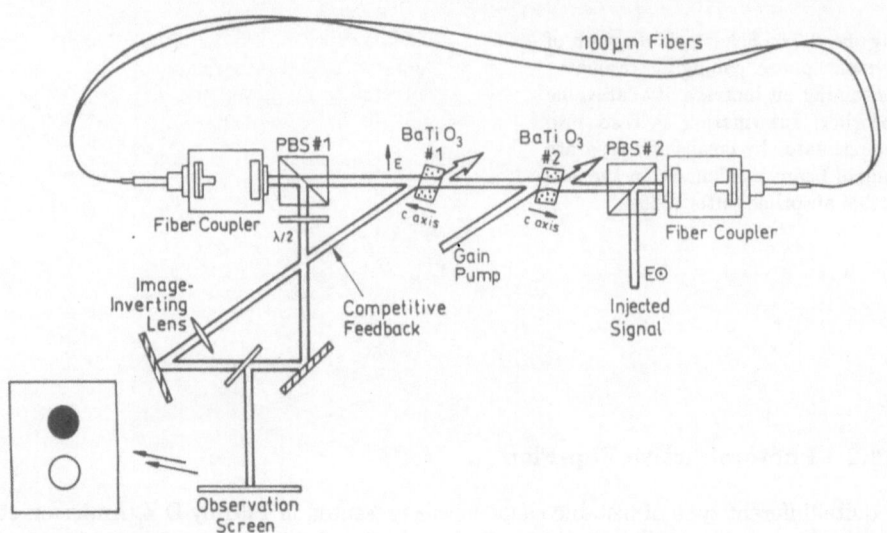

**Figure 6.18:** Experimental schematic of the photorefractive flip-flop. PBS = polarizing beam splitter (after [256]).

off. The symmetry of the system guarantees that either mode can be the oscillating one. The system can be switched from one state to the other by injecting an external signal into the off mode.

In the real experimental situation, which is depicted in fig. 6.18, the two rings are formed by two vertically displaced 100 $\mu$m multimode fibers, each of which allows for many transverse modes. At the exit of the fibers, a polarizing beam splitter divides each of the beams in half, since the multimode fiber depolarizes the light. One set of beams, which I call the signal beams, then passes through two photorefractive BaTiO$_3$ crystals before each beam is fed back into its respective fiber. The two signal beam paths are physically separate in both crystals so that there is no direct coupling between them. The beam path deviated by the first polarizing beam splitter is responsible for the mutual competition. The polarization of these competing beams is first rotated by a half-wave plate to allow for coupling with the signal beams. A lens then creates an inverted image of the two fiber modes. This image is superimposed with the corresponding uninverted image of the signal path in photorefractive crystal 1. Thus the lower mode couples to the upper one and vice versa. Crystal 1 is orientied in such a way that the signal beams experience loss depending on the strength of the competition beam. The second photorefractive crystal supplies gain for both modes through a common gain pump. The switching between the two asymmetric states of the system is achieved by injecting a light beam into only one of the modes through the port of the second polarizing beam splitter. The expected mode competition and the flip-flop switching behaviour could be demonstrated impressively by D. Anderson and coworkers [256].

With that short survey over coherent bistable switching elements implementing photorefractive devices, I want to close this section. A lot of different bistable mechanism and devices that are used for many other purposes (switches, shutters, etc.) can be found in the book of H.M Gibbs, cited in the Further Reading part of this chapter.

Bistable elements are useful devices for switching between different states in neural net-

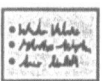   **Summary**

works. They can be realized in a variety of different configurations using nonlinear optical materials, among them photorefractive materials.

- Saturable absorbers in phase-conjugate resonators allow to operate the system in a stable or an irregularly oscillating state.
- A photorefractive flip-flop can be realized with a two-ring configuration of two photorefractive unidirectional ring oscillators that are coupled via a two-beam coupling crystal. The photorefractive coupling medium is oriented in such a way that each mode experiences a loss that depends on the intensity of the other mode.

## 6.4 Optical Data Refreshment

One of the main problems in photorefractive information storage or interconnection devices is that the recording of a new information or image partially erases previously recorded ones. Therefore, the aim to record an arbitrary number of holograms and to take account for that erasure losses in the diffraction efficiency can only be realized if special recording or exposure schemes are used. The most common and successful ones of these recording procedures have been presented in section 4.4.

Unfortunately however, as we have seen in chapter 4.4, the recording of multiple holograms according to one of these exposure schemes results in a diffraction efficiency for each hologram that is proportional to $M^{-2}$ where $M$ is the number of holograms recorded. It was this decrease in the diffraction efficiency that the authors have recognized to be the main limitation of the storage capacity in photorefractive information storage [175].

Therefore, methods to overcome these limitations, e.g. by increasing the diffraction efficiency of multiply exposure holograms, have been an attracting field of investigations in recent years, resulting in different novel techniques to ensure high diffraction efficiencies of holograms stored in photorefractive multiple image storage devices and to guarantee the readout of these data without loss of information, maintaining at the same time the capability to adapt and change the information. Of course, all refreshment devices are of important use for all sorts of optical neural network applications, because the menace of the system by degradation in reading out a photorefractive storage or thresholding device is always present.

If a prescribed, fixed interconnection device is required, several methods can be applied to prevent the system from data erasure. The most popular one is to fix the information e.g. in a thermal way by changing the electric refractive index grating into an ionic one [257, 258, 259]. In this process, the crystal is heated after recording up to a temperature where the less mobile ions can also be liberated. They will then tend to compensate the electronically written gratings. When equilibrium is reached, the crystal is cooled back to room temperature and is homogeneously illuminated in order to erase the electronic gratings. The ionic charges are no longer mobile and will remain in the grating positions, thus reconstructing a copy of the electronic grating. A different fixing process is electric domain fixing, which has been realized in SBN and $KNbO_3$ crystals [260, 261].

Another way of data conservation consists in adapting polarization, applied voltage, spatial frequency or write energies in such a way that erasure is minimized [262]. Moreover, adressing the holographic data by a wavelength different from the one which was used for writing, is also a popular means to prevent unwanted erasure during readout. However, in this case severe problems arise due to the fact that image distortion and Bragg-mismatch due to the wavelength-sensitivity of the volume storage material occurs. In order to overcome these problems, multiple reading approaches, followed by several processing steps [263], different polarization reading [264], spherical wave reading [265], and different reference angle combinations [266] have been used. For a detailed review of these methods and an analysis of these two-color adressing systems see [267]. Therefore, neural network applications have stimulated a lot of work in the area of constructing refreshment systems that are able to work in an optical feedback system with neural network architectures.

The starting point of investigation directed towards data refreshment in volatile storage and adaptive interconnection systems was the observation that a 180° phase shift in the reference beam of a single image is capable of realizing a selective erasure of the image in question. Moreover, the image could be rewritten in the crystal, this time with the phase shift of $\pi$, with a better efficiency than before. This principle has been the subject of a vast range of investigations, starting in 1975 with the first realization by J.P. Huignard et al. [103] and having experienced a renewed interest in the context of realizing neural network structures in 1993 [268, 269, 270].

Proceeding from these investigations, the most obvious solution to the problem of erasure during readout is the use of two photorefractive crystals in succession. Images are stored in the first crystal and reconstructed with low-intensity beams, then amplified in the second crystal operating in a high-gain low-noise configuration. H. Rajbenbach and coworkers [271] achieved continuous readout of angularly coded memories for more than 20 hours using the combination of $LiNbO_3/Bi_{12}SiO_{20}$ or 40 minutes, using $BaTiO_3/Bi_{12}SiO_{20}$ respectively. However, this method guarantees a better detection, but not a real update or refreshment of a grating that has been degraded by intensive read-out.

### 6.4.1  Periodic Copying Between Two Holographic Media

One of the most popular methods to recover the decrease in the diffraction efficiency while preserving the important write-erase capability of photorefractive volume storage devices is periodic copying between two holographic media. The concept is similar to an electronic dynamic random access memory, in which information stored in leaky capacitors is periodically refreshed. Already in 1971, J.C. Palais and J.A. Wise [272] demonstrated that copying a weak hologram onto a second medium, using large modulation depth, dramatically increases the diffraction efficiency of the second hologram. K.M. Johnson et al. [273] applied this technique to improve the diffraction efficiency of multiple exposed silver halogenide films. D. Brady and coworkers [274] demonstrated in 1990 that periodic copying between two dynamic media — a photorefractive crystal and a thermoplastic plate — improves the diffraction efficiency by a factor of $M$ — thus overcoming the limiting factor $M^2$ I described above. Moreover, they found that periodic copying between two holograms results in a stable diffraction efficiency when an indefinitely long sequence of exposures is performed. Although this result looks at a first glance rather academic,

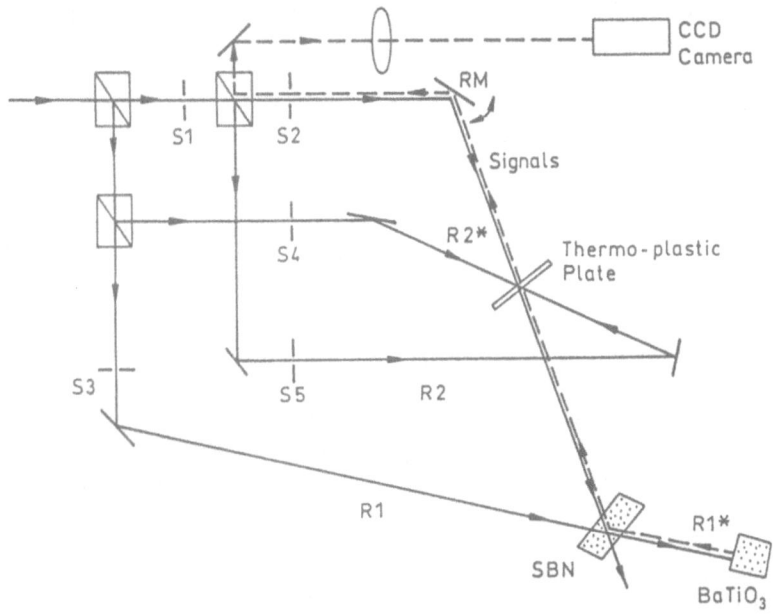

**Figure 6.19:** Experimental system for refreshment of multiply exposed photorefractive holograms implementing periodic copying (after [274]).

because "indefinitely long exposures" are as realistic as a perpetuum mobile, it is relevant for all adaptive holographic systems because they are using long sequences of exposures.

The principle architecture of the system is shown in fig. 6.19. A series of holograms between a reference plane wave and a set of signal beams (in the case of fig. 6.19 these are different plane waves, generated by rotation of the mirror $RM$) is first recorded in a strontium barium niobate (SBN):Ce crystal. During this operation, shutters $S4$ and $S5$ are closed. The diffraction efficiency of the recorded holograms is monitored by illuminating the crystal with the phase-conjugate reference beam $R1^*$. The path of the diffracted readout beam is shown as a dashed line in the figure. A self-pumped $BaTiO_3$ phase-conjugate mirror is used to generate the conjugate wave. In addition to providing automatic alignment of the conjugate beam, the phase-conjugate mirror compensates for phase distortions due to imperfections in the SBN, as I have described in detail in chapter 3.5.

When the diffraction efficiency of the holograms becomes unacceptably low because of the readout processes, the recorded holograms are copied from the SBN to a second holographic medium, a thermoplastic plate. The thermoplastic hologram is formed by using the light diffracted by the SBN hologram and a second refrence wave, $R2$. Again, shutters $S2$ and $S4$ are closed during this operation. The hologram written on the thermoplastic plate is copied back to the SBN by reference beams $R1$ and $R2^*$. The intensity of $R2^*$ is selected to make the intensity of the signal beam diffracted from the thermoplastic equal to the original signal intensity. Shutters $S1$ and $S5$ are closed during this step. The result is a rejuvenated hologram of each of the signal beams in the SBN. At this point, a new series of holograms is added to those already stored in the SBN. This point of the refreshment procedure is the most crucial point of it. To really rejuvenate the formerly stored holograms, the phase conditions for writing these holograms have to be exactly reproduced. Because this is difficult to achieve, most of the refreshment techniques rely on writing a new hologram with the same information content inside the same region of the crystal instead of exactly reproducing the former hologram structure. However, the advantage of the procedure is that when the diffraction efficiency of the holograms again falls unacceptably low, the copying process may be repeated as long as one wishes to. The method described in [275] is based on copying to a planar holographic medium, which implies that it can be copied in one exposure. The extension of periodically refreshed recording to holograms using volume degrees of freedom is more complicated because of the fact that a volume hologram can not in general be evaluated or copied in a single exposure.

Having read the principle of periodic copying for storage refreshment purposes and knowing all the advantages of photorefractive storage devices, one might argue that it should be possible and much better to use two volume photorefractive media. Unfortunately, copying between two photorefractive crystals is complicated by the decay of the hologram being copied while the second hologram is recorded. To understand this relation, suppose that we wish to copy a single hologram between two identical photorefractive crystals. As we read out the first hologram, the light intensity diffracted from it decays exponentially at a rate proportional to the readout intensity. The rate at which the second hologram builds up is determined by the overall intensity upon it. If the reference beam for the second hologram is too bright, then the modulation depth is too weak; if it is too low, then the second hologram does not build up sufficiently before the first hologram decays. To overcome this problem, we could try to set the intensity of the reference beam such that the modulation depth is constant at the second hologram. Then the overall diffraction efficiency of the copied hologram can exceed the efficiency of the original only if the strength of the perturbation produced in the crystal by a single exposure exceeds the strength required to achieve in principle 100% diffraction efficiency. This can be accomplished by selecting a crystal with the appropriate combination of electrooptic coefficients, dielectric constant, index, and thickness. For example, the critical thickness for holograms using the $r_{51}$ electrooptic coefficient in barium titanate is approximately 2 mm [274]. Because crystals of such a thickness are available, it should be possible to extend the copying technique to an all-photorefractive system although the uncertainty in the determination of the other crystal parameters makes it difficult to handle such a system in a flexible way.

H. Sasaki and coworkers [276] presented such a first all-photorefractive memory refreshment system, based on the circulation of the holograms in two SBN:Ce crystals. A schematic diagram of such an all-photorefractive dynamic memory is shown in fig. 6.20. The input images are stored sequentially in one of the two photorefractive crystals, e.g. in PCR1, until the total memory capacity of that crystal is reached. After a certain amount of readouts — the number depending on the readout intensity — the stored information from PCR1 will be degraded. Then the useful part of the information is transferred or copied into PCR2. If the storage capacity of PCR2 is not reached after the transfer, additional input images can be recorded in PCR2. The first crystal is then optically erased to prepare it for the next transfer cycle. The two crystals are functionally identical and exchange their tasks periodically. How do the authors overcome the problem of the write-erase symmetry using identical crystals? An optical amplifier is used to intensify the reconstructed output, the input to be stored, and the images transferred from one crystal to another. In that case, the number of memory acesses increases with the increase of the optical amplifier gain. Experimentally, the authors realized the transfer of a single

**Figure 6.20:** Schematic diagram of the dynamic photorefractive memory. The system consists of an optical amplifier and two functionally identical photorefractive crystals. The dashed line is the transfer path from PCR1 to PCR2 (after [276]).

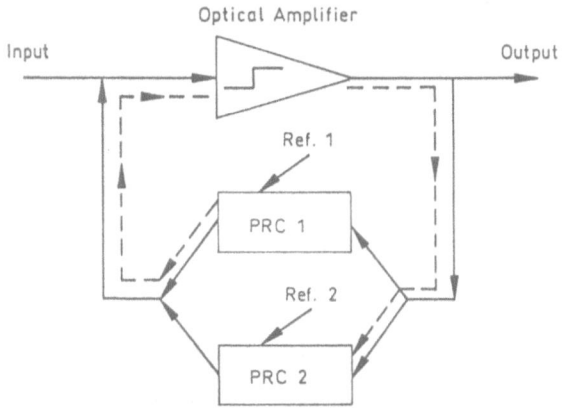

image between the two crystals in a system using two 1mm-thick SBN-crystals and a Hughes liquid crystal light valve as an optical amplifier. To avoid phase distortion in the system owing to crystal imperfections, the recording was performed in an image plane. A combination of a diffuser and a mask, placed against the writing surface of the LCLV, was used as a spatial thresholding device to reduce the system alignment requirement. The experimental setup is shown in fig. 6.21. No noticeable degradation in the pixel quality was observed after fifteen transfers, reaching an averaged signal-to-noise ratio of about 80 for both crystals.

### 6.4.2 Incoherent Feedback Systems

A scheme of refreshment and updating that is suited to be implemented in larger nets consists in adding a feedback loop to the photorefractive memory crystal. In that case, the images retrieved from the system by addressing it with the appropriate reference beams are fed back and are rewritten into the crystal after having been processed, thresholded and amplified. As an advantage of the system, it is more stable because it uses only one photorefractive holographic medium, in which the grating is renewed instead of a copying process between two photorefractive media. This process can either be realized using an electronic loop by an incoherent optical feedback or by an all-optical coherent feedback loop. In the case of an electronic loop, the images are e.g. detected by a CCD-camera, processed by a host coumputer, and then redirected to the crystal via the input spatial light modulator. For the all-optical incoherent loop, all the pixels are processed in parallel by an optically adressed spatial light modulator, which may work in a reflection geometry or in a transmissive one.

Here, one has to take into account, that the images to be written for the first time in the memory must also be written through the optical spatial light modulator so that the relative phase between the image beam and its corresponding reference beam is the same for the first recording as for the refreshing procedure. Moreover, the characteristics of the loop depends strongly on the characteristics of the optical spatial light modulator, thus requiring an intensive monitoring of the behaviour of the spatial light modulator in the resonanat configuration. Finally, it is important to prevent "walk-off" of the images

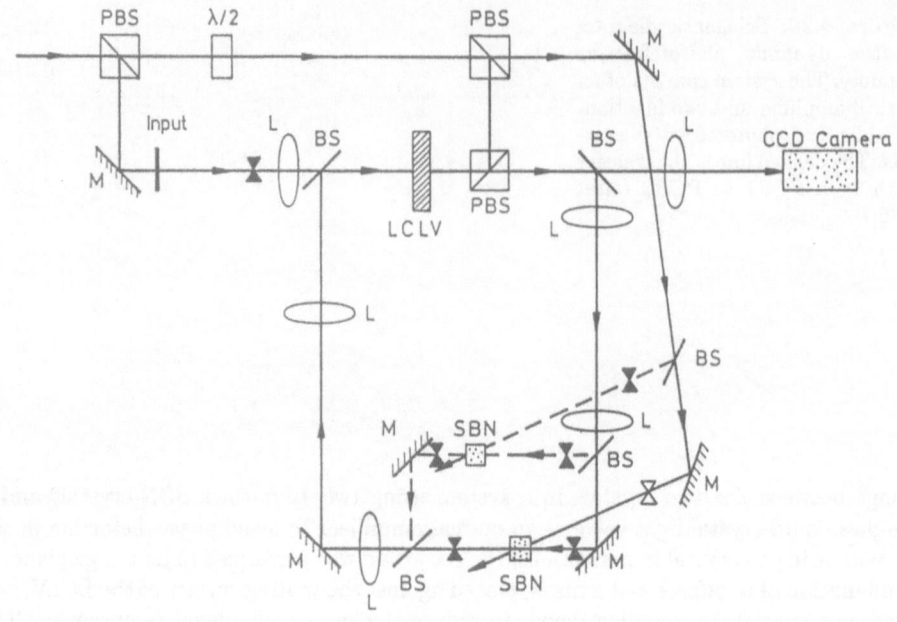

**Figure 6.21:** Experimental setup of the photorefractive dynamic memory. The two crystals are Ce-doped SBN. The image is recorded in an image plane. The output image from the LCLV-based amplifier is captured by a CCD-camera. PBS is a polarizing beamsplitter (after [276]).

from their initial position, e.g. by inserting a spatial thresholding element in the feedback loop, in order to ensure a matched feedback loop.

Experimentally, such a system implementing a spatial sampling technique has been realized using photorefractive $BaTiO_3$. In their experiment (see fig. 6.22), Y. Qiao and D. Psaltis [268] arrived at storing ten angularly multiplexed image holograms in the crystal with highly nonuniform diffraction efficiencies. Due to the copying process in the feedback system, the diffraction efficiencies could be equalized and enhanced. Moreover, they selectively erased a certain image — the fifth one of their series — by introducing a 180° phase shift in the corresponding reference beam, thus writing a hologram that is out of phase with the one to be erased (see also section 4.2.4 for the application of 180° phase shifts for implementation of bipolar weight adaptation). After that procedure, this hologram was almost erased and the other nine holograms were also partially erased. Subsequent dynamic copying rejuvenated the other nine holograms, while the fifth hologram was completely erased because it was too weak to satisfy the switching condition. Finally, a new hologram was recorded at the place of the fifth position and all holograms were again dynamically equalized.

### 6.4.3  Coherent Feedback Using a Double Phase-Conjugate Oscillator

In many storage systems that allow for a cascading of different optical processing layers, a coherent refreshment is required, leaving the phases of the recorded holograms locked. Such a coherent feedback loop with phase-matching conditions may be realized using photorefractive phase conjugation. Because these systems are the ones that are most

Figure 6.22: Schematic diagram of the sampled dynamic holographic memory. SLM = spatial light modulator, LCLV = liquid crystal light valve. (after [268]).

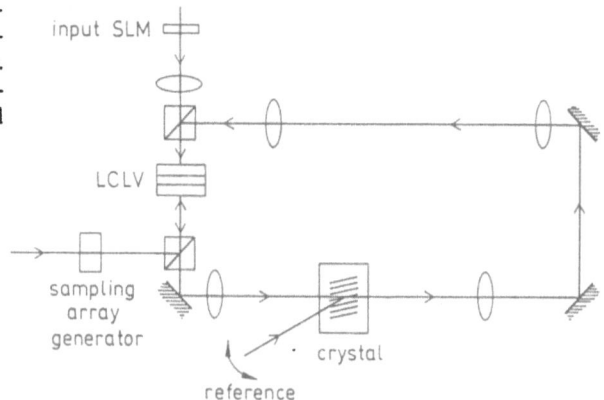

important for realizations of neural network structures, I will discuss their principle function here in more detail.

One way to realize a phase-locked system to coherently refresh photorefractive holograms is to complement the memory hologram by two phase conjugate mirrors, both realized in a single photorefractive crystal operating in the four-wave mixing configuration in order to provide gain during the feedback operation. Thus, the memory is enclosed in an oscillator bounded by two phase-conjugating mirrors. In a more practical configuration, which has been realized by the author and her coworkers, it is even possible to feed the phase-conjugate signal once again back into the same phase conjugator, where it is phase-conjugated independently from the original signal. Although the system requires a high accuracy because of its phase-preserving coherent feedback loop, the implementation of a feedback device with two phase-conjugate mirrors allows for a self-aligned loop, that may be adjusted in its performance by changing the intensity of the pump waves that pump the phase-conjugate mirror. The principle setup of such a coherent feedback refreshment device is shown in fig. 6.23.

*Mathematical Description of Refreshment Using Double Phase-Conjugate Feedback*

 The basic idea of the system is to record the holograms in the memory crystal using the interference of an image beam, $\mathcal{A}_m$, with a reference beam that can be modulated in its incoming angle, wavelength or phase in order to obtain a means to independently address different stored information pages. When such a hologram is read out with the appropriate reference beam $\mathcal{R}_m$, it can be redirected into the crystal using a double phase conjugation procedure.

For photorefractive holograms that are created by pure diffusion charge transport mechanisms, there is a phase shift of $\pi/2$ between the interference pattern and the corresponding phase hologram or refractive index grating. Thus, when the reference beam $\mathcal{R}_m$ is on and the crystal axis is oriented properly, the interference pattern formed by the reference beam $\mathcal{R}_m$ and the diffracted beam $\mathcal{D}_m$ will create a hologram that is exactly in phase with the original hologram. When this diffracted beam $\mathcal{D}_m$ is phase conjugated twice and enhanced due to externally pumped phase conjugation processes, the hologram that this diffracted beam $\mathcal{D}_m$ creates with the reference beam $\mathcal{R}_m$ is exactly in phase with the original hologram, and therefore the latter one is rewritten and updated coherently.

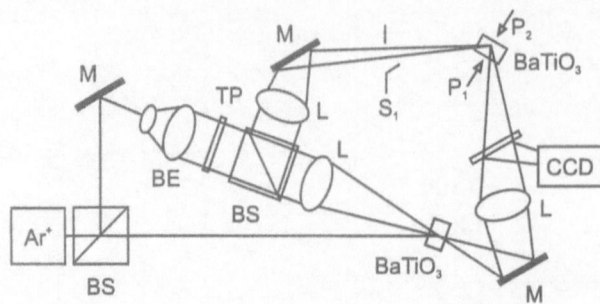

**Figure 6.23:** Scheme of the experimental setup to realize coherent refreshment using double phase conjugation in a photorefractive dynamic memory.

Let us consider here the registration and refreshment of a single hologram. During registration, shutter $S_1$ is closed. The input image $\mathcal{A}_1 = A_1 \cdot e^{i\vec{k}\vec{r}} \cdot e^{-i\omega t} + cc.$ is introduced to the system using a liquid crystal amplitude spatial light modulator (see fig. 6.23). The hologram is recorded with a plane reference wave beam $\mathcal{R}_1 = R_1 \cdot e^{i\vec{k}_1\vec{r}} \cdot e^{-i\omega t} + cc.$ in a photorefractive, nominally undoped BaTiO$_3$ crystal in the Fourier-transform plane of the input images. During readout, the crystal is illuminated with the reference beam $\mathcal{R}_1$ which reconstructs the appropriate image $\mathcal{A}_1$. The diffracted image $\mathcal{D}_1$ is in the ideal case a reconstruction of the original one, multiplied by a constant factor $\beta$ representing the losses of the system

$$\mathcal{D}_1(\vec{r}) = \beta \cdot A_1(\vec{r}) \cdot e^{i\vec{k}\vec{r}} \cdot e^{-i\omega t} + cc.. \tag{6.19}$$

This image is send to a phase conjugating mirror, set up with external four-wave mixing in a second photorefractive BaTiO$_3$ crystal. In addition to provide automatic alignment of the conjugate beam, the phase conjugate mirror compensates for phase disturbances due to imperfections in the BaTiO$_3$ storage crystal. Moreover, external four-wave mixing in these media allows for amplifications of the phase conjugated image — the author and her coworkers reached values up to 40 in the linear gain regime (undepleted pump regime) ensuring that the gain in the system is far beyond the losses and the oscillating system is able to operate. The phase conjugate signal

$$\mathcal{D}_1^*(\vec{r}) = \beta A_1^*(\vec{r}) \cdot e^{-i\vec{k}\vec{r}} \cdot e^{i\omega t} + cc. \tag{6.20}$$

passes again through the storage crystal, creating again an interference grating with the reference wave $\mathcal{R}_1$. However, this grating is not the reconstruction of the one that is erased during read out of the set of reference waves, but its conjugate. Therefore, it may be used for copying techniques in order to preserve the intensity information of the stored grating, but it is not a sufficient reconstruction in coherent, phase-sensitive systems, which require the exact reconstruction of the formerly stored refractive index grating. To achieve such a coherent refreshment, we consider the above grating as a supplementary erasure contribution and phase conjugate the signal once again after it has passed through the crystal. To ensure stability of the system and guarantee the same operation conditions for that second phase conjugation, both phase conjugation processes are performed independently in the same crystal. Both processes can be adjusted to have the same gain coefficient in the linear gain regime using two symmetric points in the angular gain-dependence of four-wave mixing. The angular configuration relative to the c-axis of the crystal can be adjusted in such a way that cross talk between the two four-wave mixing centers can be neglected (see section 3.4). The return of this phase conjugate mirror is the phase conjugate of eq. (6.20) and therefore exactly proportional to the incoming image distribution $\mathcal{A}_n(\vec{r})$.

$$\mathcal{A}_{1,\text{refresh}} = \beta'(\mathcal{D}_1^*(\vec{r}))^* = \beta'\beta A_1(\vec{r}) \cdot e^{i\vec{k}\vec{r}} \cdot e^{-i\omega t} + cc. \tag{6.21}$$

where $\beta'$ is again a constant factor representing the losses for the second phase conjugation process. Thus, $\mathcal{A}_{1,\text{refresh}}$ reconstructs and refreshes the grating in the storage crystal by rewriting exactly the former refractive index grating.

The relative dynamics of the recording and recall process in the storage crystal on the one hand and the double phase conjugation on the other hand is the key to realize an effective refreshment procedure. The temporal behaviour of these photorefractive volume holographic processes depends on the material characteristics, the applied electric field, the orientation and the period of the grating and the overall intensities in the medium. For simplicity, the index grating formation during recording, $\Delta n_r$ and the index erasure $\Delta n_e$ may be approximated by exponential relations (see also section 4.4):

$$\Delta n_r(t) = \Delta n_{\max}(1 - e^{-t/\tau_w}) \tag{6.22}$$

$$\Delta n_e(t) = \Delta n_0 e^{-t/\tau_e} \tag{6.23}$$

respectively, where $\Delta n_{\max}$ is the maximum achievable refractive index modulation in experience, $\Delta n_0$ is the refractive index modulation at the beginning of the erasure process and $\tau_w$, $\tau_e$ are the writing and erasure response times which can be determined experimentally. In eqs. (6.22) and (6.23), I have neglected absorption and will further assume that the values of $\Delta n(t)$ correspond to diffraction efficiencies that are much lower than 100%, being far from the region of saturation. For our system, eqs. (6.22) and (6.23) imply that feeding back the phase conjugate readout image into the crystal is accompagnied by the decay of the hologram being read out while the refreshment is going on. To understand the consequences this effect has on the refreshment system, suppose that we wish to refresh a single hologram implementing two completely identical photorefractive crystals for storage and both phase conjugations. As we read out the stored hologram, the light intensity diffracted from it decays exponentially at a rate proportional to the readout intensity. The rate at which the phase conjugate grating of the read out image builds up, is determined by the overall intensity incident upon it. If the two pump waves creating the first phase conjugation are too low in intensity, the phase conjugated signal does not build up sufficiently before the storage hologram decays. If they are too bright, the reconstruction of the grating in the storage crystal can be realized in a short reconstruction time, but nonlinear gain effects may decrease the reconstructed grating quality. Therefore, we have to select an appropriate combination of the time constants of both the storage and phase conjugating crystals that ensure that the time constants of the latter exceeds the one of the storage crystal in such a way that refreshment is possible without grating distortions before the stored image grating has been faded out. This can be accomplished by selecting a set of crystals with the appropriate combination of electrooptic coefficients, dielectric constants, refractive index, thickness values and adjustment of the geometrical configurations and the intensity relationship.

In order to perform a multiple image refreshment procedure, a series of cycles similar to the ones used for the incremental recording procedure (see section 4.4) can be used, but this time each image recording is followed by a time period of memory readout of all images, in which the read-out images are automatically refreshed. Additionally, in coherent refreshment using photorefractive phase conjugators as feedback devices, one has to take into account that the phase conjugate mirror gratings should be erased after every refreshment operation before the beginning of the next refreshment cycle, in order to avoid cross talk between different images.

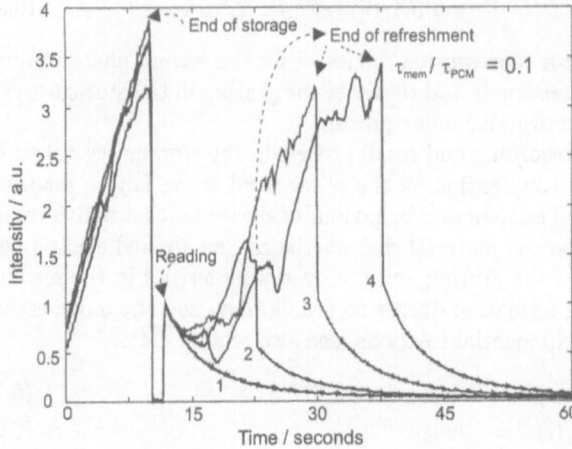

**Figure 6.24:** Dynamic behaviour of the integrated intensity of an image during coherent refreshment operation for different updating times by opening the double phase conjugate osillator. Exponential decay during read-out without refreshment (1), with 20 s (2), 26 s (3) and 33 s (4) of refreshment time. Note the enhancement in the diffraction efficiency after refreshment for (3) and (4).

*Experimental Realization of Refreshment Using Double Phase Conjugation*

In this experiment [277], which is schematically shown in fig. 6.23, both crystals, the storage as well as the phase conjugator were nominally undoped $BaTiO_3$-crystals, but differed in their thickness ($d_{st} = 5$ mm, $d_{pc} = 2.7$ mm) and in their coupling constants because of the configuration of the incoming angles and the relationship of the interacting intensities. When the pump waves of the phase conjugate mirror are chosen to have the intensities $|P_1|^2 = 17$ mW for the forward pump wave and $|P_2|^2 = 1$ mW for the backward, reading pump wave, the gain of the feedback loop becomes larger than one, compensating for the losses and allowing for a refreshment cycle. Moreover, the time constant of the phase conjugate mirror was chosen to be about 10 times smaller than the one of the storage crystal (adjusted by geometric beam configurations and the relationsship of the intensities). If these two conditions are fulfilled, the refreshment procedure as described above is operating.

To demonstrate the operation of our refreshment system, we have stored a single image, an US-Air-Force-test pattern, into the $BaTiO_3$ storage crystal and updated it using different operation times of the feedback refreshment loop. The results are shown in fig. 6.24, where we have measured the integral intensity of the read out images inside the refreshment loop (behind the storage crystal). The refreshment process was started with the beginning of the readout using the appropriate reference beam phase code and stopped after 20 (curve 2), 26 (curve 3), or 33 (curve 4) seconds of refreshment. For comparison, the dynamic behaviour of readout without refreshment by the feedback oscillator is also shown in fig. 6.24 (curve 1). The figure shows clearly that the degree of refreshment, indicated by the integrated diffraction efficiency $\eta_{ref}$ of the image after updating, depends strongly on the durance of refreshment. After 20 s of refreshment, $\eta_{ref}(20s) = 70\%$, after 26 s we get $\eta_{ref}(26s) \approx 100\%$, and finally, for 33 s, we achieve $\eta_{ref}(33s) = 120\%$. A diffraction efficiency and thus a refractive index grating modulation higher than the initial one can easily be achieved by refreshing the image long enough.

In fig. 6.25, the images corresponding to the measurements of the integrated intensity in fig. 6.24 are shown. The upper row shows the image at the beginning of the recall cycle, without refreshment, the lower row represents the refreshed image after 20 s (line 2), 26 s (line 3) and 33 s (line 4) seconds of refreshment. Because the refreshment loop

**Figure 6.25:** Coherent image refreshment of an US-Air-Force test pattern for different times of updating by operation of the double phase conjugate oscillator. Readout images according to fig. 6.24.

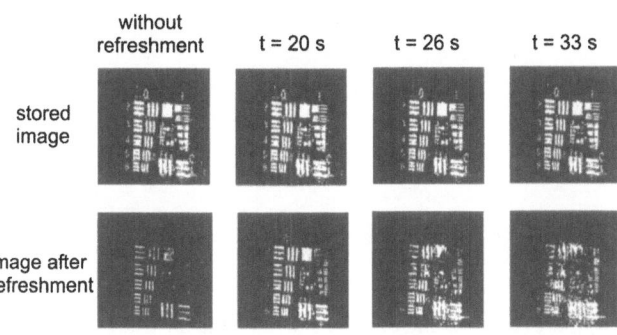

is based on a nonlinear gain oscillator build up by two phase conjugating elements, resonant self oscillation may occur during refreshment that is able to decrease the image quality. Thus, depending on the relative intensities of the pump waves of the phase conjugators, transverse effects caused by mode coupling and mode competition (see e.g. [249, 278, 279, 280]) alter the resolution of the images and contribute to distortions that destroy the exact transverse overlap of the images. Moreover, small deviations in the phase of the phase conjugate signal may lead to an oscillator setup that is not phase-matched, thus creating frequency shifts in the oscillator. The result is a decrease in resolution, contrast and signal to-noise ratio which can be seen obviously in series 4) of fig. 6.24. Fortunately, these oscillator effects occur at a time scale that is much larger than the time scale of the refreshment cycles. As a consequence, this effect can be omitted if the refreshment cycle times are chosen in such a way that transversal image distortion effects can be neglected. Refreshment times of about 24s define the region where it is possible to get an integrated intensity as high as the initial readout intensity by refreshing, with a decrease in the images' quality that is just tolerable.

For applications of coherent refreshment techniques in image updating systems, it is important to know how often an image can be refreshed with still having an acceptable decrease of its quality and diffraction efficiency. To demonstrate the ability of our setup to realize multiple refreshment cycles, we refreshed the US Air Force test pattern repeatedly. To ensure independent refreshment of each cycle, one has to erase the gratings stored in the two phase conjugators first before initializing the following refreshing procedure. This can be done by incoherent illumination of the phase conjugate crystal, erasing all grating down to zero diffraction efficiency. Every refreshment reconstructs the image with a diffraction efficiency about 100% of the initial effficiency when readout was started, showing that continous readout and refreshment is possible. With the decrease in the image quality as discussed above, we were able to realize up to 5 refreshment cycles. The decrease in the resolution and the signal-to-noise ratio is then about 25% and the reduction in the contrast of about 10%. The quality of the image can be sustained in a better way if shorter refreshing cycle times are used. To compensate for this dicrease in the image quality in multiple refreshment cycles, a threshold device as a saturable absorber [226, 281] can be integrated into the oscillator, suppressing the beginning of the self-osciclation and transversal mode coupling effects, thus allowing for an enhancement of the signal-to-noise ratio, the contrast and the resolution of the images. A similar effect can be achieved using an intracavity light pipe or a tapered fiber-optic bundle. These homogenizers significantly reduce optical turbulence and facilitate expanding the

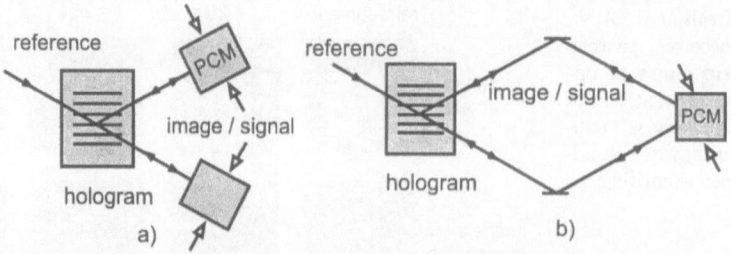

**Figure 6.26:** Scheme of coherent refreshment using phase conjugation of the information and the reference beam. (a) phase conjugation using two phase conjugators, (b) independent phase conjugation of both beams in a single crystal volume.

transverse dimension of the modes in order to support even a larger aperture or a higher number of pixels [281].

### 6.4.4 Coherent Feedback Using Single-Path Phase Conjugation

The second important method to refresh image data coherently uses a pair of phase-conjugating mirrors that reflect both the reference beam and the information carrying signal beam back into the storage crystal [282, 277]. The principle of this method is shown in fig. 6.26. If the two phase-conjugate beams are tuned — e.g. by sharing the same pair of pump waves — in such a way that the phase-conjugate beams retain the same relative phase, both phase conjugate beams will create a hologram that is exactly in phase with the original hologram, thus enhancing and sustaining the latter one. However, this technique requires two phase-conjugating elements that are tuned by shared pump waves, thus requiring a high stability in the setup and identical crystal parameters. To avoid this disadvantage, the author and her coworkers realized a setup by conjugating both beams in the same crystal volume independently from each other (see fig. 6.26 and fig. 6.27). Again, we investigated the features of the system for refreshment of a single hologram. Here, no oscillator effects disturb the image quality. However, in order to achieve the same diffraction efficiency after refreshment, a longer refreshment time has to be applied due to the lower effective amplification values of the single-pass phase conjugator. Thus, image distortion effects due to nonlinear amplification effects, especially the amplification of noise, become more important, decreasing the image quality either. In our experimental

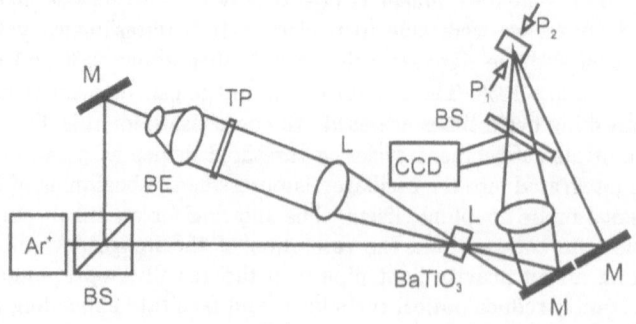

**Figure 6.27:** Scheme of the experimental setup to realize coherent refreshment by feeding back the reference as well as the image beam by a single phase conjugator.

**Figure 6.28:** Coherent refreshment of an US-Air-Force test pattern for different times of refreshment by feeding back the image as well as the reference beam. Readout images according to fig. 6.27 without refreshment at the beginning of the readout ($t = 0$,a), after 60s (b) and 120s (c). The behaviour of the integral intensity with and without refreshment is shown below.

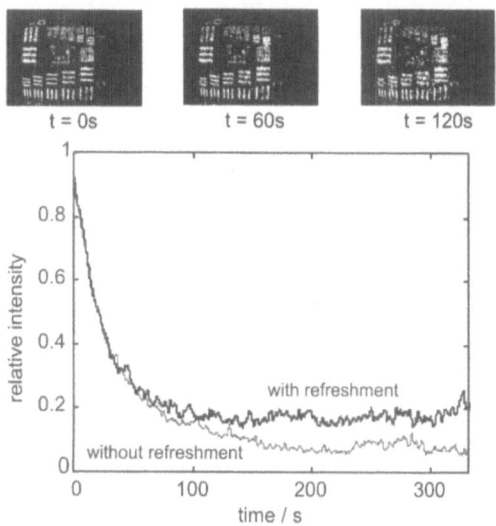

results for sustaining a single hologram, we reached about 25% of the diffraction efficiency relative to the original hologram for 200 s of refreshment. In fig. 6.28, the corresponding images are presented, showing again a decrease in the image quality that is comparable to the one of double phase conjugation refreshment. Similar results have been found in [282].

### 6.4.5  Coherent Refreshment by Alternate Readout

An improvement of the system described above can be found if an alternative readout with the reference and rewriting with the phase-conjugate of the reconstructed signal is realized. In this case, the phase-conjugate mirror is used only for the signal wave retroreflection, simulating a second readout from the rear of the medium. Such a double-side alternating readout scheme has been investigated in detail by P. Yeh and coworkers [283, 284]. It is based on the dynamic hologram redistribution that occurs during read-out in photorefractive materials. As a result, a grating that has its maximum grating amplitude near the entrance face of the medium at the beginning of the readout will be shifted towards the exit phase due to erasure processes and subsequent rewriting inside the crystal. Indeed, the index grating is always pushed towards the exit face of the medium during a single beam readout process. This effect becomes apparent for a slight Bragg-angle mismatch during readout and has been observed by the author and her coworkers in photorefractive BaTiO$_3$ [236]. If an alternate readout from both sides of the crystal is performed, the grating is pushed back and forth in the bulk of the medium. After several cycles, the resultant photorefractive grating is concentrated near the center region of the crystal. In addition, the resultant index grating tends to form a steady-state grating amplitude after several cycles of readouts, which is enhanced compared to the original grating amplitude.

The application of this principle to hologram restoration using a phase-conjugate feedback of the read-out signal beam that simulated readout from the rear side of the storage medium has been demonstrated experimentally by S. Campbell and coworkers [285]. As

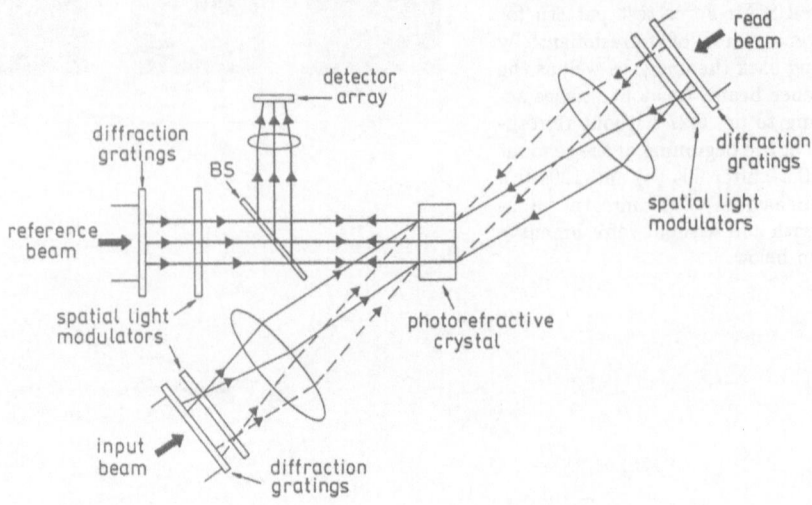

**Figure 6.29:** Schematic diagram of a dynamic interconnection based on reconfigurable volume holograms in photorefractive crystals. Recording of gratings is done between spatially multiplexed pencil beams of a reference array and angularly multiplexed collimated input beams. Reconstruction of gratings is done by read beams that are conjugates of the input beams. The configuration allows to sequentially write interconnections without erasure of previously written gratings. In addition, the gratings that are not to be reconstructed during reading are protected from erasure (after [286]).

in the cases of double phase conjugation or phase conjugation of signal and pump beam described above, the image quality was preserved during the hologram enhancement process only until saturation of the diffraction efficiency was reached. Continued toggling after saturation led to a degradation of image quality as noise levels due to internal scattering and reflections grew to match those of the signal. Again, to preserve the original hologram's image quality while allowing for a significant enhancement it is preferable to stop the enhancing process when the diffraction efficiency reaches its saturation value.

### 6.4.6   Overcoming Beam Erasure by Segregated Gratings

Up to now all methods of refreshment tried to recover the information that had been erased due to adressing the memory data by rewriting the information. Beneath this method, it is possible to avoid subsequent exposure of an interconnection location when writing new interconnection in the crystal volume, thereby minimizing erasure during the sequential writing operation.

For that purpose, K. Rastani and W.M. Hubbard [286] realized a photorefractive-based architecture that uses arrays of spatially multiplexed reference beams and angularly multiplexed input beams for recording the appropriate interconnection holograms. The configurations of the reference and the input arrays are controlled by their respective SLM's, as shown in fig. 6.29. The readout beams are conjugates of the input beams and their configuration is also controlled by a SLM array. To route one beam of the read array, to, say, three elements of the detector array, let us first consider the recording of the appropriate gratings. The interference between the input beam (conjugate of the desired read beam) and three beams in the reference array, which corresponds to the desired detectors, record three columns of spatially separated gratings within the thick photorefractive crystal. Here, SLMs have turned the appropriate beams. During reading, the desired beam is turned on (again by a SLM) and then diffracted by the three grating columns reconstructing conjugates of the appropriate reference beams. These beams are reflected from a beam splitter towards an imaging lens onto their appropriate pixels on a detector array. For routing of other read beams to their respective detectors, additional gratings are recorded as above in sequence. This is to prevent unwanted routing caused by cross recording. The angular separation between the input beams should be larger than the Bragg selectivity of each grating within the thick crystal. This condition allows simultaneous interconnection of the desired read beams (conjugates of the input beams) without cross connections.

This architecture has the advantage that read beams do not expose the gratings that are not meant for them, hence eliminating the cross-erasure problem during readout. However, the whole writing schedule is done in expense of the storage capacity of the material, because the spatially separated reference beams occupy more space than the allocated, single volume interaction region.

All the methods quoted above, beginning from periodic copying up to coherent refreshment using phase-conjugate oscillators or erasure avoiding techniques are suited to maintain information long enough in the memory crystal in order to perform operations that are necessary in higher order neural networks as e.g. error backpropagation, simulated annealing or comparisons in adaptive resonance theory. Thus, with this tools we are able to finally realize network configurations that enable to simulate a wide range of neural network architectures.

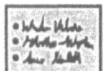

# Summary

A big problem in photorefractive information storage or interconnection devices is that the recording of a new information or image partially erases previously recorded ones. To guarantee the readout of stored data without loss of information while at the same time maintaining the capability to adapt and change the information requires updating or refreshment techniques.

- A phase shift of $\pi$ in the reference beam of a single image is capable of realizing a selective erasure of that image. Thus, arbitrary images can be rewritten with a better efficiency than before.

- Using two or more photorefractive crystals in succession, images are stored in the first crystal, reconstructed with low-intensity beams and then amplified in the second or subsequent crystals.

- Periodic copying between two holographic media, e.g. copying a weak hologam onto a second medium using a large modulation depth can dramatically increase the diffraction efficiency of the second hologram — provided that the time constants of both media are adapted.

- Adding a feedback loop to the photorefractive memory crystal using electronic feedback enables to rewrite the data in the same memory crystal.

- All-optical, phase-conjugate feedback allows to rewrite the memory with its own readout coherently, preserving the phase information of the stored data.
- Phase-conjugating both, the reference beam and the object beam and let them rewrite the grating from the opposite memory side is also feasible to refresh data coherently.

Beneath refreshment of the data due to feedback, it is possible to arrange reference and input beams in such a way that sequential recording does not erase previously written gratings.

## 6.5   Characteristics of Oscillators and Resonators

Up to now we have investigated single operations which are necessary to build up a neural network. With the concepts introduced in the previous chapters, we are able to construct a wide range of neural net architectures, as e.g. associative memories, backpropagation learning networks or self-organized structures.    To illustrate the role of oscillations or resonant effects, let me take as an example from these various models the one of associative memories. Such a memory holds in storage a large number of information entities (called objects or images), any of which may be recalled completely by addressing the memory only with a part of that information or a similar, but different information. The partial object may or may not look like one or more of the stored objects. The most striking feature of the net is that the memory process is then able to recognize the object which resembles most to the input — in other words, the system is able to adapt to an input environment that is a priori not fully determinable. Thus, one should be able to recall a dog from the dog's ear. In terms of configuration or feature space, as I have discussed it in section 2.2, objects that look alike are stored near one another — the associative memory is able to categorize what it stores.

The comparison of the input net with the stored units in these fully connected associative networks as the Hopfield net or the Boltzmann machine is realized with a feedback loop. It depends on the accuracy of that loop — its exact overlap with the incident image and its phase-preserving features — how reliable the retrieval or reconstruction of the information and therefore how powerful the associative memory as a whole will be. Therefore, the resonator plays an important role in the selection process. Because resonator theory is an own and important field in nonlinear optics, in the present book I am only able to give a small, restricted overview over the aspects of resonator theory that are relevant for applications in neural network structures using mainly photorefractive components. For a more detailed information I recommend the standard resonator literature (see also section "Further Reading" at the end of this chapter) and some overview-articles on photorefractive oscillators and their spatio-temporal dynamics as [287, 288, 289, 290].

To start with, let us first consider an empty resonator, e.g. a laser resonator with spherical mirrors. Such a resonator supports a set of Gauss-Hermite eigenmodes — optical fields that do not change their shape after one roundtrip. The modes of such a conventional resonator can be decomposed into a linear superposition of eigenmodes. Changing our point of view, the conventional resonator may be considered either as a memory for Gauss-Hermitians: the resonator memory stores objects as transverse eigenmodes of an optical resonator. Any arbitrary two-dimensional image distribution may then be represented by a weighted sum of eigenmodes in the resonator. Because the actual form of the eigenmodes of the conventional resonator is due to the choice of spherical

**Figure 6.30:** Principle of a ring resonator memory (after [249]).

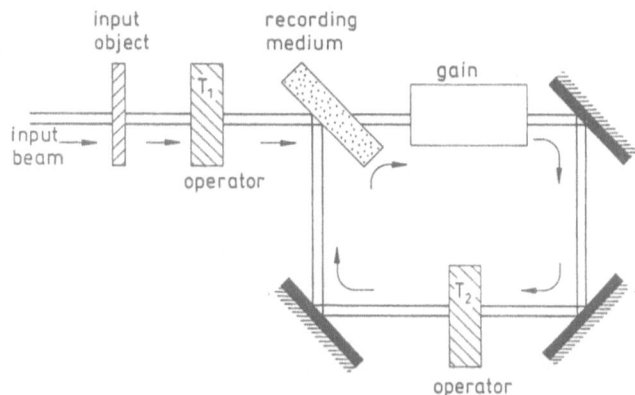

mirrors, different sets of modes can be realized by using different mirrors or intraresonator elements other than spherical ones. An example of such a resonator has been presented by D.Z. Anderson and M.C. Erie [249] and is shown in fig. 6.30. Although it is not an actual example, it shows very clearly the characteristics of resonators for neural network applications. Here, a holographic recording material plays the role of a programmable mirror. To program the resonator, an input object is illuminated with monochromatic light. The subsequent information bearing-beam is transformed by an operator labeled $T_1$ in the figure — a lens, a hologram, a fiber bundle or some other transformation operator. The beam then traverses the recording material and becomes transformed by another operator $T_2$ — the interconnection operator — before it once again traverses the recording material. This operator might be a hologram either. However, in order to get a high interconnectivity, a thin hologram is not very efficient, because after one round trip, a given ray is connected (intersected) with only one other ray. With a volume hologram in contrast, a given ray connects with all rays in a fixed plane. On a fully connected arrangement, every ray would connect with all other rays. The gain medium might be inactive during the recording process, but this is not a strict requirement. The holographic medium records the grating formed by the interference of the original beam with the return beam. The recording process is somewhat different for a real-time medium than it is for a static medium such as a photographic plate I have alreadydiscussed in section 4.1. The developed hologram now contains a grating that makes the writing beam an eigenmode of the resonator formed by the recording optics and the hologram. To understand this, suppose this beam is initially present just to the right of the hologram in the figure. It then traverses the optics in the same manner as during the writing and is transformed to the same return state. The diffraction grating of the hologram then reconstructs the supposed initial beam from the return beam. Since this optical field repeats itself after one round trip, it is an eigenmode — supposing a perfect reconstruction of the hologram. Although the eigenmode is a beam of light with a transverse distribution that varies as it travels around the resonator, I refer for the moment to the whole of the beam as the original object itself.

By multiply exposing the hologram, several objects can in principle be stored. If the set of stored objects is not an orthogonal one, a system such as the one described here will tend to orthogonalize the set; therefore, what is "memorized" is not necessarily what has been stored. In the reconstruction process, various modes of the resonator are excited according to how much the input looks like each mode. If we assume for the present, that the set of stored objects is an orthogonal set, the partial object $|P>$ excites a linear superposition of eigenmodes $|E_i>$.

$$|P> \approx \sum_i a_i|E_i> = \sum_i <E_i|P> \qquad (6.24)$$

The approximation to $|P>$ becomes exact if the set of modes $|E_i>$ is sufficient to span the space of the resonator. The memory then has to make a decision as to which stored mode the input resembles most. This is done through a competitive process in the gain medium. If the gain in the resonator is activated and sufficiently great, oscillation will commence, favourizing one or more eigenmodes. these modes compete with one another for gain; in general, the more one mode uses up the available gain, the less there is for another mode. If mode competition is strong, then the presence of one mode can suppress the oscillation of all other modes. When an injected signal is present, the competition is biased in favour of the mode that most resembles the input [291]. The nature of the competition depends upon the chosen gain medium. A. Herden and F.E. Ammann [291] used two-beam coupling in photorefractive $BaTiO_3$ — an easy-to-implement process providing very large gain coefficients. Because the photorefractive beam coupling process is tuned by the pump beam, the overlap of the frequency of the pump beam and the frequency-gain-response of the resonator will define the modes of the photorefractive ring resonator that can be excited. Due to the narrow line width of the photorefractive gain medium, a strong frequency pulling will result in the resonator, thus allowing several modes that are near the frequency of the pump wave to be excited simultaneously in the resonator [292].

Although the complex physics of two-beam coupling in an optical resonator has not yet been fully analyzed — especially in its transverse dimension, a perturbative treatment of the evolution of the fields inside a resonator has been carried out by D.Z. Anderson and coworkers [293]. The main result of that analysis concerning the competitive behaviour in the resonator can be shortly summarized as follows. In the absence of an injected signal, the equation of motion for the intensity $I_n$ of mode $n$ can be written in the form [249]

$$\dot{I}_n = \alpha I_n - \Theta_{nn} I_n^2 - \sum_{m \neq n} \Theta_{nm} I_n I_m \qquad (6.25)$$

where $\alpha_n$ is a linear gain coefficient, $\Theta_{nn}$ is a self-saturation coefficient describing how much the presence of a mode reduces the net gain for itself, and $\Theta_{nm}$ is a cross-saturation coefficient indicating how much the presence of mode $m$ reduces the net gain for mode $n$. This last term is the one describing the effect of competition. It is important to note that $\Theta_{nm}$ is proportional to the mode intensity overlap integral, taken over the gain volume and not depending on the amplitude. In that case, two objects can be "orthogonal", as expressed by a vanishing amplitude overlap integral, but still have a finite intensity overlap.

If we describe the amplitude distribution of mode $n$ as $A_n(r)$ and write the overlap integral as [249]

$$\Theta_{nm} = \int_{\text{gainvolume}} |A_n(r)|^2 \cdot |A_m(r)|^2 \; d^3r, \qquad (6.26)$$

we may describe for the case of e.g. two modes the competition by the ratio of cross- to self-saturation [249]

$$C = \frac{\Theta_{12}\Theta_{21}}{\Theta_{11}\Theta_{22}} \tag{6.27}$$

This expression is a good basis to explain the different cases of competition. If $C \ll 1$, then competition is weak, indicating that the presence of one mode does not influence the presence of the other. If $C \gg 1$, competition is strong and only one mode will oscillate. The case $C = 1$ (neutral coupling) is perhaps the most interesting one for applications in associative or neural network systems. Then the injection of an external signal will be decisive to select the winning mode. In the experimental operation conditions, such a situation can be adjusted by operating the resonator slightly below the threshold of self-oscillation. In that case, an extra input or object beam pushes the system above threshold and initializes the competitive process. It will end up with the image distribution which resembles most the input image distribution. We may call the resonator in that case an associative selector of eigenmodes or a coherent optical eigenstate memory.

If a volume hologram is introduced into the system instead of a plane hologram, the number of oscillating modes is reduced. We will get a programmable, selective memory, which allows only for special superpositions of eigenmodes. To analyze these processes in more detail, we now have to take attention to the special coupling and amplification processes we are using in our system. I will focus here on photorefractive two-beam coupling and phase conjugation via four-wave mixing and will give an overview over the conditions for resonators with these elements by investigating the stability conditions as well as the longitudinal and transversal interaction of the modes.

At last, let me emphasize one important point. The equations of motion for the mode intensities are similar, for example, to S. Grossbergs field equations for neural networks using the adaptive resonance theory (see section 2.7. However, S. Grossbergs equations of motion describe the evolutions of the states of individual neurons and the interconnection strengths, whereas here I describe the evolution of the states of the stored objects (the eigenmodes).

### 6.5.1 Problems of Real-Time Programmable Resonators

Writing into a real-time memory in a configuration as in fig. 6.32 is somewhat different than for the static memory. For the static case, one can simply multiply expose the photographic plate while taking into account factors such as reciprocity failure and the limited dynamic range of the photographic emulsion. The number of objects that can be stored is limited by the required diffraction efficiency and the film's dynamic range. If the film is properly exposed, one can have a hologram that essentially contains the sum of the sequentially exposed interference patterns. In a photorefractive medium, however, writing one grating tends to erase partially any gratings written previously. Therefore, the recording mechanism has to be adapted appropriately — e.g. via the sequential or the incremental recording method I have introduced in section 4.4.

Another characteristic of real-time resonators which is connected to that dynamic modification of gratings is that as they recall an object, they will also forget this object since the object's gratings tend to become erased. This effect becomes important for the case of a process called *daydreaming*. It takes place in the absence of an injected signal in the resonator. Then the state of the resonator is not static — the output is seen to wander, apparently randomly, from mode to mode, it changes continually (see e.g. [294, 295, 292]). Periodically, a given state will remain for a considerable length of time, then it will suddenly yield to wander again. As one mode begins to appear and becomes strong, an existing mode begins to fade and then disappears. The time scale for these transitions is governed by the response time of the photorefractive gain medium, which is about some milliseconds to seconds, depending on the operating intensity. So the fluctuations in the resonator occur at a "linear time scale". When a signal is injected, the wandering immediately ceases and the resonator collapses into the desired state. The state is not, however, a stable one: as soon as the injected signal is turned off, wandering commences again.

What causes "daydreaming"? Quantitatively, a mode that arises during daydreaming presumably has the highest net gain. As it begins to oscillate more stongly, a given mode may cause self-gain modification or cross-gain modification other than the simple third-order saturation effects included in the weak-field theory. In such a case, a different mode may become more highly favoured. Nonstationary mode competition has even been observed with apertures restricting the number of active modes [294]. The time dependence of the mode emission can be attributed to drift of the cavity length or the optical length (e.g. by changing the propagation length in the resonator in resonators with two-beam coupling or the relative arrangement of the pump waves in four-wave mixing configurations, tuning the cavity modes in or out of resonance with the gain line. In several experiments implementing resonators with two-beam coupling gain, the author and her coworkers could show that all competitional behaviour can be explained by changes in the cavity length, mainly due to thermal effects [292]. Consequently, by stabilizing the setup properly, photorefractive oscillators have been shown to be tunable like a a laser with a narrow gain line to particular transverse mode families. Here, the overlap of the gain line of the photorefractive medium and the one of the resonator may allow for competition between different modes. This competitive behaviour is completely different from the one that is observed without stabilizing the resonator, where modes may wander around randomly caused by resonator length changes.

How does daydreaming influence the resonator's performance? In a resonator, the most likely objects to be recalled are those having the lowest losses — or, in other words, those that have a strong grating in the photorefractive medium. The arising process then reduces the strength of this particular grating. Thus, daydreaming in the resonator induces the various grating strengths to equalize. J. Hopfield et al. [10] demonstrated already in 1983, that an "unlearning process" such as the one of daydreaming has a stabilizing effect in neural network memories. In the absence of any input, daydreaming continues until no grating is no longer sufficiently strong to support oscillations in the resonator. A consequence out of this behaviour is, that if mode competition has to be established before a grating becomes erased, the response time of the recording medium has to be slower than that one of the gain medium.

For applications of resonant structures in optical neural networks, these phenomena can be avoided using stabilized and specially prepared systems (see e.g. [226, 281]) or they can be exploited in order to use features that match perfectly the requirements of neural network structures. Among these features, pattern competition in resonators can be used e.g. to do unsupervised search for states in self-organized systems, sampling of input data or for state selection by temporal pattern search [296].

## 6.5.2  Stability Analysis

To analyze if a resonator system with photorefractive or other nonlinear processing elements will have a stable steady state operation mode, we may use the ray transfer matrix algorithm. This algorithm is useful if the transformation of a beam through different optical systems has to be evaluated in the paraxial beam approximation. In that case, a beam is completely described by its tranversal displacement $x$ and its slope or angle $\alpha$ relative to the optical z-axis. The output parameters $(x_2, \alpha_2)$ depend on the input values $(x_1, \alpha_1)$ by the transfer matrix $M$.

$$\begin{pmatrix} x_2 \\ \alpha_2 \end{pmatrix} = \begin{pmatrix} A & B \\ C & D \end{pmatrix} \cdot \begin{pmatrix} x_1 \\ \alpha_1 \end{pmatrix} \tag{6.28}$$

where $A, B, C, D$ are the parameters characterizing the change of the beams. All periodic sequences can be analyzed with that formalism with regard to their stability. A sequence is called stable, if the trace $(A + D)$ is able to fulfil $-1 < 0.5(A + D) < 1$ [297].

If we apply that analysis to a resonator that uses nonlinear two-beam coupling, one round-trip of the optical beam is given by the product of the matrix of propagation for the beam development of the free-space translation of the resonator length and the matrices of the reflection on the conventional mirror with curvature $R$. Because the photorefractive crystal is small compared to the length of the resonator, the passage through the crystal can be neglected. Therefore, we get the same beam transfer matrix as for the case of a passive resonator.

$$\begin{aligned} M &= M_{prop} \cdot M_{mir} \cdot M_{prop} \cdot M_{mir} \tag{6.29} \\ &= \begin{pmatrix} 1 & L \\ 0 & 1 \end{pmatrix} \cdot \begin{pmatrix} 1 & 0 \\ 2/R & 1 \end{pmatrix} \cdot \begin{pmatrix} 1 & L \\ 0 & 1 \end{pmatrix} \cdot \begin{pmatrix} 1 & 0 \\ -2/R & 1 \end{pmatrix} \end{aligned}$$

The stability condition requires in that case, that $|L/R| < 1$. Thus, not all sorts and shapes of resonators are suited for feedback with amplification via two-beam coupling.

If we want to use resonators with phase-conjugate mirrors instead of one or both conventional mirrors, we need an expression for the transfer matrix describing phase conjugation to obtain the stability analysis. Phase conjugation can be described as the reversal of an optical beam to its origin with the same inclination and phase. Therefore, the transfer matrix is given by [298]

$$M_{pcm} = \begin{pmatrix} 1 & 0 \\ 0 & -1 \end{pmatrix} \tag{6.30}$$

One roundtrip in a resonator with one phase-conjugate element is then given by the product of the matrix for propagation of the resonator length $L$, the matrix for the conventional mirror with curvature $R$ and the matrix describing phase conjugation as mentioned above.

$$\begin{aligned} M &= M_{prop} \cdot M_{pcm} \cdot M_{prop} \cdot M_{mir} \tag{6.31} \\ &= \begin{pmatrix} 1 & L \\ 0 & 1 \end{pmatrix} \cdot \begin{pmatrix} 1 & 0 \\ 0 & -1 \end{pmatrix} \cdot \begin{pmatrix} 1 & L \\ 0 & 1 \end{pmatrix} \cdot \begin{pmatrix} 1 & 0 \\ -2/R & 1 \end{pmatrix} \\ &= \begin{pmatrix} 1 & 0 \\ 2/R & -1 \end{pmatrix} \end{aligned}$$

The condition for stability resulting from that equation is fulfilled in any case, because the sum of the diagonal elements just gives zero. In analogy, the transfer matrix for resonators containing two phase-conjugate mirros gives

$$M = \begin{pmatrix} 1 & 0 \\ 0 & -1 \end{pmatrix} \qquad (6.32)$$

which again results in a completely stable system.

To summarize, resonators with at least one phase-conjugating element instead of a conventional mirror are stable for all possible setup conditions — in contrast to conventional resonators, where resonator length and mirror curvature have to be adapted properly.

### 6.5.3 Longitudinal Mode Analysis

To analyze the longitudinal modes of a resonator, the extension of the active medium can be considered to be small relative to the length of the resonator. Thus, the conventional analysis of Fresnel and Kirchhoff as it is used for passive resonators can be applied. The modes in the resonator are defined by their Fresnel number $F \propto k(x^2 + y^2)/z$. If we have the case of $F \ll 1$ (farfield diffraction), the diffraction effect for modes of higher orders is very strong, leading to an attenuation of these modes. Here, only those modes having low orders are able to "survive". The behaviour changes, if we allow higher Fresnel numbers. The coexistence of many modes of higher orders with similar energy-volumes is possible. Therefore, mode competition may take place.

*Unidirectional Ring Resonators*

The theory of multimode oscillation in a ring resonator that has photorefractive gain developed in the following is based mainly on the derivations of H. Kogelnik [297] and D.Z. Anderson [293]. I will retain only the gratings formed by each mode with the pump beam and ignore the intermode gratings describing the interaction of signals among each other in the weak-field limit (see also sec. 3.4.4). Moreover, I assume negligible pump beam depletion and a small modulation index of the intensity pattern in the photorefractive medium. The nonlinear polarization induced in the photorefractive medium is obtained by considering the Bragg scattering of the pump field from the index gratings in the direction of the resonator field. An overview over the complex interaction of the modes with the external fields and their feedback into the resonator is given schematically in fig. 6.31. Because from a calculational point of view, the unidirectional ring oscillator is the simplest, I have chosen it as a model to analyze the transversal mode behaviour (see fig. 6.32). In that case, the photorefractive gain medium is pumped by a plane reference wave $\vec{P}(\vec{r}, t)$ (see eq. 3.19). The oscillating beam that establishes itself in the ring oscillator is denoted as $\vec{S}(\vec{r}, t)$ and may be expanded in the complete set of the normal eigenmodes of the cavity.

$$\vec{S}(\vec{r}, t) = 0.5 \sum_n S_n(t) e^{-i(\omega_n t + \phi_n(t))} \cdot \vec{U}_n(\vec{r}) + cc. \qquad (6.33)$$

The spatial eigenmodes are also taken to be plane waves for the unidirectional ring resonator. If we neglect the effect of transverse distributions for the moment, the longitudinal part can be written as

$$\vec{U}_n(\vec{r}) = \hat{e}_n \cdot e^{i(\vec{k}_n \vec{r})} = \hat{e}_n \cdot e^{ik_n z} \qquad (6.34)$$

**Figure 6.31:** Schematic flow diagram of the self-consistent analysis of oscillation in resonators with photorefractive gain.

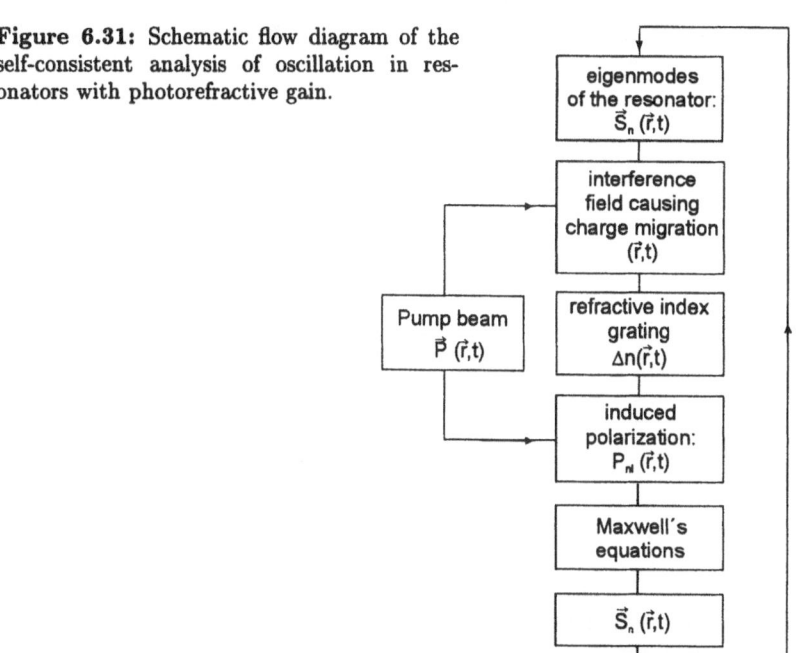

with the wavevector $\vec{k}_n$ for the $n$-th mode. Note, that the magnitude of the wavevectors for the pump and the resonator fields differ from the free space propagation value in the photorefractive medium. The orthogonality of the eigenmodes is given by

$$\int_0^L \vec{U}_m^*(z) \cdot \vec{U}_n(z)dz = \delta_{mn}L \tag{6.35}$$

Because the electric field amplitude $S_n(t)$ and the phase $\phi_n(t)$ vary little in an optical frequency period, they are regarded as real quantities. The optical fields that develop in the resonator and the comparatively static field in the photorefractive medium are connected through an induced optical polarization (see eq. (3.32)). The spatially periodic intensity distribution, that is created by the interference of the resonator field with the pump beam within the crystal, may be written for the case of negligible intermode interaction as

$$I(\vec{r}, t) = |\vec{P}(\vec{r}, t) + \vec{S}(\vec{r}, t)|^2 = I_0 \left( \frac{1}{2} + \sum_n M_n e^{i(\vec{K}_n \vec{r} + \omega_p t + \omega_n t - \phi_n(t))} + c.c. \right) \tag{6.36}$$

with the incoming intensity $I_0(t) = I_p + I_s(t) = 0.5|P|^2 + 0.5 \sum_n |S_n(t)|^2$, the modulation terms $M_n(t) = 0.5 I_0(t)^{-1} \cdot (\hat{e}_n \cdot \hat{e}_p^*) P S_n(t)$ and the grating vectors for the different index modulations $\vec{K}_n = \vec{k}_n - \vec{k}_p$. Using Kukhtarev's equations (eqs. (3.14) — (3.17)), the space charge field results in

$$E_{sc}(\vec{r}, t) = \sum_n E_{sc,n}(t) e^{i\vec{K}_n \vec{r}} + cc. \tag{6.37}$$

and the refractive index modulation is given by

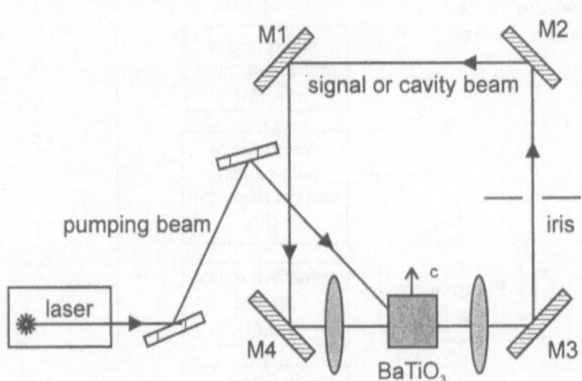

**Figure 6.32:** Schematic of a unidirectional ring oscillator with photorefractive gain via two-beam coupling.

$$\Delta n(\vec{r}, t) = \sum_n \Delta n_n(t) e^{i\vec{K}_n \vec{r} + \nu_n t + \Theta_n(t)} + cc. \tag{6.38}$$

with $\nu_n t + \Theta_n(t)$ being the actual frequency of the $n$-th fourier component of the refractive index grating. The equations of motion of both terms are (see also eqs. (3.26), (3.35)) given by

$$\frac{dE_{sc,n}}{dt} = -\frac{1}{\tau_0} \frac{E_d + E_q}{E_d + E_m} \cdot E_{sc,n} - iM_n \frac{E_q}{\tau_0} \frac{E_d}{E_d + E_m} \cdot e^{i(\omega_p t - \omega_n t - \phi_n(t))} \tag{6.39}$$

and

$$\frac{d\Delta n_n(t)}{dt} = -\sigma \Delta n_n(t) + \beta \sigma M_n \sin \Psi_n(t) \tag{6.40}$$

$$\nu_n + \frac{d\Theta_n(t)}{dt} = -\beta \sigma \frac{M_n}{\Delta n_n(t)} \cos \Psi_n(t) \tag{6.41}$$

with the grating constant $\beta = 0.5 n^3 r_{eff} E_q \cdot E_d / (E_d + E_q)$, $\sigma = \tau_0^{-1} \cdot (E_d + E_q)/(E_d + E_m)$ and the instantaneous phase difference between the interference and the refractive index grating, $\Psi_n(t) = \omega_p t - (\omega_n t + \phi_n(t)) - (\nu_n(t) + \Theta_n(t))$. Bragg scattering of the optical fields from the index grating creates a nonlinear polarization in the photorefractive medium. I will neglect effects that arise because of scattering of the resonator field back into the pump field in the weak field limit. The pump field incident upon the index grating scatters only in the direction of the resonator field. Here, Bragg's condition is automatically satisfied, and this acts as a source of polarization for the resonator field.

$$\vec{P}_{nl}(\vec{r}, t) = 2\epsilon_0 \Delta n(\vec{r}, t) \vec{P}(\vec{r}, t)$$
$$= \epsilon_0 \hat{e}_p P \sum_n \Delta n_n(t) e^{-i\Psi_n(t)} \cdot e^{i(\omega_n t + \phi_n(t) - \vec{k}_n \vec{r})} + c.c.$$

where I have dropped all phase mismatched terms that will average to zero over the crystal volume. Having now obtained the nonlinear polarization induced in the photorefractive medium for a multimode resonator field, we are able to derive the equations of motion for the amplitudes and frequencies of oscillation from the self-consistency equations.

If we apply the above equations for only *one oscillating mode*, we find an instantaneous and typical beat frequency phenomenon: the frequency of the oscillation will be shifted relative to the pump frequency. The direction of the shift depends on the relative change of the cavity length — whether there has been a rise or a drop in the cavity length relative to the length at the resonator frequency.

For *two modes* there is no longer the possibility of mode locking [293]. For the stationary equilibrium state of the system, which is given by the phase angle

$$\Psi_{i,s} = \frac{\Omega_i - \omega_p}{\gamma} = \Delta_i \qquad (6.42)$$

and is equal to the normalized cavity detuning ($\Omega_i = k_i/\sqrt{\epsilon\nu}$ is the frequency of the passive resonator), we find the two possible mode intensities at steady state given by [293]

$$I_{1,s} + I_{2,s} = \frac{\beta_1}{1 + \Delta_1^2} - 1 \qquad (6.43)$$

$$I_{1,s} + I_{2,s} = \frac{\beta_2}{1 + \Delta_2^2} - 1 \qquad (6.44)$$

using the normalized amplification factors $\beta_i$. These equations show that only in the case of identical gain factors, losses and detunings of the cavity, simultaneous oscillation of both modes is possible at steady state. In all other cases, the dynamics of the system leads always to a steady-state situation where one mode completely inhibits the oscillation of the other regardless of the initial conditions, thus being the mode with the larger gain or the mode with the smaller magnitude of cavity detuning. The steady-state intensity of the favoured mode is that which would have been obtained if only one mode had been oscillating in the cavity.

The analysis for *three or more modes* leads to the same result. The mode with the highest amplification will win the competition and its intensity value in the stationary state is equal to the one of single-mode operation. As a conclusion, there is no possibility for different longitudinal modes to coexist in a resonator with photorefractive two-beam coupling gain — except for those modes that have exactly the dame gain factors in the photorefractive amplifier.

*Phase-Conjugate Oscillators*

To analyze the conditions of the existence of longitudinal modes for the case of oscillators with one or two phase conjugating mirrors (see fig. 6.33), I choose once again the situation of plane waves. Then, the problem can be solved in analogy to the case of an unidirectional ring resonator. Therefore, I do not want to repeat the calculation once again, but instead present a simple argumentation to understand the phase relationships that are most important for the existence of longitudinal modes. Let us assume a plane wave that is incident on the oscillator with a constant phase $\phi_0$. The reflection at the conventional mirror gives the additional phase term $\phi_{mir}$, the free propagation phase is $\phi_{prop} = kL$. The phase-conjugate mirror creates its inherent phase change $\phi_{pcm,0}$, which is $\phi_{pcm,0} = \pi$ for ideal phase conjugation and beneath that a second propagation in free space in the opposite direction: $\phi_{pcm} = \phi_{pcm,0} - \phi_{prop} - \phi_{mir} - \phi_0$. Thus, the resulting phase shift after one roundtrip in the resonator is only given by the phase change of the phase-conjugate

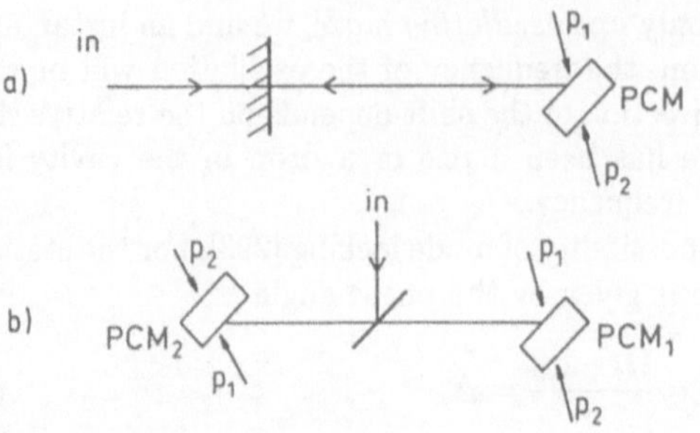

a)

b)

Figure 6.33: Schematic of phase-conjugating oscillators with photorefractive gain via four-wave mixing.(a) Oscillator with a single phase-conjugated mirror, (b) oscillator with two phase-conjugated mirrors.

mirror, which may be a multiple of $\pi$ for an ideal phase conjugator. Consequently, there is only a condition on the phase of the wave, but not on the length of the resonator. This is different to conventional resonators, where a longitudinal phase shift of $2kL$ is created for every roundtrip. The sensitive dependence of the resonator output on the phase of the wave will be discussed in section 6.5.5.

As a conclusion, there are no longitudinal mode conditions for oscillators with at least one phase-conjugating mirror — provided that the phase conjugation is an exact one. If an additional phase shift is impressed on the ocillator wave by the phase-conjugation process, this phase plays an important role in tuning the oscillator. Therefore, a continuous range of frequencies can be existent in the cavity as long as they are coexistent with the pump waves. If the frequency of the pump waves is different from the one of the incident waves on the phase-conjugate mirror e.g. by a detuning of $\Delta\omega$, the reflected signal is as well shifted about $-\Delta\omega$ and therefore it its no longer the phase-conjugate of the incident wave. This frequency detuning has been investigated in [299] and allows the system to be tuned in order to support different longitudinal modes. Thus, such an oscillator can be continuously tuned by changing the frequency of the pump waves as the relevant control parameter. For the case of an oscillator with two phase conjugating mirrors, an analogous argumentation gives the result that the two nonlinear phase shifts of both mirrors add in such a way that $\phi_{pcm1,0} - \phi_{pcm2,0} = 2\pi n$. Therefore, phaseshifts play an important role in these type of oscillators. A complete analysis of them has been given by M.S. Cohen [300].

### 6.5.4 Transversal Mode Analysis

To get an insight in the behaviour of the transversal case, we have to extend the description of the eigenmodes and include the transversal distribution of the modes.

$$\vec{U}_n(\vec{r}) = \hat{e}_n A_n(x,y) \cdot e^{ik_n z} \tag{6.45}$$

As before, these eigenmodes are taken to be orthogonal, with a normalization factor given by

$$\int_{V_{res}} |U_n(\vec{r})|^2 d^3r = L \int_S \int |A_n(x,y)|^2 dx dy = LN_n \tag{6.46}$$

where the double integral is taken over the transverse dimensions of the resonator. Thus, the modulation of the refractive index as well as the polarization become dependent on the position in space. If we proceed the same calculation as in the longitudinal case, we are able to compare the intensity in the quasi-stationary case to the intensity for the case of plane waves [293]

$$I_{1,s} = \frac{\beta_1 - \Delta_1^2 - 1}{\beta_1} \frac{\int \int_S |A_1(x,y)|^2 dxdy}{\int \int_S |A_1(x,y)|^4 dxdy} \qquad (6.47)$$

Thus the transversal distribution changes the stationary intensity of the modes. For the case of two modes we get

$$I_{i,s} = \frac{\alpha_i}{B_i} - \frac{T_{ij}}{B_i} \cdot \frac{\alpha_j}{B_j} \cdot (1 - C)^{-1} \qquad (6.48)$$

with $i, j = 1, 2$ and the coupling constant $C$ between both modes

$$C = \frac{T_{12}T_{21}}{B_1 B_2} = \frac{\int \int_S |A_1(x,y)|^2 |A_2(x,y)|^2 dxdy}{\int \int_S |A_1(x,y)|^4 dxdy \cdot \int \int_S |A_2(x,y)|^4 dxdy} \qquad (6.49)$$

For the case $T_{12} = T_{21} = 0$, when there is no overlap between the modes, the two modes are no longer coupled with each other and both modes acquire the intensity of the single-mode case. If $C \gg 1$, the competition is weak, the modes do not influence each other. If in contrast we have $C \ll 1$, the competition is as strong that only one mode is able to oscillate. Finally, the case of $C = 1$ describes the most important case for our applications in neural networks. It is the case of neutral coupling, where competition is already strong, but both modes are still able to oscillate simultaneously as long as their cavity detuning, gains and losses are equal. Any difference in cavity detuning or the gain factors (e.g. by injecting a signal from outside or by changing the losses with transversally modulated intensity absorbers) leads to complete inhibition of oscillation of one mode by the favoured mode. In absence of an external driving signal, the competition processes described above result in a dynamic state, in which the modes are changing arbitrarily. This describes exactly the case of "daydreaming" or "mode dancing" I discussed above.

What do these derivations implement for our neural network applications? For the case of e.g. an associative memory device, a lot of images or information pages have to be stored in the photorefractive crystal using one of the multiplexing techniques described previously. A given, partially erased or noisy input image then creates a reference wave, that has to be thresholded and decoded to result in the right image information. In unidirectional ring resonators, these conditions for stable oscillations are difficult to fulfil for all image distributions and reference beam configurations — especially if a phase-matched feedback is required. Therefore, the implementations of phase-conjugating resonators are much more promising for these applications, because they guarantee a priori an easy adjustment and a selfregulating stability. Moreover, they allow for the correction of intracavity distortions and at the same time for energy optimization via amplification.

### Phase-Conjugate Oscillators

 To investigate the transversal mode structure for the case of oscillators containing one or two phase-conjugating mirrors, C. Huygens' principle in the approximation of J. Kirchoff can be used. In that case, the field can be expressed as the product of $E(x, y, z) = u(x, z) \cdot u(y, z)$, where

$$u(x, z) = \frac{\exp\left(i(\frac{\pi}{4} - \frac{k|z - z_0|}{2})\right)}{\sqrt{\lambda|z - z_0|}} \cdot \int_{-a}^{a} u(x_0, z_0) \exp\left(-i\frac{k}{2}\frac{(x - x_0)^2}{|z - z_0|^2}\right) dx_0 \qquad (6.50)$$

and $2a$ is the size of the aperture in the oscillator. The condition of self-consistency for transversal modes, which requires that the field has to reproduce itself after one roundtrip, results for the case of resonators with one phase-conjugating mirror in the Fredholm-integral equation [301]

$$\gamma u(x) = \int_{-a}^{a} u^*(x_0) \exp\left(-i\frac{k}{2L}g(x^2 - x_0^2)\right) \cdot \frac{\sin(\frac{k}{2L}b(x - x_0))}{\pi(x - x_0)} dx_0 \qquad (6.51)$$

with the complex constant $\gamma$, which defines the loss and the phase shift per round trip, the length $L$ of the cavity, the degree $g = 1 - L/R$, which measures the non-confocal symmetry and the transversal dimension $2b$ of the phase-conjugating mirror. In that equation, the conjugation of $u(x, z)$ takes place by the process of phase conjugation.

For the confocal case $(R = L)$, we get for that integral exact analytical solutions, which are exactly those for normal symmetric confocal resonators [301]. The wavefront of the transversal modes is in that case equal to the curvature of the conventional mirror. For a nonconfocal symmetry $(R \neq L)$, a numerical analysis done by C. Wilbur and cowork-ers [301] gives, that the phase fronts of the transversal modes are not in conformity with the curvature of the conventional mirror. In that case, diffraction effects grow important to define the phase of the wave. For high Fresnel numbers, the phase deviations of the conventional mirror and the medium can be compensated by the phase-conjugating mirror. However, for low Fresnel numbers, this can no longer be guaranteed. In other words, the phase conjugator is a real conjugator as long as its Fresnel number is sufficiently high.

For the case of oscillators with two phase-conjugate mirrors, we get from the condition of self-consistency

$$\gamma u(x) = \int_{-a}^{a} u(x_0) \exp\left(-i\frac{k}{2L}g(x^2 - x_0^2)\right) \cdot \frac{\sin(\frac{k}{2L}b(x - x_0))}{\pi(x - x_0)} dx_0 \qquad (6.52)$$

which can be solved analytically. The solution is given by

$$u_m(x) \quad \propto \quad S_m(x, ab/\lambda L)e^{-ikx^2/2L} \qquad (6.53)$$

$$\gamma_m \quad \propto \quad (R_m(1, ab/\lambda L))^2 \qquad (6.54)$$

with the radial and the spherical wave functions $R_m, S_m$ respectively, as they can be found in conventional confocal systems. It is important to note that the field $u_m(x)$ contains an effective radius of the curvature, which corresponds to the cavity length $L$. Thus, a change in $L$ automatically induces a change in the radius of the curvature.

To summarize, transversal modes in resonators with one or two phase-conjugating mirrors exist for all combinations of mirror curvatures and resonator lengths. Phase deformations of the conventional mirror can be compensated as long as the phase con-jugating mirror is sufficiently large. For our implementation in optical neural nets this means, that because of the self-adjusting features and the possibility to couple many modes transversally, resonators with phase-conjugating elements have to be preferred compared to unidirectional photorefractive ring resonators.

**Figure 6.34:** Transfer function of the phase of the backward pump beam into the phase of the phase-conjugate beam in photorefractive four-wave mixing using BaTiO$_3$ for different beam geometries.

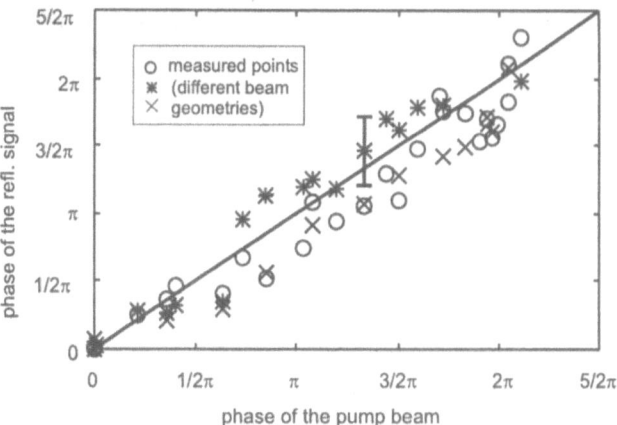

### 6.5.5 The Role of the Phase in Phase-Conjugate Oscillators

From the above derivations concerning stability as well as longitudinal and transversal mode analysis, phase-conjugate resonators may be described as the experimentalist's dream — they do not need to be adjusted at all, but are self-aligning systems — at least in a theoretical treatment. In reality, this property of self-aligment in resonators with one or more phase-conjugating elements depends sensitively on the alignment of the relative phases of the pump waves. The author and her coworkers showed, that the relative phase of the backward or reading pump wave is transferred to the phase-conjugate signal, thus defining the phase offset in the resonator. It is that phase that can be used to tune a resonant system with a phase-conjugate mirror to its proper oscillation state. In fig. 6.34, the transfer function of the phase of the backward pump wave, changed by a pure phase modulator that was introduced into the reference beam arm, in a four-wave mixing photorefractive phase conjugation configuration to the phase-conjugate signal is shown. For different pump wave geometries, the effect is always the same: The phase of the pump wave defines the phase of the conjugate wave. Consequently, "conjugation" is a relative description, depending strongly on how the pump wave is tuned compared to the signal wave. Exact phase conjugation can only be achieved if the two pump waves — the backward and the forward, writing pump wave — are exact phase conjugates of each other. For all other cases, the two pump waves define a reference system, which allows to tune the phase of the phase-conjugate signal beam. Thus, the backward pump waves can be used to vary the complex reflectivity coefficient in photorefractive four-wave mixing in its whole complexity — in its absolute value by changing the pump beam angle (see section 3.5) or in its phase by tuning the phase of the pump beam. The same behaviour can be observed when changing the phase of the forward, writing pump wave. However, because this wave is the dominant one in writing the refractive index grating in a transmission grating configuration, it results also in a shift of the grating in the crystal volume, thus giving a temporally-dependent change in the phase of the phase-conjugate wave. What are the consequences of that phase shift transfer for phase-conjugate oscillators? In fig. 6.36, the output intensity of an oscillator containing a single phase-conjugate mirror (Fabry-Perot type) is shown in dependance on the phase of one of the pump waves. The oscillatory, cosinus-like behaviour suggests, that the phase of the

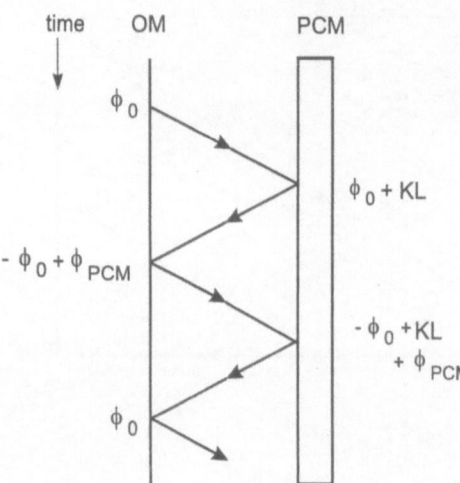

**Figure 6.35:** Schematic illustration of the phase matching conditions of a phase-conjugate Fabry-Perot resonator consisting of an ordinary mirror (OM) and a non-ideal phase-conjugating mirror (PCM).

pump wave "tunes" the resonators, thus giving in-phase (maximum) and out-of phase (minimum) values for the resulting intensity. These results show clearly that the phase-conjugating Fabry-Perot oscillator can be tuned using the phase of the backward pump wave. By changing this phase, the intensity output of the resonator can be maximized or minimized, leading to a high-gain system or a non-oscillating system, if the resulting intensity is below the threshold value of oscillation.

This behaviour can be explained if we estimate an oscillator that is tuned only for every two round-trips, as shown in fig. 6.35. Imagine a wave with a phase (or phase difference) of $\phi_0$ emerging from the conventional or ordinary mirror (OM) into the oscillator. After reflection on the phase-conjugate mirror (PCM), the resulting phase is given by $\phi_0 + kL$. The return to the OM gives a phase of $-(\phi_0 + kL) + kL + \phi_{pcm}$. Only for the second round trip, all phases are compensated, leaving the initial phase of $\phi_0$. For $\phi_0 = \phi_{pcm}/2$, the input field will be reproduced after a single roundtrip, giving the results we discussed already for the case of an ideal phase-conjugate resonator (see sections 6.5.3 and 6.5.4). An explanation for this behaviour can be found by estimating two fields in the resonator, giving by the input amplitude and its phase-conjugate. These two fields lead to the cosinus-dependence shown in fig. 6.34, which is typical for a two-beam interferometer.

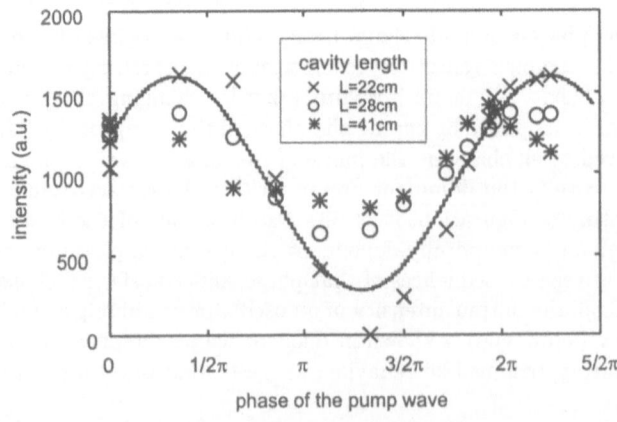

**Figure 6.36:** Behaviour of the output intensity of a phase-conjugate oscillator when changing the phase of the backward pump wave.

However, from our experience we know that the phase-conjugate oscillator resonates. Thus, we can interpret the behaviour as a superposition of a resonator that is always tuned with a two-beam interferometer. Because the contribution of the resonator part is not transferred into the transmission function of the resonator, we observe the typical two-beam interference behaviour. This interpretation has already been theoretically given in [302] and is in good agreement with our observations.

With these special resonator conditions and the results from the previous chapters, we are now finally able to look at optical realizations of feedback systems and thus at realizations of almost all of the neural network architectures I have discussed in the second chapter. I will begin with the most popular sort of networks, associative memories, and will go into detail of the association process and then proceed to more complex neural network implementations.

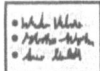 ## Summary

In fully-interconnected networks, the comparison between an input and the stored units is realized with feedback loops. The reliability of the retrieval is strongly dependent on the accuracy of that feedback. The features of resonators are therefore important for the construction of a well-working optical neural net.

- The conventional resonator may be considered as a memory for Gauss-Hermite eigenmodes: the resonator memory stores objects as transverse eigenmodes of an optical resonator. Any arbitrary two-dimensional image distribution may then be represented by a weighted sum of eigenmodes (linear superposition) in the resonator.
- If a recorded hologram is introduced in the resonator, it contains a grating that makes the writing beam an eigenmode of the resonator formed by the recording optics and the hologram.If the set of stored objects is not an orthogonal one, the system will tend to orthogonalize the set. Therefore, what is "memorized" is not necessarily what has been stored.
- For image comparison with an input object, the memory has to make a decision as to which stored mode the input resembles most. This is done through a competitive process in the gain medium.
- Resonators with at least one phase-conjugating element instead of a conventional mirror are stable for all possible setup conditions — in contrast to conventional resonators, where resonator length and mirror curvature have to be adapted properly.
- In a resonator with photorefractive two-beam coupling gain, there is no possibility for different longitudinal modes to coexist. In contrast, there are no longitudinal mode conditions for oscillators with at least one phase-conjugating mirror. Therefore, a continuous range of frequencies can be existent in the cavity as long as they are coexistent with the pump waves. The phase of the backward pump wave of PCM can be used to tune the resonator.
- In unidirectional ring resonators, conditions on the curvature and the reflectivity of the mirrors for stable oscillations are difficult to fulfil for all image distributions and

reference beam configurations — especially if a phase-matched feedback is required. Thus, transversal competition exists.

- Transversal modes in resonators with one or two phase-conjugating mirrors exist for all combinations of mirror curvatures and resonator lengths. Phase deformations of the conventional mirror can be compensated as long as the phase conjugating mirror is sufficiently large.

 **Further Reading**

1. F.T.S. Yu, S. Jutamulia, *Optical Signal Processing, Computing and Neural Networks*, Wiley Series in Microwave and Optical Engineering, New York (1992).
   A basic and easy-to-read introduction to optical signal processing, combined with actual application examples. The chapters on *Photorefractive signal processing* and *Optical neural networks* give a good overview over the topic.
2. B. Javidi, J.L. Horner (eds.), *Real-time Optical Information Processing*, Academic Press, San Diego (1994).
   Among the important chapters in that overview, you will found *Fixed hologram neural networks* by D. Casasent, *Holographic neural networks based on multigrating processes* by Y. Owechko, B. Soffer, and *Optical Phase Conjugation for Interconnection and Image Processing* by P. Yeh, A. Chiou.
3. H. M. Gibbs, *Optical Bistability: Controlling Light with Light*, Academic Press, New York (1985).
   An easy-to-read introduction to all possibilities of light-light interaction with special emphasize on optical bistability.
4. A.D. McAulay, *Optical Computer Architectures*, Wiley & Sons (1991).
   This book provides the theory, hardware design techniques, and lot of examples of the application of optical concepts to computing.
5. B. Fischer, S. Sternklar, S. Weiss, *Photorefractive Oscillators*, IEEE QE 25(3) (1989).
   A concise review article on photorefractive oscillators, the mutual light-crystal interactions and oscillator applications.

# Part III

# Concepts, Architectures and Complex Systems

# 7    Associative Memories

After having introduced the different possiblities to realize the most important components of optical neural networks — interconnection and storage systems, threshold devices and optical processing configurations — it is time to put these components together to complete optical neural networks. The following chapters will present examples of realizations of the different neural network types following the schematic presented in chapter 2. The emphasize of the next chapters is again on those systems that use the components and algorithms I have described in the first two parts of the book. I have focussed on those realizations that are based on all-optical components, using optoelectronic devices to support the optical function of the system. Neural network implementations using mainly electronic elements that are supported by optical interconnections are only mentioned rarely for some special realizations, but are mainly left out in this part.

In order to understand the principle function of the different examples of optical realizations of neural networks presented in the present part, I have chosen among the existing proposals and realizations of optical neural networks a limited number of interesting, promising and well-developped systems that are described in detail. The concentration on such a limited number, but profundely described systems implies often a more detailed technical description of the system performance, but will give a deeper insight in the performance, state-of-the-art and limitations of these systems. Moreover, numerous citations are given for those who become interested in a special system and want to explore its function in all its details. Beneath these systems that are described extensively, in all chapters several aspects of supplementary realizations are mentioned shortly, giving the reader the possibility to go into details using the cited references.

For the first chapter of this part, I want to start these descriptions with optical realizations of associative memories, I have already introduced in the second chapter, because they represent on the one hand applications of simple neural networks as perceptrons or Hopfield networks. On the other hand, associative memories have basic roots in optics because of their similarity to correlation and image comparison, making them since several decades to an attractive research topic in nonlinear optics.

Associations are a common experience of our everydays life. We are easily able to put names to faces, recall that someone looks familiar because we work with him or we have met him before and so on. We form links between people, events and places or between shapes, objects and concepts. It is essentially this ability that allows to match an input to and output pattern and to build our own representation of the world as we see it. Inputs to our senses usually cause a cascade of associations and recollections, each one prompting the next. Human memory therefore works in an associative fashion - a memory system such that an input specifically evokes the associated response.

Computational models of associative memories have been studied for many years in the framework of realizing neural network models, and much of the work in neural networks draw on ideas developed in this field. The distinction between an "associative memory" and a "neural network" therefore is imprecise and not clearly defined. Often, it is a matter of personal preference how a network is named, since many networks operate as associative memories (as e.g. the Hopfield network, in which patterns are associated

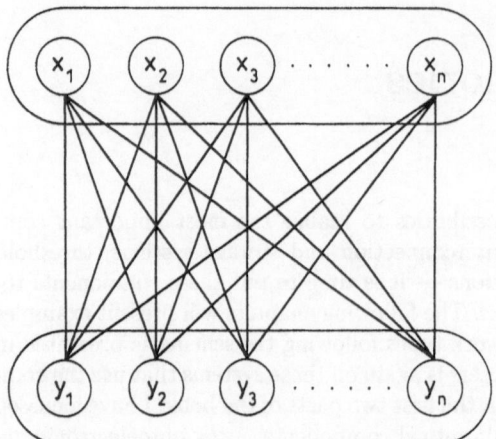

**Figure 7.1:** The bidirectional associative memory (BAM) seen as a two-layer network. Each input $x_i$ is connected to every output $y_i$.

with themselves), whilst some associative memories perform the same processing as the network. It can be shown that neural network models for associative memories can be derived from a simple optimization principle without resort to any 'biological' assumptions such as the Hebbian learning rule, the threshold and the updating rule. All the features of the Hopfield model for neural networks can be obtained with that derivation, including the inner product algorithm, the symmetric and zero-diagonal interconnection matrix and bipolar als well as binary realizations. This interesting derivation, which is in detail described in [303] shows again the close connection of associative memories and Hopfield neural networks. In other words, the Hopfield neural network model can be described in terms of the inner product, e.g. as a matched Vander Lugt filtering step followed by a pattern synthesis step, whereas associative memories can be described in terms of the Hopfield net algorithm.

On the other hand, some of the adaptive processing algorithms for pattern recognition and image processing, in terms of neural networks, can be seen as adaptive heteroassociative and autoassociative memories, respectively.For example, a certain class of image processing problems, as e.g. image restoration, is closely related to autoassociative memories, while pattern recognition is closely related to heteroassociative memories.

An improvement of associative memories that allows to see the close connection of association and neural network models was proposed in 1988 by B. Kosko [304]. His bidirectional associative memory (BAM) can be seen as a simple form of a two-layer nonlinear feedback network, as shown in fig. 7.1. It consists of two fields of processing elements, each receiving inputs from all elements in the opposite field, but none from elements in its own field. Each of the neural processing elements forms a weighted sum of its inputs. A threshold — hard-limited or sigmoidal — is applied to this sum to give the output value of the element. Inputs to the BAM are brought in as extra input lines to the processing elements in one or both, typically both, fields. System outputs are taken as the input lines of some or all of the processing elements. BAM operation begins with the presentation of initialized values on the input lines. The state of the processing elements may subsequently be updated one at a time in a randomly ordered fashion, or the fields may update all their elements simultaneously. During these processes, patterns sweep from one neuron layer to the next, and then back again, slowly relaxing into a stable state that represents the network's association of the two patterns. The weight in the forward pass can be represented as a connection matrix $M$, whilst those in the backward pass are given by the transpose of this matrix, denoted $M^T$. The use of the connection matrix's transpose makes the bidirectional associative memory interesting, since this distinguishes it from other systems which use a different matrix of connections for the backward pass.

The BAM stores pairs of patterns $X_i, Y_i$. It is autoassociative if $X_i = Y_i$ and heteroassociative if $Y_i$ is different from $X_i$. In a standard heteroassociative memory, $X$ is presented to $M$, then thresholded to produce output $Y$ that is hopefully closer to the stored pattern $Y_i$ than to all the other patterns $Y_j$, if $X$ was closer to $X_i$. However, this assumption is not always valid and we would like to have a procedure that allows us to increase the accuracy of the final recall. The BAM achieves this by passing the output $Y$ back through the system to produce a new value, $X'$, which should be closer to the stored pattern $X_i$ than was the original pattern $X$. This new value is passed forward again, producing a better estimate $Y'$, and the process repeats until it settles down to a steady resonance between the stored patterns $X_i$ and $Y_i$. The advantage of using the transpose of the matrix $M$, $M^T$, is that it requires no additional information, and this information is locally available to each node. B. Kosko has proved that the BAM converges to a fixed pair of stored patterns by extending Hopfield's argument, and demonstrates that the Hopfield case of autoassociation is simply a specialized case of the BAM, when $X = Y$. In other words, the sequence $X \rightarrow M \rightarrow Y$, followed by $Y \rightarrow M^T \rightarrow X$, which continues producing a series of approximations $(X, Y), (X', Y'), (X'', Y''), \cdots$, will converge to a steady resonant state that reverberates between the fixed pairs $(X_f, Y_f)$. Having proved that any matrix $M$ is bidirectionally stable in this way, B. Kosko goes on to show that patterns can not only be recalled from a fixed matrix $M$, as in the Hopfield net, but that if small changes are made to $M$ in accordance with a Hebbian learning rule, it will learn to associate two patterns. In this case, as the patterns oscillate back and forth, pattern information is allowed to sweep into the weights, resulting in the learning of an association between two patterns.

You will see in the following sections, that most of the all-optical realizations of associative memories refer to the Hopfield model as a fully-connected network that takes the output as the new input for the next iteration. As in the Hopfield model, most of these architectures display their solution as a steady-state output, when there is no longer a change from cycle to cycle.

## 7.1 Linear Optical Associative Memories

The good match between the parallelism and interconnectivity of optics and the requirement of associative memory paradigms has not gone unnoticed over the years. In contrast, it was the principle of association that attracted researchers to exploit optics for neural network purposes. D. Gabor, the inventor of holography, appreciated its associative properties. He found that the correlation signal of a traditional optical correlation setup can be used as a metric to compare an unknown input image with a library of programmed reference images in a shift-invariant fashion [305]. Such a system therefore operates as an autoassociative memory. Therefore, the correlation-based associative memory is today dated back to D. Gabor. P.J. Van Heerden [100] predicted in 1963 that a hologram would produce a "ghost image" of a stored image upon illumination with a fragment of the original image. This was subsequently confirmed by G.W. Stroke et al. [306]. These early holographic associative memories (often named ghost image holography systems) suffered from distortions, poor signal-to-noise ratio and low storage capacity. However, the development of off-axis holography, invented in 1962 by E.N. Leith and J. Upat-nieks [307] greatly improved the signal-to-noise ratio by angularly separating the desired signal term from the undesired noise due to self-interference among scattered waves form the original image. R.J. Collier and K.S. Pennington [308] demonstrated only a short time later ghost image reconstructions using this off-axis approach. The next important step in the development of associative memories was the introduction of off-axis holograms for matched filter recognition of objects. The idea is to use a setup writing a Fourier hologram, i.e. that the hologram is placed in the focal plane behind a lens. If a second lens is now placed behind the hologram in the associative reconstruction process, the correlation of the input image with the stored image appears in the back focal plane or correlation plane. Moreover, the location of this spot corresponds directly to the location of the matching image in the input plane. The *Van der Lugt linear optical correlator* has found many applications in pattern recognition, signal processing and optical associative memories.

The perfect analogy between a general neural network for associative purposes and a neural network with an optical Van der Lugt correlator can be easily seen. In a two-layer neural network, the input vector is multiplied with the weight vector connecting the input neurons to the first neuron in the hidden layer. The product of the input and the weight vector is then fed into the neuron and then thresholded to produce the output. Thus we see that the transfer function between the input and the output of a neuron can be governed by the product followed by thresholding (see section 2.1 for further details). In a correlator setup, the input is first Fourier-transformed and then multiplied with a holographic filter function in the matched-filter plane. The product of these two complex spatial functions is then inversely Fourier-transformed. The output plane spatial light distribution is the correlation between the input and the spatial neural-weight function. The center light intensity of the correlation signal is proportional to the product of the input and the neural-weight function. Therefore, it is feasible to use an optical correlator as a generic building block for an optical neural network.

### 7.1.1 Page-Oriented Linear Associative Memories

One of the earliest applications of the optical correlator for optical associative memo-ries was in the page-oriented holographic associative memory (HAM) [309] for digital computers. In this application, memory data are stored in a large number of spatially multiplexed (shifted) holograms. During recording, different data planes or "pages" are recorded in each hologram sequentially by shifting a plane wave reference from hologram to hologram. In the readout phase, the light from the input data page illuminates the entire set of holograms, allowing for an associative search of all the stored data simul-taneously. A detector matrix determines the location of the resultant correlation peak which in turn determines the location of the hologram containing the matching data. This information is used to shift a readout reference wave to the proper hologram for readout of the associated data.

These investigations marked the starting point for the development of page-oriented

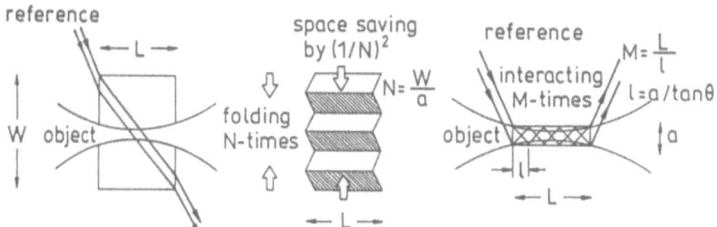

**Figure 7.2:** Interaction of object and reference beams in holographic recordings: in bulk crystals (left) and in a photorefractive optical fiber (right). The multiple interaction beam configuration can be explained by an *n*-times folding of space, reducing the space occupied by the photorefractive medium by $1/n^2$ (middle).

associative holographic memories which are capable of large storage capacities. Up to present, a lot of improvements have been made to make optical associative correlation systems compact, practicable for industrial applications and as adaptive as possible to introduce them in larger setups. Among them, Van der Lugt correlators that use photorefractive crystals or organic polymers have found a lot of interest because of their ease of alignment and especially their feature for real-time optical correlations that can be implemented in robotic applications (see e.g. [310, 311]) or in applications of identification (finger prints, faces) [312]. An interesting and impressingly well-working application that combines this correlation technique with training by gradually adapting photorefractive holograms is a face recognition system realized by H.L. Sidney Li and coworkers. They used faces from different positions as the training data and realized the discrimination of faces from three different persons by correlation with input data with almost arbitrary face position and inclination. Turning of the head could be tolerated up to 30° in each direction and the tilt of the head could be about 12° in either direction [313].

### 7.1.2 Photorefractive Fiber-Based Linear Associative Memory

As an example of a linear associative memory using the concept of photorefractive holographic storage, I want to present a system based on photorefractive fibers. In these devices, photorefractive holographic gratings may be formed between a signal and a reference beam inside a fiber via total reflections at the fiber faces. Because of the multiple reflection scheme of the fibers, a large interaction region between the two beam can be created (see fig. 7.2), ensuring thus large energy transfer between the beams. In such a fiber, even an axial hologram can be recorded. For that purpose, the fact is exploited that a linearly polarized plane wave that is coupled into a fiber using a condensing lens will exit at the other end of the fiber as a cone of light beams, having different polarizations, caused by the anisotropy and the Fresnel reflections inside the fiber. Thus, as the light beam propagates along the fiber, the spatial information of the incident light beam is scrambled. If the incident light beam is an object beam that interferes with a reference beam which is coupled into the opposite end of the fiber, a hologram similar to that of a reflection hologram can be recorded in the fiber. However, this object beam can not be read out by merely using the reference beam. In contrast, if the recording is made using a conjugate reference beam, which can be derived with a polarization-preserving phase-conjugate mirror, a transmission-type hologram can be recorded in the fiber. In

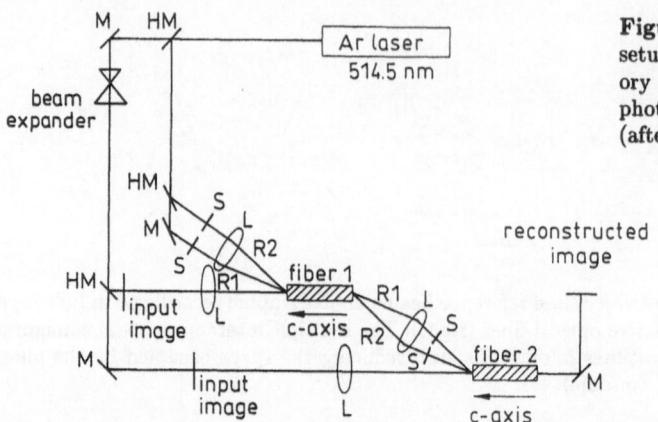

**Figure 7.3:** Experimental setup of an associative memory system using cascaded photorefractive optical fibers (after [314]).

this case, if the recorded photorefractive fiber is illuminated by the reference beam, the phase conjugate object beam will be reconstructed.

An experimental setup for performing linear association with these fiber devices has been realized by H. Yoshinaga and coworkers [314] and is shown in fig. 7.3. The two LiNbO$_3$ fibers are connected in cascade. For recording of multiple images, angular multiplexing is achieved by changing the incident angle of the reference beam to the fiber axis for each image. All images are Fourier-transformed by a lens and recorded by being launched on the axis in the fibers. The reference beam used for recording in the first fiber propagates through it and is successively used for recording in fiber 2. In the experiment, auto- or heteroassociation can be performed by using the same or another image as the input image for fiber 2. The authors could show storage and associative recall from a partial input image for two images. Although this setup is ingenious because of its compactness and ease of adjustment, it suffers from great degradations in the quality of the reconstructed images since the aperture of the fiber is too small to record the entire Fourier spectrum of the image.

*The Problem of Position Invariance*

Although these linear page-oriented holographic memories have some outstanding properties as quickness and ease of implementation, most of them are limited because they are not shift invariant. The problem of position invariance is a global one, also being a limitation to nonlinear holographic associative memories. They work best if the matching patterns always appear in the same position. However, many associative processing tasks as the detection of faults in production sequences or quality inspection require such an insensitivity especially to size and location of the probe. Therefore, methods have been developed to preprocess data in such a way that the associative recognition then can take place in a unit invariant to these effects. Some methods of preprocessing have already been described in chapter 6 as optical arithmetic operations (section 6.1), temporal data reduction processes as optical novelty filtering (section 6.2) or data refreshment techniques (section 6.4). Moreover, there is the possibility to reduce the data to be associated in such a way that an information invariant to scale, rotation and shift results, that may be further processed.

Y.-K. Lee [315] recently presented a method of optical angular feature mapping of an object for pattern recognition that uses self-pumped phase conjugation followed by a vertical projection. In this method the vertical contour features of an object are most enhanced and vertically projected at each angle as the object is rotated. The vertical enhancement of the contour and the projection are performed optically with self-pumped phase conjugation in $BaTiO_3$. As the crystal is located in the back focal plane of a Fourier-transforming lens, the vertical contours of an object orthogonal to the plane of incidence become most significantly enhanced. Others that deviated from the orthogonal direction gradually decreased in brightness. These angular features are vertically and then horizontally integrated using a cylindrical lens in order to obtain angular signatures at an angle. The resulting signatures form an angular signature function. A cross correlation between two angular signatures may then be performed by different means, e. g. electronically calculated or optically realized.

Although a real Fourier-hologram correlation setup shows shift-invariance, its signal-to-noise ratio strongly depends on the shift distance. Consider a conventional matched filter correlation configuration. Since the overall operation of writing and retrieving is a cascade of correlation and convolution, the system is shift invariant. However, the output contains the correct recall term, but combined with cross-correlation noise. Consequently, the signal-to-noise ratio is inversely proportional to the number of pages to be stored. When $N$ is the number of pixels in a stochastic two-dimensional image distribution, and $M$ is the number of vectors to be stored, the signal-to-noise ratio in the correlation plane is given by

$$SNR = \sqrt{1}2M \tag{7.1}$$

So, even if $M = 1$, the $SNR = \sqrt{1/2}$, which implies that the reconstruction of even a single, stored association is noisy and thus being ineffective for shift-invariant operation.

In order to reduce this noise, spacial encoding technique can be applied — using in the most simplest case an aperture that can be placed in the correlation plane. This aperture allows to transmit correlation peaks that are at a distance $L$ from the zero position, resulting still in a shift invariant reconstruction to a certain amount. Then, SNR is given by

$$SNR = \sqrt{\frac{N}{(2L+1)M}} \tag{7.2}$$

J. Hong and D. Psaltis [316] implemented a more elaborated encoding scheme, allowing a shift-invariant memory that can store multiple one-dimensional vectors. The performance of the system could be shown to approach that of the outer-product scheme, which is not shift invariant. Moreover, nonlinear interconnection may also be used through the use of a squaring nonlinearity in the correlation plane in order to overcome the probem of shift-invariance. I will discuss implementations using this nonlinear holographic memory approach in the following section.

In response to the need for highly-parallel archtitectures for neural network models and nonlinearities in the correlation plane, a new class of holographic associative memories has been developed recently which is also based on the optical correlator principle. These new devices also perform associations using correlation as a measure of similarity. However, they use the storage in volume holographic media instead of planar hologram devices. Thus, a higher storage density and the independent addressing of each stored page is possible without cross talk (see also section 4.4).

The second change in the configuration compared to page-oriented linear holographic devices is the use of nonlinear gain and feedback provided by e.g. phase conjugation to implement competition between stored memory pages. This competition is used to perform associations with error correction, allowing to filter out of the correlation the one with the best match to the input and an improved signal-to-noise ratio on multiple inputs in parallel.

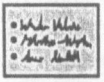 **Summary**

Beneath the Hopfield algorithm, the correlation signal of a traditional optical correlation setup can be used as a metric to compare an unknown input image with a library of programmed reference images in a shift-invariant fashion. The most important items of the resulting linear holographic associative memory are:

- In a correlator setup, the input is first Fourier-transformed and then multiplied with a holographic filter function in the matched-filter plane. The product of these two complex spatial functions is then inversely Fourier-transformed. The output plane spatial light distribution is the correlation between the input and the spatial neural-weight function. The center light intensity of the correlation signal is proportional to the product of the input and the neural-weight function.
- Correlators can be realized using fixed holographic plate filters, holographic fibers, optical disks in combination with volume holograms or volume holographic memory crystals with the ability of adaptive change of the gratings or filters stored.

## 7.2   Nonlinear Holographic Associative Memories

The nonlinear holographic associative memory — which I will refer to in the following as NHAM — is an optical associative memory which combines the fully-parallel image-to-image heteroassociative capabilites of ghost image holography whith the high signal-to-noise ratio, processing gain and storage capacity of thresholded Van der Lugt correlators. In addition, nonlinearities allow an NHAM to select a particular stored datum over all others on the basis of incomplete input data.

A schematic diagram of a representative NHAM system is shown in fig. 7.4. I have chosen a system that is based on double phase conjugation to achieve a resonant process because it seems to be the most promising and easy-to-build system among the various architectures that have been proposed. The author of this book as well as a number of different authors have investigated the properties of that configuration in detail [317, 146, 277] (see also section 6.4). Other architectures will be described below after having explained the principle mechanism of that system. The heart of the system is a volume hologram in which Fourier transforms of objects $a^{m,0}$ are recorded sequentially or incrementally using in this case angularly multiplexed reference beams $b^m$ (see section 4.4).   For readout of the NHAM, phase-conjugate mirrors or other means of forming retroreflected time-reversed beams are positioned on both sides of the hologram, forming a phase-conjugate resonator. As I have discussed in section 5.2, linear as well

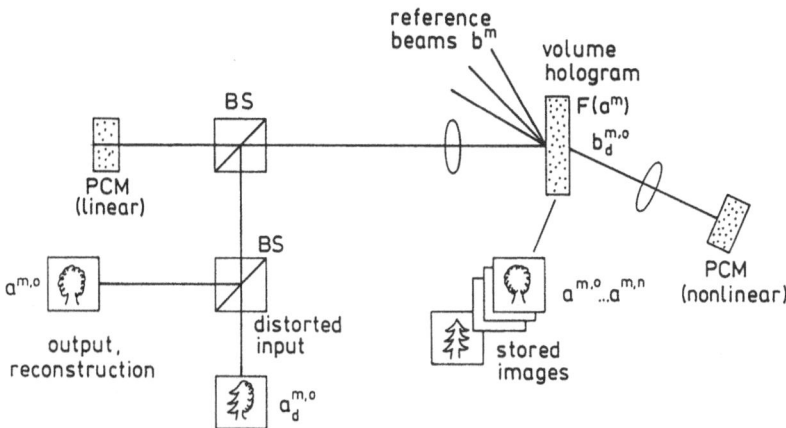

**Figure 7.4:** Principle configuration of a nonlinear holographic associative memory. Images $a^{m,i}$ are stored with angularly multiplexed reference beams $b^m$ in the memory crystal. When adressing it with a distorted input $a_d^{m,0}$, the reconstructed reference will be nonlinearly reflected by the phase-conjugate mirror and reconstructs the best-matched exemplar. This reconstruction may be retroreflected again by a linear phase conjugator in order to enhance the reliability of the system response (after [318]).

as nonlinear thresholding in combination with beam amplification may be achieved implementing this sort of retroreflection. The volume hologram divides the resonator into the object and the reference arms. When a partial or distored version of an object $a_d^{m,0}$ addresses the hologram via the beamsplitter, a set of partially reconstructed reference beams $b_d^{m,0}$ is generated. Each reconstructed reference beam is convolved with the correlation of the input object with the stored object associated with that particular reference beam. This part of the system is identical to a matched filter Van der Lugt correlator. The distorted reconstructed reference beams are phase-conjugated by the reference arm phase-conjugate mirror and retrace their paths to the hologram. These beams then reconstruct the complete stored objects. The reconstructed objects are phase-conjugated by the object arm phase-conjugate mirror. The process iterates until the system settles into a self-consistent solution or eigenmode, assuming the gain of the phase-conjugate mirrors is sufficient for oscillation. As I have discussed in section 6.5, in the absence of the hologram, the phase-conjugate resonator could support a continuum of different resonator modes. Because of the volume hologram interaction with the resonator, the eigenmodes of the NHAM resonator are defined by the stored wavefronts in the hologram.

An important common feature of most different configurations of that principle NHAM is — as I document in the name of that associative memory — the introduction of nonlinearity. Without it, NHAM's could not "choose" a particular datum over all others and the output would be a linear superposition of multiple recalled memories. If the stored objects are considered to be vectors in state space, then NHAM nonlinearities form regions of attraction around the stored object vectors in a manner analogous to neural network formulations of associative memories. A lot of different nonlinearities can be implemented to realize such a nonlinear associative memory, beginning with the most simple one, the step- or hard-limited nonlinearity, up to complicated ones as the sigmoidal nonlinearity. The nonlinear response and multiple stable states of the NHAM allow selections between patterns to be made on the basis of incomplete data since

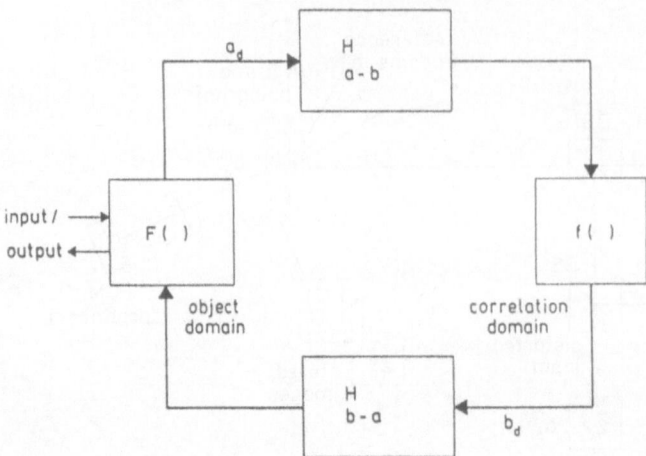

**Figure 7.5:** Schematic block diagram illustrating the resonant nature of a nonlinear holographic associative memory. f() = nonlinear reflectivity of the reference arm in the correlation domain, F() = output plane point nonlinearity in the object domain. H(a-b), H(b-a) = forward and backward paths through the hologram (after [318]).

gain will exceed loss only for the stored pattern with the largest overlap with the input pattern. Nonlinearities also improve the signal-to-noise ratio and storage capacity over ghost image holography or linear matched filter correlators. The output association is available in two forms depending on where the output is coupled out. The reference side of the NHAM is essentially a Van der Lugt correlator where a correlation peak marks the location of the recognized object in the input plane. In the object arm, an undistorted version of the associated stored object is superimposed over the partial input object. The output can be separated from the input with a beamsplitter.

### 7.2.1 Mathematical Description

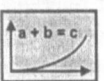 Now let us have a more detailed look at the effects of nonlinearities in the reference arm on the signal-to-noise ratio and storage capacity of an NHAM. The resonant nature of the NHAM is illustrated in fig. 7.5. Assuming for the moment thin Fourier transform holograms and using the approach of ghost image holography, the NHAM output is given by the iterative equation [318]

$$a_n^{m,0} = F\left[\sum_{m'}\left(f\left\{\sum_m(a_{n-1}^{m,0} \odot a^m) * b^m\right\}\right) \odot b^{m'} * a^{m'}\right] \tag{7.3}$$

where $a_n^{m,0}$ is the amplitude in the object arm after the $n$th round-trip in the resonator, $a^m$ are the objects stored in the hologram, and $b^m$ are the reference beams used in holographically recording the objects. f() represents the nonlinear reflectivity of the reference arm, F() represents an output plane point nonlinearity, $*$ and $\odot$ represent convolution and correlation, respectively. $a_0^{m,0}$ is the input "seed" for the resonator that represents a partial or distorted version of the input object $(m,0)$. The output of the $n$th round-trip consists of a double sum of cascaded correlations-convolutions. The double sum over the object indices $m$ and $m'$ is due to the double pass through the hologram. Assuming angularly multiplexed plane waves as reference beams, the $b^m$ functions are spatially displaced delta functions, $b_m = \delta(x - x_m)$. This results in the new iterative equation [318]

$$a_n^{m,0}(x) = F\left\{\sum_{m'}\sum_m[C_{n-1}^m(x - x_m + x_{m'})] * a^{m'}\right\} \tag{7.4}$$

with $C_{n-1}^m(x) = a_{n-1}^{m,0} \odot a_m$ is the correlation between the stored object $m$ and the resonator amplitude distribution in the $n$th iteration. In eq. (7.4), it is assumed that the angular separation between reference beams is large enough to separate the correlation-convolution terms in the reference arm, allowing to disregard cross terms due to the nonlinear reflectivity f(). To compare such a system with outer product neural network models of associative memories, let us assume objects consisting of binary $N$-dimensional vectors, uniformly distributed and equally spaced reference functions $B^m$. If these spacings are wider than the width of the objects, let us place an aperture over the output plane, thus preventing any ambiguity in the output plane occuring otherwise for thin holograms. For that case, only terms for which $m = m'$ are retained, giving [318]

$$a_n^{m,0} = F\left[\sum_m \sum_p f(C_{n-1}^m(p))a^m(i-p)\right] \tag{7.5}$$

Now we are able to calculate the signal-to-noise ratio in the first iteration (or at the end of the first roundtrip) from [318]

$$SNR = \frac{\mid f[C_0^{m,0}(0)] \mid}{\sqrt{\sum_{p\neq 0} \mid f[C_0^{m,0}(p)] \mid^2 + \sum_{m\neq(m,0)} \sum_p \mid f[C_0^m(p)] \mid^2}} \tag{7.6}$$

If we assume that the objects are random, not orthogonalized distributions, they can be described by a balanced binary phase diffuser model [319] with

$$C_0^m(p) = \begin{cases} N\delta(p) + \sqrt{2(N-\mid p \mid)/3} & \text{if } m = (m,0) \\ \sqrt{2(N-\mid p \mid)/3} & \text{if } m \neq (m,0) \end{cases} \tag{7.7}$$

where $N$ is the size of the stored object vectors. If we finally assume that the point nonlinearity in the reference or correlation domain has the form $f(x) = x^n$, which means that the effects of arbitrary nonlinearities can then be estimated by using a polynomial approximation and superposition, we arrrive for $N \gg 1$ at [318]

$$SNR_{\text{NHAM}} = \beta(3/2)^{n/2}\sqrt{(n+1)/2}\frac{N^{(n-1)/2}}{\sqrt{M}} \tag{7.8}$$

where $\beta$ is the fraction of $a^{m,0}$ used as the input object and $M$ is the number of stored vectors. From that expression, we can estimate that for an SNR that is small enough to achieve successful recall, $M$ is proportional to $N^{n-1}$. Therefore, we can conclude that the storage capacity of an NHAM can be increased by increasing the nonlinearity in the correlation domain.

A similar analysis for the outer-product associative memory results in

$$SNR_{\text{outer-product}} = \mid 2\beta - 1 \mid \sqrt{n/M} \tag{7.9}$$

indicating that $M$ is proportional to $N$. This is the same result as reported by J.J. Hopfield [10] who estimated the storage capacity as linear in $N$ based on empirical evidence for small $N$ values. This estimation was supported by Y.S. Abu-Mostafa and J.N. St.Jacques [320] who used hyperplane counting arguments to show that the outer-product model is bounded from above by $N$. Other authors (see e.g. [321, 322]) applied techniques from coding theory to the outer-product model and showed that for random objects the maximum asymptotic value of $M$ for which all objects can be recognized is $N/(4\log N)$ as

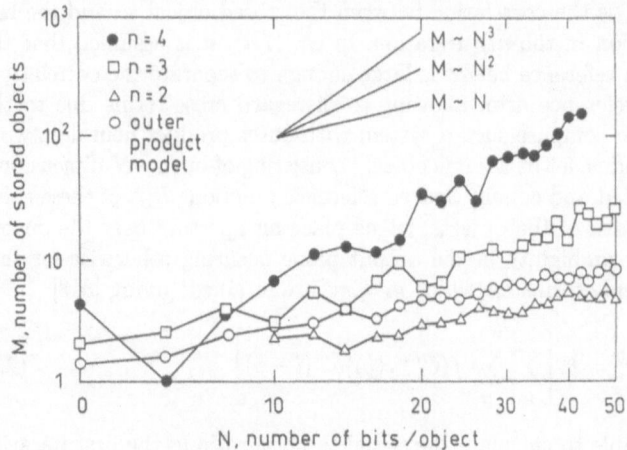

**Figure 7.6:** Comparison of the storage capacity of reference-based nonlinear holographic associative memories in the correlation domain of the form $f(x) = x^n$, assuming error-free input objects (after [323]).

$N$ approaches infinity. In fig. 7.6, computer simulations performed by Y. Owechko et al. [323] are shown, comparing the storage capacity of the outer-product model and the NHAM presented in fig. 7.4 for the number of vectors stored as a function of $N$ for power law nonlinearities with $n = 2, 3$ and 4.

NHAM implementations can be categorized based on the resonator geometry and the method used for generating the reference beams used in recording the holograms. They can be further differentiated by the form and implementation of the nonlinearities. Most of the systems reported up to date have been based either on a unidirectional feedback using photorefractive two-beam coupling or on a double phase-conjugate configuration similar to the one described above in which a separate independent reference beam is associated with each object beam. In general, most NHAM implementations up to now have not relied on the nonlinearity of the phase-conjugating mirror. Instead, various external nonlinear mechanisms (as e.g. saturable absorption) have been used. Beside these multipass resonant configurations, ring resonator geometries have been proposed and demonstrated which derive the reference beam from the object beam, relying on the eigenmode selection of the resonator itself. Although these systems lack some of the discrimination obtainable using separate reference beams, they do incorporate competition between stored modes using nonlinear gain saturation. Moreover, several nonresonant configurations, that realize association in a single pass through the memory, have also been presented, using storage in photorefractive volume materials and nonlinear thresholding in liquid crystal devices. In the following sections, some specific and promising implementations of these categories of NHAM will be described.

## 7.2.2 Nonlinear Photorefractive Associative Memories

*Thresholding Using Phase Conjugate Mirrors*

B.H. Soffer et al. [323] were the first who demonstrated in 1986 a first step towards the system described in figure 7.4. Basically, their system is identical to the one shown in fig. 7.4 - an idea that has been discussed by several authors [291, 324], too. Instead of volume holograms, they used thin thermoplastic Fourier-transform holograms as the storage medium. This allows to include shift invariance and the capability of programming heteroassociations by manipulating the correlation plane. A disadvantage of that system however is the lack of Bragg selectivity which results in low information storage capacity compared to volume holograms. Experimentally, the group demonstrated a single iteration nonresonant configuration without implementation of thresholding, whether due to competition or to nonlinearities in the phase-conjugate reflectivity. Compared to fig. 7.4, the second phase-conjugating mirror at the left is missing in that demonstration. However, even with this reduced device, discrimination of two objects was demonstrated using black-and-white binary as well as gray-scale images.

A lot of improvements have been proposed after that first demonstration. D.M. Pepper [325] discussed an alternative method for thresholding and conjugating an optical wavefront in such an NHAM using a phase-conjugate mirror for conjugation and a liquid crystal light valve for controllable external thresholding. In 1989, K.H. Fielding et al. finally arrived at implementing nonlinear thresholding to the setup using a self-pumped $BaTiO_3$ phase-conjugate mirror. The nonlinear intensity thresholding properties of such a self-pumped phase conjugator have been shown in section 5.2. Moreover, they achieved a position, rotation, and scale invariant correlation system implementing a polar representation of the square of the Fourier transform, which is realized optically using a coordinate transforming computer-generated hologram.

In 1990, H. Xu et al. realized a similar system, using photorefractive Vanadium-doped KNSBN as the storage material and Cupper-doped KNSBN as the phase-conjugator in a semi-self pumped configuration [326]. H. Kang [327] and coworkers finally presented an all-optical coherent system that realized the storage hologram as well as the phase-conjugating element providing feedback, thresholding and gain using photorefractive $LiNbO_3$. Their system possesses content addressing ability even when the input is only 25% of the stored information. Moreover, the resolution as well as the storage capacity of their system is greatly improved using two photorefractive volume storage devices.

*Electrooptic, Liquid Crystal-Based Nonlinear Thresholding*

In order to address the issues of implementing controllable arbitrary nonlinearities in the correlation plane, making diffusers unnecessary and increasing at the same time the optical gain in order to achieve resonator oscillation, Y. Owechko [328] suggested and implemented a hybrid optical-electronic NHAM, based on liquid crystal light valves. A

**Figure 7.7:** Block diagram of an optoelectronic nonlinear holographic associative memory (after [328]).

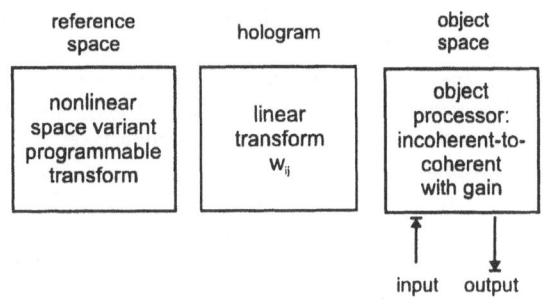

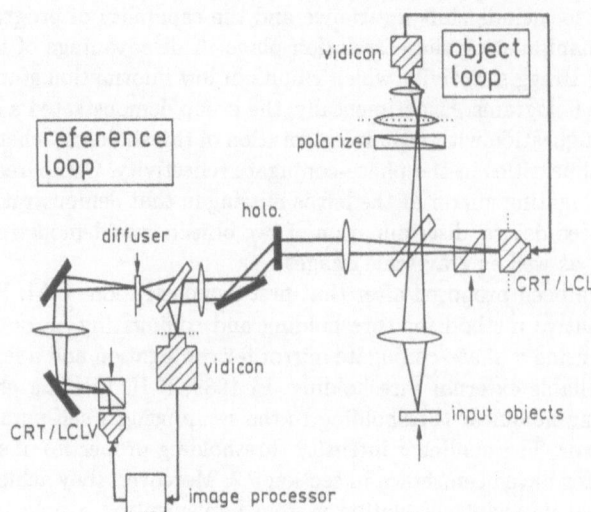

block diagram showing the principle function of such a hybrid NHAM is shown in fig. 7.7 and a detailed schematic in fig. 7.8.

Although the basic principles of the hybrid and the all-optical nonlinear holographic associative memory are the same and the division in an object and a reference loop is preserved, the implementation of the input and the feedback mechanism are quite different. Instead of using degenerate four-wave mixing in photorefractive materials to create true phase-conjugates of the reconstructed reference and object waves, a system using video detectors and CRT-addressed liquid crystal light valves (LCLV) was used. As shown in fig. 7.8, the partial input image is focused onto an object loop video detector and transfers the image to a CRT-addressed LCLV. The optical output of the object loop LCLV addresses the thermoplastic hologram and reconstructs the correlation plane which is focused on the reference loop video detector. A one-to-one mapping is performed between the detector and the output of the reference loop LCLV, which is positioned at the back focal plane of the correlation lens. Thresholded correlation peaks on the reference loop LCLV are converted into backward propagating plane wave reference beams by the correlation lens. These beams address the hologram, reconstructing recorded objects which are in turn focused on the object loop video detector, closing the resonator loop. Because the gain of the combined detector-CRT-LCLV loops is sufficient to overcome optical losses, a feedback signal can be installed. The advantage of the system is, that general nonlinear feedback functions can be easily programmed by the look-up tables in a PC board level image processor driving the correlation plane. With that device, the image processor can also be programmed to realize heteroassociations or multilayer optical neural networks by shuffling subregions of the correlation plane (see section 7.4 for a more detailed example of a multilayer network based on that hybrid nonlinear holographic associative memory). Moreover, the method can be used to classify patterns on structured backgrounds. A model for such an adaptive optical neural network has been presented by A.V. Pavlov [329], where the pattern and the background are regarded as the implementation of one steady-state random process and are distinguished from each other by their temporal stability. The difference to the hybrid NHAM presented by Y. Owechko exists in the functional modes of the neuron layer in the correlation plane and

**Figure 7.9:** Recording of multiple objects in a thin Fourier transform hologram using spatial multiplexing of the objects (after [330]).

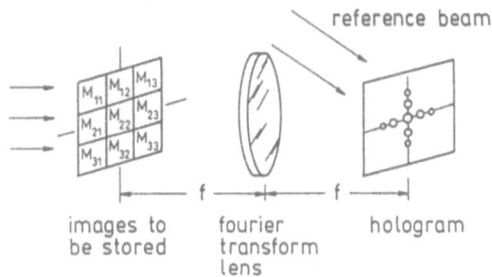

images to        fourier        hologram
be stored        transform
                 lens

in an operative (during operation) variation of the weights of the interconnections in the spectral plane. In other words, the distinction is in a wider use of the set of eigenmodes of the system's cavity.

However, the combination of electronics and optics introduces bottlenecks of transformation units into the system, destroying the coherence, parallelism and ease of processing that are the inherent advantages of all-optical systems.

### Pinhole-Array Mirrors as Thresholders Combined with Conjugate Feedback

As an alternative to the all-optical feedback using phase conjugation and the optoelectronic systems that combine optical correlation with electronic feedback and thresholding, D. Psaltis and coworkers [121, 330, 2] implemented threshold utilizing pinhole array-mirrors and unidirectional or conjugate feedback. As a big advantage, these pinhole arrays sharpen the correlation peak of the system, but on the other hand destroy the shift-invariance nature of the holographic systems described above. On that basis, they could demonstrate two different NHAM systems.

In the first system, a single-pass passive system, a set of spatially multiplexed objects are simultaneously recorded as Fourier holograms in a thin holographic medium using a single reference beam. This method of multiplexing, shown in fig. 7.9, is similar to spatial multiplexing introduced in section 4.1. It can be interpreted as the registration of a single "macro object", which consists of many subregions, each containing a single object. During readout (see fig. 7.10), an aperture equal in size and shape to the subregions in the macro object is centered in the input plane and input objects are placed inside it. This approach is equivalent to sequentially recording objects centered in the same aperture but with angularly shifted plane wave reference beams. Both approaches divide the correlation plane into subregions. During readout, the presence of a correlation peak in a particular subregion is a unique label for which the stored object is being recognized. The location of the correlation within the subregion has a one-to-one correspondence to the location of the object in the input aperture.

In the corresponding setup, thresholding is implemented using a pinhole array in contact with a mirror placed in the back focal plane of a correlation lens. This correlation lens in turn is positioned in such a way that its front focal plane coincides with the Fourier transform hologram. Thus, the correlation lens and the mirror act as a pseudo-conjugator which retroreflect the reconstructed reference beams back to the hologram for readout of the hologram. The appropriate stored object is reconstructed centered on the input aperture. Because the pinhole array passes only the peaks of the correlations, suppressing the sidelobe noise — one could call this a sort of hard-limiting threshold device (see

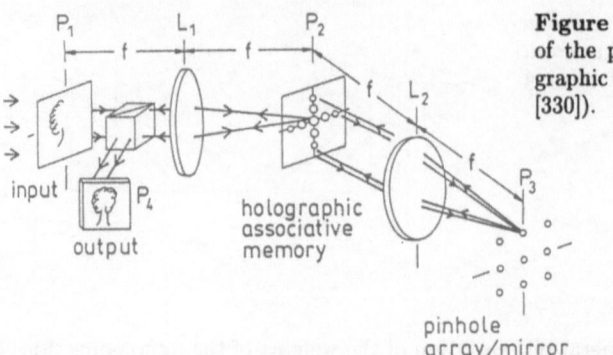

**Figure 7.10:** Schematic diagram of the pinhole array mirror holographic memory system (after [330]).

chapter 5) — , the image quality of the reconstruction was improved. Experimentally, the authors stored four objects in their system, which could be reconstructed with partial inputs. An application of this system has been presented in [331], where individual words were identified in a concatenated word string.

Several similar systems have been realized on the basis of these pinhole arrays mirrors combined with pseudo-conjugate feedback of the thresholded signals, extending the system to different storage materials and wavelengths [332] or using electrooptic switches as reflective thresholding neurons [333].

In 1988, H.J. White and coworkers [334] constructed a nonlinear holographic memory which utilizes a pinhole array and a photorefractive phase-conjugate mirror in combination for thresholding the correlation plane (see fig. 7.11). In their experiments, two objects were sequentially recorded using angularly multiplexed plane reference beams in a dichromate gelatine hologram. During recording, the reference-object beam ratio was adjusted to enhance the high spatial frequencies of the object, resulting in edge enhancement. This endge enhancement sharpened the autocorrelation peaks and improved object discrimination. They implemented the pinholes in the pathway at the correlation plane to filter out cross-correlation terms and enhance the quality of the autocorrelation function ($\delta$-function similarity). The restored reference beam is then retroreflected back along its path to the hologram using a phase-conjugate mirror rather than an ordinary mirror. Their experimental results are shown in fig. 7.12.

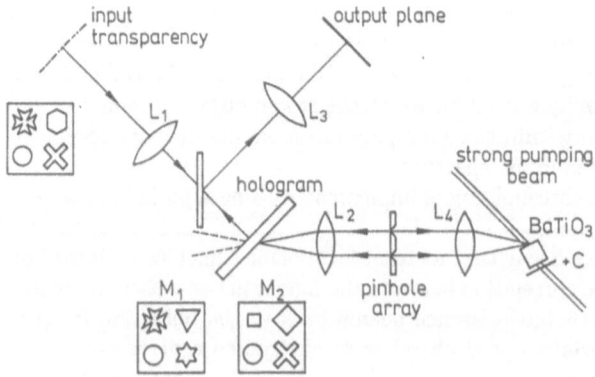

**Figure 7.11:** Associative memory using a pinhole array and phase-conjugate mirror based on four-wave mixing in photorefractive BaTiO$_3$ (after [334]).

**Figure 7.12:** Associative recall (right) for pairs of test inputs (left) using the associative memory system of fig. 7.11: circle and Maltese cross (a), circle and diagonal cross (b) (after [334]).

a)

b)

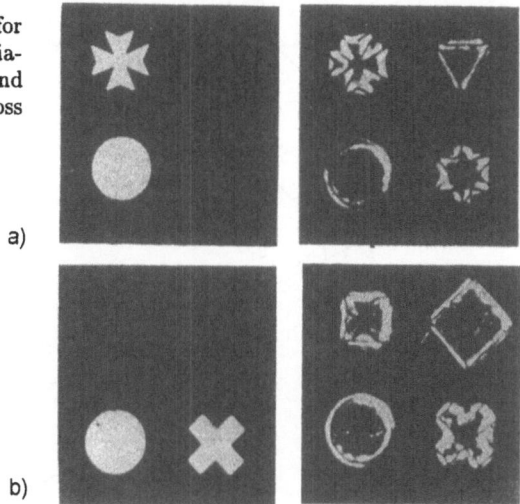

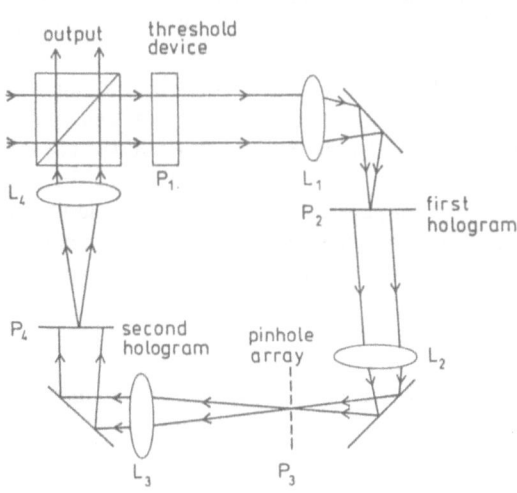

Although the system allowed to enhance the memory with the larger correlation, the discrimination between the objects was not complete as a faint image of the other memory can still be seen. The authors attributed this to a lack of nonlinearity in the phase-conjugate mirror, since the reflectivity of the four-wave mixing process is essentially linear for low probe beam intensities (no pump beam depletion, see section 3.5).

*Pinhole-Array Mirrors in Unidirectional Feedback Memories*

Although these approaches to correlation plane nonlinearities have the advantage of simplicity, they destroys the natural shift invariance of the Fourier transform holograms due to the fixed spatial position of the pinholes. In order to preserve this feature, E.G. Paek and D. Psaltis separated the functions of identification and reconstruction and used a separate hologram for each function. This second NHAM is shown in fig. 7.13. The combination of the first Fourier transform lens, the hologram, the correlation lens and the

**Figure 7.13:** Pinhole array optical associative loop (after [330]).

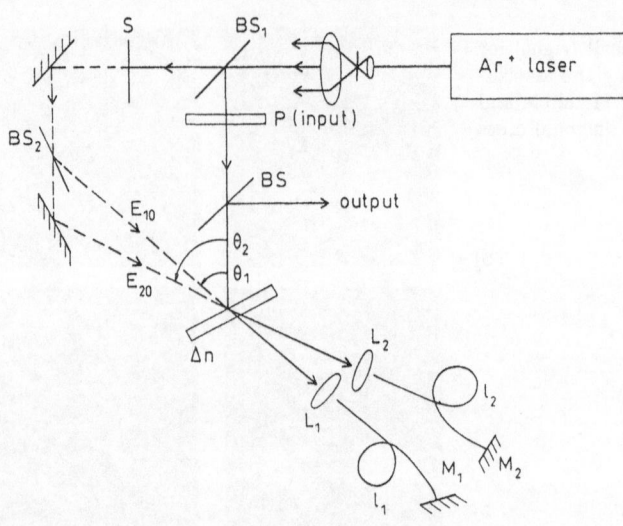

**Figure 7.14:** Experimental arrangement of image comparison using feedback from optical fibers (after [335]).

pinhole array is identical to the thresholded Van der Lugt correlator portion of their first system. But instead of retroreflecting the correlation peak back towards the first hologram, it is passed on to a second correlation lens, converting it to a plane wave reference beam which reads out a second hologram. Because this second hologram is identical to the first one (it is recorded in the same setup as the first using the same objects and reference beam), it reconstructs the associated object when addressed by the thresholded reference beam.

How is thresholding performed in that setup? It is a nonlinear microchannel spatial light modulator in the reflection mode, which reflects most strongly from its front surface the pattern that appears brightest on its back. The function of the whole setup is therefore as follows. The input is passed to a beamsplitter which sends one copy of the input to the front of the thresholding device, and passes another to the first hologram. Here, the input pattern is correlated with the stored images. The correlations are a measure of the similarity between the patterns, and the pattern that is most similar to the input is the brightest. This is passed through the pinhole array wich separates the images, and via a lens and a mirror through the second hologram, swhich is identical to the first. This correlates the new images and passes the results to the rear of the tresholding device. The back of the thresholder therefore receives a set of images corresponding to the stored images in the system.

The brightest one of these will be the one that the original image was most similar to, and this means that this pattern will be most strongly reflected from the front of the device. This new enhanced pattern will then pass round the loop for further enhancement and the system will quickly settle into a state in which the pattern most likely to the input pattern goes round and round the loop until stopped. The combination of object thesholding, emphasize of high spatial frequencies in the object during recording, and display optimization by separation of recognition and reconstruction functions between the two holograms greatly improved the reconstruction quality and enabled the system to reconstruct without failure four stored images when only a very small portion of the original was presented.

Meanwhile, several theoretical and numerical investigations of these configurations have shown that a feedback holographic associative memory becomes a pinhole array

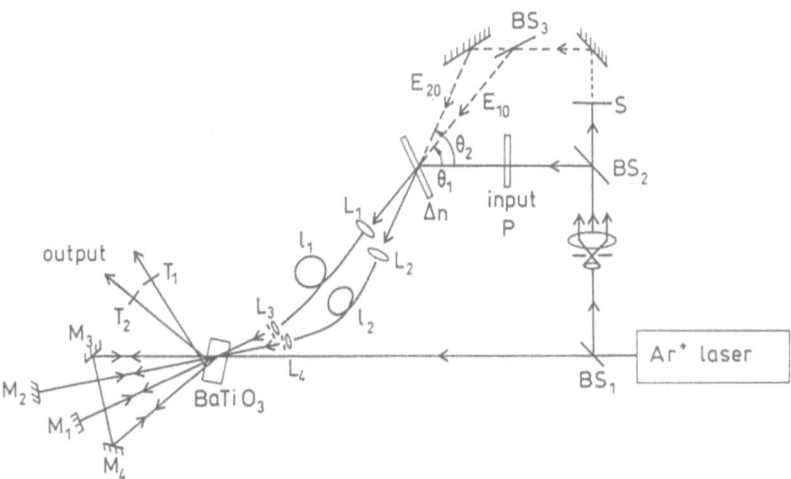

**Figure 7.15:** Associative memory using a bistable oscillator based on a passive ring phase-conjugate mirror (after [335]).

associative memory under certain conditions, and thus is a generalization of the pinhole array holographic associative memory [336].

*Thresholding Using Optical Fibers and Mirrors*

An alternative, but closely-related approach to thresholding the correlation plane using pinhole array mirrors is the use of optical fibers coupled to mirrors to retroreflect the central peak of the correlation function back to the hologram. This approach was demonstrated by A. Yariv and coworkers [335] and is shown in fig. 7.14.

In their configuration, two objects are recorded in a volume holographic material using angularly multiplexed reference beams. Optical fibers are used to sample the peak in the correlation plane and retroreflect the beam by the mirrors that were butted against the other end of the fibers. Since the fiber ends are located in the back focal plane of correlation lenses, reconstructed plane wave reference beams illuminated the hologram. This is another conceptual idea of spatial filtering that is identical to the pinhole-mirror technique used by E.G. Paek, D. Psaltis or the pinhole phase-conjugate mirror technique realized by J. White (see above). The authors could store in their system two overlapping, spatially nonorthogonal objects with only a poor resolution and a quite bad image quality for the reconstructed objects because of the lack of thresholding devices.

In a modification of the system, the mirrors are replaced by a conjugating-thresholding element. This element consists of a bistable oscillator using a ring resonator passive phase-conjugate mirror — an idea that will be discussed in more detail in the following section. The bistable oscillator utilizes mode competition to selectively enhance the strongest mode at the expense of weaker ones and retroreflect it back to the hologram. The bistable oscillator is added to the NHAM, as shown in fig. 7.15, to further enhance the discrimination between reconstructed reference beams.

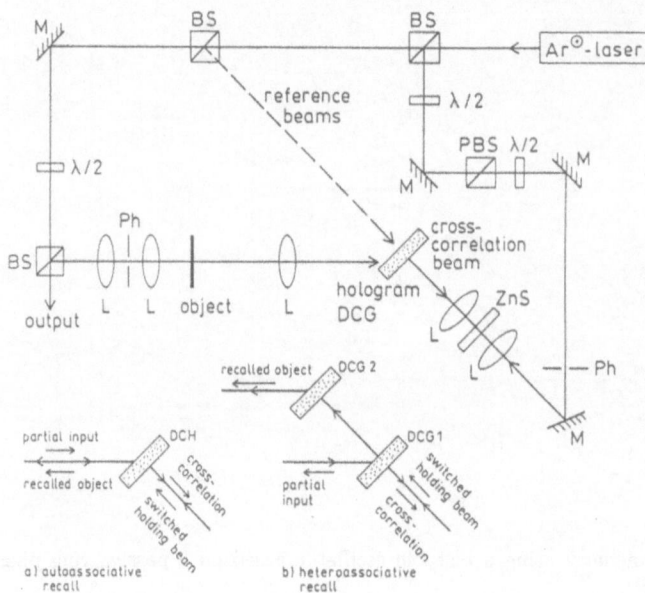

**Figure 7.16:** Associative memory apparatus using a dichromated gelatine hologram as a storage device and nonlinear ZnS etalons as thresholding devices (after [337]).

## Thresholding Utilizing Bistable Etalons

A novel method of thresholding the correlation plane in an NHAM was demonstrated by L. Wang et al. [337]. Their system is similar to the single-pass systems described above using a dichromate gelatine holographic thin storage element, but as the th2resholding element they used a ZnS bistable etalon. The principle function of those coherent thresholding devices has already been described in section 5.2. As in these systems and as shown in the configuration of the device in fig. 7.16, holding beams are used to bias the etalon just below the threshold point where it would switch from nontransmitting to transmitting. The etalon is positioned in the back focal plane of the correlation lens. If the peak of the autocorrelation function is sufficient to switch the etalon, the holding beam at that point will be transmitted. Since the holding beams are aligned to be counterpropagating with the reference beams, the transmitted holding beams read out the hologram and reconstruct the associated image. The need for a phase-conjugate mirror is therefore avoided. Both auto- and heteroassociation can be implemented by directing the holding beams to the same or different holograms. Associations of two stored fingerprint images have been demonstrated. Moreover, L. Wang et al. have discussed various practical limitations of this approach, including the high power requirements and nonuniformity of the ZnS etalon. Finally, a phase-conjugator or pseudoconjugator would have to be added on the object side of the hologram in order to form a multipass resonator.

## Associative Memories Using Second-Order Diffraction

To simplify nonlinear holographic associative memory setups, E.G. Paek and E.C. Jung [338] introduced two new concepts to their system. The system utilizes now the second-order diffraction instead of the first-order one. To enhance the second order diffraction efficiency, a novel angle-tuning recording using a thermoplastic plate was employed. A

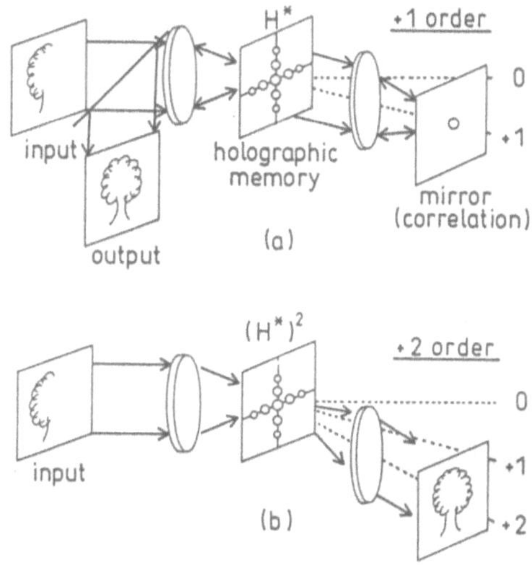

**Figure 7.17**: Comparison of a conventional associative memory using first-order diffraction (a) and a memory using second-order diffraction (b) (after [338]).

comparison between the system described above and that new idea is schematically shown in fig. 7.17. As a result, the size of the system is reduced by one half with a remarkable increase (about 100 times) in light efficiency compared with that of conventional systems. Moreover, the need for critical alignment is eliminated.

However, as in the conventional first order associative memory, because of the passive nature of the system and the resultant lack of gain, a resonator architecture can not be implemented optically, but only using electronic amplifying devices [338].

### 7.2.3 Nonresonant Memories

Another all-optical associative memory implementation based on a nonresonant cavity containing a photorefractive storage crystal and a nematic liquid crystal thresholding cell has been presented by the group of P. Günter [339].

As shown in fig. 7.18, the permanently stored images, which are stored in the system

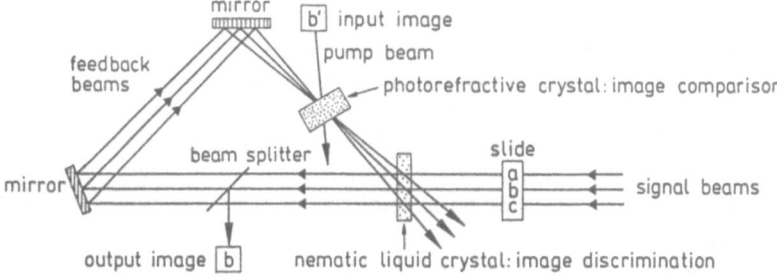

**Figure 7.18**: Schematic of an associative, nonresonant cavity containing a photorefractive storage crystal and a nematic liquid crystal thresholding cell. Three images (a-c) stored on a slide are injected into the cavity. Upon addressing the photorefractive amplifier with an input image (b'), the most similar among the stored images (b) is read out selectively by bistable switching in the NLC cell (after [339]).

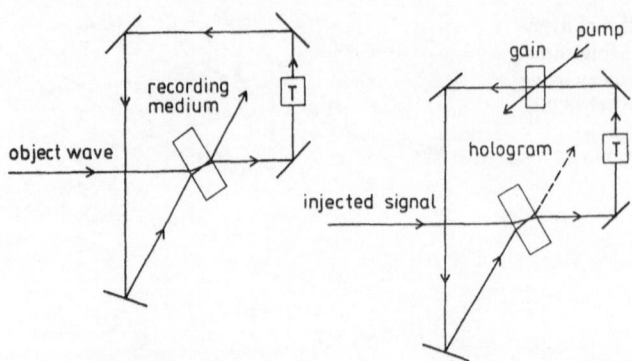

**Figure 7.19:** Holographic ring resonator memory. Recording of the hologram (left) and recall by an injected signal. Gain is supplied by a pumped photorefractive medium (after [278]).

on slides, modulate the intensities of the signal beams circulating in the cavity. In a photorefractive $KNbO_3$-crystal, which is used for dynamic image storage, these signal images are compared with an arbitrary input image on the pump beam by multiple two-beam coupling. However, two-wave mixing does not permit enough intensity variations among the signal beams after amplification. Therefore, external thresholding is performed by the nematic liquid-crystal cell, where the images are spatially separated and thresholded independently of each other. The transmission of the NLC between crossed polarizers is a bistable function of the amplification of every feedback intensity. Consequently the differences in the amplifications allow the NLC to discriminate the strongest beam containing the most similar image by bistable switching. In the output, which is extracted from the cavity by a beamsplitter, the associated image is bright and the other stored images remain dark.

### 7.2.4   Ring Resonator Associative Memories

An alternative type of optical associative memories implementing photorefractive devices is the ring resonator NHAM described and demonstrated by D.Z. Anderson [278]. In such a configuration, the reference beam for recording the hologram is derived from the object beam in a ring configuration, as shown in fig. 7.19. After the hologram is recorded, each stored pattern defines an eigenmode of the resonator in the same manner as for the linear resonator NHAM's described previously (see also section 6.5). An association is made by injecting a portion of the original pattern. A gain medium inside the resonator amplifies the eigenmode with the largest overlap with the injected field. The other eigenmodes are suppressed by a gain competition mechanism. As derived in section 6.5, for the case of two stored eigenmodes, the gain competition between modes is described by the ratio of cross-to self-saturation. For associative functions, the competition between modes can then be biased with an injected signal. D.Z. Anderson and M. Erie [249] have demonstrated this concept using both simple plane waves and printed characters as the recorded eigenmodes.

M.S. Cohen [340] took advantage of that concept to realize a resonant system based on a double phase-conjugating resonator that is loaded with a volume-holographic recording medium in which multiplexed images have been stored. That system is — for special operating conditions — able to associate the input to the appropriate stored pattern. If the only nonlinearity present in the system is the saturating gain of the conjugators, there is no multistability. The most deeply imprinted image always wins the competition for saturation. If several images are imprinted with equal depth, any slowly varying superposition of these images is neutrally stable. However, if a nonlinearity, arising from a space-charge dependent cubic polarization, is present, the resonator does exhibit multiple basins of stability that correspond to the faithful reconstruction of each stored image. The momentary injection of an initial image will then tip the system into the basin that corresponds to the stored image that it most closely resembles, even if that image is not the most deeply imprinted one. The symmetric double phase-conjugating resonator has two qualitative properties that make it a good candidate for attempts in realizing an associative memory. First, the unloaded resonator has — in an ideal case — an infinite number of degenerate eigenmodes at the pump frequency $\omega$ (see section 6.5 for further details on that topic). In practice, a finite, but large number $n$ of nearly degenerate eigenmodes is expected. This gives a high-dimensional set of basin vectors in which to encode input images, yielding greater resolution and higher storage capacity than conventionally bounded resonators. Second, phase shifts that are due to intracavity refraction are automatically canceled in the double phase conjugator. The same holds true for phase shifts on Bragg reflection [340]. Introducing a volume hologram recording medium into the cavity will not produce added phase shifts that can throw the cavity off resonance; it merely forces the normal modes of the cavity to become the eigenvectors of the holographic mode-coupling matrix. As these modes grow because of the gain in the conjugators, any nonlinearities present create a competition among modes for saturation. The modal spectrum at saturation depends on the form of the nonlinearities and on the initial state. However, the system lacks at the moment any chance of realization because there is no cubic Kerr-effect material available. Such an effect could be looked for in semiconductors that are capable of imprinting two-photon gratings and that possess quadratic potential wells. It may be hoped that the utility of such materials for optical image processing will spur their development, because several other applications of such tensorial couplings besides associative memories are possible, as many-to-one associations, strings, cycles, branching trees of associations and group-invariant pattern recognition.

This approach is completely different from the previously described NHAM architectures in that the reference beam is derived from the object beam for recording the hologram. During readout no separate thresholding is performed on the reconstructed reference beam. Instead of a nonlinear gain, a competition mechanism is relied on to favour one reconstruction over other possible ones. This results in a simple design and automatic generation of reference beams for recording, but at the expense of losing the flexibility and storage capacity advantages of the plane wave reference based NHAM's described above.

Among the various incoherent models to realize all-optical associative memories, which I will not mention here, the ones using microchannel spatial light modulators (MSLM) revealed to be the most promising ones. MSLM's have already been introduced in sections 4.1 and 5.1. They are used in associative memories with their special functions of storage, control of the surface charge with a mesh electrode and real-time hard-clipping thresholding. With that basis elements, M. Ishikawa and coworkers realized a new architecture for a nonlinear optical associative memory, called the optical associatron [341] . For those interested in these systems, references [341, 342] give a detailed insight in the function and application of these associative memories for the recognition of handwritten characters.

The experimental systems discussed in that chapter demonstrate the potential of NHAM's in a quite impressing variety, but they are also evidence of the immature state of NHAM implementations to date. There are still questions open in finding the optimum nonlinear mechanism for coherent thresholding. Also, there is still a need for demonstrating better image quality, larger storage capacity, and programmability of the systems as well as showing the capability of minituarization of the setups to create demonstrators for real-time applications in technical and industrial situations. Moreover, it is still neccesary to develop interfaces to conventional electronic host computers and incorporate the systems in general purpose optical neural network architectures in order to implement higher order tasks such as rotation and scale invariance and recognition of patterns. Nevertheless, these first generation systems have demonstrated several design principles which will doubtless be incorporated in future optical associative memories and neural network processors.

 **Summary**

Nonlinear holographic associative memories (NHAM's) combine ghost image holography with thresholded Van der Lugt correlators. In addition, nonlinearities allow an NHAM to select a particular stored memory over all others on the basis of incomplete input data. The function and configuration of these NHAM's are as follows:

- The heart of the system is a planar or volume hologram in which Fourier transforms of the objects are stored using a multiplexing technique of section 4.4.

- NHAM implementations can be categorized based on the resonator geometry (uni- or bidirectional), the method used for generating the reference beams (different multiplexing techniques) and the nonlinearity applied.

- For thresholding of the NHAM different devices can be used, as e.g. electronic thresholding via look-up tables, pinhole arrays as frequency domain nonlinear filters, nematic liquid crystals, nonlinear Fabry-Perot etalons, and phase conjugation (e.g. self-pumped phase conjugators exactly display the sigmoidal intensity threshold function).

- The feedback oscillator can be built using two phase-conjugating mirrors, multipass resonant systems utilizing retroreflecting or phase-conjugating mirrors in combination with pinhole arrays or optical fibers, and unidirectional ring resonators that obtain gain via a pumped photorefractive two-beam coupling medium.

- When a partial or distorted version of an object addresses the hologram, a set of partially reconstructed reference beams is generated. Each reconstructed reference beam is convolved with the correlation of the input object with the stored object associated with that particular reference beam. This part of the system is identical to a matched filter Van der Lugt correlator.

- The distorted reconstructed reference beams are fed back and retrace their paths to the hologram, reconstructing the complete stored objects. The reconstructed objects are phase-conjugated by the object arm phase-conjugate mirror.

- The process iterates until the system settles into a self-consistent solution, assuming the gain in the system is sufficient for oscillation.

## 7.3 Bidirectional Associative Memories

As in the precedent sections for conventional linear and nonlinear associative memories, optical implementations of bidirectional associative memories (BAM) can be realized in two main categories — matrix-vector multipliers (see also section 4.3.3) based on Hopfield-like networks or holographic correlators (section 7.1). Most of the BAM-like structures have already been described in the precedent chapters, although not for their properties as BAM's. It is therefore now easy to realize from these concepts some BAM implementations just by allowing the system to propagate from both directions in the system. In the following, I will introduce the concepts of some of these modifications, relating them closely to the appropriate unidirectional systems.

### 7.3.1 Matrix-Vector Multiplier Bidirectional Associative Memory

Optical BAM's using a matrix-vector multiplication approach are all similar in their design to the first optical neural net system presented in 1985 by N. Farhat and D. Psaltis. It is based on the idea to realize a Hopfield neural network and thus its concept is extensively discussed in section 7.2.2. The basic element of this net, the matrix-vector, is also a standard one and has been presented in section 4.3.3. Based on this system, R. Athale et al. have demonstrated a small autoassociative system. The BAM differs from these systems in that there are sources and detectors on both sides of the connection matrix. The two fields of this optical BAM are implemented as linear arrays of paired detector and source devices. The light falling on a detector is converted to an electronic signal, amplified, passed through a threshold circuit and ultimately drives its associated optical source. The connection matrix can be a photographic film or holographic plates for fixed connections, two-dimensional, electrically addressable spatial light modulators or photorefractive volume holographic devices. As with all optical, incoherent matrix-vector multipliers, bipolar input values and connection weights must be handled by doubling the number of elements, by wavelength multiplexing or some other scheme as discussed in section 4.2.4. For compact devices, it is desirable to have both fields of the BAM in the same plane. This is possible if the connection matrix spatial light modulator is operating in the reflective mode. Because in this case light passes twice through the connection matrix mask in this configuration, the single pass mask transmittance must be the square root of the desired value.

A similar setup using a spatial light modulator as the interconnection device, but realizing the bidirectionality of the neural network by a loop-back feature and orthogonally polarized counterpropagating waves, where the element is processed and fed back to a different location on the SLM, has been proposed by J.M. Kinser et al. Details of this "big BAM" design can be found in [343].

### 7.3.2 Photorefractive, Correlation-Based Bidirectional Associative Memory

A BAM system design using a photorefractive reflection volume hologram to implement the interconnection device has been suggested by C.C. Guest and R. TeKolste [344] and is shown in fig. 7.20. The two BAM fields exist side by side on a single device. The device provides amplification of an input image followed by thresholding. A liquid crystal light valve, a microchannel spatial light modulator, or two-wave mixing in a photorefractive

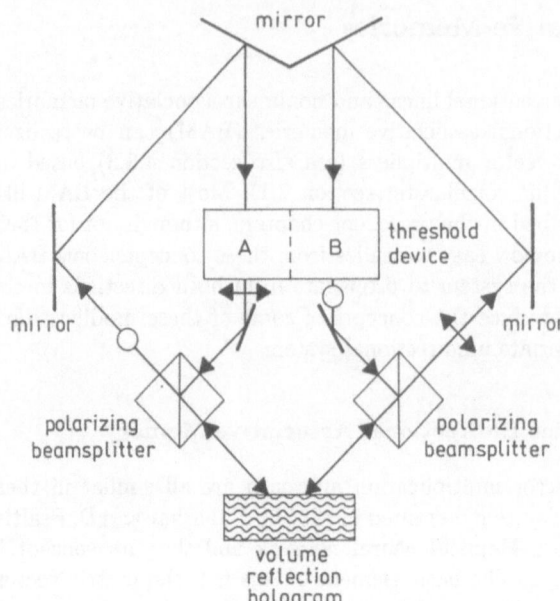

**Figure 7.20:** Interfield connections in a bidirectional associative memory, realized with a volume reflection hologram (after [344]).

crystal are candidate implementations for this function. Light from the two BAM fields ($A$ and $B$) are again polarized orthogonally so that they may follow different paths around the system. Light from the output side of the field $A$ is reflected from the hologram and passes to the input side of the $B$ field. Light from the output side of the $B$ field is reflected from the hologram and passes to the input side of the $A$ field. Note first that the beams used for reading the holographic gratings are not in a configuration to write new reflection gratings. Thus the operation of learning is decoupled from data processing. This is not a liability since the adaptation characteristics of the holographic material is rarely what is desirable from an algorithmic standpoint. For learning to occur in this system, separate beams must be brought into the other side of the hologram; their intensity and duration may be adjusted to give a controlled learning characteristic. The second point to note is that this system does not require phase conjugation or any sort of resonant cavity. This can be expected to improve the system stability and reduce system complexity.

A similar design for a BAM using a transmission hologram — also proposed by C.C. Guest et al. [344] — is shown in fig. 7.21. Again, some type of nonlinear optical amplifier is used for the BAM fields. In this case the hologram reading beams are in a configuration to write new gratings, provided the polarizations of the beams are aligned. The confocal mirror configuration is used only as a convenient way of performing Fourier transform operations and folding the optical path; the optical cavity is not used as a resonator.

The above description is a very schematic one, giving an illustrative example of the possible form that holographic BAM's may take, but leaving outside all practical details as the form of the threshold device and the coding of the data fields to allow an appreciable fraction of the storage capacity of the volume hologram to be utilized. With the information in the previous chapters however and the quick development, photorefractive interconnections are subject to actually, the next future will surely bring the experimental realization of some of these bidirectional associative memories.

**Figure 7.21:** Photorefractive bidirectional associative memory using a volume transmission hologram as the interconnection device (after [344]).

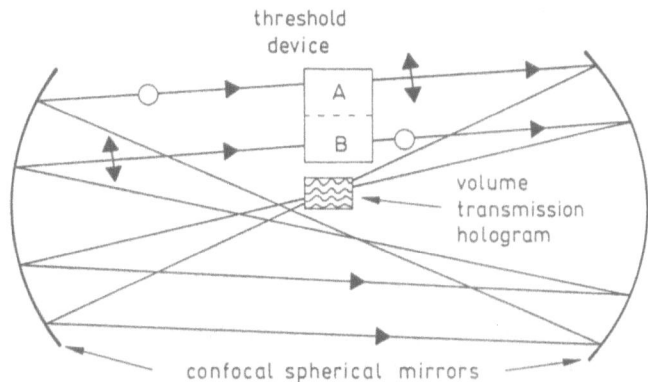

 ## Summary

Optical implementations of bidirectional associative memories (BAM) can be realized in two main categories — matrix-vector multipliers (see also sections 4.3) and holographic correlators (section 7.1). A BAM differs from other associative systems in that there are sources and detectors on both sides of the connection matrix. Thus, the system does not require any sort of resonant cavity. BAM's can be realized using

- linear arrays of paired detector and source devices. The network signal incident on a detector is converted to an electronic signal, amplified, passed through a threshold circuit and ultimately drives its associated optical source for the processing way in the inversed case.
- a loop-back feature and orthogonally polarized counterpropagating waves, where the element is processed and fed back to a different location on the SLM.
- a photorefractive reflection volume hologram to implement the interconnection device.
- a device that lets the two BAM fields exist side by side on a single device, both processing light with different polarizations. The device provides amplification of an input image followed by thresholding. A liquid crystal light valve, a microchannel spatial light modulator, or two-wave mixing in a photorefractive crystal are candidate implementations for this function.

## 7.4 Associative Memories for Complex Multilayer Networks

Because associative memory approaches have the capability for tackling the large parallel symbolic-processing demands of problems from image understanding, image synthesis and reconnection, robotic manipulation or locomotion, speech understanding and synthesis, language processing, expert-system problem solving, and other difficult artificial

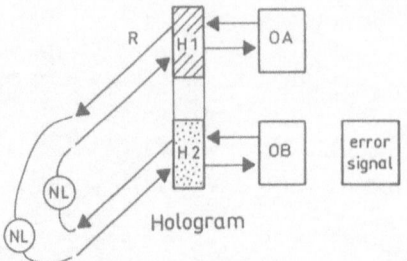

**Figure 7.22:** Block diagram of a proposed optoelectronic two-layer neural network (after [328]).

intelligence domains, a major aim of present investigations is to generate systems that are able to realize multilayer, adaptive associative architectures. For that purpose, a modular approach is the most promising one, where a number of smaller adaptive, associative modules (which may also be termed slabs or layers) are nonlinearly interconnected and cascaded under the guidance of a variety of design guidelines or organizational principles to construct larger architectures for solving specific problems. Because each individual module is a complete, associative memory which can be modified by exposure to sets of associated information patters (e.g. feature vectors, encoded symbols, images, ...), all the systems described in the previous sections of that chapter are in principle well suited for such a multilayer modular architecture.

For the case of optoelectronic associative memories, several attempts have already been made to realize these modular systems. The most comprehensive one, realized by A.D. Fisher and coworkers [216], includes optoelectronic Hebbian, Widrow-Hoff and differential learning. For all-optical realizations, the optoelectronic nonlinear holographic associative memory presented in the previous chapter (see fig. 7.8) can also be extended to implement an adaptable and reconfigurable multilayer optical neural network with large storage capacity and parallel weight update capability. Moreover, the possibility to use several wavelengths and gray-levels in order to realize multilayer associative networks is discussed.

### 7.4.1   Multilayer Nonlinear Holographic Associative Neural Network

A block diagram of the system is shown in fig. 7.22. A two-dimensional neural activation pattern object ($OA$) addresses subhologram $H_1$ and reconstructs another activation pattern (reference $R$). This in turn is nonlinearly processed and then shifted so that it addresses a second subhologram $H_2$ and reconstructs a third pattern (object $OB$). The two subholograms $H_1$ and $H_2$ are physically adjacent on the same substrate and form the link weights between the input and the output layers $OA$ and $OB$, and the hidden layer $R$. The hologram substrate is a volume photorefractive crystal such as LiNbO$_3$ or BaTiO$_3$ in which link weights can be continuously reinforced or inhibited. The optical pathways or links are bidirectional so that light can propagate not only in the direction $OA$-$R$-$OB$ but also $OB$-$R$-$OA$. An error signal is backpropagated through the net with its phase shifted by 0 or $\pi$ so that gratings which contribute to a large error signal can be enhanced or inhibited. Thus this system can implement an optical version of the backpropagation algorithm for associative purposes. A detailed drawing of the proposed optical backpropagation system is shown in fig. 7.23. I will describe it in that section in more detail to show the virtual identity between holographic associative memories and

**Figure 7.23:** Detailed schematic of an opto-electronic two-layer neural network (after [328]).

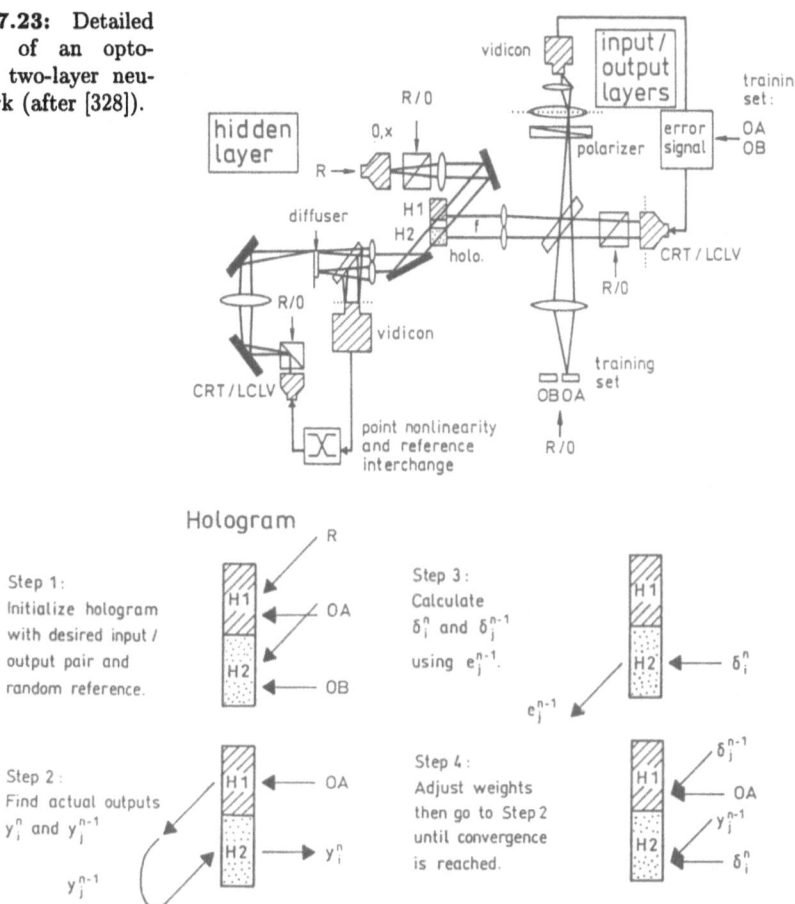

**Figure 7.24:** Optoelectronic backpropagation learning procedure with two layers (after [328]).

some neural network algorithms. More nonlinear optical implementations of the back-propagation algorithms or multilayer networks are presented in chapter 9. In that setup, the top and the bottom activation pattern $OA$ and $OB$ are located side by side in the input plane. An incoherent-to-coherent conversion is performed in the object loop using a vidicon detector and LCLV. The optoelectronic learning procedure is outlined in fig. 7.24. In the first step, the connection weights in subholograms $H_1$ and $H_2$ are initialized to random values. This is done by recording $OA$ and $OB$ with random references in $H_1$ and $H_2$, respectively. In the second step, the actual output for an input of $OA$ is found. $OA$ appears on the object LCTV, is transformed by a Fourier transform lens, and directed to subhologram $H_1$. This reconstructs $R$ which is detected by the reference loop vidicon. $R$ is processed electronically using a look-up table and shifted spatially before it is converted back into an optical form by the second LCTV. This spatial shifting causes it to read out subhologram $H_2$, which reconstructs the actual ouptut $Y$ in the vidicon detector plane. In the third step, two error signals for holograms $H_1$ and $H_2$ are calculated electronically using a local error signal generation rule (see section 2.3). Therefore, it is sufficient to use a single point transformation as it can be implemented using digital

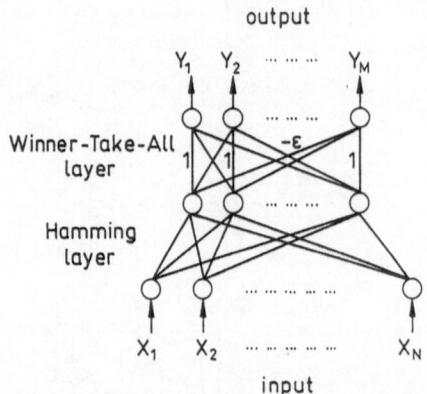

output

Winner-Take-All layer

Hamming layer

input

**Figure 7.25:** Schematic diagram of an associative neural network based on two layers, the first one being a Hamming net and the second one a winner-takes-all net (after [345]).

look-up tables to generate the error signal. In step 4, an outer product is formed at $H_1$ between the $H_1$ error signal and $OA$. In accordance with the backpropagation algorithm, the gratings formed in the $LiNbO_3$ crystal during this step act to inhibit or enhance gratings formed in previous iterations in such a way as to minimize the error signals. Step 2 is then repeated and the cycle continues until convergence is reached.

With that configuration and the use of spatially multiplexed holograms in $LiNbO_3$, the implementation of a multilayer optical neural network with about $10^6$ neurons and with potential processing rates of $10^9$ interconnections per second were estimated to be possible [328].

### 7.4.2   Multilayer Associative Networks Based on Multiple Color Encoding

Up to now, multilayer networks with content adressable memory features for association are based on a single color or wavelength approach. In order to enhance the storage capacity of such a system, it is possible to use different colors for image classification. By introducing such a color encoding technique in conjunction with a conventional white light source, a polychromatic Hamming net can be sythesized. In the following, I will describe the principle of the net realized by F.T.S. Yu and coworkers in more detail, because it represents a good example for two-level associative neural networks cascading a single layer based on a Hamming net and a second layer based on the winner-takes-all principle to construct a hetero-associative memory [345]. Its basic construction is shown in fig. 7.25. A schematic optical setup of the polychromatic neural network part is shown in fig. 7.26. Its heart consists of two liquid crystal television (LCTV) panels tightly cascased for displaying the input pattern and the interconnection weight matrix. If we denote the light intensity of the red, green and the blue components in color pixel element as $I_R(x,y)$, $I_G(x,y)$, and $I_B(x,y)$, respectively, the color intensity field from the LCTV panel can be written as

$$I(x,y) = I_R(x,y) + I_G(x,y) + I_B(x,y) \qquad (7.10)$$

where $(x,y)$ represents the spatial coordinate of the liquid crystal panel. This equation allows to introduce the concept of color decomposition and composition, by which each image pixel can be decomposed and composed with the primary colors. These colors can be used as the training sets for the construction of the three color interconnection weight matrices. By combining these set of weight matrices, a multicolor interconnection weight matrix can be displayed on the second LCTV panel. If a color pattern is fed into LCTV1, the summed signal detected by the color CCD would be the output of the color Hamming net. If this output is sent to a color winner-take-all layer and then selecting the maximum input node, a classification of the input color pattern can be made.

**Figure 7.26:** Experimental setup of a polychromatic two-level content-adressable neural net based on the combination of a a color liquid crystal television-based polychromatic Hamming net as the first step and a winner-takes-all unit using a photorefractive memory crystal (after [345]).

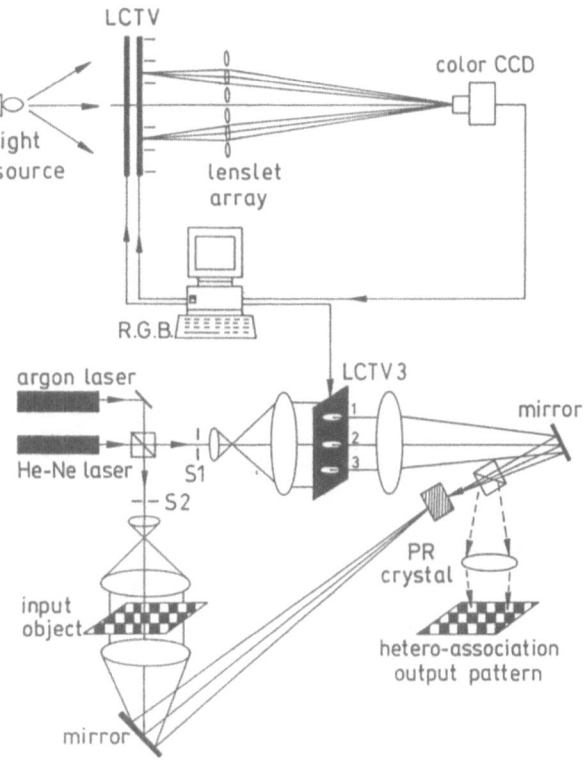

In order to include the usual associative memory features into the system, a mapping net has to be introduced as a second layer into the system, which has the ability of mapping the maximum output from the Hamming net to the desired output pattern. These mapping processes can be done by using either auto-association or hetero-association, depending on the application, as e.g. pattern recognition (auto-association) or pattern translation (hetero-assocation). The second mapping level proposed by F.T.S. Yu and coworkers is essentially a holographic interconnection system, in which a set of Fourier holograms is recorded using the outputs from the polychromatic Hamming net as the read-out training sets (see fig. 7.26). In the readout process, the object beam is blocked while the reference beam will be used to reconstruct the stored exemplars. Since a different input position corresponds to the different reference beam angle, an angular multiplexed crystal hologram can be used. Thus, a one-to-one correspondence between the reading angles and the output stored exemplar can be established in the second level. since the different input pattern correspont to a different output position from the Hamming net, the content-adressable network can be developed using this two-level neural net.

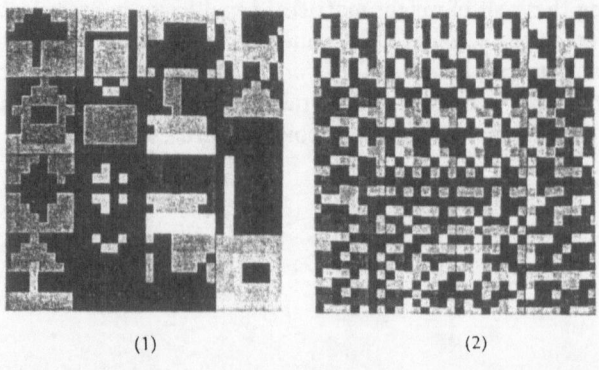

**Figure 7.27:** Experimental results (1) 16 training object patttern; (2) interconnection weight matrix of the first level; (3), (4), (5) three input object patterns; (6), (7), (8) the corresponding output results from the winner-takes-all net; (9), (10), (11) the corresponding hetero-association output pattern obtained from the second level (from [345]).

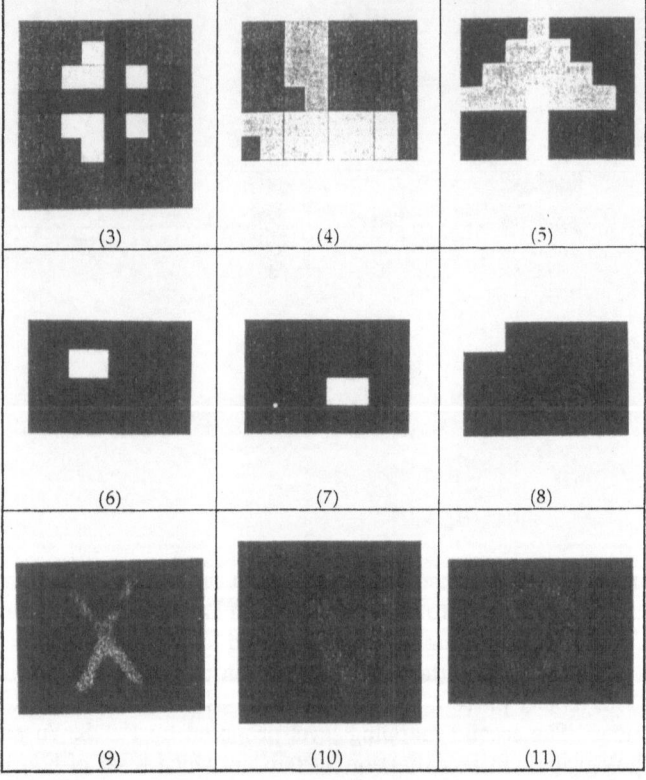

In their experimental demonstration, F.T.S. Yu et. al. used a Ce,Fe-doped LiNbO$_3$ crystal as the recording medium, being sensitive to the entire visible-light spectrum. A training set of 16 patterns as shown in fig. 7.27(1) has been used for the construction of the interconnection weight matrix (fig. 7.27(2)). Subsequently, the input objects of figs. 7.27(3,4,5) are sequentially presented to the neural net (on LCTV1) and the interconnection weight matrix is displayed on LCTV2, obeying size matching of the two patterns. By a lenslet array, the Hamming net algorithm can be realized on the color CCD camera. By applying the winner-take-all algorithm carried out by a computer, the right output of the Hamming layer can be obtained by taking about two to three cycles of iteration using a simple winner-takes-all system. Figs. 7.27 (6,7,8) show a set of winner-take-all outputs, corresponding to the input objects of figs. 7.27 (3,4,5). To demonstrate the application of this net to hetero-association, an associative training set of color patterns is recorded in the photorefractive crystal — the red letter X, the green letter V and the blue letter D, using different light sources and angular multiplexing. The net works in the readout phase of the memory crystal. If the input pattern of fig. 7.27(3) is presented to the LCTV1 of the neural net, a transparent spot (fig. 7.27(6)) will be registered on LCTV3, for which the transmitted coherent light is used to read out the hetero-associated photorefractive hologram, giving fig. 7.27(9) as the output answer. Similarly, figs. 7.27(4,5) result in reconstruction of figs. 7.27(10,11). Thus, the system allows pattern translation or heteroassociative neural net functions in color. Moreover, the system can also be used for gray-level pattern recognition, because gray levels can be coded with different colors. This is an important feature of image processing using associative memories.

The associative memories I discussed up to now were discrete ones, mostly based on the Hopfield model in which the outer-product operation for the construction of the interconnection weight matrix is used, or on the interpattern association, in which the interconnection weight matrix is constructed by using simple logic operations. Obviously, by mapping (e.g. binarizing) a gray-level object into a bipolar/ binary pattern, a great amount of information is lost. For example, a binary pattern would lose the spatial-frequency content, and hnce its special features would be lost. These characteristics are sometimes critical for pattern recogition. In these cases, gray-level neural networks are needed.

### 7.4.3 Multilayer Associative Memories Based on Gray-Level Discretization

In recent years, several techniques for gray-level neural networks have been developed. M. Taketa and J.W. Goodman [346] have presented a number-representation scheme. They encoded the gray level on binary neurons and then used an group-and-weight scheme to represent the gray level. This method permits the network to operate in a largely redundant mode and offers more chances to reach the correct solution. However, this scheme needs a much greater number of neurons and interconnection links, which increases the programming complexity.

G.Y. Sirat and coworkers [347] proposed another technique called a discrete state phasor neural network, in which they mapped the gray level to the phase domain. Although this technique is a promising one for a couple of neural network models, it is not applied easily to associative calculations. More recently, W. Zhang et al. [348] adopted a multistate neuron to reduce the huge number of binary neurons and interconnections required by gray-level processing and introduced the simulated annealing technique into the Hopfield model. Of course, with the use of multistate neurons, the dramatic increase in the number of neurons and interconnections as well as in the program complexity, can be avoided. However, the annealing process in usually time consuming.

In 1993, F.T.S. Yu and coworkers [349] combined a gray-level pattern decomposition-composition technique with the discrete Hopfield and interpattern association models to extend their use to gray-level objects. By decomposing a gray-level pattern into sets of bipolar/binary subpatterns, each subpattern set can be processed based upon the discrete-model algorithm. A preprocessing step is necessary for removing dc bias and normalizing the gray-level scale, done on the input gray-level pattern. This eliminates mismatch and saturation problems caused by the bias level, which shifts the pattern gray levels throughout the pattern. This net has an information content of $G^M$, where $G$ is the number of gray levels and $M$ the number of nodes, compared to $2^M$ for the bipolar/binary pattern systems. They could also show that this gray-level network can be applied to the interpattern association model.

We have seen in the last sections, that associative memories can be realized in a number of different configurations, including nonlinear feedback by phase conjugation, ring resonator configurations, bidirectional information transmission and multilayer structures. The applications of these associative memories range from autoassociation from partially available inputs, heteroassociation from one special pattern to another up to pattern classification. The vast range of application examples discussed in this chapter shows impressively the power of neural network processing in optics as well as the actual need in those pattern processing tools. In the following section, I will describe optical realizations that fit the algorithms described in the second chapter, beginning with realization of simple perceptrons up to implementations of complex adaptive resonance systems.

 ## Summary

First implementations of associative functions in larger multilayer neural networks show the possible realization of backpropagation networks with
- nonlinear holographic storage and thresholding using subholograms representing the different network layers
- a combination of Hamming learning and winner-takes-all selection using different colours
- the realization of gray-level networks by decomposing a gray-level pattern into sets of bipolar/binary subpatterns, each of them being processed based upon the discrete-model algorithm.

 ## Further Reading

1. Y. Owechko, *Nonlinear Holographic Associative Memories*, IEEE Journ. of Quant. El., QE-25 (1989) 619.
   A well-written overview over the most promising nonlinear holographic associative memory systems with a basic introduction into the features of nonlinear associative memories.
2. G. Hinton, J. Anderson, *Parallel Models of Associative Memories*, Lawrence-Erlbaum, 1981.
   An overview over models of linear associative memories.
3. J.J. Hopfield, *Neural networks and physical systems with emergent collective computational abilities*, Proc. Nat. Acad. of Sciences, 79 (1982).
   The original work of Hopfield on the concept of associative memories based on the Hopfield outer-product algorithm.

4. Y. Abu-Mostafa, D. Psaltis, *Optical pattern recognition*, Scientific American, March 1987.
   An excellent introduction into the features and basic principles of associative pattern recognition devices. It describes the system presented in 7.2 in detail.
5. K. Kyuma, *Optical neural networks — a review*, Nonlinear Optics 1 (1991), pp. 39-49.
   A short and compact overview over current implementations of photorefractive neural networks, especially of associative memories.

# 8 Optical Realizations of Perceptron-like Neural Networks

The perceptron learning and network algorithm is one of the most simple and popular structures in optical neural network relizations. Although it allows only for processing of linear decision problems because it is based on a simple outer product summation of the inputs followed by a nonlinear thresholding operation (see section 2.2), it is the basis element for all further multilayer nets and easy to implement optically using different interconnection techniques. An important feature of perceptron algorithms is that they allow both additive and subtractive changes to the weights (see section 4.2.4). This bipolar weight realization allows for excitatory as well as for inhibitory interconnections and therefore for bipolar weight changes. It is the realization of that feature that is required in most of the multiple layer networks such as backpropagation nets and will play an important role for all-optical realizations

The basic components necessary to implement a perceptron-like network are an input device to convert patterns into the appropriate form (e.g. electric to optical or incoherent to coherent optical), an interconnection device, and a thresholding device for the output unit. Input devices may be electronically driven spatial light modulators that convert electronic input data into an optical form or optically addressable light modulators that allow to transform an optically incoherent input (e.g. a diapositive) into an optical coherent form. Moreover, liquid crystal screens that can be placed in the coherent beam path allow to impress an image information on the beam. The function of the interconnections in this context is to simply compute the outer product between the input patterns and the weights. Thresholding then is accomplished to achieve a nonlinear selection of the stored patterns in order to obtain a target output that resembles most the input.

Volume holograms are well suited to implement the interconnection functions in a way that is extendable to multiple category cases (e.g. multiple outer products). In the multiple category a number of different products need to be computed simultaneously, requiring the third dimension of volume memories using e.g. angular, phase or wavelength multiplexing techniques (see section 4.4) or spatial multiplexing of planar holograms (see section 4.1). However, for single-layer perceptrons, this is clearly an overkill since the same function could be achieved with a planar hologram, although it makes sense to implement even in that easy case volume holographic interconnections in order to leave open the door to the extension to the multiple category case.

Finally, it is difficult to achieve shift-invariant pattern recognition in volume holographic devices, because of the volume property of these storage media. In contrast, planar storage devices have the advantage to be shift-invariant.

As a consequence, in the present chapter I will divide the implementations of simple single-layer percepetrons using nonlinear optics in those who realize them using planar interconnection devices and those using volume memories. Moreover, attention is paid to the fact how the bipolar weight requirement of perceptrons described in detail in section 4.2.4 is realized.

## 8.1   Perceptron Pattern Classification Using Planar Interconnection Devices

A linear, one-layer perceptron can be used in pattern classification for linearly separable decision problems or, in other words, to correlate the input to two different classes. The purpose of training the perceptron before it is operating (prescribed learning) is to adjust the interconnection weights of the system such that the system produces correct responses for all training patterns. For the two-category pattern classification, the patterns of class one are specified as those producing a high response and the patterns of class two are those producing a low response. In the learning process, the interconnection weight is initially set to be zero, and responses of the training patterns are checked one by one by the perceptron. The outer product of the input pattern with the interconnection weight is calculated and then compared to the desired response. If the response is correct, then another pattern is presented for the next check. Otherwise, an error signal is generated for modifying the interconnection weight according to the perceptron learning algorithm (see section 2.2). The process of weight updating is repeated until all training patterns are correctly classified. Then learning is complete.

Since during the training process the response of the input pattern is determined by the product of the weight matrix with the input vector, and since in optical systems the value of the product can be obtained by detecting the signal at the center of the output plane of an optical correlator, a high response implies a sharp correlation peak. Hence, the trained interconnection weight matrix can be used as a correlation filter for pattern classification. This possibility has been used by O.V. Dubrovskaya and E.I. Shubnikov to realize correlators for optical perceptron implementations [350]. Consequently, a lot of optical perceptron learning implementations are based on that possibility to use correlator functions — thus connecting perceptron realizations closely to applications for associative memory tasks. The interconnection device has often been realized using spatial light modulators, with the weight adaptation in the prescribed learning process being performed electronically before transfer into the SLM or optically using specific SLM architectures.

### 8.1.1   Spatial Light Rebroadcaster-Based Perceptron

As a classical example for an electrooptic perceptron realization, where thresholding is realized in an electronic way and the interconnection is done optically, I have chosen a system based on spatial rebroadcaster interconnections (see section 4.1.1). They can store the weights for a neural network on their two-dimensional surface and permit them to be individually increased by writing with one wavelength and individually decreased by writing with another, different wavelength. This is an excellent match to the perceptron algorithm, which requires small modifications to the weights both in a positive and a negative direction. In the configuration proposed by A.D. McAulay and coworkers, the weights are thus updated optically on two-dimensional arrays. A summing lens forms the product between the elements of the array. Thus the electronic computer is not required to compute weights, but has to compute a single threshold. As a result, the amount of electronic computation is of the order $n^2$ less than that performed in optics. However, still most of the time is spent in loading the next image onto an electronically settable spatial light modulator as an input device. This limitation may be overcome by using microlaser arrays that have fast modulation times.

The principle function of the bipolar net is to process two optical similar setups, one for learning with the original patterns and one for learning with the complementary patterns. Both systems operate as perceptrons with positive weights. The bipolar character is then introduced by adding the outputs of the two detectors when using the system for classification. One system is composed of the SLM as the input image device, the SLR for the interconnection weight learning process, a lens to perform summation, a detector-computer system to perform electronic thresholding and feedback in a read and a write channel to the input SLM. The process is started by writing uniformly with low intensity on the electronic SLR. Then the first pattern is set on the SLM. This pattern is multiplied pixel by pixel by the weights on the SLR. The lens integrates the array information spatially to a detector, perfoming thus the product computation between the weights and the input vector. The computer reads the detector. If the output is correct during training, the computer controls the feedback to the write channel. It opens the write channel in that case only long enough to compensate for losses from the SLR during reading. This corresponds to no change in the perceptron algorithm. If the detector output is opposite to that desired during training for class one, then the computer opens the write shutter for a long time in order to add information selectively on the SLR. If the detector output is opposite to that desired during training for class two, then the computer opens the read shutter to remove some of the information selectively on the SLR. Here, the electronic threshold is not a bottleneck in the system because the optics perform a two-dimensional product while only a single number needs to be thresholded electronically and the response time of the SLR is in the range of 100 ns.

In their experimental procedure, A.D. McAulay and coworkers trained four patterns to be classified in two groups to the system by letting the computer rotate through all patterns until there are no more changes in the weights stored on the SLR, using an Ar-ion laser wavelength for the write and an infrared laser for the read channel. After 10 cycles through all the patterns — in both the original and the complementary setup with the same learning rate and its final combination —, the system answered and classified correctly to the presentation of one of the four patterns.

### 8.1.2 Spatial Light Modulator-Based Perceptron Using Four-Quadrant Multiplication

In the next example, a separation of the weight channels into the inhibitory and the excitatory ones has to be done, followed by an electronic summation of the outputs (see section 4.2.4 for more details on bipolar interconnections) in order to realize a bipolar network in an incoherent interconnection system. However, if a liquid crystal spatial light modulator is divided into four quadrants, symmetric bipolar activation function can be simulated, allowing faster convergence. The main principle permitting this four-quadrant multiplication is based on two adjacent twisted-nematic liquid crystal displays (LCD's) with a polarizer on one side and an analyzer on the other side. The polarization direction of the input light is rotated by the joint LCD's according to the electric field applied to the two LCD's. The transmitted light is thus affected by the patterns written into both LCD's. In twisted nematic LC's, if no electric field is applied, the polarization plane rotation is 90°. If $\theta$ is the polarization plane rotation under the maximal electric field applied to the LCD and if $\phi$ is the angle between the polarizer and the analyzer

**Figure 8.1:** Complete four-quadrant matrix-vector multiplier for perceptron learning (after [351]).

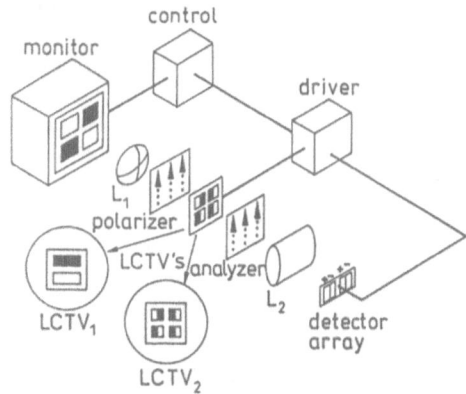

polarization axes, then a two-variable transparency function for unpolarized light is given by

$$T(X,Y) = \begin{cases} I_p\cos^2(\phi) & \text{for } X = -1, Y = -1 \\ I_p\sin^2(\phi + \theta) & \text{for } X = -1, Y = +1 \\ & \text{or } X = +1, Y = -1 \\ I_p\cos^2(\phi + 2\theta) & \text{for } X = +1, Y = -1 \end{cases} \qquad (8.1)$$

where the +1 and -1 values of $X$ and $Y$ represent the presence or the absence of an electric field, and $I_p$ is the transmission intensity of two parallel polarizers. The above definition of $T(X,Y)$ can also be written in a polynomial representation as

$$T(X,Y) = aXY + b(X + Y) + c, \qquad (8.2)$$

where the constants $a$, $b$, and $c$, resolved from a linear equation set, are

$$\begin{aligned} a &= (I_p/4) \cdot \left(\cos^2(\phi + 2\theta) - 2\sin^2(\phi + \theta)\right) \\ b &= (I_p/4) \cdot \left(\cos^2(\phi) - \cos^2(\phi + 2\theta)\right) \\ c &= (I_p/4) \cdot \left(\cos^2(\phi) + \cos^2(\phi + 2\theta) + 2\sin^2(\phi + \theta)\right). \end{aligned} \qquad (8.3)$$

Fig. 8.1 represents a complete four-quadrant multiplication machine between a continous-value multiplicand $W$ and a unity bipolar multiplier $X$, realized by S. Abramson and coworkers [351]. The light source intensity represents the absolute value of $W$, the first LCD represents $X$ and the second LCD is separated into two subcells for each cell of the first LCD, which represent the sign of $W$ and the opposite sign of $W$, respectively. The light source is imaged upon the joint LCD's, and the LCD's are imaged by a cylindrical lens upon a photodetector array, so that each detector receives the light that passed through the first LCD and the respective part of the second LCD. The signals from the photodetector are summed separately for each polarity and are then subtracted to produce the bipolar results. Each pixel on the LCTV that holds the matrix signs consits of two contrast subpixels. Two detectors are used for each output vector item. It should be emphasized here that the partition of each element into two subpixels in not similar to the separation of negative and positive matrices used for implementing multiplications between bipolar weights and unipolar neuron outputs in the two-channel multiplication method. Instead, this partition is a convenient technique for achieving four-quadrant multiplication.

The basic machine described above can perform a feed-forward information flow from one layer to another. Therefore, beneath multilayer and single-layer percpetron networks, networks that require several iterations can be implemented by writing the output-vector results directly to the input-vector display LCTV, for further iterations.

Experimentally, S. Abramson and coworkers realized a partial model having an input layer of 8 × 8 neurons and an output layer of 6 neurons that was capable of classifying six digits. The training set consisted of 15 out of 30 sets of six digitized handwritten digits written by three people. The system converged successfully after about 10.000 iterations — each iteration included a recall of a training vector and a weight-matrix modification according the the perceptron learning algorithm — to 0% error for the training set and to about 15% rejected and misclassified vectors for the entire test set.

Current limitations on the computational abilities of the system are the resolution and the time response of the LCTV. However, developments in technology and the use of a combination of several LCTV's and high-resolution monitors, as well as optimal tuning and adjustment of the hardware components could yield in a promisingly high performance.

### 8.1.3   Perceptron Learning Using Computer-Generated Holograms

In optical implementations of the perceptron learning algorithm, the Van der Lugt correlator architecture (see section 7.1) is well suited to perform binary weight adaptation. After the hologram is trained by recording the appropriate interconnection weight, it can be used as a matched filter for pattern recognition. Here again, the close connection of the network structure to associative memories becomes apparent.

The learning procedure can be performed in media that are able to adapt weights or — as proposed and realized by K.Y. Hsu and coworkers [71] — in a hybrid way by electronically training the perceptron network to obtain the desired interconnection weights and then transfer the trained weights into a hologram by using the technique of computer-generated holograms, thus resulting in a prescribed learning procedure with finally fixed interconnection weights. Afterwards, the "trained" computer generated hologram (CGH) is used as a correlation filter for optical pattern recognition.

Compared to other, real-time optically adaptive interconnection devices, this method has the advantage that exposure constraints as in photorefractive multiple grating formation are avoided. Thus the learning behaviour of the traditional perceptron can be fully implemented. Second, because the CGH is a fixed one which is not erasable, the weight decay problem during operation of the learned network that arises e.g. in volume holographic photorefractive interconnections is omitted. However, we obtain a fixed hologram for a special pattern classification problem, losing adaptivity and therefore the possibility to use this interconnection device for other pattern classification tasks.

Another optical possibility to train a CGH for implementation in a perceptron-like network structure was proposed by Y. Kajiki et al. using CGH's and a dot liquid crystal display as a shutter array. The CGH's are used for weights and summation. Since the CGH used is a Fourier transform hologram, the position of the reconstructed image does not change even if the CGH is shifted. In the optical perceptron-like system, the weights correspond to the intensity of the images reconstructed by the CGH. The weight modification can be performed by selecting some CGH's having several intensities and by superimposing the selected CGH reconstructed images. When several CGH's are reconstructed in parallel — using a dot matrix liquid crystal device as a shutter to address them — the reconstructed images are superimposed at the same position. The sum operation is performed by this superposition of images. The intensity distribution of the reconstructed images can be specified for each CGH, corresponding to the connection weight. Since the subtraction for negative weight is impossible in the summation by optical superposition of intensity-reconstructed images, two regions representing the positive and negative terms are arranged in the reconstructed image. This corresponds to the dual-rail coding or double-channel methods described in section 4.2.4. After each summation for positive and negative terms is performed optically, the subtraction between them, as well as the threshold operation, is performed by electric circuits.

The optical system realized with that concept by Y. Kajiki et al. allows to construct a trainable optical pattern classification system which has five outputs, in which ten connection strengths from one input to ten sum regions can be varied.

In all the systems metioned up to now, learning is performed with the help f electronic summation, thresholding or feedback, thus destroying the advantages that originate from the inherent parallelism, speed and possibility to realize bipolar weights. Compared with these rather conventional correlation techniques for pattern recognition using perceptron-like structures, the volume holographic perceptron approach has the important advantage of learning ability. Because several techniques of volume grating recording have been developed in the last years, all being possible to realize simple perceptrons, recently, several remarkable optical architectures for implementing perceptron-like networks using photorefractive dynamic holographic techniques have been presented. Therefore, the following section will focus on their realization as examples for perceptron implementations using coherent, all-optical processing.

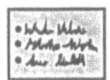 ## Summary

A single-layer perceptron can be used in pattern classification for linearly separable problems, mostly two-category pattern classification. The preceptron is trained before operating (prescribed learning), thus adjusting the interconnection weights of the system. During operation, the product of the input pattern with the interconnection weight is calculated and then compared to the desired response. There are several possibilities to realize perceptrons in an electrooptical way. Among them, the most important concepts are:

- Interconnection weights can be realized using spatial light modulators (SLM's, e.g. spatial light rebroadcasters), liquid crystal or optoelectronic devices and plane (computer generated) or volume holograms.
- Summation is done optically by superimposing or matrix-vector multiplying the input with the stored interconnection matrix (see also section 4.3).
- When SLM's, liquid crystal or optolelectronic devices are used, learning or weight adaptation is performed in the prescribed learning process electronically before transferring the final weights into the device.

## 8.2   Perceptron Pattern Classification Using Volume Holograms

The classsical holographic setup utilizing a photorefractive crystal to record the interconnection strengths allows to implement single-layer perceptrons that have multicategory
classification capabilities. Almost all methods of beam coupling in photorefractive materials can be used to generate the single-layer perceptron learning procedure. Among
them, photorefractive two-beam coupling is the most widely used process. Beneath, perceptrons have been realized exploiting photorefractive beam fanning as well as phase
conjugation based on four-wave mixing or self-pumped mechanisms. Here, I will present
several important and exemplary realizations that demonstrate the principles and the
abilites of photorefractive volume holograms for the realization of powerful perceptron
devices. However, as in the precedent section, most implementations utilize optoelectronic threshold devices in order to complement the volume holographic interconnection
devices to a perceptron system.

### 8.2.1   Electrooptic Two-Beam Coupling Perceptron

A possible setup to realize a perceptron based on two-beam coupling interconnections
was realized by E.G. Paek and coworkers [352] [353] using photorefractive LiNbO$_3$ as
the interconnection volume hologram. Fig. 8.2 shows a diagram of the experimental optical learning machine, which is even suited for multicategory classification. Each of the
input training patterns, stored on an optical memory disk recorder, is loaded onto a two-
dimensional spatial light modulator, using a monitor and a lens. The liquid crystal SLM
has about 500×500 resolution elements, representing about 250,000 input neurons. After
the input image is loaded onto the liquid-crystal light valve, it is read out by a collimated
coherent beam from an argon-ion laser and focused onto the photorefractive crystal. The
light diffracted from it is collected and focused onto a detector array of 20 detectors representing 10 bipolar outputs using the dual-channel technique (see section 4.2.4), which is
located at the image plane of an array of liquid-crystal modulators (LCM's). The outputs
from each pair of detectors are subtracted, thresholded and latched. The error signals are
generated by comparing the output with the desired target signal electronically, loaded
to the array of 20 LCM's, representing as well 10 bipolar error values. The light through
the LCM array is focused to the photorefractive crystal by lenses and interferes with the
input to reconfigure the information recorded on the crystal. Cylindrical lenses are added
to increase the light efficiency. After each reconfiguration of the hologram the error signals are disabled before detection, thus allowing the measurement of only the diffracted
beam intensity.

Using that system, the authors could demonstrate the learning of five standard capital
letters (A, B, C, D, E). Each training pattern was exposed for an equal time (7 s) at
the same optical power level (18.5 mW/cm$^2$). When the system is trained starting from
arbitrary small initial values of the interconnection strengths, it converges to the non-
error state after 10 iterations, although the training sets are far from being orthogonal.
After the convergence to the non-error state, the system stays at the stable state for more
than 30 iterations — enough time to fix the interconnection strengths to their learned
values using temperature or electric field methods before the system eventually diverges
owing to the gradual decay of interconnection strengths inside the photorefractive crystal.

**Figure 8.2:** Schematic of the holographic learning machine for multicategory classification (after [352]).

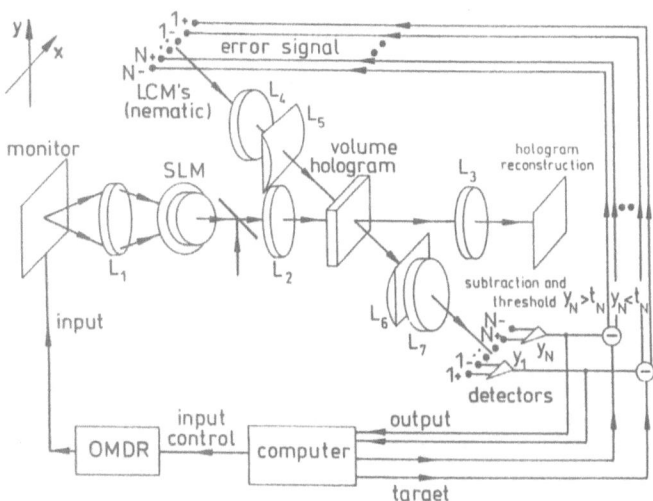

## 8.2.2 All-Optical Two-Beam Coupling Perceptron

In contrast to that optoelectronic realization of bipolar weight adaptation and feedback, the model of H. Yoshinaga and coworkers is an all-optical one that allows for two-category classification [354]. The interconnection storage device here is a BSO crystal and the holographically written gratings are changed in their strength or weight using coherent enhancing or erasing at the same wavelength. For a learning pattern with a "low" output that should be classified as "high", the corresponding gratings must be enhanced. For a learning pattern with a "high" output that should be classified as "low", the corresponding gratings must be erased. The learning rate is related to the exposure time of the hologram. To determine it, the temporal characteristics of recording and erasing must be investigated. Experimentally, it can be found that the higher the beam intensity, the shorter the recording and erasing times [355]. The number of iterations required for a learning convergence depends on the amount of weight change required at any single learning step. The larger the learning rate, the faster the learning converges, but the output then becomes oscillatory. Conversely, the smaller the learning rate, the slower it converges, but in this case the output is stable. Therefore, the learning process must be optimally scheduled, depending on the selected storage probe. In the present implementation this is done by manual setting — the most simple method of thresholding and feedback. This procedure could also be done electronically, which is equivalent to incorporate electronic thresholding and feedback to the electronically addressable input device. However, by incorporating an all-optical error signal generation system based on the generalized delta rule, as they were presented in section 5.1.3 using e.g. microchannel spatial light modulators, the construction of all-optical multilayer neural networks with backpropagation learning will be possible. Therefore, we will find this concept again in the next chapter describing these multilayer perceptron architectures.

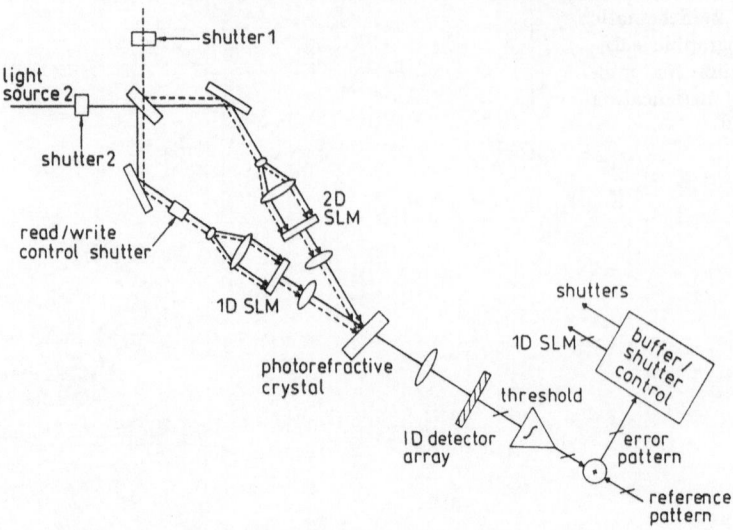

**Figure 8.3:** Multiple category pattern classifier using photorefractive interconnection storage. Bipolar weights are realized using holographic phase control in a Mach-Zehnder-like input setup. The perceptron algorithm here is: if error = 1, then open shutter 1; if error = −1, open shutter 2; if error = 0, then go to next pattern (after [105]).

### 8.2.3    Interferometric Beam Coupling Perceptron

The third realization of a photorefractive perceptron is based on coherent bipolar weight adaptation using $\pi$ phase-shifted recording beams for erasure in a Mach-Zehnder interferometric setup [105, 356, 357]. The principle of that bipolar weight realization has already been presented in section 4.2.4. The system shown in fig. 8.3 is able to realize multiple category classification when the reference for each pattern is no longer a single plane wave but a set of plane waves dictated by the collection of opening in the one-dimensional SLM. Initially, the crystal contains no hologram and interconnections are built by simply exposing the hologram with light source 1 with the pattern in the two-dimensional SLM and its associated reference pattern in the one-dimensional SLM. The process is repeated for each pattern in the training set. After the initialization, the first pattern is loaded into the two-dimensional SLM and the reference beam shutter is closed to interrogate the system. The reconstructed ouptut pattern is then compared with the desired output pattern to yield the error vector. The algorithm must now be performed in two steps: First, only those portions of the one-dimensional SLM corresponding to the positive portions of the error vector are opened. Light source 1 is turned on to strengthen certain interconnections following the perceptron recipe. Then, only those portions corresponding to negative elements of the error vector are loaded into the one-dimensional SLM and light source 2 is turned on to weaken the appropriate weights.

For a training set of 10 patterns that was to be separated into two classes, the solution was reached within 10 iterations. However, further reading or testing of the holograms causes the gratings to decay owing to incoherent erasure, leading eventually to errors. If the algorithm is further applied, these errors are regarded by the system as non-settled solutions and are corrected again automatically, thus resulting in an oscillatory behaviour between the final state and its surroundings. The incoherent erasure of previ-

**Figure 8.4:** Simple photorefractive learning system. PB: polarizing beam splitter, $L_1$, $L_2$: imaging lenses, BS: beam splitter, WP: quarterwave plate, SH: shutter, P: polarizer, D: detector, TV: television monitor, M: mirror (after [101]).

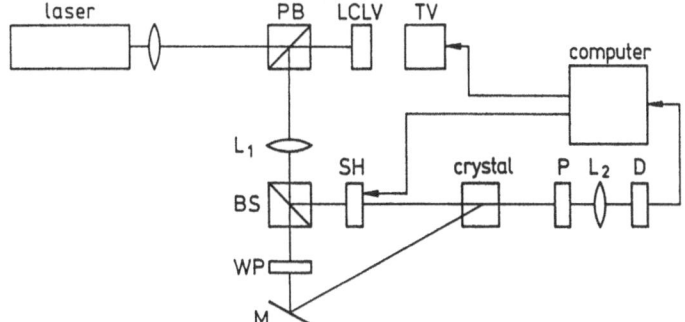

ously recorded holograms in exposing a new hologram poses a problem in the learning implementation of a perceptron using photorefractive crystals. In terms of neural network language, as new holograms are recorded, the photorefractive crystal slowly forgets its history. To avoid this problem, first an exposure schedule can be incorporated into the training procedure to yield equal-strength contributions from individual patterns in a training cycle (see sections 4.4.2 and 4.4.2. However, the total number of patterns that can be accomodated (the number of possible exposures in a cycle) is limited by the minimum diffraction efficiency required for registration in the system. If the patterns used in the exposures are statistically independent, the diffraction efficiency of the composite hologram can be shown to decrease in inverse proportion to the total number of exposures in the training sequence [99]. The number of training patterns that can be accomodated is then limited by not only the dynamic range of the photorefractive crystal but also the overall system noise, including noise in the detector and the associated electronics. As a solution, hologram strength normalization techniques using photorefractive phase-conjugate mirrors to refresh and amplify weak photorefractive holograms that are volatile owing to the photorefractive erasure can be implemented (see section 6.4).

### 8.2.4 Polarization-Selective Coherent Beam Coupling Perceptron

A very similar system was proposed and proved experimentally in a first, rudimentary way in 1988 by D. Psaltis and coworkers (see fig. 8.4) [101]. The aim of their setup was to realize a single-layer of what might be a possible net for a multilayer architecture that allows for backpropagation. Therefore, to allow for forward and backward beam path in the same holographic interconnections and threshold units, they used the polarization as an encoding means.

In their implementation they used two-dimensional patterns recorded on a liquid crystal light valve from a video monitor as stored vectors. The output vectors correspond to a single bit output of the detector. An input vector is imaged onto a photorefractive crystal via two separate paths. The strength of the grating between the image of the input along one path and the image along the other path is read out by light propagating along the path of one of the write beams in the orthogonal polarization, i.e. while the write beam incident on the detector is linearly polarized, the other write beam is circularly polarized. The polarizer blocks the linearly polarized light and one component of the diffracted circularly polarized beam, passing only the orthogonally polarized diffracted beam. This allows readout of the grating as it is being recorded. The diffracted light is imaged onto the detector. This system classifies input patterns presented to it into two classes according to whether the output of the detector when the pattern is presented is high or low. If during training a pattern we would like to classify as high yields a low response, the hologram is reinforced by exposing the crystal to the interference of the two beams, each carrying the image of that pattern. This exposure continues until the diffracted output increases by a fixed amount. If a pattern which should be classified as low is found during training to yield a diffracted output that is too high, the hologram diffracting that pattern is weakened by a fixed amount by exposing the crystal with only one of the imaging beams (incoherent erasure, compare to section 4.2.4). Experimentally, four patterns were trained to the system. The most limiting problem that arised during the experiment was the fact that overlap of patterns is well defined only in the image plane, meaning that the crystal must be thinner than the depth of focus of the images. To utilize the full capacity of photorefractive volume holograms it will be necessary to move beyond this implementation to architectures utilizing the full three-dimensional capacity of the crystal.

### 8.2.5   Perceptron Based on Beam Fanning- and Self-Pumped Phase Conjugation Interconnections

Finally, I want to discuss an implementation of the single-layer perceptron that is based on photorefractive interconnections that are angularly and spatially multiplexed in the registration volume using self-pumped or stimulated photorefractive gratings. A diagram of such a *stimulated photorefractive optical neural network* is shown in fig. 8.5.

In principle, the network contains a single photorefractive crystal, a spatial light modulator (SLM) and a CCD detector. Multilayer networks may be implemented by superimposing different weight layers in the same crystal. During operation, a plane wave readout beam from an argon-ion laser is modulated spatially by the SLM, which is driven by a host computer. The most important trick of the system is that the input plane impressed on the input light consists of two portions: the reference plane (R) and the object plane (O). A system of mirrors separates the O and R planes. The R beam enters the photorefractive crystal (in his experiment, Y. Owechko used $BaTiO_3$ [131]) at an angle relative to the crystal's c-axis such that it experiences high two-wave mixing gain, resulting in fanning. The O beam is directed at a low-gain angle and does not fan. Fanning of the object beam would degrade the optical quality of reconstructed images since the original O beam would be distorted. Pixels in the O and R planes represent neurons in the various network layers.

After a hologram is written, shutter $s_1$ is closed by the host computer, and the R beam reads out the hologram. Closing $S_1$ ensures that no light leakage owing to the finite contrast ratio of the SLM is passed through from the O input plane to the detector during readout. The CCD camera detects the holographic reconstruction of the O plane. Due to the fact that Bragg degeneracy is greatly reduced owing to the fanning of the R beam, the beam quality of the reconstructed image is improved compared to conventional two- or four-wave mixing interconnnection setups. The image is then stored into the computer memory by an image-processor-frame-grabber card. Because the optical neurons are oversampled by the CCD pixels, a simple automated alignment procedure is used for each output neuron, allowing for reduction of noise in the stored images. Moreover, two computer-controlled liquid crystal cells (LC-1, LC-2) are used to control the optical

**Figure 8.5:** Experimental configuration of the stimulated photorefractive optical neural network. The fanned reference beam R reads out the unfanned object beam O (after [131]).

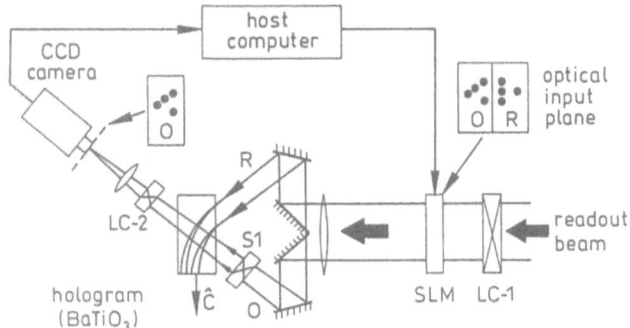

power during the hologram writing (high power R) and hologram reading phases (low power R).

The topology of the network being implemented is determined by the organization of the input plane. A single layer network with only forward connections is implemented by devoting the entire R plane to the input layer of the network, $L_0$. Similarly, output layer $L_1$ occupies the entire O-plane. Turning pixels on in the R and O planes forms connections between $L_0$ and $L_1$. $L_0$ then reads out the weights connecting the R and O planes. Multilayer neural networks are implemented by dividing the input plane into sectors. For example, to implement a net with input layer $L_0$, hidden layer $L_1$, and output layer $L_2$, the input plane is divided into four sectors. Both possibilities are shown in fig. 8.6. Connections between $L_0$ and $L_1$ are formed by writing values to the $L_0$ and $L_1$ sectors in the R and O planes, respectively. Similarly, connections between $L_1$ and $L_2$ are formed by writing values to the $L_1$ and $L_2$ sectors in the R and O planes, respectively. The two weight layers are exposed separately to avoid unwanted links between $L_0$ and $L_2$. To read out the network, first an input pattern is written to $L_0$ in the R plane. An optical matrix-vector product is then detected by the CCD camera at $L_1$ in the O plane and processed electronically to form the neuron value in the hidden layer $L_1$. These values are then written to the $L_1$ sector in the R portion of the optical input plane. Another optical matrix-vector product is formed, detected at $L_2$ in the O plane, and processed to form the output neuron values in $L_2$.

**Figure 8.6:** Stimulated photorefractive neural network implementation of multilayered neural networks by spatial organization of the optical input plane (after [131]).

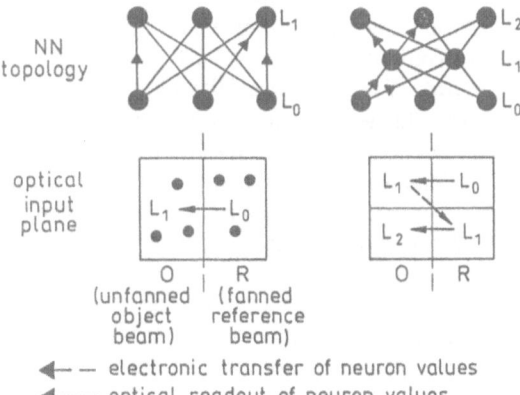

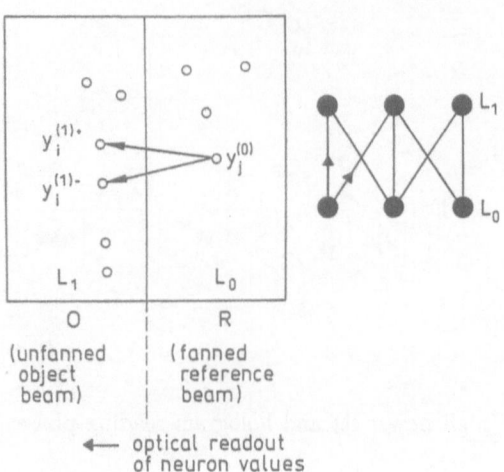

**Figure 8.7:** Stimulated photorefractive neural network input plane for a single-layer perceptron (after [131]).

If more complicated networks with additional features such as backward connections or additional layers are required, they can be implemented simply by dividing the optical input plane into more sectors and by changing the software. It is not necessary to adjust the hardware or to add more crystals. This easy reconfigurability is due to the capability of arranging neurons in arbitrary patterns in the optical input plane, a feature made possible by elimination of Bragg degeneracy.

Bipolar weights in these nets can be achieved using either the coherent approach of direct phase modulation of the phase of the written gratings or to incoherently use spatial multiplexing and electronic subtraction (see section 4.2.4). A diagram of the stimulated photorefractive optical neural network optical input plane for implementation of a perceptron network is shown in fig. 8.7. The R plane contains the pattern in the input neuron layer $L_0$, and the O plane contains the output neuron layer $L_1$. It is possible to use both single output and many output neurons. The latter case corresponds to many perceptrons operating in parallel on the same input pattern but with different classification goals. Y. Owechko [131] realized a system that was able to learn 96 random patterns in 29 iterations.Each pattern consisted of 1920 pixels. Increasing the number of pixels tended to reduce the number of patterns that could be learned. With 7680 pixels for each pattern, the system could still learn 42 patterns. The system could separate patterns containing as few as 0.5% differing pixels.

Moreover, Y. Owecko was also able to implement multiple-output-neuron perceptron networks. The networks learned to perform a one-to-one transformation of a given set of random binary patterns (values 0 and 1) into another set of random patterns (values -1 and 1). This network contained 1740 input and 870 output neurons, giving 1.5 million weights. Nevertheless, learning took place after about 20 iterations, showing the quick adaptation of the network. This type of network was also applied to handwritten digit recognition application, using exemplars from a database of handwritten digits from the U.S. Post Office. The network consisted of 676 input neurons and 10 output neurons. It was trained to label inputs as one of the 10 output digits. The network achieved a 92% correct recognition rate on the training set and a 75% correct recognition rate on 40 test digits the system had not seen before, although it contained only a single layer of weights and the caracter problem is not linearly separable.

The realization of backpropagation networks in the same setup will be discussed in the following chapter.

### 8.2.6   The Problem of Weight Decay

 Although my descriptions of the perceptron learning algorithm implemented with photorefractive volume holography seem to be fairly simple and easy-to-implement, there is a fundamental problem we have to consider: The decay of the weight vector during the training stages. If you look back e.g. at the previous implementation using interferometric phase control to realize bipolar weigths, a set of weights is learned for every input pattern sequentially. However, changing the weights for the second and following patterns implies that previously set values will be affected too. Thus the weight decay imposes a limitation on every photorefractive perceptron network, in such a way that convergence of the net may be jeopardized. Significant decay of the interconnection weights may even lead to a divergence of the learning process. In mathematical expressions, the perceptron algorithm in photorefractive media is modified compared to the original algorithm (see box 1 in section 2.2) including weight decay as

$$w_i(k+1) = w_i(k)\cdot\exp\left[-|a(k)|t/\tau\right]+a(k)[1-\exp\left(-t/\tau\right)]\cdot x_i(k) \quad \text{for} \quad 0 \le i \le N-1 \quad (8.4)$$

and

$$a(k) = \begin{cases} 0 & \text{if output is correct} \\ +1 & \text{if output 0, should be 1 (class A)} \\ -1 & \text{if output 1, should be 0 (class B)} \end{cases} \quad (8.5)$$

Here, in contrast to section 2.2, $t$ denotes the exposure time, $\tau$ is the hologram decay time constant of the photorefractive material, $k$ is the integer registering the number of interrogations, $x_i(k)$ is the $i$th element of the $k$th input vector and $\sigma(k)$ represents the perceptron threshold function.

K.Y. Hsu and coworkers theoretically proved that the photorefractive perceptron learning algorithm according to eq. (8.4) will converge, provided that the exposure time is sufficiently small relative to the decay time constant [358]. This requirement is based on the fact that the decay of photorefractive holograms during the process of interconnection weight changes. As a result of the hologram erasure, the convergence of the learning process is dependent on the exposure time during the weight changes.

To overcome the hologram or weight decay problem, the interconnection weights must be strengthened by optical techniques. The most promising of these weight restoration methods — or in other words, refreshment or updating techniques — have already been discussed in section 6.4. The best suited method for weight updating is to read out the weight vector and then rerecording it by using the technique of double phase conjugation during each exposure for update by previous records in holographic associative memories. This method has been realized by the author and her coworkers and has also been presented in detail in section 6.4.

Mathematically, the equation for the weight update becomes

$$w_i(k + 1) = w_i(k) \cdot \exp\left(-t/\tau_e\right) + [1 - \exp\left(-t/\tau_r\right)] \cdot (y_i(k) + Aw_i(k)) \quad (8.6)$$

where A is a constant. In other words, a previous weight vector is added to the input vector to compensate for weight decay. Writing the equation in a slightly different way, we obtain

$$w_i(k+1) = \exp\left(-t/\tau_e\right) + A[1 - \exp\left(-t/\tau_r\right)]w_i(l) + [1 - \exp\left(-t/\tau_r\right)]y_i(k) \qquad (8.7)$$

The constant A can be properly chosen to tailor the gain factor that in turn affects the convergence of perceptron learning of the system. K.Y. Hsu and coworkers found in a theoretical analysis, that for short exposure times $t$ with

$$0 < t < \tau_e \cdot \ln\left(\frac{p_m + 1}{p_m - 1}\right) \qquad (8.8)$$

($p_m$ being the number of updating steps till the $m$th cycle), A is given by [359]

$$A \approx \tau_r/\tau_e. \qquad (8.9)$$

With this basic knowledge about the construction and constraints of single-layer perceptrons, we are on the one hand able to optically realize linearly separable classification problems. On the other hand, we have obtained the tools to construct more complex, larger multiple layer perceptron nets by combining the principles of these single layer operations to several, cascaded layers and introduce nonlinear thresholding capable of error backpropagation into the systems. Examples of how multilayer perceptrons can be realized all-optically, using especially photorefractive interconnections, will be the contents of the next chapter.

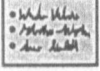 ## Summary

The classsical holographic setup utilizing a photorefractive crystal (BSO, LiNbO$_3$ or BaTiO$_3$) to record the interconnection strengths allows to implement single-layer perceptrons that have multicategory classification capabilities. The components and performance of such are network are the following:

- Training patterns as well as later on input patterns are loaded into the system using spatial light modulators.
- Learning and operation or readout of the photorefractive memory are based on pure matrix-vector multiplication (see section 4.3), realized by classical two- or multiple beam coupling interconnection configurations.
- Beam-fanning and self-pumped photorefractive processes in contrast realize interconnections that are angularly and spatially multiplexed in the registration volume.
- Learning or weight adaptation takes place using coherent enhancing or erasing at the same wavelength (e.g. erasure by $\pi$ phase-shifted recording beams). The learning rate is related to the exposure time of the hologram.
- During storage or learning, special writing and refreshment algorithms have to be performed to account for the erasure effects and thus for the weight decay in photorefractive materials (see sections 4.4 and 6.4).

- Provided that the exposure time is sufficiently small relative to the decay time constant, convergence of the photorefractive perceptron learning algorithm is still possible.
- Operation of the system can be done using orthogonal polarization in order to preserve the grating strengths of the learned interconnections.
- Thresholding is again done electronically in most cases.
- A simple extension to multiple layer networks for the case of self-pumped or beam-fanned gratings is to divide the input plane into sectors. For example, to implement a net with an input layer, a hidden layer, and an output layer, the input plane is divided into four sectors. This easy reconfigurability is due to the capability of arranging neurons in arbitrary patterns in the optical input plane, a feature made possible by elimination of Bragg degeneracy.

 **Further Reading**

1. B.K. Jenkins and A.R. Tanguay, *Photonic implementations of neural networks*, in *Neural networks for signal processing*, B. Kosko, ed., pp. 287 - 382, Prentice Hall, Englewood Cliffs, NJ (1992).

# 9   Optical Realizations of Multilayer Perceptrons

I have shown in section 2.2 that, if a nonlinear decision problem has to be solved, a single-layer perceptron fails to give the right response because of its linear separation of the classes to be distinguished. However, multilayer perceptrons that allow for sigmoidal threshold in combination with the possibility to adapt the slope of this decision function are well suited to solve these problems (see section 2.3 for a detailed explanation). These cascaded systems may be easily realized using optical devices, because cascadability is an inherent feature of most of the simple optical implementations I have discussed up to now. The critical point in realizing all-optical multilayer perceptrons however is to realize the error backpropagation function which has to be the derivative of the forward propagating sigmoid threshold function. Many of the all-optical realizations of the sigmoidal threshold function have derivatives that do not simulate the requirement of the error backpropagation to have its maximum in the region where the slope of the sigmoidal function is steepest.

Therefore, two solutions to realize optical multiple layer implementations are mainly suggested. The first is to realize that part of the system either with electronic or electrooptic devices. The second method is using functions with error backpropagation curves that deviate slightly from the one prescribed by the error backpropagation algorithms. For the first case, it is obvious that the potential of coherent processing that allows for an easy way to cascade the system all-optical, is destroyed, thus introducing again a limitation in the speed of data processing. In the second case, in contrast, one has to investigate carefully how the deviation from the proper gradient descent backpropagation algorithm will influence the performance and the convergence of the net's response. Several investigations have shown [131, 223] that the network performance is relatively insensitive to the precise values of these functions as long as the general sigmoidal and humplike form of the threshold function and its derivative, respectively, are preserved.

Based on these possibilities to realize optical multilayer perceptrons, I will begin my description with simple models that use electronic or electrooptic implementations of the threshold unit. Among them, you will find multilayer perceptrons with optoelectronic learning as the optical connectionist machine, a system based on the neocognitron paradigm using holographic interconnection devices and a system realizing near-field learning using photorefractive volume holography.

Then, alternative implementations of nonlinear classification problems and optical perceptron-like networks will be described. Instead of cascading several single-layer perceptrons with nonlinear thresholding units as in multilayer perceptron realizations, the method of higher-order neural networks consists of calculation the higher order correlations and to use them as input to the network (see section 2.3.4). This concept is mainly based on autocorrelation — a concept that can be easily realized optically. An alternative to the problem that the backpropagation algorithm requires knowledge of the weights in the second layer in order to modify the ones in the first layer — thus being a nonlocal learning algorithm — is the application of local learning rules for multiple layer perceptrons. I will present an implementation of optical local learning for multilayer percpetrons based on photorefractive interconnections. Finally, I will also describe more complex, but more elegant all-optical nets that allow for coherent processing, but are more difficult to handle and stabilize. Here, I will focus my attention on a polarization multiplexing technique for photorefractive interconnection-based networks.

In order to analytically determine the conditions that are necessary for photorefractive backward-error propagation learning, G. C. Petrisor and coworkers [360] investigated both cases of realizations, a fully coherent system that uses electric-field amplitude encoding of signals and weights and an incoherent one using intensity encoding of the signals and weights. They analytically determined that the backward-error-propagation learning algorithm has a well-defined region of convergence in neural learning-parameter space, depending strongly on system gain and exposure energy. Whereas the first one determines the learning-rate coefficient, the second one is responsible for weight decay. The convergence condition gives for both cases — coherent and incoherent realization — an expression for a minimum gain that will ensure convergence. The optimal gain, in terms of maximizing the signal intensity incident upon the detectors, is the minimum gain required for solving the problem. Generally speaking, convergence in guaranteed by using a sufficiently high gain and a sufficiently low exposure energy per weight update — thus giving for all photorefractive implementations a region in parameter space that allows for successful implementation of backpropagation perceptron realizations. The details of that analytical derivation can be found in [360].

## 9.1 Optoelectronic Realizations of Multilayer Perceptrons

Among the optoelectronic multilayer implementations of neural networks with adaptive learning capabilities, systems using interconnection storage in optically adressable spatial light modulators or in holographic plane or volume devices, combined with optoelectronic or incoherent threshold function generation are the most widely spread ones. Consequently, I will present in the following some examples that are based on these realizations of the interconnection unit. The detailed mechanism of interconnecting information in spatial light modulators or in holograms has already been described in chapter 4. Several single-layer perceptrons units have been presented in chapter 8. The reader may refer to the appropriate sections in that chapter for detailed explanations.

### 9.1.1 The Bipolar Optoelectronic Connectionist Machine

The optoelectronic connectionist machine presented by K.M. Johnson and coworkers [361] allows bipolar synaptic weight values using polarization states and is able to implement the backward propagation of errors for multilayer perceptrons. The light transmitted by each pixel in the one-dimensional spatial light modulator (SLM) represents the activation value of an artifical neuron. This optical beam is then incident upon a two-dimensional SLM. Pixels in the two-dimensional array encode the weighted interconnects between input and output neurons. As described in section 4.2.4, polarization logic is used to perform the multiplication of the input activation values by the synaptic weights. One drawback of the optical connectionist machine shown in fig. 9.1a is that a controlling computer is required to calculate the output error and to program the calculated weight modifications on the two-dimensional electrically addressed SLM, which is a time-consuming

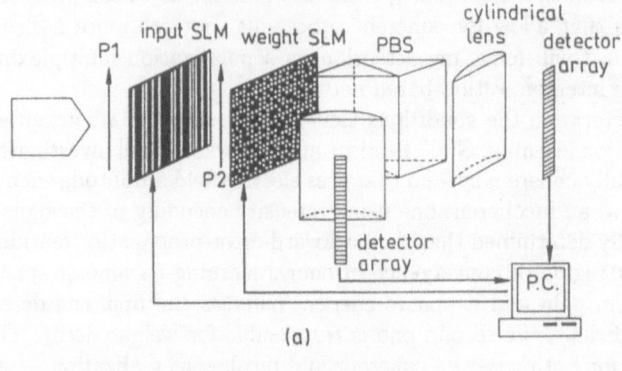

**Figure 9.1:** Optoelectronic connectionist machine architectures (a) with weights stored on an electrically adressed spatial light modulator and (b) with weights stored on an optically adressed spatial light modulator. $P_1$ and $P_2$: polarizers, SLM: spatial light modulator, PBS: polarizing beam splitter, CRT: cathode-ray tube (after [361]).

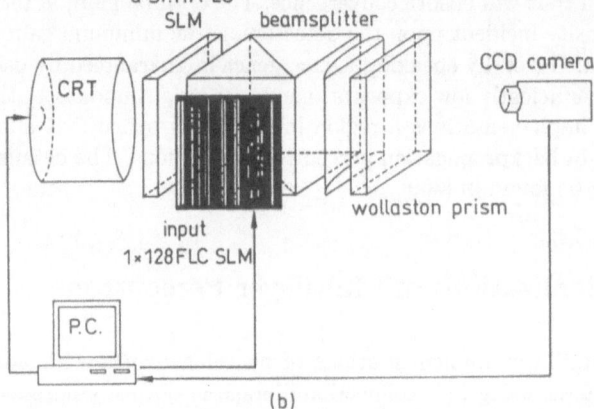

task when done in a sequential electronic way. A more time-efficient algorithm adapted to the problem of that electrooptic implementation was proposed by P.J. de Groot and R. Noll [362]. They substituted the backpropagation algorithm, which can also be interpreted as a gradient descent in energy space towards the nearest minimum, by an algorithm that subjects all weights to a random fluctuation similar to the one used in simulated annealing (see section 2.6). In their implementation, the error function is evaluated at each clock cycle. If it is nonzero, then all of the weights are subject to a random fluctuation. When the global minimum is reached, the fluctuation ceases. This method allows the network to learn new patterns at any time, without requiring a change in the system parameters.

In addition, the trained weights in the connectionist machine need to be refreshed from a nonvolatile memory stored in the external computer. The optical connectionist machine version in fig. 9.1b in contrast replaces the two-dimensional addressed SLM with an optically adressed ferroelectric liquid crystal SLM. This device exhibits bistability, or in other words, a memory. Hence the trained weights no longer need to be refreshed by an external computer. Once the network is trained, pattern classification is accomplished in the time that it takes to program the input SLM with the neuron activation values and to detect the result of the polarization-based, optical vector-matrix multiplication. The training time, however, is still limited by the data acquisition and the clock speed of the external controller that generates the error and determines the weight modifications.

**Figure 9.2:** Schematic diagram of a vector feature extracting optical multilayer perceptron without any inhibitory synaptic connections (after [364]).

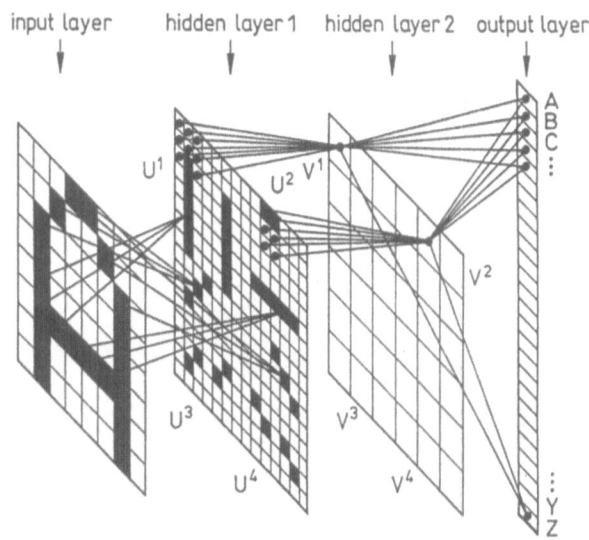

The heart of the system is the ferroelectric liquid crystal modulator in very-large-scale-integrated (VLSI) fashion. It permits the optical connectionist machine to operate without the limitations of an external controlling computer both in training the network and in using it to perform pattern classification. The device computes the changes in the weights according to the delta-rule learning algorithm and encodes the result on a pair of optical beams. It was specifically designed to make the connectionist machine work as an all-optical learning machine, and it can be used to modify the weights stored in optical media such as optically addressed SLM's, nonlinear planar interconnection devices as e.g. optically stimulable phosphor [363] and photorefractive volume holograms.

### 9.1.2 Vector-Feature Extraction Multilayer System

Most perceptron systems described up to now are based on bipolar weight value realizations, either by dividing the interconnection weight matrix into two parts, one for positive and the other for negative values, or — in the case of the connectionist machine — by implementing bipolar weight values by use of linearly polarized light. The subtraction to achieve bipolar weightening is then executed on the electrical signals. It is also possible to realize a multilayer perceptron without any inhibitory synaptic connections. Such a network has been introduced by S. Kuratomi and coworkers [364], and named vector-feature extracting optical neural network. It can correctly recognize hand-written letters by extracting a vector feature such as its direction, its existing position, and the length of line segments in an input pattern.

Fig. 9.2 shows the schematic diagram of the vector feature extraction network. It is a hierarchical multilayer feed-forward network consisting of four layers and can be divided into two parts: a vector feature extraction part (the input layer, hidden layer 1, and hidden layer 2) and a recognition part (hidden layer 2 and the output layer). In this network, a hand-written letter is given to the input layer with $n_i \times n_i$ meshes. In the input layer the hand-written letter is converted into a binary mesh-featured pattern by the following process: the output value of the neuron, including a part of the hand-written letter, is set

feature-extraction
layer $u^l$

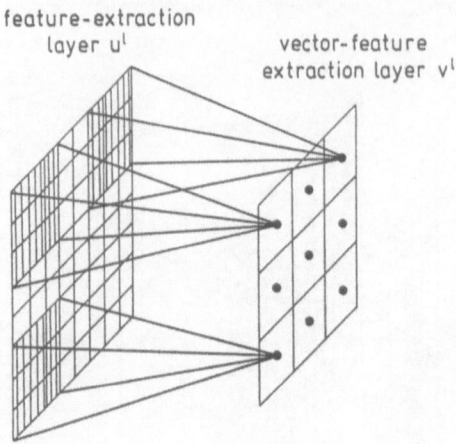

vector-feature
extraction layer $v^l$

**Figure 9.3:** Schematic diagram showing the connection between the $u^l$ layer and the $v^l$ layer. A neuron in the $v^l$ layer integrates outputs from neurons in a restricted area of the $u^l$ layer. Its weighted sum corresponds to the length of a line segment existing in its restricted area (after [364]).

to one; otherwise, it is set to zero. The hidden layer 1 consists of four feature-extracting layers ($u^l$ with $l = 1, \ldots, 4$). Each $u^l$ layer has again $n_i \times n_i$ neurons, and extracts a specific line segment, such as vertical, horizontal, left-oblique, and right-oblique line segments included in the input pattern without any inhibitory synaptic weights. Hidden layer 2 consists of four vector feature extracting layers, denoted as $v^l$. Each $v^l$ layer has $n_v \times n_v$ neurons and extracts a vector feature. Fig. 9.3 shows the schematic diagram of the connection between the $u^l$ layer and the $v^l$ layer. The $u^l$ layer is divided into $n_v \times n_v$ areas, which do not overlap each other. A neuron in the $v^l$ layer is designed to receive the signals from neurons that belong to one corresponding area in the $u^l$ layer. The recognition part of the network is a two-layer perceptron, which can be easily implemented by an all-optical system. Each neuron in the output layer is fully connected with all of the neurons in hidden layer 2. During the learning process, the synaptic weights between hidden layer 2 and the output layer are renewed. The learning process is performed by use of the error-correction multilayer perceptron training rule (see section 2.3). Compared to a simple multilayer system, this feature-extracting network allows to reduce the number of iterations until convergence dramatically, in combination in enhancing the recognition rate. However, because there is one more hidden layer involved in the process, the iteration time for one iteration enhances dramatically.

Actually, a lot of efforts are done to realize these multilayer feedforward systems in a rather compact way for real-time pattern recognition. Due to the introduction of specialized electronic neural network processors and VLSI neural chips to the commercial market, several compact systems have been realized using the same principle as the connectionist machine. The system of S. Kuratomi was developed into a feature-extraction optical neuron chip using aluminium thin-film segments operating as neuron electrodes arranged between a hydrogenated amorphhous silicon (a-Si:H) layer as a photoreceptor and a ferroelectric liquid-crystal layer in a chiral smectic C phase used as a light modulator. The processing speed of the system is typically $300\mu s$. Details of the configuration can be found in [364].

In alteration to the backpropagtion configuration in the connectionist machine and similar to the three-layer net based on vector feature extraction in the second layer, several multilayer feedforward net are not using error backpropagation, but a two-layer perceptron in which the first layer is based on a modified Hamming net and the second is performing a winner-takes-all realization. These system have fewer interconnections and a higher storage capacity than a fully interconnected Hopfield network. In comparison with pure multilayer perceptrons, they have advantages such as easy and direct training and unipolar binary interconnection weights. Especially the last point makes them more suitable for implementations with currently available optoelectronic devices. Z. When and coworkers [365] realized an optical system on that basis employing Dammann gratings in combination with liquid crystal spatial light modulators. T. Lu and coworkers [366] realized recently an impressing holographic optical neural network on that basis that uses a holographic interconnection device and is capable of performing massive three-dimensional interconnection of tens of thousands of neurons. The system is packaged in a small attache case ("lunchbox system"), having a size of $9 \times 12 \times 5$ cubic inches and being portable.

### 9.1.3 The Neocognitron Paradigm Using Holographic Interconnections

The general automatic target recognition problem involves both data fusion and pattern recognition. An algorithm that is able to perform both of these functions is the neocognitron neural network paradigm first proposed by K. Fukushima and coworkers [367]. The neocognitron uses a feature-fusion approach to discrimination that is carried out with a multilayer feedforward hierarchy, namely a multilayer perceptron. The evidence provided by individual image features is combined in stages, and at each stage, compensation is made to scale and aspect angle variations. As a result, the neocognitron is able to robustly classify input images over a large range of scales and aspects.

To implement an optical neural network based on that paradigm, T.H. Chao and W.W. Stoner [368] used a multichannel optical correlator as the building block of a generic single layer of their network. The idea behind this implementation is the strong analogy between a general neural network and an optical neural network with an optical correlator (see e.g. chapter 7), here used as a multichannel device. The multichannel correlator permits the parallel processing of a large number of selected features. To further implement the multilayer processing, the authors built an electronic feedback loop to connect the output plane of the thresholding detector to the input plane of the SLM. Thus, with the proper update of the previously stored holographic weight filter array for each layer of operation, a multilayer neocognitron-type neural network is implemented. A schematic diagram of the optical neural network applied to the recognition of laser radar images is shown in fig. 9.4. The idea of the experimental investigation was to built a system for discrimination of laser-radar images of reentry vehicles (cones) and decoys (beachballs). The authors used specifically the angle-angle images of both the cones and the beachballs for their experiments. They selected eight different cone images with various shapes, sizes and orientations. In order to have more intraclass fault tolerance, these eight objects were further partitioned into two different groups. A careful selection of characteristic features was made so that the objects could be recognized simultaneously.

In operation, the two-dimensional input, consisting of multiple targets, is fed into the input SLM of a multichannel correlator. A two-dimensional binary Dammann grating and a cascaded Fourier-transform lens are used to generate an array of spatially replicated Fourier spectra of the input objects. A holographic Fourier-plane filter performs matched spatial filtering between the input and selected characteristic features. This holographic filter plays the role of the interconnection weight array for the neural network. A Fourier-transform lenslet array is placed behind the Fourier hologram array to perform inverse

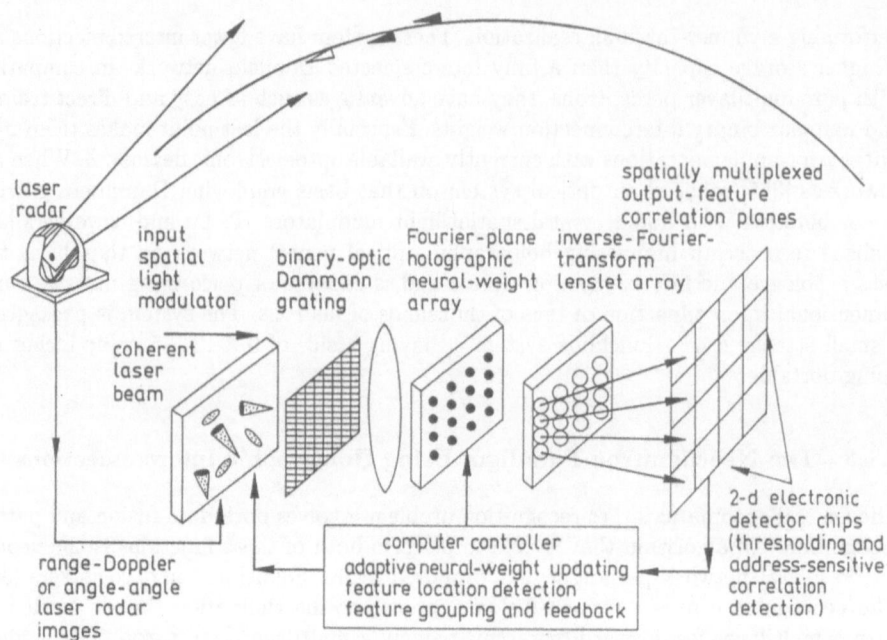

**Figure 9.4:** Schematic of a multichannel-correlator-based optical multilayer neural network for applications in laser radar image recognition (after [368]).

Fourier-transformation so that correlation peaks are obtained in the output plane. The mapping between the lenslet array and the Fourier-hologram can be one-to-one or one-to-many, depending on the input pattern and the processing requirements. If a specific mapping relationship is required, a custom lenslet array can be built using holographic or binary-optics technology. A photodetector array is placed in the output plane of each lenslet so that a cluster of spatial features can be extracted from the input pattern and converted into the corresponding cluster of correlation peaks. These correlation peaks are in turn detected by the photodetector array.

In this optical neural network, off-line-training and on-line recognition are adapted. Characteristic features of training targets are selected by a computer simulator. The criterion of this feature selection is similar to that of the interpattern association model (see section 2.5.2 for further details). Using this model, it is possible to identify representative features easily and use them effectively for pattern recognition. To ensure intraclass fault tolerance, common features from each class of targets are extracted. To achieve intraclass discrimination, one need to ensure that features extracted from different classes have minimum cross correlation. These characteristic feature patterns are then recorded upon the Fourier hologram array. In other words, with that method a two-dimensional input pattern, upon passing through the multichannel correlator, is reduced to a cluster of feature correlation peaks. Each cluster is detected simultaneously and thresholded by a different photodetector array.

The spatial resolution of the photodetector array is sufficient to preserve the position of the feature-correlation peaks. For complicated input patterns a single pass of the multichannel correlator may not be adequate to complete the entire recognition and classification process. Thus it is necessary to pass the feature-correlation patterns along for processing in a second layer. A time-multiplexing scheme was devised to accomplish this feedback using one input SLM only. The output of each photodetector array is fed back to the input SLM, and at the same time the neural-weight holograms are switched for a second layer processing. This process is continued until the final recognition and classification is finished in the deepest level.

It is feasible to store all the neural weights that are required for use in each processing layer on the same piece of holographic film or volume. Then it is necessary to develop an effective addressing scheme that is fast and accurate for this multilayer optical neural network. In principle, holographic neural weight array addressing can be accomplished using all methods described in chapter 4, especially spatial-, angular-, phase- and wavelength-multiplexing. The spatial multiplexing can be obtained by partitioning the available holographic neural-weight arrays into a number of groups, with each group assigned to a specific processing layer. However, in that case the available number of feature-correlation channels for each processing layer is reduced. Moreover, crosstalk may occur due to the global interconnection between the input and the Fourier-planes. Wavelength multiplexing as another example can be achieved by incorporating the use of a tunable laser source and a diffractive Damman grating. The latter one enables that the scale of the Fourier diffraction pattern is proportional to the laser wavelength. Thus when the wavelength of the laser is tuned incrementally, the scale of the Fourier spectra array is varied accordingly, such that a different Fourier-holographic neural-weight array can be addressed. This scheme permits the sequential addressing of a multiple number of holographic neural-weight arrays with a holographic filter in each array. By synchronizing the wavelength tuning, we can synchronize the feedback from the output detector to the input SLM such that multilayer processing can be achieved.

In their preliminary experimental investigations, W.W. Stoner et al. did not implement noise and clutter to be present in the input target imageries. However, this might be a problem for real-life images. Thus an image cleanup preprocessor for noise and clutter removal might be useful, especially for networks with a larger number of layers.

### 9.1.4 Beam coupling Mean-Field Theory and Backpropagation Learning

Volume holography in combination with electronic feedback was proposed in 1990 by C. Petersen et al. to implement a supervised learning multilayer neural network [369]. It is based on spatial multiplexing rather than the more commly used angular multiplexing of the interconnection gratings (see fig. 9.5). Among the various multilayer supervised learning algorithms, the single-crystal architecture allows for implementation of mean-field theory, backpropagation and Marr-Albus-Kanerva-style algorithms (see section 2.3 for further details).

The system configuration is shown in fig. 9.5 and has two principal optical paths, a reference path (1) and an object path (2). Each path has a spatial filter, a beamsplitter, a SLM and an imaging lens system. The object path ends with a CCD array. The photorefractive crystal is SBN and an Argon ion laser is used as a coherent light source. Thresholding using the binary sigmoidal function ($\tanh(x)$) and loading of the SLM's take place electronically. Both mean-field theory and backpropagation learning algorithm implementations have distincitve read ($a_1$ plus $a_2$) and write ($a_2$ plus $b$) phases that use this generic architecture. As described in chapter 5, the temperature annealing procedure that is necessary to implement the mean-field algorithm (see section 2.6) can be realized in photorefractive materials by varying the gain (which is equal to the inverse

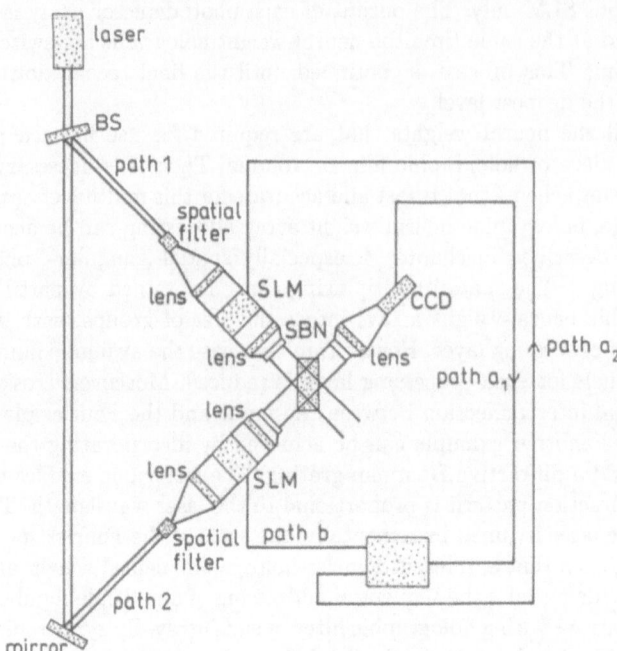

**Figure 9.5:** Optoelectronic implementation of a multilayer neural network in a single photorefractive crystal (after [369]).

temperature). Computer simulations of C. Peterson and coworkers showed a good convergence of such a net using the gain as the variation parameter to perform simulated annealing using photorefractive interconnection devices.

One of the big advantages of the system has been shown in simulations that include "real world" effects of photorefractive recording such as rescattering, beam depletion, beam absorption and grating decay. Normally, rescattering, absorption and beam depletion cause asymmetries in the weights, thus tending to distort the matrix-matrix multiplication (see section 4.3.2) and limiting the performance of the interconnection device severely. Grating decay, however, exerts a symmetrizing influence. Consequently, grating decay actually aids the learning process, typically reducing the number of training presentations required. Generally, the weights become more symmetric the longer learning proceeds. In addition to supervised learning, a key ingredient for handling these physical phenomena is a geometric "temperature gradient" for the output intensities. The detailed function of that net and its performance for the different neural net implementations can be found in [369].

### 9.1.5  Beam-Fanning Backpropagation Network

Another scheme using photorefractive volume holographic interconnections has been realized by Y. Owechko: It takes advantage of the possibility to use stimulated photorefraction that distributes the gratings among a series of angularly and spatially multiplexed gratings, thus eleminating the Bragg degeneracy and allow for a higher storage density. The principle of that method for a single layer of such a network implementation has already been presented in the previous chapter, in section 8.2 (see figs. 8.5, 8.6, 8.7).

**Figure 9.6:** Optical input plane for optical backpropagation with a single hidden layer (after [131]).

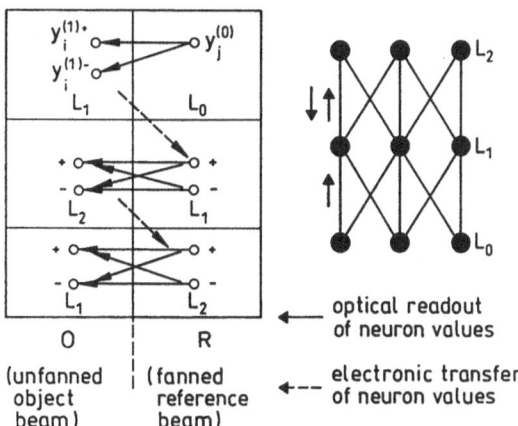

This principle can be extended to the case of backpropagation. For that case, Y. Owechko used the fact that the network performance is relatively insensitive to the precise values of the sigmoidal function and its derivative, as long as the general sigmoidal and humplike forms of both functions are preserved [223, 131]. In his optical system, he actually implemented a variation of backpropagation in which the input error signals were trinary quantized to $+1$, $0$, $-1$ according to an algorithm reported by P.A. Shoemaker et al. [370]. They found that trinary quantization improved the convergence speed of backpropagation for a wide range of problems. Moreover, trinary quantization also helps to avoid amplitude-dependent phase errors in our liquid-crystal-based spatial light modulators.

The optical input plane for backpropagation is shown in fig. 9.6. The network consists of three neuron layers ($L_0$, $L_1$, $L_2$) and two weight layers. The $L_1 \leftrightarrow L_2$ weights are actually implemented as two seperate sets of weights, one for the forward pass $L_1 \to L_2$ and another one for the backward pass $L_2 \to L_1$. As explained previously (see section 8.2), this is done because the fanned reference beam (R plane) is always used to read out the unfanned object beam (O plane). In order to implement forward and backward passes through the same set of effective weights, two sets of photorefractive weights must be exposed. In this case they are exposed so as to make the forward and backward weights as nearly equal as possible (symmetric connections) although they can also be made different (asymmetric connections) if required for certain networks. Although the $L_1 \to L_2$ and $L_2 \to L_1$ sections of the input plane are shown separated spatially for clarity, actually they are interleaved spatially in order to make the connections as symmetric as possible. The solid arrows in fig. 9.6 indicate optical connections between neurons be the hologram. Dashed arrows denote electronic transfer of detected outputs from one layer (O plane) to the inputs of the next layer in the R plane. This electronic operation is an order-$N$ operation, while the optical connection is of the order of $N^2$, where $N$ is the number of neurons.

Experimental results for the problem of transforming one random binary pattern into another using optical backpropagation were realized for a network with 320 (1140) neurons. The number of neurons in the three layers was 160-80-80 (570-285-285). For the smaller network, the error dropped below 0.1 after about 40 cycles, for the larger network it took about 200 cycles for convergence. An application of this system to the recognition of handwritten digits with a system having 676 input neurons, 90 hidden neurons and 16 output neurons gave a performance on novel test digits with 97% correct classification [131].

---

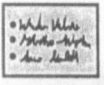 ## Summary

---

Among the optoelectronic multilayer implementations of neural networks with adaptive learning capabilities, systems using interconnection storage in optically adressable spatial light modulators or in holographic plane or volume devices are the most widely spread ones. Possible architectures and functions are:

- The optoelectronic connectionist machine allows bipolar synaptic weight values using polarization states in an optically adressed ferroelectric, bistable liquid crystal SLM or photorefractive volume holograms. It is able to store and compute the changes in the weights according to the delta rule, the backward propagation of error or higher-order neural network supervised learning algorithms.
- The neocognitron multilayer feedforward perceptron is able to perform both data fusion and pattern recognition using a multichannel optical correlator (holographic interconnections and filters) and an electronic feedback loop to connect the output plane of the thresholding detector with the input plane (SLM). In operation, a two-dimensional input pattern, upon passing through the multichannel correlator, is reduced to a cluster of feature correlation peaks.
- Multiple volume holography (spatial multiplexing) in a single photorefractive crystal and a two-beam coupling or two-arm configuration allows for implementation of mean-field theory, backpropagation and Marr-Albus-Kanerva-style algorithms. A key ingredient for handling rescattering and grating decay is a geometric "temperature gradient" for the output intensities, which can be exploited in realizing a simulated annealing algorithm.
- Volume holography using distributed, angularly and spatially multiplexed gratings resulting from beam fanning or self-pumping can be used to realize a network with error backpropagation. The latter one can be realized electronically.

## 9.2   Optical Implementations of Higher-Order Neural Networks

In section 2.3.4 I have already discussed the suitability of second-order neural networks for optical implementations, especially because of their feature to realize translation-invariant pattern recognition. Thus, higher-order neural networks are an alternative implementation to allow for the solution of nonlinear classification problems. Instead of cascading several single-layer perceptrons with nonlinear thresholding units, the method of higher-order neural networks consists of calculation the higher order correlations and to use them as input to the network (see section 2.3.4). For optical implementations, the simple form of a second-order-only network with the processing algorithm (see also eq. (2.30 in section 2.3.4)

$$y_i = f\left(\sum_j \sum_k w_{ijk} x_j x_k\right) \tag{9.1}$$

can be broken into a series of steps to be performed sequentially. The initial input $x_j$ is correlated with itself to form the autocorrelation matrix $x_j x_k$. The terms of the matrix are then weighted by $w_{ijk}$. The weighted terms are summed and the total is then thresholded to give the final output. The weights are calculated during a training phase, when good examples of patterns to be recognized are presented. As in most of the neural network algorithms, starting from some initial condition e.g. small random values, the weights are iteratively adjusted according to a supervised learning rule, such as the perceptron delta rule, until the training pattern are correctly classified. Unseen patterns will then be classified according to the training pattern they most resemble – a simple learning rule that can be used in the case of second order networks because they are still single-layer networks.

In an optical implementation, it is possible to generate all the terms of the auto-correlation matrix simultaneously. Translation invariance requires that the network is transparent to translated versions of the training pattern. This is implemented if $w_{ijk}$ is the same for all $j$ and $k$ such that $k - j = $ constant [371]. In terms of the two-dimensional matrix this means that all terms on any one diagonal of the matrix should be multiplied by the same weighting. The resulting number of independent weights is reduced from the order of $n^2$ to the order of $n$, and the simple symmetry can be exploited to simplify optical implementation.

### 9.2.1 Cascaded Liquid Crystal Interconnections for Second-Order Networks

P. Horan and coworkers [372] realized several configurations of second order networks using liquid crystal displays that are able to perform on-line learning and updating of weights. Their first realization consisted of a pair of two crossed binary liquid crystal modulators with uniform illumination. Weight multiplication is achieved using a one-dimensional, variable transmission modulator, included in the system along one diagonal direction. The diagonal symmetry and classical optics (lenses) to achieve weight summation are exploited to simplify this part of the implementation to allow the use of a linear rather than a planar modulator, which is technologically much easier to manufacture and to address. The terms from each diagonal are imaged onto the modulator using a cylindrical lens orientated at right angles to the main diagonal, i.e. at 45° to the matrix. A cylindrical lens focuses all the terms on any diagonal to a single point allowing the use of a linear modulator. Positive and negative weights are realized exploiting the symmetric nature of the autorcorrelation matrix of the system. They are instituted on either side of the main diagonal. Where a positive weight is used, the corresponding entry on the opposite side is zero, and vice versa for negative weights. Translation variance requires the same weight along each diagonal. After multiplication the negative and the positive terms are imaged to separate photodetectors, thus implementing summation. The photodetector giving the largest output indicates the classification. This scheme requires that the weight matrix is calculated separately offline. The system operated with an expanded He-Ne laser beam, processing 8 bit words. It revealed to be 100% translation invariant and to be highly tolerant of misalignment.

A second implementation by P. Horan and coworkers, now realized with 20 inputs ($x_j$, binary patterns or bars) and a single output ($y[i = 1]$ in eq. (2.30)), acted as a binary

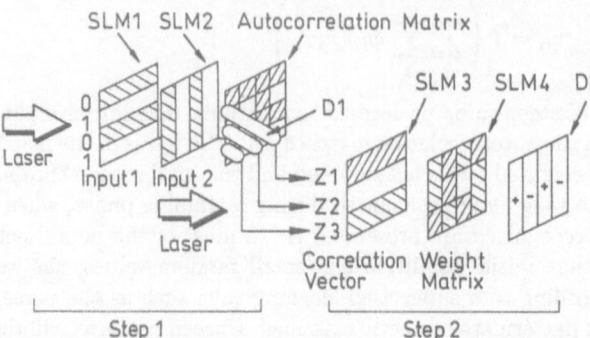

**Figure 9.7:** Optical implementation stages of a second-order neural network with a LCD spatial light modulator and a CCD array (after [4]).

classifier. Spatial light modulators are used to generate the input vector and the weight matrix ($w_{ijk}$). Both function have been combined in a single LCD screen in transmission, using three areas of the screen. The three windows are kept at separate locations of the screen. An input He-Ne beam passes through the first array of input vectors, is then translated and retroreflected by a large right- angle prism to impinge on the second image of the input array, which is rotated through 90° with respect to the first. The net transmission is the autocorrelation matrix $x_j x_k$. The resultant image is then once more translated and retroreflected by a second large prism to pass through the screen a third time to carry out the weight multiplication. This simple optical scheme uses the large area available, minimizing the optics used. Again, positive and negative weights are implemented as described above using the symmetry of the autorcorrelation matrix.

In 1994, S. Kakizaki and P. Horan extended their idea and proposed a system that is capable of multicategory or multiple pattern classification [4]. A scheme of that configuration is shown in fig. 9.7. As in the earlier approach with a single-output system, a pair of crossed binary spatial light modulators (SLM1 and SLM2) with uniform illumination is used to obtain the autocorrelation matrix directly. The matrix is symmetric about the main diagonal and thus gives two parallel results, one of which is chosen to obtain a correlation vector. The diagonal terms are summed with a detector array (D1) to form the correlation vector of the input. In the second step, two analog modulators (SLM3 and SLM4) that can perform gray-scale operation are used to display the correlation vector and the weight matrix. the correlation vector is displayed again, the transmission of each element being proportional to the power detected by the array (D1). With suitable efficient optics, the light from the first stage could also be summed up as in the first realization with a cylindrical lens and passed directly to the weighting modulator. In this case, uniform illumination of the two modulators executes the multiplication of the correlation vector and the weight matrix. Each row consists of the multiplication of a correlation vector by the relevant weights in a conventional vector-matrix multiplier. Again, positive and negative terms are displayed side by side. The result is summed with a detector array (D2), and the subtraction of negative terms from positive terms is executed. Finally, the total is thresholded to give the final output.

Experimentally, the system was realized using a conventional LC-display as shown in fig. 9.8. Uniform light, again from a He-Ne laser, illuminates the input vector, and this pattern is imaged onto input vector 2 in the same plane with lenses and a prism. The resultant image is the autocorrelation matrix, which is recorded with a CCD-camera. The summation as well as the second step of the system, the multiplication of the corre-

**Figure 9.8:** Schematic of the experimental system to realize an optical second-order neural network for multiple pattern classification (upper part). An example of the input vector (a), the rotated input (b) and the resultant autocorrelation matrix (c) (after [4]).

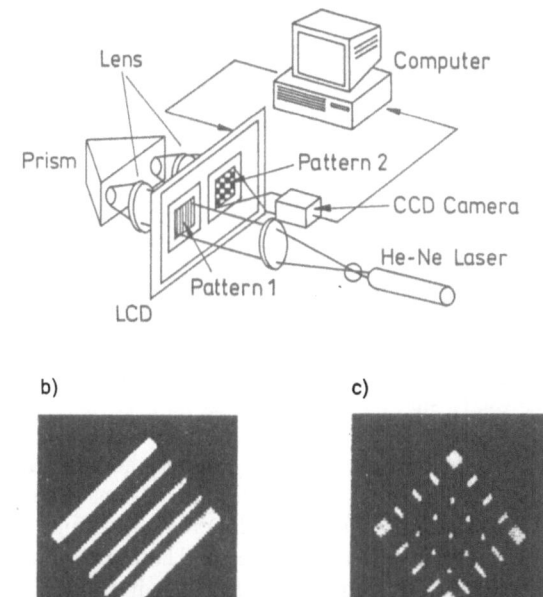

a)          b)          c)

lation vector with the weight matrix, is done electronically. To face the problem of the requirement of high resolution during the training phase, an adaptive training rule was introduced in the system, whereby the optical power is adjusted during training.

In the systems described above, the weighted interconnections are implemented by varying the optical power with liquid-crystal SLM's. Many alternative geometries and devices can be readily envisaged, such as changing the outputs of optical sources or sensitivities of optical detectors, along with various available SLM technologies. Devices such as surface-emitting laser-diode arrays, variable-sensitivity photodetectors, self-electro-optic effect devices with analog function or high-constrast asymmetric Fabry-Perot modulators are capable of implementing gray-scale weighted interconnections for the purposes of second-order and many other optical neural networks.

Thus, a fourth possibility suggested by P. Horan et al. [373] is the use of an asymmetric Fabry-perot modulator (AFPM), operated as an electroabsorptive device that uses the quantum-confined Stark effect in gallium arsenide multiple quantum wells. These devices can be used to carry out the operations of optical input to a network and the calculation of the autocorrelation matrix that is required for any second-order neural network. They have already been described when discussing the possiblities of realizing optical thresholding elements (see section 5.1). The potential of such a system lies in its ability to be integrated easily in fully integrated networks. It has recently been demonstrated successfully for the classification of binary input vectors derived from 12 input neurons [374]. Convergence was achieved in 4 - 8 iterations of the weights when an optimized threshold function was applied. Also, they demonstrated the invariance to translation as expected for the algorithm. Similar systems using optoelelectronic implementation of the interconnection part by two liquid-crystal modulators have been realized by A. von Lehmen et al. [375] and A. L. Mikaelian et al. [376].

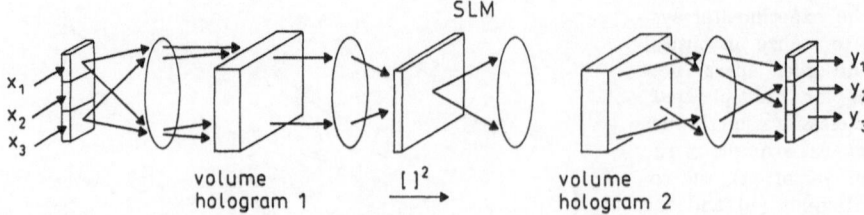

**Figure 9.9:** Optical higher order associative memory implemented with two volume holograms and an SLM that performs the square-law nonlinearity (after [3]).

### 9.2.2 Cascaded Second-Order Network with Volume Holographic Interconnections

Finally, D. Psaltis and J. Hong [3] showed the suitability of second order neural networks to holographic optical implementation, because the autocorrelation matrix as well as the weight calculation are both inherent to volume holographic storage principles as I have discussed in previous chapters. The outer product quadratic associative memory requires three basic components for their implementation: interconnectectivity weights, a square-law device and a threshold nonlinearity. For the implementation of quadratic memories, volume holograms can be used to fully interconnect 2-D pattern to a 1-D pattern ($N^2 \rightarrow N$ mappings) and also the reverse ($N \rightarrow N^2$ mappings). The procedure of recording and reconstruction of the weights is the same as those we have already discussed in section 4.3, because this part of the system is basically a matrix-vector multiplier. By using two of these volume holograms, a complete quadratic associative memory can be realized. The first hologram is prepared with a multiple exposure scheme where for each exposure a memory vector in the one-dimensional input (vector) and one point in the two-dimensional input (matrix) training array are turned on simultaneously. The second hologram is prepared by a similar procedure except that the associated output vectors are recorded in correspondence to each point in the two dimensional training plane. After the holograms are prepared in this way, an input vector is loaded into the one- dimensional input array and the correlations between it and the $M$ memory vectors are displayed in the output plane. In this plane, an optically addressable SLM can be used to produce an amplitude distribution which is the square of the incident correlation amplitudes. The processed light then illuminates the second hologram which servers as a $M \rightarrow N$ interconnection, each correlation peak in the SLM plane reading out its corresponding memory vector and forming a weighted sum of the stored memories on the one dimensional output detector array. This configuration, shown in fig. 9.9, is a direct optical implementation of the system shown in fig. 2.12, with the SLM performing the square-law nonlinearity at the middle plane and the two volume holograms providing the interconnections to the input and output. Thresholding is then performed at the end of the system on the outputs in an electronical or optical way.

### 9.2.3 Cascaded Second-Order Network by Photon-Echo-Interconnections

This two-step interconnection method can also be adapted to the photon echo effect, by using its temporal properties. In this case, the systems consists of two inhomogeneously broadened media, one being a 2-D matrix for the interconnections of the input vector with the first template and the second one being a 1-D array for the second-order multiplication. The principle function of this device introduced by E.A. Manykin and M.N. Belov [377] is shown schematically in fig. 9.10. Suppose that an image $E_i$, $(i = 1, 2, \ldots, N)$, should be processed in the optical quadratic neural network. In this case, the SLM is programmed with the help of the input vector $\vec{E}_i$. This allows an optical signal with the 2-D amplitude distribution $E_i E_j$ to be obtained. Then, this pulse interacts with the nonlinear medium of the $M$ matrix. Due to the preliminary storage for this medium (fig. 9.10b), the temporal sequence of stimulated photon echo signals occurs in the correlation domain. After summation in the transverse plane, their amplitudes have the form

$$S(m) \sim \sum_{I=1}^{N} \sum_{J=1}^{N} E_i^{(m)} E_j^{(m)} E_i E_j = \left( \sum_{j=1}^{N} E_i E_j^{(m)} \right)^2 \qquad (9.2)$$

with the corresponding time constants $t_m = T + t_0 + (M - 1)\tau$, $\tau$ being the separation time between the first and the second pulses of the photon echo sequence and $t_0$ being the separation time between the excitation pulse sequence and the stimulated photon echo signal.

The first pulse amplitude of the echo sequence is equal to unity and therefore omitted in eq. 9.2. The pulse sequence with amplitudes given by eq. 9.2 weights images stored in the 1-D array A (fig. 9.10c). This causes a large number of stimulated echo-responses which propagate into the output plane. The useful response should be detected at the

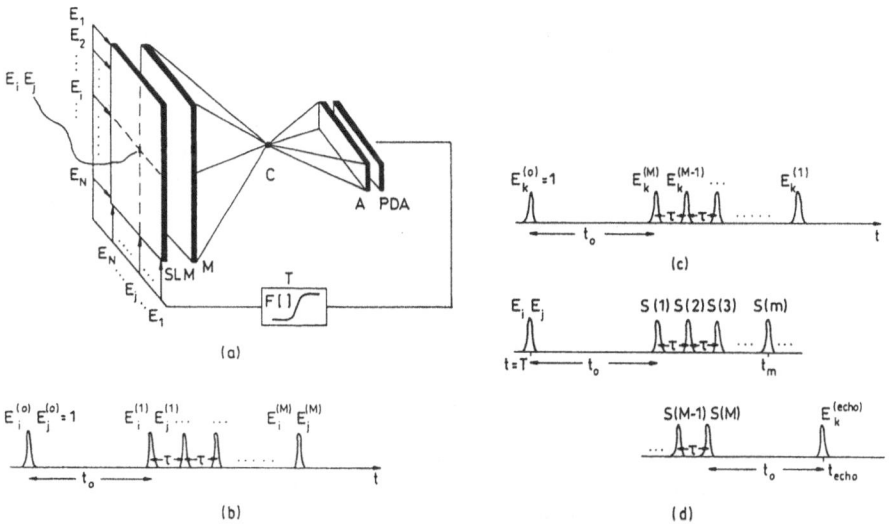

**Figure 9.10:** (a) Optical scheme of the second-order neural network, implemented using the photon echo effect. SLM = electrically or optically adressed 2-D spatial light modulator, M = 2-D storage matrix, A = 1-D storage array, C = correlation domain, PDA = photodetector array, T = threshold device. (b) The pulse sequence for memory storage in the 2-D matrix M, (c) the pulse sequence for memory storage in the 1-D array A, (d) the temporal sequence of optical pulses during the recognition phase. (after [377]).

time instant $t_{\text{echo}} = T + 2t_0 + (M - 1)\tau$. For this case, it is possible to separate the signal of the stimulated photon echo with the amplitude on the time axis

$$E_k^{(\text{echo})} \sim \sum_{m=1}^{M} S(m) E_K^{(m)} \left( \sum_{j=1}^{N} E_j E_j^{(m)} \right)^2 . \tag{9.3}$$

Generation times of other echo-signals differ from $t_{\text{echo}}$. If a threshold device is available in the system, as shown in fig. 9.10a, this system is equivalent to a second-order neural network.

Higher-order neural networks as third-order ones can also be realized optoelectronically using the effects mentioned above, when e.g. a 3-D interconnection is realized by combinatory methods of 2-D matrices or masks [378].

---

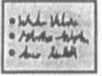 ## Summary

---

Higher-order neural networks are an alternative implementation to allow for the solution of nonlinear classification problems. Their characteristics are :

- Instead of cascading several single-layer perceptrons with nonlinear thresholding units, higher order correlations are calculated and used them as the input to the network.
- The initial input is correlated optically with itself to form an autocorrelation matrix directlyusing e.g. liquid crystal displays, Fabry-Perot modulators, volume holograms or photon-echo devices. The terms of the matrix are then weighted, summed and the total is then thresholded to give the final output.
- The weights are calculated during a training phase as in conventional multilayer perceptrons by presenting examples and adjust the best match.
- They are able to realize translation-invariant pattern recognition because of their correlation approach.

## 9.3    Optical Implementations of Local Backpropagation Algorithms

The big disadvantage of the backpropagation algorithm for optical systems lies in the fact that knowledge is required of the second layer weights in order to realize a backpropagation function that changes the weights in the first layer. As an alternative to that situation, which is often difficult to realize in optics, it is possible to use a local learning rule, in which the weight updates are calculated from the signal at the input of each connection, the signal at the other end of the same connection and a global scalar error signal. For that purpose, an anti-Hebbian learning rule has to be applied (see section refsec:multilayerpercep), resulting still in a descent rule, although not performing a steepest descent to the solution. Although the convergence rate of the backpropagation algorithm is 6 times faster than the Anti-Hebbian learning [5], it allows an easy way of

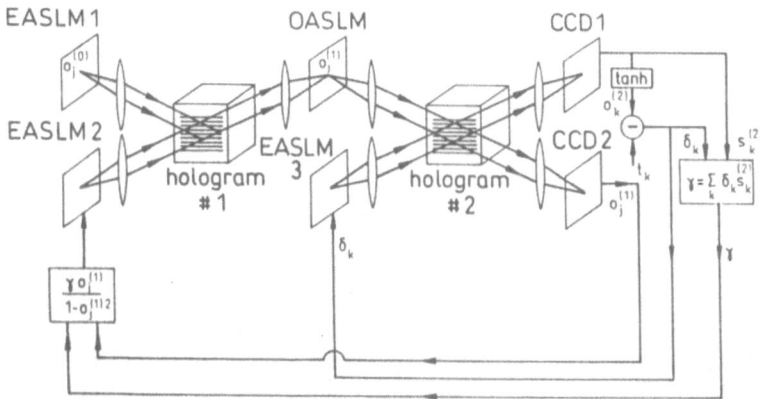

**Figure 9.11:** Schematic of the optical architecture that implements the anti-Hebbian local learning algorithm (after [5]).

realizing optically a descent rule, maintaining the same low error quote as backpropation error systems.

An example of an all-optical realization of local learning using photorefractive interconnections is shown in fig. 9.11. The input images are recorded in an electrically adressable spatial light modulator, and hologram no. 1 interconnects the pixel at the input plane to the pixels at the intermediate layer or hidden layer plane. The nonlinear response of the neurons at the hidden layer is simulated by an optically adressed spatial light modulator. The second layer is similar to the first one, with hologram no. 2 interconnecting pixels from the optical adressable SLM to the output plane, where a two-dimensional CCD detector is placed to detect the light. At the input stage there is a second electrically adressed SLM on which the reference signals that are necessary for the adaptation of hologram no. 1 are recorded. Similary at the hidden layer there is a third electrically adressed SLM to record the reference for hologram no. 2, which is simply the error signal $\delta_k$. This error signal is produced by subtracting the network output from the desired target signal, a point operation that can be accomplished either optically (see section 6.1) or electronically. The reference for hologram no. 1 involves the global error signal $\gamma$ and the response of the hidden layer. $\gamma$ can be calculated with point operations from signals that are already available at the output of the system, but a second CCD detector is needed at the output to record the vector $o^{(1)}$ as it is imaged through hologram no. 2. Once the reference signals are calculated and recorded on the second and the third SLM, hologram no. 1 is exposed to the interference between the signal recorded on SLM2 (SLM3) and the vector $o^{(0)}$ ($o^{(1)}$). This requires that a latching device, such as the microchannel spatial light modulator, should be used as the optically adressed SLM. An exposure schedule (see sections 4.4.2 and 4.4.2) must be used to ensure that each holographic exposure contributes equally to the overall hologram.

The key difference between the anti-Hebbian learning architecture and all-optical backpropagation algorithm architectures as I will also present in one of the next sections, is that the light always travels in the same direction throughout the system. This simplifies the construction and the alignment of the system, and, most importantly, does not require a device at the hidden layer that operates in both operations and has a different response function in each direction. This is a big advantage for optical realizations of backpropagation perceptrons.

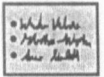   **Summary**

Local feedback to realize an optical multilayer network with a descent rule can be accomplished by a cascade of volume holograms. The reference for the second layer is the error signal, derived from the first layer, thus allowing to adapt the interconnection from the second layer to the output locally. In contrast to nonlocal backpropagation, the light travels in a single direction throughout the system.

## 9.4   All-Optical Realizations of Multilayer Perceptrons

In order to realize an all-optical backpropagation network, one has to ensure that the interconnection can be accessed bidirectionally in both forward and backward directions and that the nonlinear thresholding units are able to realize the shape equal or similar to the sigmoidal function and its derivative. Moreover, the thresholding element should enable to pass back the error signal.

Therefore, volume holographic interconnections that enable read-out from both forward- and backward directions and phase-conjugating elements that are inherently able to pass back a beam from where it came from without multiplying noise and disturbances, are well-adapted candidates for that task.

Their potential for implementation of backpropagation networks was first investigated in 1987 by D. Psaltis and K. Wagner. They were the first to propose several interesting models of all-optical multilayer perceptrons that had a pioneer function and formulated the most important characteristics of volume holography for these purposes. Thus, I will discuss in the following some of these early examples. Although some of these concepts revealed to be difficult to realize with the present technology, the ideas of D. Psaltis and coworkers represent pioneer ideas that are still important for the development of the field.

These examples use multiple coherent optical beams to represent signals, holographic gratings for interconnection (weighting), and nonlinear optical or optoelectronic arrays for processing. In such architectures, opticals signals and weights are approriately described using complex numbers, when both magnitude and phase information are included — resulting in complex optical functions to be processed in these neural networks. In order to consider complex interconnection weights, the phasor model of neural networks [379, 380] has been introduced. It will be described in more detail for the example of a complex Hopfield neural network in section 11.2.3.

### 9.4.1  Modifications of Backpropagation for Complex Optical Fields

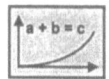 For the case of the backpropagation model, where often the interconnection weight is a complex one, whereas the nonlinear operation in a neuron is a function only of the optical intensity at the neuron — which is the case for optoelectronic processing or the use of all-optical nonlinear processes like a saturable absorber — I will present such a generalization to include complex weights and thresholds based on a derivation by G.D. Little and coworkers [381].

The system described here has $I$ external inputs, $J$ neurons in a hidden layer and $K$ neurons in the output layer. The $i$th external input will be named $y_i$ while the outputs of the $j$th ($k$th) neuron in the hidden (output) layer are $y_j$ ($y_k$). The superscripts $r$ and $i$ are used to index the real and the imaginary parts of a parameter, and a nonsubscripted $i$ is the complex value $\sqrt{-1}$. In this notation, $w_{ij}^r + iw_{ij}^i$ represents the complex interconnection weight between input $i$ and hidden layer neuron $j$ while $w_{jk}^r + iw_{jk}^i$ represents the complex weight between hidden layer neuron $j$ and output layer neuron $k$. I will use the symbol net $j$ (net $k$) to represent the total complex input to the $j$th ($k$th) neuron in the hidden (output) layer, i.e.

$$net_k^{r/i} = \sum_{j=1}^{J} w_{jk}^{r/i} y_j + w_{ok}^{r/i}$$

$$net_j^{r/i} = \sum_{i=1}^{I} w_{ij}^{r/i} y_i + w_{oj}^{r/i}. \tag{9.4}$$

Here, $r/i$ indicates that either the $r$ or the $i$ superscript is to be used consistently in the expression, $w_{ok}^{r/i}$ and $w_{oj}^{r/i}$ weights represent thresholds or offsets. with these conventions, the neuron outputs are

$$y_k = f\left[(net_k^r)^2 + (net_k^i)^2\right],$$

$$y_j = f\left[(net_j^r)^2 + (net_j^i)^2\right]. \tag{9.5}$$

Here, the nonlinearity is described as a two-stage process, the explicit coherent-to-incoherent conversion operation and the as-yet-unspecified function $g$. Clearly, one form of $f$ which is of interest is the sigmoid nonlinearity $f(x) = (1 + \exp(-x))^{-1}$ commonly used in standard backpropagation. However, the first stage nonlinearity allows useful network computation even if a linear function ($f(x) = x$) is used in eqs. (9.5) and (9.5).

Suppose that when the neural network inputs are $y_i$, the desired outputs are $d_k$. Then the weights $w_{jk}^{r/i}$, $w_{ok}^{r/i}$, $w_{ij}^{r/i}$, $w_{oj}^{r/i}$, are to be found such that a quantity $E$ proportional to the sum of squared errors is minimized, where

$$E = \frac{1}{2} \sum_{k=1}^{K} (d_k - y_y)^2. \tag{9.6}$$

If the weights have nonzero values (e.g. due to nonprior training or due to the selection of small random initial values), the problem reduces to finding optimum weight changes. For weights terminating on output neurons, the weight change is given by

$$\Delta w_{jk}^{r/i} = -\eta \frac{\partial E}{\partial w_{jk}^{r/i}} = \eta \delta_k^{r/i} y_j, \tag{9.7}$$

where $\eta$ is a gain parameter and

$$\delta_k^{r/i} = -\frac{\partial E}{\partial net_k^{r/i}} = 2(d_k - y_k)net_k^{r/i}f_k'. \tag{9.8}$$

Here, $f_k'$ represents the derivative of the nonlinearity function as evaluated for the $k$th neuron. For weights terminating on hidden neurons, the change is

$$\Delta w_{ij}^{r/i} = -\eta\frac{\partial E}{\partial w_{ij}^{r/i}} = \eta\delta_j^{r/i}y_i, \tag{9.9}$$

where

$$\delta_j^{r/i} = -\frac{\partial E}{\partial net_j^{r/i}} = \sum_{k=1}^{K}\delta_k^r w_{jk}^r 2net_j^{r/i}f_j' + \delta_k^i w_{jk}^i 2net_j^{r/i}f_j'. \tag{9.10}$$

Results collected from the equations above are

$$\Delta w_{jk}^{r/i} = \eta\delta_k^{r/i}y_j$$
$$\text{with} \quad \delta_k^{r/i} = 2net_k^{r/i}(d_k - y_k)f_k'$$
$$\text{and} \quad \Delta w_{ij}^{r/i} = \eta\delta_i^{r/i}y_i$$
$$\text{with} \quad \delta_i^{r/i} = 2net_i^{r/i}\sum_{k=1}^{K}(\delta_k^r w_{jk}^r + \delta_k^i w_{ik}^i)f_j'. \tag{9.11}$$

A notable feature of these results is that both $\delta_j^r$ and $\delta_j^i$ depend separately on both $\delta_k^r$ and $\delta_k^i$. Note also that the last equations require that the total input to each neuron has to be available during computation of the delta values. This is deemed a relatively minor cost in that it requires additional storage equal to the number of neurons $(I + J)$ rather than the number of interconnections $(IJ + JK)$.

G.R. Little and coworkers realized tests of that complex weight backpropagation algorithm using computer simulations, using a linear function for $f$ and investigating the exclusive-or logic problem. In their simulation, they adapted the weight adaptation to include the momentum term commomnly used in backpropagation (see section 2.3):

$$\Delta w_{ik}^{r/i} = \eta\delta_k^{r/i}y_j + \alpha\delta_{jk}^{r/i},$$
$$\Delta w_{ij}^{r/i} = \eta\delta_j^{r/i}y_i + \alpha\delta_{ij}^{r/i}. \tag{9.12}$$

Small gain values ($\eta \leq 0.05$) are necesary to prevent learning from being unstable. Optimium performance, as measured by learning speed, can be achieved for $\eta = 0.01$ and $\alpha = 0.9$. In comparison with the standard backpropagation algorithm, an order of magnitude improvement in learning speed is gained for the complex weight case. However, such dramatic improvement seem to be not persistent for other choices of algorithm paramters or for larger or more complex problems, but giving also a certain improvement of the standard backpropagation algorithm.

With these mathematical derivations I want to close the introductions to optical back-propagation models, and present several examples how to realize bipolar multilayer perceptrons, using photorefractive interconnection devices and polarization encoding as well as spatial light modulator technology.

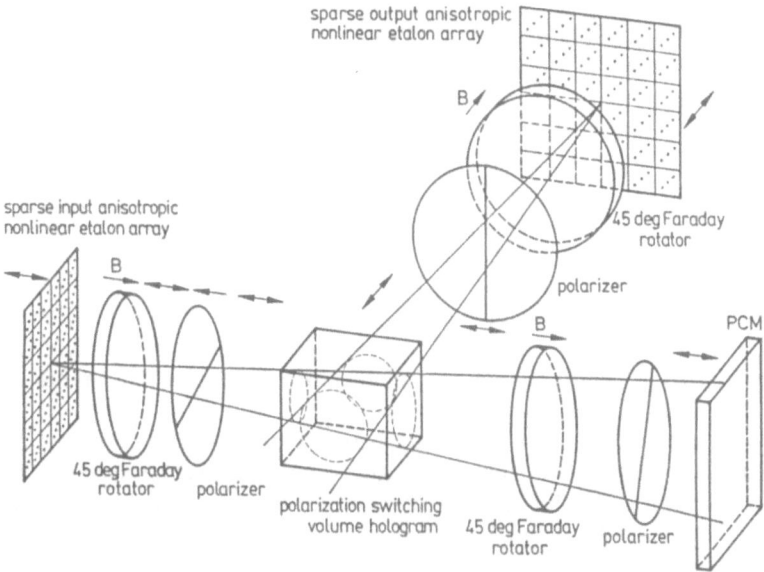

**Figure 9.12:** One layer of an optical backward error propagation architecture with polarization multiplexed forward and backward waves, nonreciprocal polarization filtering and a self-aligning polarization switching volume hologram (after [223]).

### 9.4.2 Polarization-Multiplexed Photorefractive Multilayer Perceptron

The polarization multiplexed multilayer optical learning network is based on the volume holographic interconnection systems described in sections 4.3 and 4.4 and the nonlinear Fabry-Perot thresholding elements presented in section 5.2. To realize proper separation of the two channels in bipolar networks, polarization multiplexed forward and backward waves are implemented, using nonreciprocal polarization filtering and self-aligning polarization switching volume holograms. A single layer of that multilayer network is shown in fig. 9.12.

This system may operate without lenses, because the volume hologram can perform the desired weighted interconnection imaging by exposing it with the proper expanding image and focusing reference beam to form a Fresnel volume hologram. If Fourier lenses are inserted between the etalon arrays and the volume holographic crystal, the exposed hologram will be a Fourier hologram with planar fringes, and the momentum space analysis will be simplified, but the processor learning and self-aligning operations will be similar.

The operation of this model is as follows: The forward propagating pattern vector transmitted by the anisotropic nonlinear etalon array on the left-hand side of fig. 9.12 is polarized at a -45° angle and is rotated clockwise by 45° as it passes through the nonreciprocal Faraday rotator so that it becomes horizontally polarized. This aligns the forward propagating beam with the polarizer allowing it to pass and illuminate the polarization switching volume hologram. The diffracted beam consists of a weighted interconnection of the forward propagating pattern vector by the current state of the holographically represented interconnection matrix, stored as a superposition of curved and chirping space charge gratings within the photorefractive crystal. The diffracted beam is polarization switched by the birefringent diffraction mechanism to an orthogonal polarization to the input, and this vertical polarization state is rotated clockwise by 45°

through the following Faraday rotator so that it falls on the next etalon array with the same -45° polarization as the forward propagating transmitted beam from the previous state. The undiffracted beam passes straight through the volume hologram and has its vertical polarization rotated by 45° as it passes through the Faraday rotator, so that it falls on the phase-conjugate mirror (PCM, see section 3.5) with a polarization angle of 45°, the same as the counterpropagating pump beams (for simplicity they are not shown in fig. 9.12), producing an identically polarized phase-conjugate beam, which is composed of an array of beamlets that are retroreflected towards the etalon sources that each originated from. This phase-conjugate beam passes through the nonreciprocal Faraday rotator picking up another 45° rotation (instead of unwrapping the rotation as it would occur with a reciprocal optical activity based rotator) emerging vertically polarized to act as the reference beam array for the self-aligning holographic outer product exposure with the backward propagating error signal.

The backward propagating error signal emerges from the backside of the output etalon array with a 45° polarization that is orthogonal to the forward propagating beam. This allows for the polarization filtering based separation of the reflected forward propagating beam as well as the independent tuning of the relative Fabry-Perot resonance position of the forward and backward propagating beams. The backward propagating error signal is rotated to a vertical polarization by the Faraday rotator so that it interferes in the volume hologram with the vertically polarized phase-conjugate reference beam and not with the horizontally polarized undiffracted forward propagating signal. The interference of a back-propagating error signal emerging from a particular etalon at the output with the phase-conjugate forward propagating beam emerging from a particular etalon at the input produces a self-aligning volume Fresnel holographic interference pattern that interconnects these two etalons for both forward and backward propagating beams with exactly the same diffraction efficiency, or weight, due to the reciprocity of linear electromagnetic systems. The interference of the backward propagating error beam with the phase-conjugate of the forward propagating beam records a Fresnel grating due to each pair of beams that is present, perturbing the weighted interconnection matrix represented by the hologram by the outer product of the signal and error vectors and thereby pushing the matrix towards the desired interconnection solution. The backward propagating beam is polarization switched by the volume holographic diffraction mechanism, producing a horizontally polarized beam which is the appropriate weighted summation of the error signal by the transpose of the interconnection matrix seen by the forward propagating beam. This passes through the polarizer and is Faraday rotated by 45° to be incident on the etalon array with a 45° polarization angle, orthogonal to the forward propagating beam, and the same as the backward propagating beam which emerged from the previous output layer. The undiffracted phase-conjugate of the forward propagating beam needs to be blocked so that it is not confused with the copropagating diffracted backward error propagation signal, and this is accomplished by the polarizer which blocks the vertical polarization of the undesired phase-conjugate reference beam. The indicated nonreciprocal polarization filtering will also remove the unwanted diffraction terms produced by the hologram. The diffracted phase-conjugate reference and undiffracted backward error signal emerge at different angles and will not focus on the etalon; thus they can be ignored, or they can be examined to determine intermediate states of the hidden neurons.

Each layer is completely compatible with the previous and the following layers. Therefore, this type of learning network can be stacked up to form a complex multilayer learning machine. A complete two-layer system is illustrated in fig. 9.13.

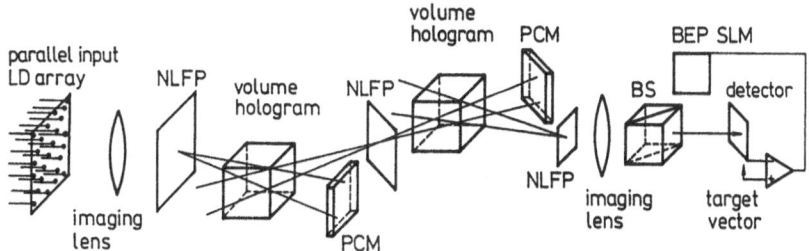

**Figure 9.13:** Complete system for two-layer backpropagation optical learning including a massively parallel input laser array and electronic error detection at the output. The error detection can either be performed optically. LD: laser diode (or fiber optics), NLFP: nonlinear Fabry-Perot etalons, PCM: phase-conjugate mirror, BEP SLM: spatial light modulator for backward error propagation (after [223]).

To realize such a network, a high speed method is required to enter the data for pattern transformation processing and to introduce the backward propagating error signals for the learning phase. The most promising approaches to high speed data entry are sparse parallel laser diode or micro-laser arrays or a fiber-optic input array, demagnified on the first layer nonlinear etalon array which is operated in a bistable regime. In this way, the subthresholded coherent bias beams transmitted by each addressed device can be modulated by the data signals, thereby using the input nonlinear Fabry-Perot etalon array as high speed incoherent to coherent converter with memory. At the final layer of the system, error signals need to be generated (or computed) and injected back into the system with the appropriate wavelength and the phase shift or timing needed to represent the sign of the error. As shown in fig. 9.13, the system can be designed with either optical or electronic error detection and generation circuitry at the output to introduce the backpropagating error.

Optical subtraction techniques can be considered for an optical approach to teaching the system. Image subtraction using a phase-conjugate Michelson interferometer (see section 6.1) or phase-encoded multiplexing (see section 6.1.2) seems to be the most promising approaches for this application since they produce subtracted fields with the appropriate phase shift to represent the sign of the error, without the accurate phase adjustments required by other interferometric approaches to image subtraction. Alternatively, since the computational load required at the output is relatively small, optical detectors can be combined with electronic subtraction from the target vector to generate the bipolar error vector, which can be applied to a spatial light modulator at the output to introduce the backpropagation error. However, in this case, all the problems arising when electrooptic or electronic components have to realize bipolar weight values (see section 4.2.4) have to be taken into account.

Finally, the disadvantages and limitations of that system should also be mentioned. The fan-out capability of each layer is determined by the gain of the nonlinear devices, the holographic diffraction efficiency and the polarization components throughput, and is essential for the information throughput of the architecture. For example, if the product of optical efficiencies is about 3%, a network with 30,000 bit input pattern vectors might be processed by 1000 hidden units that communicate with thirty output devices, which simplifies the error generation process at the output.

The ability of the system to process large amounts of data in parallel at very high speed is limited by the electronic addressing of the input array, and the output photodetector readout time, and not by the intervening optical system because of the extremely fast response that can be achieved using nonlinear etalons and the almost instantaneous optical interconnection delay.

The optical power requirements depend primarily on the requirement of the first layer of nonlinear etalons, which are not bidirectional and can be optimized to have switching energies of several picojoules [223]. However, even this value leads to a power requirement of about 0.4 mW/etalon or 12 W for 30,000 input etalons. Backpropagation neurons will require even more power because the dual functions of the bidirectional cavities. Thus, power supply (e.g. pulsed laser beams), heat dissipation, and interconnection crystal damage intensities are problems arising concerning the point of optical power flow in the network.

Nevertheless, the investigations by D. Psaltis and K. Wagner showed the possibilities that can be envisaged using all-optical multilayer backpropagation networks. Some of the problems of the components of that network even seem to help the whole network to operate: Error driven learning operations, such as backpropagation should be able to compensate for many of the technological faults inherent to an optical implementation by adaptively sensing the misbehaviour of the system and driving it in the appropriate direction necessary to overcome these imperfections. The nonideal optical implementation of a backpropagation network — the backpropagation characteristic is not exactly the derivate of the sigmoidal function (see also sections 5.2 and 9.1.4) — may actually have improved performance over that of an idealized digital simulation because noise will always be present in the system, helping it to avoid shallow local minima (see sections 2.3.3 and 9.1.4) and pushing the interconnection matrix away from solution boundaries. Imperfections of the holographic interconnection and effects of erasure while addressing the system will help the system perform symmetry breaking which the idealized model can not perform spontaneously (compare to section 9.1.4, where I have discussed the influence of grating erasure to backpropagation network performance). The simultaneous self-aligning and learning of the optical system make this approach to multilayer optical neural processing experimentally feasible and allow for the implementation of complicated systems that could not be completely specified a priori but can be learned and modified as the desired processing operation slowly changes. The slow learning of the holographic gratings — actually often cited as one of the big disadvantages of photorefractive interconnections — combined with the extremely high speed processing of the nonlinear etalons gives this system an enormous throughput potential and the capability for solving complicated cogitive problems. Therefore, it is worth being studied in more detail in spite of the experimental difficulties that many of the scientists working in that field are focussing actually.

### 9.4.3   Multilayer Perceptron with SLM Thresholding

A variation of the system discussed above has been presented one year later, in 1988, by D. Psaltis, D. Brady and K. Wagner. The example system they presented consisted of two layers, but an arbitrary number of layers can be implemented as a straightforward extension. The investigation of the appropriate single layer of that network has already been presented in section 8.2 (see fig. 8.4). As shown in fig. 9.14, an input training pattern is placed at plane $N_1$. The pattern is then interconnected to the intermediate (hidden)

layer $N_2$ via the volume hologram $H_1$. A two-dimensional spatial light modulator —
and this is the important difference to the previous model — placed at $N_2$ performs
a soft thresholding operation on the light incident on it, simulating the action of a
two-dimensional array of neurons, and relays the light to the next state. Hologram $H_2$
interconnects $N_2$ to the output plane $N_4$ where a spatial light modulator performs the
final thresholding and produces a two-dimensional pattern representing the response of
the network to the particular input pattern. This output pattern is compared to the
desired output and the appropriate error image is generated, as in the previous model,
in an electronic or optical way, and impressed on the spatial light modulator $N_4$. The
undiffracted beams from $N_1$ and $N_2$ are recorded on spatial light modulators at $N_3$
or $N_5$ repectively. The signals stored at $N_3$, $N_4$, and $N_5$ are then illuminated from
the right so that light propagates back toward the left. The backpropagation algorithm
demands a change in the interconnection matrix stored in $H_2$ equivalent to (compare to
the derivation in section 2.3.1)

$$\Delta w_{ij}^{(2)} \sim \delta_i f'(x_i^{\text{in}}) x_j^{\text{out}} \tag{9.13}$$

where $\delta_i$ is the error signal at the $i$th neuron in $N_4$, $x_i^{\text{in}}$ is the input diffracted onto the
$i$th neuron in $N_4$ from $N_2$, $f'(x)$ is the derivative of the threshold function $f(x)$ which
operates on the input to each neuron in the forward pass and $x_j^{\text{out}}$ is the output of the $j$th
neuron in $N_2$. Each neuron in $N_4$ is illuminated from the right by the error signal $\Delta_i$ and
the backward transmittance of each neuron is proportional to the derivative of the forward
output evaluated at the level of the forward propagating signal. As described above, the
hologram $H_2$ is the outer product of the activity pattern of the activity patterns incident
from $N_4$ and $N_5$. Thus the change obtained in the holographic interconnection is stored
in $H_2$ and is proportional to the change described in eq. (9.13).

The change that is required in the interconnection matrix stored in $H_1$ under the
backpropagation algorithm is

$$\Delta w_{lm}^{(1)} \approx -\sum_i \Delta_i f'(x_i^{\text{in}}) w_{il}^{(2)} f'(x_l^{\text{in}}) x_m^0 \tag{9.14}$$

where $x_m^0$ is the activity on the $m$th input on $N_1$. The error signal applied to $N_4$ produces
a diffracted signal at the $l$th neuron in $N_2$ which is proportional to

$$-\sum_i \Delta_i f'(x_i^{\text{in}}) w_{il}^{(2)}. \tag{9.15}$$

If we assume that during the correction cycle for $H_1$, $N_5$ is inactive, the backward transmittance of the $l$th neuron is proportional to $f'(x_l^{\text{in}})$, the change made to the hologram
by the signals propagating back from $N_2$ and $N_3$ is proportional to the change prescribed
in eq. (9.14).

A critical key element in this architecture is the assumption that the spatial light
modulators at $N_2$ and $N_4$ may have transmittances which may be switched between a
function $f(x)/x$ for the forward propagating signal and $f'(x)$ from the backpropagating
signal. As I have described in the previous section and above, the uncritical behaviour of
the networks relative to the exactness of both functions allows to implement nonlinear
etalon switches or electrooptic spatial light modulators.

Experimentally, D. Psaltis and coworkers could realize in 1988 only the one-layer version of the architecture already described in section 8.2. However, the ease of dynamic
holographic modification of interconnections in photorefractive crystals allows up to date
the successful implementation of a large class of outer product learning networks.

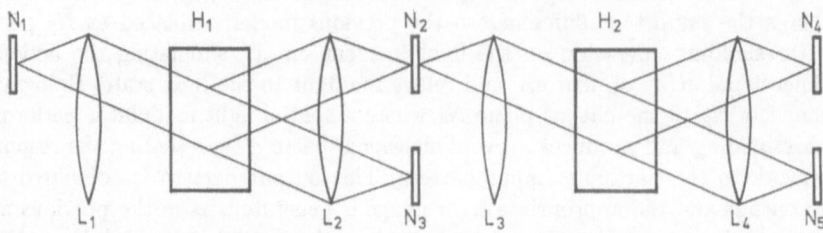

**Figure 9.14:** Optical architecture for backward error propagation learning (after [101]).

Although most of the experimental realizations up to now have renounced to implement all-optical backpropagation algorithms because of stabilization problems in phase-sensitive setups, the models of D. Psaltis have a pioneer function in implementations of all-optical, coherent neural networks using photorefractive volume holographic interconnection and phase-conjugate threshold and feedback units.

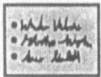 ## Summary

Volume holographic interconnections that enable readout from both forward and backward directions and phase-conjugating elements that are inherently able to pass back a beam to where it came from without multiplying noise and disturbances are well-adapted candidates for multilayer perceptrons with error backpropagation. They can be realized using the following concepts:

- To realize proper separation of the two channels in bipolar networks, polarization multiplexed forward and backward waves can be implemented, using nonreciprocal polarization filtering and self-aligning polarization switching volume holograms.
- Spatial light modulators can perform soft thresholding on the light incident on it, with a sigmoidal forward and a derivative backward threshold characteristic.
- Error calculation can be performed optoelectronically using Fabry-Perot etalons or optically using subtraction techniques from section 6.

# 10 Optical Realizations of Self-Organizing Neural Networks

Among the architectures and algorithms suggested for neural network implementations, the self-organizing map (see section 2.4 and [382]) has the property of creating spatially organized internal representations of various input signals or patterns. It does this by taking the spatial neighbourhood of the cells into account during learning. In its basic version a cell or a group of cells becomes specifically tuned to various input patterns or classes of input patterns through an unsupervised learning process. The spatial location of a cell in the network corresponds to a particular set of input patterns. The spatial clustering of the cells and their organization into topologically related subsets then result in a high degree of classification efficiency. This is particularly relevant in cases in which no a priori information about the scaling of the problem is known, a fact that argues strongly in favour of the use of an optically addressable, continuous adaptive medium for implementation of the neural activity.

In the previous sections, I have described various optical implementations of different neural net architectures with prescribed learning procedures. However, in many cases, the actual, technical limitations of optics become apparent when one tries to implement experimentally realizations of complex learning algorithms as e.g. the error backpropagation rule. In that case, the all-optical loop is often opened to introduce electrooptic devices or digital computers for performing basic tasks such as weight updating and decision making. In most cases this introduction is done at the expense of destruction of the former coherent nature of the interconnection and severe reduction of the adavantages of speed and high parallelism of the systems. However, the category of resonator memories (see section 6.5 for details on these memories), involving a three-dimensional optical storage medium, e.g. volume holographic elements, escapesthis problem. Corresponding architectures were already proposed as early as 1963 by P.J. van Heerden [100] to simulate information recording in the brain. Most of the optical models that realize Hopfield-like models take advantage of these ideas of resonator memories. I will discuss several of these approaches in the next chapter. Moreover, some of the few concepts of all-optical realizations of multilayer perceptrons with error backpropagation have exploited the features of resonator memories, as e.g. the model of D. Psaltis desribed in the previous chapter. Also, some of the models of associative memories are based on resonators that implement photorefractive or phase-conjugating nonlinearities (see chapter 7). These implementations emphasized the volume storage capability and the gain selection of photorefractive materials, relating them naturally to the category of analog self-organizing systems.

In this chapter, I want to discuss several different models to realize self-organization, among them mostly implementations using photorefractive volume holograms in resonant configurations, but also alternative approaches based on bistable, optically adressed liquid crystal spatial light modulators.

## 10.1   Electrooptic Competitive Learning and Self-Organization Systems

For optoelectronic realizations of winner-takes-all or competitive learning systems, optically addressed ferroelectric liquid crystal spatial light modulators are used for interconnection as well as for winner-takes-all selectors. However, in many cases, the interconnection device is realized using a volume holographic memory, whereas all fedback and thresholding functions are implemented in an electrooptic way. With liquid crystal or conventional holographic storage devices, information is not engraved in the volume, as in the case of photorefractive materials, but in a plane. The drawback of a reduction in storage capacity may be partially compensated by a better control of the memorizing and learning procedures. Thus, real Kohonen-like self-organizing networks can be constructed.

Here, I will first introduce different all-optical maximum element or winner-takes-all selectors for implementation in self-organized Kohonen networks. Then, I will focus my attention to the realization of a complete all-optical Kohonen networks implementing self-organization, competitive learning and winner-take-all modules using liquid crystal spatial light modulators.

### 10.1.1   All-Optical Maximum Element Selector for Competitive Learning

The closest-vector selection necessary for Kohonen maps can be viewed as a matrix-vector multiplication, followed by a maximum or minimum selection. For a matrix-vector multiplication, one of the schemes presented in section 4.3.3 can be used. The more important part of the system is the operation of the maximum element selector, which has several steps. If the optical intensity is chosen to be the measure of magnitude, the task is to select the brightest spot of a set of given light spots. This can be accomplished in a very conventional way using linear optics by means of a gray wedge, a spatial thresholding detector and devices for optical logic manipulations in combination [383].

Let the input to the system be an array of horizontal sources of different intensities. These sources may correspond to the output of a matrix-vector multiplication. The light is passed through an elliptical lens to spreach it out vertically and to image it horizontally. The vertical streaks thus formed are applied to a gray wedge, a device whose transmission increases monotonically with distance from bottom to top. The output of the gray wedge is a set of vertically streaks whose intensity decreases with height. Next to the gray wedge is a point-by-point spatial thresholding device. This can be accomplished using spatial light modulation or optical flip-flops (see e.g. section 6.3). Past the threshold device, the more intense spots create longer streaks, the longest streak corresponding to the brightest spot. Now the selection of the brightest spot can be accomplished. To isolate the longest line, the pattern are optically smeared horizontally, using again an elliptical lens. Then, the smeared pattern is shifted upward by one image element, and is combined with itself using a dot product operation. This can be performed using a semitransparent mirror in combination with a polarization-selective superposition. A single spot then remains at the location of the tip of the longest streak. The horizontal position indexed the location of the brightest spot, and the vertical position describes the magnitude of the intensity.

**Figure 10.1:** Competitive optical learning network using the transmission from the competitive etalon network as the holographic reference beam for learning the appropriate interconnections (after [110]).

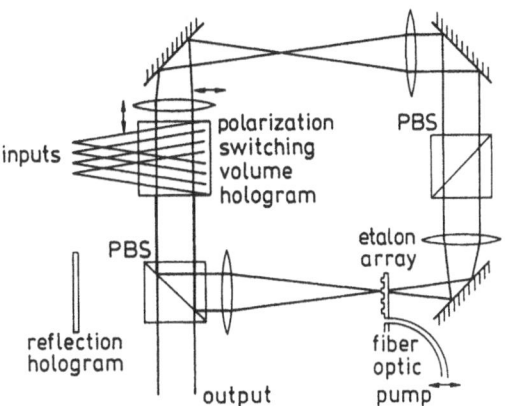

## 10.1.2 Competitive Learning with Fabry-Perot Etalons

The idea of using a dynamic volume hologram for interconnection weight storage and learning in a self-organizing network, a very-large-scale-integrated (VLSI) liquid crystal spatial light modulator (LC-SLM) to perform the optical winner-takes-all neuron function in combination with linear feedback has already been proposed in 1987 by K. Wagner and D. Psaltis [110]. Photorefractive phase conjugation to guarantee proper feedback in the configuration has been proposed and partially realized by K. Wagner and T.M. Slagle in 1993 [6]. It is a further development of the idea of using polarization switching holograms for the realization of backpropagation neural networks (see example in section 9.4). The first prototype of such a system using unidirectional, conventional feedback with dielectric mirrors is shown in fig. 10.1. As I have discussed already above, the neurons within a single competitive patch must be inhibitorily interconnected in order to produce a dynamical system that converges to the winner-takes-all output that implements the desired maximum operator without resorting to a sliding threshold. Thus, negative interconnection must be implemented somehow. From the concepts of bipolar weightening discussed in section 4.2.4, we know that destructive interferometric detection is one way to realize negative interconnection values coherently, with each of the feedback interconnections beeing out of phase with the reference beam. Although the coherent nature of this process makes it well suited for phase-preserving, coherent implementations of optical neural networks, it is actually regarded for many robust applications as not being stable and practicable enough. Here, alternative incoherent approaches to inhibitory interconnections have to be applied.

One approach is to use an optical neuron with an inverted sigmoidal transfer characteristic, so that an increase of incident optical intensity produces a saturation decrease of the output intensity. This type of negative differential threshold characteristic can be obtained over a limited operating regime from a properly detuned nonlinear etalon. It can be used to implement an optical flip-flop by biasing two single-valued nonlinear etalons up to the regime at which negative differential reflection begins and feeding the reflected output of each device into the input of the other. In a certain region, two stable states of the two etalon flip-flop can be achieved, corresponding to one of the devices producing a large reflection which pushes the other device into the low reflection regime, which is consistent with the original device producing a large reflection of the bias beam.

However, the etalon that started with the largest optical input is pushed further along the negative differential reflected intensity curve producing a smaller reflection, and becoming — at stability — the device with the lowest reflected intensity. For a two-neuron competitive network this is just the opposite of the behaviour desired, the neuron with the largest input has won the competition and thereby produces the smallest reflected output. This is a consequence of the fact that the interconnections are actually positive and an inverting sigmoidal reflection transfer characteristic is utilized. This situation can be inverted by using the transmitted output of the pair of competing etalons as the competitive network output and noting that when the reflection is low, the transmission is high. The winning etalons, which had a low reflection, will have a high transmission and conversely the losing etalon, which had a high reflection, will have a low transmission, which is of the appropriate form for the output of a competitive network. The same form has already been discussed for the output of photorefractive operations using multiple beam interaction (see section 3.4.4). The advantage of such a two-neuron competitive network in contrast to coherent photorefractive multibeam coupling is that only positive interconnections are involved, which can be more easily achieved with a number of incoherent feedback techniques.

A multidimensional extension of this flip-flop behaviour can be constructed by mutually interconnecting the reflections from an array of $N$ etalons, as long as the fan-out of the reflection from each device is large enough. One possible arrangement is a Fourier transform reflection hologram that is fabricated with a space invariant impulse response so that each point source is interconnected to $N$ etalons on either side, but not to itself (see also fig. 10.1). An array of $N$ etalons is placed in a resonator structure with the etalon in the zero position replaced by an optical point source (such as a fiber optic feed through). The light from the fiber is interconnected via the reflection hologram to provide a bias beam for each etalon in the array and a polarizing beam splitter cube does not reflect any of this polarization. The reflection produced by each etalon acts as a point source that is recollimated by the lens, and is incident on the hologram in order to produce $2N$ reflected plane waves, which are focused onto $N$ spots on either side of the initial etalon, thereby providing feedback for the other $N - 1$ etalons and waisting the power in $N + 1$ beams. Meanwhile, the development of space variant interconnection schemes has proceeded a lot (see section 4.1), so that more efficient schemes can be derived using thick holograms.

The operation of this competitive resonator is initiated by sending an orthogonal polarized array of collimated input beams into the side port of the beamsplitter, so that it is reflected into the resonator and focused on the $N$ etalons. At this point the fiber optic pump is turned on and the competition begins. The pump beam provides enough power to bias all $N$ etalons up to the point that negative differential reflection begins and the etalon with the largest orthogonally polarized input is pushed the furthest towards the minimum reflectivity. The reflection from the winning etalon is small and does not push the losing etalons towards a smaller reflectivity as well as the losing etalons push the winning etalon towards a low reflectivity.

The competitive etalon network described above can be used as the basis of a competitive optical learning network. An input array of coherent light sources is interconnected through a polarization switching volume hologram to produce the orthogonally polarized inner product array which reflects off the polarizing beamsplitter into the competitive resonator to initiate the competition. In this unsupervised learning procedure the hologram is initially in a randomly exposed, saturated diffraction efficiency state, so the diffraction onto the etalons is essentially random, but for a given input one of the etalons must win the competition and produce a transmitted beam. This transmitted beam is imaged around an optical loop with a four lens (8f) noninverting imaging system exactly back onto itself, except that the final loop mirror is the polarizing beamsplitter within the competitive resonator, so the wrongly polarized light is actually transmitted by this beamsplitter and does not reilluminate the etalon array. However, within the hologram the transmitted beam from the winning etalon produces a collimated plane wave that is collinearly polarized with the input beam, so it will expose a holographic interconnection grating that will strengthen the interconnections to the winning etalon. The next time that the same input pattern is presented to the network, the polarization-switched diffraction to the winning etalon will even be larger and it will be even more likely to win the competition. Cycling through the set of input patterns will result in different statistical pattern clusters becoming tuned to different output neurons through holographic adaptive filtering in a completely unsupervised fashion.

Up to day the system that was proposed already in 1987 has revealed to have several difficulties to be realized experimentally, but its important ideas are still valid. The most difficult problems the system has to focus on are the effects of weight vector explosion or domination by a single etalon, which makes normalization effects necessary. Fortunately, a lot of unavoidable artifacts in the optical system and the forgetting characteristics of photorefractive holographic gratings due to their dark conductivity allowed the system to saturate. However, the fact that previously written gratings can be erased by new registrations or adaptations, is an additional, unwanted effect that has to be taken into account (see also section 6.4). These limitations have finally led to a lot of improvements in the feedback technique and the winner-takes-all etalons and will be discussed in the following section.

### 10.1.3  Competitive Learning with Liquid Crystal Winner-Takes-All Modulators and Photorefractive Phase Conjugation

In order to understand the operation of the modified unsupervised optical learning machine using optical phase conjugation, let us trace again the flow of optical signals through the system in fig. 10.2. Input patterns are applied one at a time to the input SLM. The modulated light illuminates the hologram, which is erased initially and only scatters a little of the light randomly through photorefractive beam fanning (see section 4.3.5). Some of the light is diffracted by the hologram and passes through the polarizer towards the array of reflective winner-takes-all modulators. The winner-takes-all array is subdivided into several competitive patches of different size. The units in the arrays are arranged on a sparse-grid topology in order to eliminate crosstalk from Bragg degeneracy (see section 4.3). Each unit consists of a reflective modulator placed directly on top of a photodetector that detects a small fraction of the incident light for use in the electronic decision-making operation. The winner-takes-all electronic circuitry causes the units with the largest inputs in each competitive patch to switch the liquid crystal above that unit's detector or modulator to the nonpolarization rotating state, while the remaining units rotate the polarization of the reflected light by 90°. As a result, the reflected light from the losing units is blocked by the polarizer, while the light reflected from the winning units passes back through the polarizer and reilluminates the volume hologram. Meanwhile, the undiffracted light from the input SLM that has passed through the volume hologram records a dynamic grating in the PCM, which is read out by a strong counterpropagating pump

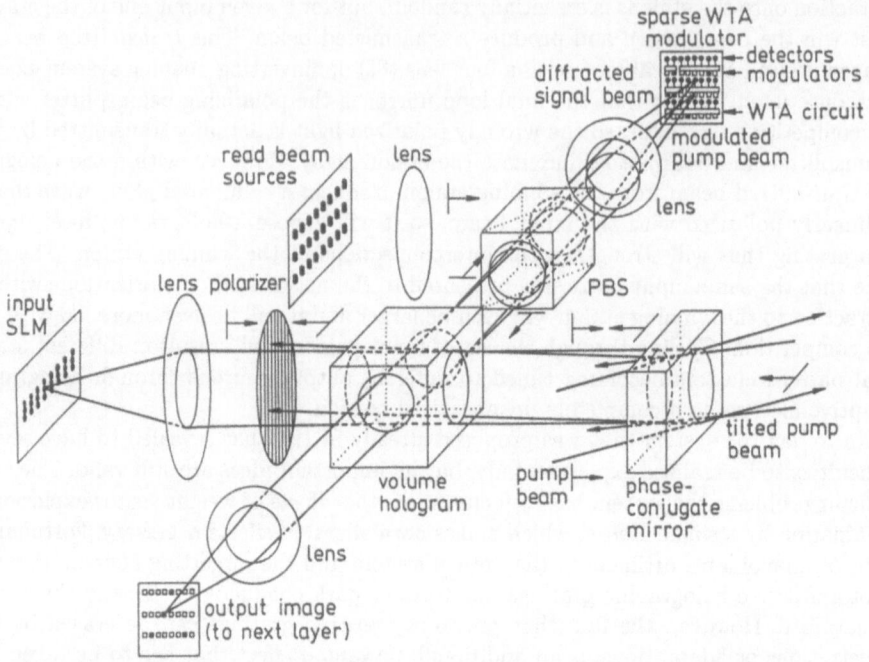

**Figure 10.2:** Fourier-hologram competitive learning architecture with the inclusion of spatially separated modulators and detectors on the winner-takes-all device in order to achieve winning-unit power normalization and gain (after [6]).

beam. The diffracted pump produces a phase-conjugate wave that focuses back through the volume hologram towards the input SLM pixels from which it originated. Inside the dynamic volume hologram the phase-conjugate wave and the reflections from the winning pixels interfere, adding a grating perturbation to the existing hologram. This performs the outer-product weight matrix update required by the competitive learning algorithm, strenghtening the interconnections between the input pattern and the winning pixels in each competitive patch. The next time that a similar pattern is applied to the input SLM, it is even more likely to produce the largest diffraction towards the same winning node locations. After cycling through a statistically clustered set of input patterns many times, individual nodes in the different competitive patches become tuned by the different clusters through the adaptation of the holographic weight interconnection. Since the number of clusters in the input statistical distribution is unknown, many different sizes of competitive patches should be included in the sparse competitive winner-takes-all device. The different-size patches produce clustering of the input-pattern space with different levels of detail resulting in the emergence of different types of feature detectors.

There are several difficulties in that setup concerning the realization of the winner-takes-all unit. First, the reflected light by the winner-takes-all unit is initially very weak. Therefore, as shown in fig. 10.2, the winner-takes-all unit may be read out by strong uniform read beams rather than by reflecting the incoming light. This can be accomplished by adding a polarizing beamsplitter to reflect an array of vertically polarized read beams upon the modulator array and to black any reflected losing beams from reilluminating the hologram.

When the winner-takes-all devices are used with the detector under the modulator, the read beams must be perfectly uniform, and the winner-takes-all detectors must have identical sensitivities. Otherwise, the additional light from the read beams could disrupt the operation of the winner-takes-all units. Realistically, the read beam illuminating the LC modulators must be separated from the winner-takes-all detector inputs in either time or space. Temporal multiplexing using the technique of time-sequential gain [384] (compare to section 4.4.3 for methods of temporal recording) can be used to maintain the self-aligning property of the system. With this technique, the winner-takes-all circuits first select the maximum based on the weak input light diffracted from the hologram. Then the winner-takes-all outputs would be latched elelctronically, and a strong read beam would be turned on to read out the state of the modulators, giving an effective gain in the modulators outputs. The modulated read beams would then interfere with the phase conjugate of the input to record the appropriate incremental perturbation to the hologram.

Alternatively, the detectors and modulators can be separated spatially, as shown in fig. 10.2. In this case the strong read beams illuminating the modulators do not produce any photocurrents in the detectors, so they do not affect the decision making of the winner-takes-all circuits. However, the self-algining character of the learning is then destroyed because of the lack of spatial coincidence of the detectors and the modulators. Proper operation can here be regained by producing a phase-conjugate of the input beam with a vertical tilt equal to the angular separation of the modulators and their corresponding detectors [236, 299]. This idea requires an accurate adjustment of the waves to ensure proper interference of the waves in the volume hologram. A detailed description of the function and inconveniences of that method can be found in [6].

After learning converges, input of a particular pattern produces a sparse pattern of activity across the array of competitive patches, with only one unit on each. This distributed representation can be used by subsequent layers of optical processing for further categorization. Multiple layers of these clustering networks can be cascaded to produce an unsupervised working classifier that is able to solve nonlinearly separable decision problems. However, there are indications that an even more promising technique would be to follow a first layer of competitive learning with a second layer of a supervised single-layer learning network such as a perceptron and to use the competitive learning system only as an unsupervised feature recognizer. Such a combination of an unsupervised clustering layer and a supervised classification layer permits the computation of nonlinearly separable decision surfaces without requiring the backpropagation of error signals through the system.

When the input to the second stage of a multilayer network is the output from a previous competitive learning layer, the phase-conjugate beams could propagate back through the hologram to the input modulators and affect the detectors under the modulators from which they originated. For a winner-takes-all detector or modulator this could result in latching of the winners in the on state owing to the feedback from the phase-conjugate mirror. In order to block the phase-conjugate wave from reaching the input, it is possible to use polarization multiplexing, where the phase-conjugate reflection is produced with a polarization orthogonal to the forward-proapgating beam. This can be readily accomplished using orthogonally polarized pump waves. Because these phase-conjugated beams should interfere with the reflection from the winning modulators it is necessary that they have the same polarization. This requires that the volume hologram in turn is operated in a polarization-switching diffraction mode (see section 9.4). In optically isotropic polarization-switching holographic materials this mode of diffraction is compatible with self-aligning holographic learning. However, for optically anisotropic materials — as are the most appropriate photorefractive materials for learning with long storage times — polarization-switching diffraction presents some phase-matching difficulties when using large space-bandwidth images.

To summarize, I have shown that the combination of a dynamic volume hologram as the adaptive interconnections between an input spatial light modulator array and a winner-takes-all reflective modulator based on liquid crystal spatial light modulators is suitable to realize learning processes without the aid of a teaching algorithm. The hologram is exposed repetitively to the phase-conjugate of the input and to the reflections from the winner-takes-all spatial light modulator, and this exposure perturbs existing interconnections in a way analogous to an outer-product update rule.

### 10.1.4  Self-Organizing Maps Using Gray-Level-Based Liquid Crystal Devices

Although ferroelectric liquid crystal binary and optically adressable SLM's are intrinsically bistable devices, by operation near the threshold voltage and by use of spatial integration techniques, they can be made to show a gray-level behaviour [385], resulting from the phenomenon of multidomain switching of the ferroelectric liquid crystal. In this manner the value of the gray level may be stored, increased or decreased and easily thresholded, thus allowing to use these SLM's for self-organization realizations. Additionally, by variation of the SLM's electrical drive parameters as amplitude and duration of the applied voltage pulse, the apparent sensitivity of the device can be influenced. In this way it is possible to vary the learning rate during the learning phase. The third function to be realized, the neighbourhood function, can be realized in the case of simple local neighbourhood using the fact that neural firing can be dilated (or eroded) by an increase in the write voltage of the storage SLM. Combination of this neighbourhood operation with the operations of accumulation and subtraction leads to an original implementation of the Kohonen activation and inhibition neural functions.

The optical setup resulting from that concept is shown in fig. 10.3. The setup can be described instructively if we consider the forward and the backward phase separately. In the forward phase, the formation of the neural activity takes place. The inner product is implemented by means of crossbar switching with a lenslet array and a collimating lens. The transparent pixels of the input pattern select the corresponding weight maps. The neural activity results from the spatial summation of these activated weight maps. Each selected weight map of the array of weight maps is imaged onto the neural map (materialized by the storage SLM) in the focal plane of the collimating lens. For better control of the internal states of the neural network, the neural firing is memorized on the thresholding device. In the backward phase, the weight updating is performed. The operation to be realized is the Hebbian outer product learning. The theoretical solution to this learning problem consists of implementing the inverse function of interconnection network. This can also be carried out optically by use of a 4-f confocal configuration with phase-conjugate mirrors. Similar systems for associative memory applications has been presented in sections 7.2 and 7.4. Although this solution is theoretically very satisfying, because it guarantees a perfect correspondence between the neural map and the weight maps and moreover limits alignment problems, the introduction of a nonlinear decision step (see chapter 5) considerably complicates such an implementation of the inverse

**Figure 10.3:** Experimental setup to realize an optical implementation of a Kohonen self-organization resonator scheme (after [386]).

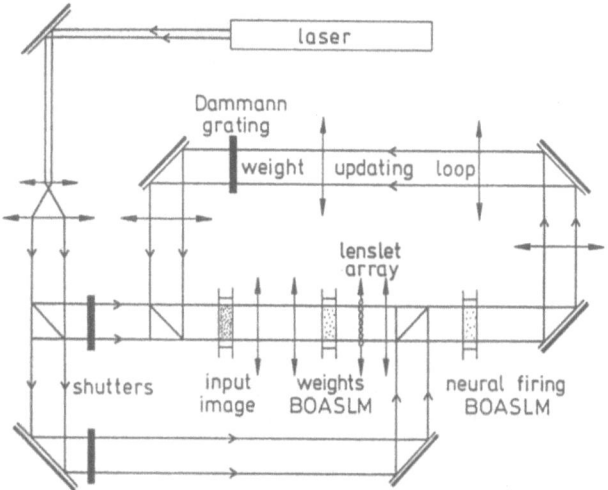

operation. The reasons are that — if one wants to realize a phase-preserving, coherent thresholding — phase restoration is very sensitive and difficult to achieve experimentally or that — if one stores only the intensity in the device, — the optical path is not entirely retrievable.

The solution proposed by J. Duvillier et al. [385] to this problem consists of duplication of the neural firing with a holographic Dammann grating and projection of the duplicated images through the input pattern onto the array of weight maps that contribute to creating the neural activity. In this way the duplicated images are modulated by the input pattern $x_{ij}$, and only the activated weight maps are modified. The resulting optical setup, shown in fig. 10.3, is based on the principle of a ring architecture. Experimentally, however, this simple approach leads to a problem. If the weights are only increased and never decreased, they soon become saturated. In the original self-organizing map algorithm (see section 2.4), the weights are renormalized to prevent saturation, but such a global renormalization is difficult to perform experimentally in optics. However, there are other possibilities to avoid possible weight saturation, e.g. by using an extension of the algorithm that weakens weights maps that have not been activated. The most popular methods to realize this bipolar weightening have been discussed in section 4.2.4. J. Duvillier and coworkers [386] used the method of subtraction of a supplementary set of weights by changing the voltage of the SLM. The selection of the activated and unactivated weight maps is determined by alternately complementing the input SLM. First, the input pattern is displayed, and the neural firing is added (reinforcement) to the active weight maps through the input pattern, as described above. Second, the complementary pattern is displayed, and the neural firing is then subtracted (weakening step) from the unactivated maps in the same way. The expected consequence of this procedure, at the next iteration, is the emphasis of the link between the pattern and its neural firing. In this manner the architecture preserves a spatial correspondence between neighbourhoods on the neural map and the weight maps. In fig. 10.4, the principle effect of lateral inhibition by use of voltage control is shown. Although the system is a possible alternative to actual models using photorefractive materials, they suffer from several limiting factors. The discriminating capability of the system is strongly related to its ability to

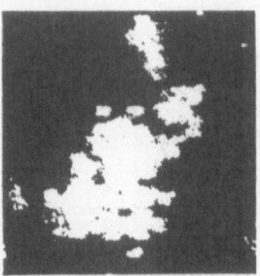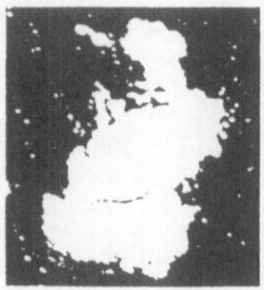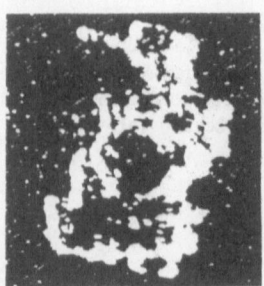

**Figure 10.4:** Photographs showing the principle effect of lateral inhibition by use of voltage control: (a) close-up of an updated weight map showing a single replication of the neural map carried out at a voltage $V1$; (b) result of the updating operation at voltage $V2 > V1$; (c) result of an updating at $+V2$, followed by an update at $-V1$ with $|V2| > |V1|$ (from [386]).

address fine details on the weight map. i.e. details close to the domain size. This fact requires that the optical system is able to transmit such fine details. If this requirement is not met, the patterns become increasingly indistinguishable. Other limiting parameters such as the presence of noise and device inhomogeneities can also be considered as capacity-limiting factors. However, the major problem of such a solution is related to the correspondence (mainly of scale and position) between the imaging performed by the lenslet array and that performed by the Dammann grating. Particular attention must be paid to the quality and regularity of the lenslet array. Out of these reasons, J. Duvillier et al. [385] realized experimentally only local neural inhibition or activation functions and the neighbourhood function as single units, but not a complete learning cycle that combines the basic operations.

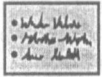 **Summary**

A self-organizing neural network has the following properties:
- It creates spatially organized internal representations of various input signals or patterns.
- It does this by taking the spatial neighbourhood of the cellls into account during competitive or winner-takes-all learning.
- Optically, this can be realized using resonant configurations combined with optically addressed spatial light modulators.

Winner-takes-all or competitive learning for maximum selection as one of the crucial functions of such a self-organized net can be realized using
- a gray wedge, a spatial thresholding detector and devices for optical logic manipulations in combination.
- nonlinear Fabry Perot etalons that enable to realize an inverted sigmoidal transfer characteristic, so that an increase of incident optical intensity produces a saturation decrease of the output intensity. This type of negative differential threshold characteristic can be obtained over a limited operating regime from a properly detuned nonlinear etalon.

- Fourier-hologram competitive learning architecture with the inclusion of spatially separated modulators and detectors on the winner-takes-all device in order to achieve winning-unit power normalization and gain.

A complete electrooptic Kohonen feature map-like neural network can be realized with two spatial light modulators, one of which is devoted to the information storage and the other one to the decision-making unit. Its main features are:

- Gray levels may be stored in a liquid crystal binary and optically addressable SLM near the threshold voltage using spatial integration techniques.
- By variation of the SLM's electrical drive parameters (amplitude, duration of the applied voltage pulse), the apparent sensitivity of the device can be influenced and thus the learning rate can be varied during the learning phase.
- The local neighborhood function can be realized using the fact that neural firing can be dilated or eroded by an increase in the write voltage of the storage SLM.
- Combination of the neighborhood operation with the operations of accumulation and subtraction leads to an original implementation of the Kohonen activation and inhibition neural functions.

## 10.2 Coherent Self-Organized Kohonen Networks

If we identify the adjustable interconnection network with the index grating in the volume of the photorefractive gain crystal which adaptively connects spatially complex input signals with an optical feedback that realized the winner-takes-all or competitive behaviour, the result is a dynamical evolution of the input pattern that implements the functionality of Kohonen's algorithm. Therefore, self-organizing memories are ideal candidates to be realized with resonator memories involving a three-dimensional storage medium as a photorefractive volume hologram. These implementations generally rely on the volume storage capability and the gain selection of the material, relating them to the category of analog self-organizing systems. In these configurations, the different resonator modes correspond to the different neural firings produced by the various classes of input patterns. Here, self-organization takes place naturally as the modes of the resonant configuration compete in order to learn from a set of patterns presented at the network input. The aim of the algorithm is to build a spatial representation such that two similar patterns correspond to two related modes.

I will discuss in the following sections systems using photorefractive interconnection units, that are based on coherent optical feedback in photorefractive ring oscillators.

### 10.2.1 Topological Map by Photorefractive Self-Organization

As an electrooptic implementation of photorefractive self-organized learning, T. Galstyan and coworkers demonstrated experimentally feature extraction in a set of images using photorefractive beam coupling and electronic feedback [387], . Consider a photorefractive recording scheme using angular multiplexed reference beams. If the amplitudes of the signal beams are made proportional to the neuron input vector components, the diffracted amplitudes of the memory represents the neuron outputs, whereas the index modulations in the volume holographic memory are proportional to the components of the weight vectors. In order to implement an adaptive neuron, these index modulations must be adapted according to the value of the neuron output by using a feedback loop. With such a feedback system, the different adaptation laws (see section 2.4, Box 5) can be implemented using simple hologram formation in photorefractive materials. In a variation of the formulation used for the Kohonen map, a general vectorial equation can be formulated for the weight adaptation rule:

$$\Delta \vec{w} = (ay^q \vec{x} - by^p \vec{w}) \cdot \Delta t \tag{10.1}$$

The potential and performances of the neurons considerably differ according to the values of the parameters $p$ and $q$. For $p = q = 0$, this adaptation law makes the weight vector $\vec{w}$ become proportional to the "moving average" of the input vector $\vec{x}$ according to T. Kohonen. Thus, this learning law does not show any interesting self-organizing characteristics. The situation changes completely for $p = q = 1$. Because the learning rate is proportional to the output $y$ through the term $ay\vec{x}$ in eq. (10.1), the neuron becomes selectively adapted to one of the input vectors if they are orthogonal. Furthermore, it decreases the scalar product for other input vectors through the forgetting term $-by\vec{w}$. This learning is a competitive one known as *winner-takes-all* behaviour.. If the set of input training vectors contains subsets of correlated vectors, the system can extract their common features and adapts following vectors to these general features. Finally, for $p = 2$, $q = 1$, the learning law also leads to adaptive characteristics such as feature extraction and competitive learning. However, the norm of the weight vector is now independent of $x$, which is known as weight vector normalization (see section 2.4). In eq. (10.1), there is no threshold adjustable with time. Thus, competitive learning also exists after the neuron has been sensitized to one specific input vector. If this vector is then suppressed from the set of input vectors, the neuron will learn to recognize another input vector. This new learning will force the neuron to forget the previously learned vector. These characteristics of *active weight forgetting* are maintained when the input vectors are presented in a random sequence.

In the photorefractive realization, a discrete time updating method can be realized with the following procedure:

- For each input vector $\vec{x}$ the diffracted amplitude (with the reference beam switched off) provides the neuron output $y$.
- The weights are updated by reinforcing the gratings during a time period $\Delta t$ with the reference beam switched on. Both the amplitude of the reference beam and the time period $\Delta t$ are adjusted according to the neuron output and the desired adaptive law.

If we consider the electric field change in the photorefractive materials, this procedure gives [387]

$$\Delta E_i = \frac{2 \mid S_i \mid \cdot \mid R \mid}{\tau_0} \cdot E_{sc} \Delta t - \frac{I_t}{I_0} \cdot E_i \Delta t. \tag{10.2}$$

with $R, S$ being the amplitudes of the reference and the signal beams, respectively.

By comparison with eq. (10.1), the following rules can be derived [387]:

- Case $p = q = 0$: This case is achieved by setting the amplitude of the reference beam and the time period $\Delta t$ to a constant value, whatever the output result $y$ gives.
- Case $p = q = 1$: This case is easily implemented by setting the amplitude of the reference beam to a constant value but by making the recording time $\Delta t$ proportional to the output value $y$. Consequently, there is a total correspondence between the updating law and the kinetics of the photorefractive effect.

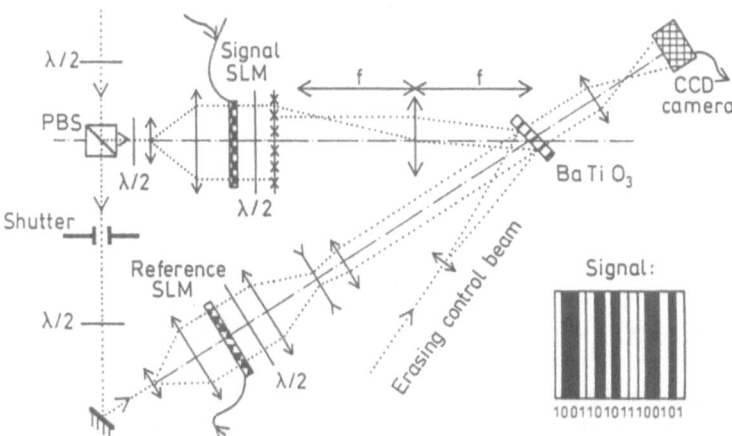

**Figure 10.5:** Experimental setup to realize a photorefractive self-organizing neural network with feature extraction and competitive learning potentials. PBS = polarizing beam splitter, SLM = spatial light modulator, $\lambda/2$ = half-wave plate. The inset shows an example of the input vector (after [388]).

- Case $p = 2$, $q = 1$: To implement this case, the different beam intensities have to be adjusted so that the reference beam intensity is much larger than the sum of the signal beam intensities. Moreover, the recording time $\Delta t$ is a constant, but the reference beam amplitude is made proportional to the neuron output $y$.

In their experimental realization of the self-organized learning rules, T. Galstyan and coworkers [387] used an Ar-ion laser and a liquid crystal television for realization of the input vectors (stripes). Each of these vectors is expanded to be incident on the whole surface of the $BaTiO_3$ crystal, allowing to superimpose all beams in the crystal volume. A photodiode detects the values of the output vectors $y$ and is coupled with the SLM in order to adapt the weights. With that setup, competitive learning between two vectors as well as feature extraction and active forgetting could be realized. In an improved version, which is shown in fig. 10.5, a whole opto-electronic self-organizing neural network was realized, being able to create a topological map from a learning set of vectors. After an iterative presentation of all vectors, the input of one vector activates a few neurons of the two dimensional output whose positions indicate the correlation of this vector with the other vectors of the learning set. Again, the system benefits from the specific kynetics of the photorefractive effect to holographically record and update the neural interconnection weights in the photorefractive crystal.

### 10.2.2 Winner-Takes-All Behaviour in a Photorefractive Ring Oscillator

A winner-takes-all system based on a photorefractive ring oscillator is a coherent multi-stable system in which the different modes are coupled in a purely competitive way. This leads to dynamics in which the modes compete with each other until one mode wins the competition and suppresses the oscillation of the remaining modes. Any mode can be the winner and the actual outcome of the competition is determined by the initial mode intensities or it can be weighted through some internal bias of the modes. The system will then start to suppress the oscillation of all modes other than the chosen one. In the final state all the energy is concentrated in the only oscillating mode which constitutes a localized center of activity.

In addition to providing gain for a multimode oscillator, two-beam coupling in photorefractive media can be used to establish mode coupling.

Three features of photorefraction are particularly important for the realization of competitive feedback systems. First, enourmous gains are possible, rising up to experimentally observed factors of more than 10.000 [255]. Second, ferroelectric photorefractive crystals are very slow by nonlinear optics standard in the formation of the nonlinear effect. Time constants are typically in the range of milliseconds to seconds. Third, increasing the intensity of the pump beam does not increase the gain for an unsaturated signal, because the photorefractive effect is based on the intensity modulation depth or on the intensity ratio between the interaction beams and not on the total incoming intensity. However, increasing the intensity decreases the response time of the material. The response remains out of these reasons nonlinear even at very low light intensities, also below $1 \, \mathrm{mW/mm^2}$.

C. Benkert and D.Z. Anderson [389] realized an optical circuit based on these principle for winner-takes-all competition, that is rather similar to the system descibed in section 6.3. It is a unidirectional ring oscillator with three photorefractive $BaTiO_3$ crystals. Five ring paths are formed with five optical fibers, each carrying many modes. In the following however, I will ignore these multimode fiber outputs and refer to the collection of fiber modes as a single mode. The first photorefractive crystal is oriented in such a way that the oscillator experiences loss. The second and the third crystal provide gain the the oscillation signal. Moreover, the third crystal is pumped by an external laser source and supplies gain to all five modes of the system. The other two crystals allow to program the interaction matrix between the five fiber beams. The resonator is fed by the output of the optical fiber. A polarizing beam splitter provides two copies of the five modes. The beam passing straight forward is the "resonator" beam, the ohter one is used as an interaction beam. A lens is used on each of these two beams to produce the Fourier transform of the output of the fibers. The first photorefractive crystal is placed at the intersection of the two Fourier planes and its axis is arranges so that the interaction beams derive energy from the resonator beams. If we think of the modes of optical rays, then in the Fourier-plane junction the two sets of rays cross. Any of the interaction rays can deplete every resonator ray of its energy. This has the effect of increasing all saturation coefficients, including self-saturation. A pair of lenses reimages both sets of beams: they intersect again, but this time in an image plane. The second photorefractive crystal is arranged in such a way to cause energy tranfer from the interacting modes into the resonator modes. here, each mode intersects only its corresponding twin. Hence, an interaction beam gives back whatever energy it obtained in the first crystal to its own resonator mode in the second crystal. This reduces self-saturation relative to the cross saturation. As a result, only one of the five modes is stably oscillating at any one time. The constrast ratio between the oscilating mode and the other suppressed modes is greater than 100:1, The system is optically switched between the different modes by injecting a signal into the respective mode.

### 10.2.3 Topology-Preserving Mapping in a Photorefractive Ring Resonator

Self-organization can also be realized using a resonator architecture that learns to recognize features in complex, information bearing beams. The resonator supports a continuous family of spatially localized transverse modes. When the resonator is pumped by an information bearing beam, the tranverse modes self-organize to generate a mapping of input signals onto localized resonator modes due to the dynamics of competition between the transverse modes of the resonator (see section 6.5). Thus, different input features are mapped onto different groups of resonator modes. This mapping is topology-preserving. In the recall phase, where new data are presented to the resonator, groups of modes turn on, depending on the degree of similarity between the training set used in the learning phase, and new inputs.

The recall phase of the feature extractor is fundamentally different from the recall phase in an associative memory. In the associative memory the signal that is injected in the recall phase biases the nonlinear mode competition to favor the stored resonator eigenmode that most closely matches the input (see section 7.2.4). Thus, a mode containing the complete information is excited associatively when the resonator is presented with partial information. All other resonator modes are suppressed. In the feature extractor in contrast, the recall phase is purely linear. When partial information is presented, all linearly dependent resonator modes are excited, not only the mode corresponding to the training pattern that bears the greatest resemblance to the new input.

The resonant structure introduced by M. Saffman et al. [390] is a self-imaging resonator where the transverse mode structure is strongly influenced by the nonlinear gain and loss, and only weakly affected by the linear boundary conditions. It consists of a unidirectional, self-imaging ring in which saturable gain by two-beam coupling in one plane of the resonator is combined with saturable loss in a spatially distinct plane. The transverse mode profile in a high Fresnel number self imaging optical cavity is not well defined. The observed transverse structure in self-oscillation is a complex pattern, due to a continously changing superposition of many metastable transverse modes [391]. By placing saturable photorefractive gain and loss in spatially distinct resonator planes, this continuum of transverse modes collapses into a localized, single-mode-like oszillation, at an arbitrary transverse location in the resonator [392]. This corresponds to a modified winner-takes-all behaviour. Feedback provided by the optical cavity serves to connect this element with the interconnection gratings stored in the photorefractive gain medium. The result is a dynamical evolution that inplements the features of T. Kohonen's algorithm.

In order to realize this configuration, M. Saffman and coworkers influenced the dynamics of the localized modes. They depend strongly on the cavity misalignment. For small misalignments, such that the transverse phase mismatch across a limiting iris is smaller than $\pi$, the transverse location of the localized mode becomes unstable. The cavity misalignment causes a linear feeding of energy from the oscillating mode to neighbouring locations. This results in a continuous drifting of the cavity spot in the transverse plane of the resonator. In order to eliminate this drift motion, the cavity was misaligned is such a way that the transverse phase mismatch is several $\pi$. In this case the oscillating pattern, with no loss pumping, is a set of fringes that represents the cavity equiphase contours. When the pump is turned on, localized modes still form, but they are restricted to locations on the bright fringes. The dark fringes act as barriers to the spot motion. Although using this approach modes can no longer form at arbitrary transverse locations, modes can still form along an arbitrary location along a single fringe. Thus, the resonator is suitable for implementing mappings from a two-dimensional input space to a one-dimensional output space.

Another advantage of photorefractive ring resonators comes into play here. The pump beam to the photorefractive gain crystal need not have a smooth Gaussian transverse profile. Since the photorefractive gain results from diffraction in a volume hologram, arbitrary spatial-pump profiles may be transformed into arbitrary resonator-mode profiles. When a Gaussian beam is propagated through a length of multimode fiber it emerges as a speckle pattern. Pumping the resonator with a speckle pattern also leads to a mode with a smooth, localized envelope. When the resonator is pumped with two spatially distinct speckle patterns, each one leads to a localized mode. When the speckle patterns are spatially and temporally orthogonal, the resulting resonator modes are also spatially and temporally orthogonal. Thus, the resonator can be considered as a feature extractor.

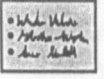 ## Summary

Coherent optical competitive learning self-organizing networks can be built using resonator configurations. The neurons within a single competitive unit must be inhibitorily interconnected in order to produce a dynamical system that converges to the winner-takes-all output that implements the desired maximum operator without resorting to a sliding threshold. This can be accomplished by

- implementing a photorefractive recording scheme using angular multiplexed reference beams. If the amplitudes of the signal beams are made proportional to the neuron input vector components, the diffracted amplitudes of the memory represents the neuron outputs, whereas the index modulations in the volume holographic memory are proportional to the components of the weight vectors. In order to implement an adaptive neuron, these index modulations are adapted according to the value of the neuron output by using an electrooptic feedback loop. Thus, competitive learning with winner-takes-all features, feature extraction and active weight forgetting can be implemented.

- self-imaging unidirectional ring oscillators with nonlinear gain and loss units supplied by two photorefractive beam-coupling devices. The different resonator modes correspond to the different competing neural firings produced by the various classes of input patterns, resulting in self-organized competitive learning. The functionality of the resonator modes is equivalent to that of Kohonen's algorithm for a self-organizing topology-preserving map.

 ## Further Reading

1. R.A. Athale, J. Davis (eds.), *Neural network models for optical computing*, Proc. SPIE 882 (1988).
   An older conference proceeding on optical neural network realizations, with several important contributions on self-organized and competitive learning.

# 11 Optical Realizations of Hopfield and Boltzmann Neural Networks

One of the major reasons why Hopfield neural networks attract much interest as candidates for optical realizations is the fact that they easily provide schemes for pattern recognition and associative memories. Beneath the possibility to exploit the associative memory properties of classical Van der Lugt filters as they have been discussed in chapter 7, models based on directly transferring the Hopfield model from their digital electronic manifestations into optical systems are the most widely spread and successfully realized ones.

When J. Hopfield [10] described in 1982 his simple model for the operation of fully interconnected feedback networks, his aim was to construct a simple network that is able to solve complex decision problems. Its basic structure (see section 2.5) suggests the action of individual neurons to be modeled as a thresholding operation, whereas information is stored in the interconnections among the neurons. Computation is performed by setting the state (on or off) of some of the neurons according to an external stimulus and, with the interconnections set according to the recipe that Hopfield described (see box 6 in section 2.5), the state of all neurons that are interconnected to those that are externally stimulated spontaneously converges to the stored pattern that is most similar to the external input. The basic operation performed is nearest neighbour search. Because it is a fundamental operation for pattern recognition, associative memory and error correction, it is obvious why the Hofield model has become so important for the realization of associative pattern recognition tasks. The advantages of the model — as there are the fact that its computational power lies in very simple identical logic elements (the neurons), its storage capacity (approximately $N/4 \cdot \ln N$ bits/neuron), the possibility of concurrent, distributed processing that enables a massively parallel structure and does not require synchronization and finally its insensitivity to local imperfections such as variations in the threshold level of individual neurons or the weights of the interconnection — recommend the investigation of optical implementations.

Already in 1985, N. Farhat and coworkers discussed the possibility to use Hopfield nets to realize associative memories [121, 393]. I have introduced the basic principles of the analogies of optical characteristics derived in these publications and neural network models like the Hopfield model at the beginning of chapter 4. In the present chapter my aim is to introduce to some of the novel concepts to realize Hopfield-like optical neural networks rather than to repeat these basic principles. Because many of the realizations of optical Hopfield networks are closely connected to the realization of associative memories, I will concentrate here on the basic realizations that do not purely intend to realize pattern recognition systems.

In order to classify the different optical implementations of Hopfield-like neural networks, I will distinguish two different approaches. In the first one — the incoherent approach — the states of the neurons are represented by the intensities of the optical fields. In previous chapters, we have already got to know several of these models (see chapters 8.1, 9.1 and 10.1). Because the intensity-modulation devices such as spatial light modulators can not have negative intensity transmittance or reflectance, the incoherent approach requires several modifications or optoelectronic hybridization to achieve bipolar synaptic weights. I have discussed already in section 4.2.4 several general methods of realizing bipolar weights in that situation — among them the most widely spread principle of channel doubling.

In the coherent approach, the states of the neurons are represented by optical fields. The photorefractive resonator memories I have presented throughout the precedent chapters are examples for these coherent neural network realizations (see sections 8.2, 9.4, and 10.1). The coherent approach has — as I have discussed throughout the book — the possibility of an all-optical implementation, because negative synaptic weights can be realized simply by a change in the phase of the transmitted or reflected amplitude by $\pi$. In section 5.2, the amplitude characteristics of photorefractive materials have been discussed to realize all-optical coherent thresholding devices. Thus, in the coherent approach, the states of neurons take not only bipolar values but also complex values whose amplitude and phase correspond to those of the optical fields (see e.g. section 9.4). The same is true for the synaptic weights, which are represented by complex amplitude transmittance or reflectance. The fact that both the states of the neurons and the synaptic weights can take complex values is one of the most important features that distinguish coherent optical implementations from other implementations. It is this fact that requires some changes in the Hopfield model in order to be adapted to complex weights. However, only several authors have recognized that feature in its whole importance up to now. Therefore, in the following I will review exemplarily some of the most important optical incoherent Hopfield networks and then come to these more important and promising coherent approaches, which I will discuss in more detail. Moreover, optical realizations of interpattern associations allowing to discriminate similar patterns that are not orthogonal or independent of each other, will be described.

One of the most important and restricting limitation of Hopfield nets is their tendency to converge to local minima without having a chance to escape from it. I have discussed extensively in section 2.6 that the model of statistical physics can provide an excellent tool to overcome this situation: simulated annealing. This method of providing energy to a system that stucks in a local minimum in order to allow to escape from it and then slowly reducing the amount of supplementary energy in order to enable the network to make its way towards a global solution, has attracted many researchers in completely different fields of physics.

Again, as in the previous section, optoelectronic feedback is the solution in most of the incoherent optical realizations of fully-interconnected networks. Here again, the combination of neural net modeling, Boltzmann machines, and simulated annealing concepts with high-speed optoelectronic implementations promises to produce high-speed artificial neural net processors with stochastic rather than deterministic rules for decision making and state update. However, cascading these systems in order to obtain multilayer networks becomes difficult because of the operation time necessary for electronics-to-optics and back layer-to-layer processing.

On the other hand, if one wants to maintain a coherent, all-optical processing network, it is difficult to realize the features of the methods of Boltzmann machines as simulated annealing in coherent optics. In contrast to prescribed learning procedures as the outer-product learning rule, induced self-organized or self-teaching nets have not yet been extensively investigated in coherent optical neural net implementations. Consequently, there are only a few attempts in optics to realize Boltzmann machines, based on volume holographic interconnections, combined with computer controlled simulated annealing processes. I will discuss them in the last section of this chapter.

## 11.1    Incoherent Optical Realizations of Hopfield-like Networks

The following incoherent optical realizations of Hopfield-like networks are simple combinations — as in the case of the single-layer perceptron – of the elements discussed in chapter 4 and chapter 5, combined with electronic feedback. Because I have discussed the basic elements — interconnection device and thresholding operation — already in detail, in this chapter I will describe a selection of the most promising models combining these devices.

The earliest model was proposed in 1985 by N. Farhat and coworkers [121, 394], using an array of light emitting diodes (LED's) to represent the logic elements or neurons of the network. Their state (on or off) can represent unipolar binary vectors that are stored in the memory matrix. Global interconnection of the elements is realized through the addition of electronic nonlinear feedback (thresholding, gain and feedback) to a conventional vector-matrix multiplier in which the array of LED's represents the input vector and an array of photodiodes is used to detect the output vector (see fig. 11.1). Some of these schemes of matrix-vector and matrix-matrix interconnection devices using liquid crystal devices or spatial light modulators as well as photorefractive volume holographic media have been presented in section 4.3.

An optical feedback becomes more attractive when we consider that arrays of nonlinear optical light amplifiers with internal feedback, as e.g. the nonlinear Fabry-Perot etalons described in section 5.2 or optical bistable devices can be used to replace the photodiode/LED arrays. This can lead to simple and compact structures of Hopfield networks, that are able to realize content-addressable memories or associative memories.

Here, I will describe some extensions of these ideas to more actual realizations of Hopfield models. However, the basic ideas still remain the same. Their simplicity and ease of combination is one of the reasons for the popularity the Hopfield network still gains in optical neural network realizations.

**Figure 11.1:** Concept for optical implementation of adding nonlinear feedback to an optical vector-matrix multiplier (after [121])

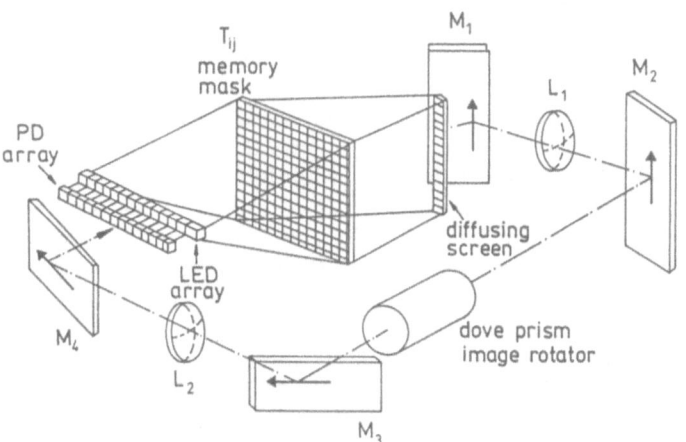

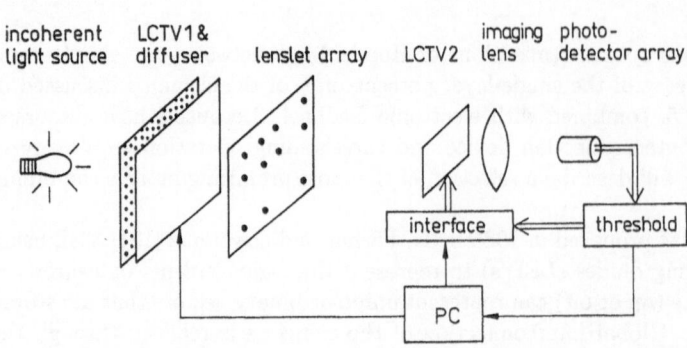

Schematic diagram of the setup of a compact optical neural network based on the Hopfield model using a pocket-sized liquid crystal television as interconnection device (after [395]).

## 11.1.1   Dual Channel Realization Using Compacts LCTV's

In order to make the optical realizations of Hopfield networks compact without reducing the space-bandwidth product, F.T.S. Yu and coworkers [395] realized in 1990 a Hopfield network using a pocket-sized liquid crystal television (LCTV) as the interconnection device, allowing to interconnect $8 \times 8$ or 64 neurons (see fig. 11.2). The LCTV (LCTV1) realizes the generation of the interconnection weight matrix, which consists of submatrices, each submatrix having $8 \times 8$ elements. This interconnection matrix is then displayed on a fine diffuser immediately behind the first LCTV. A second LCTV, LCTV2, is used as an input device for the generation of input patterns. During learning, the lenslet array, consisting again of $8 \times 8$ lenses, provides the interconnections between the interconnection weight matrix and the input pattern. Here, positive and negative weights are realized using the concept of two channels for each polarity as presented in section 4.2.4, which are sequentially displayed on LCTV1 to realize the interconnection matrix, followed by an electronic combination in order to realize bipolar neurons. Each lens of the lenslet array images each of the interconnection weight matrix submatrices onto the input LCTV2 to establish the proper interconnections. Thus the input matrix is superimposed onto and multiplied (or interconnected) by all the interconnection weight matrix submatrices.

This represents the connectivity part of the Hopfield net equation (see section 2.5.1). The output result from LCTV2 is then imaged onto a charge-coupled device (CCD) array detector. The signals collected by the CCD camera are then sent to a thresholding circuit, and the final results can be fed back to LCTV2 for a second iteration. Note that the data flow in the optical system is controlled primarily by a personal computer. Thus, although an optical setup, the interconnection weight matrices and the input patterns are written onto LCTV1 and LCTV2 through the PC and the PC can also make decisions based on the output results of the neural network. Consequently, this optical Hopfield network proposed by F.T.S. Yu and coworkers is programmable and adaptive in an electronic and not an optical way. In the recall or operation mode, an input pattern, e.g. a noisy pattern similar to one of the stored patterns, is presented on LCTV2. The input pattern is then multiplied with the submatrices of the interconnection weight matrix using the lenslet array. The positive and negative parts of the output results are captured by the CCD camera, and the subtraction and thresholding operations are performed by the PC. A

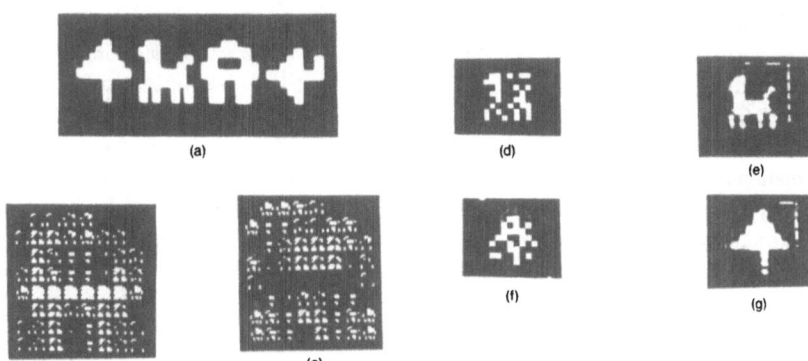

**Figure 11.3:** Experimental results of the compact optical neural network. (a) Four images used as the reference patterns. (b) Positive and negative parts of the interconnection weight matrix. (d), (f) Input pattern embedded in 25% noise, and (e), (g) its output results obtained in one single iteration (from [395]).

noisy image with 25% random noise could be recognized properly in these experiments in only one single iteration. Thus, the system is able to perform pattern recognition and image reconstruction operations based on the Hopfield neural network algorithm (see fig. 11.3).

### 11.1.2 Solely Inhibitory Interconnection Weights

A common solution to the bipolar weight value problem in incoherent optical network implementations is the doubling of the interconnection channels as I have described it in the example above. Another practical solution may be to implemented the Hopfield model optically by adding a constant to the interconnection matrix, so that all its elements become positive. This results in an additional dynamic variable for the network, the neuron's threshold, which has to be electronically detected and corrected at each time step. This concept of removing negative numbers in the Hopfield model will be exemplarily described for a coherent realization of a Hopfield network in the next section.

However, if using only inhibitory interconnections, it is possible to realize at network that performs similarly to networks with both inhibitory and excitatory interconnections. This is due to the fact that although the network is composed of both positive and negative elements, the stored information is not equally divided between the positive and negative elements. Specifically, it has been shown that for small networks with around 100 neurons the performance of the system is essentially unchanged by discarding (setting to zero) all positive interconnections $w_{ij}$. In contrast, when only using the positive interconnections, setting the negative ones to zero, the Hopfield network loses its ability to function as an associative memory [396].

This behaviour can be explained by the following arguments. In a network with only positive elements, each neuron is forced to be in a state that is equal to the states of all the other neurons. This occurs only when all neurons are 1 or all are −1, resulting in two highly stable states that do not match the stored states. Consequently, poor associativity is achieved. In a network with only negative interconnections, each neuron tends to be ins a state opposite to that of the other neurons. This can not be accomplished for all neurons simultaneously. At best the neuron can be in a state in which half of the neurons are 1 and the other half are $-1$. Such a situation occurs in a Hopfield-type network where the stored states are chosen at random, so that the distributions of the 1's and the $-1$'s in these stored states are nearly equal. Some other learning algorithms, such as the perpectron, can also be modified to yield interconnections that are solely inhibitory. It can be shown that the maximal number of stored states in a Hopfield network with only inhibitory connections would be reduced by 10 - 20% when compared with the original Hopfield network [396]. Thus, the $w_{ij}$'s can be replaced by $w_{ij}$ values that are defined as (compare to section 2.5)

$$w_{ij} = \begin{cases} w_{ij} & \text{when } w_{ij} \le 0 \\ 0 & \text{otherwise} \end{cases} \qquad (11.1)$$

to form a new dynamics rule

$$\mu_j = f\left(\sum_{i=1}^{N} w'_{ij}\mu_i\right) \qquad (11.2)$$

Substitution of a new set of variables, $\eta_j$ that are related to the old set $\mu_j$ as $\eta_j = (1 - \mu_j/2)$ yields

$$\eta_j = \frac{1}{2}\left(1 - f\left(-2\sum_i w'_{ij}\eta_i + \sum_i w'_{ij}\right)\right) \qquad (11.3)$$

Using the fact that the function $f(x)$ is a hard-limiting bipolar function as

$$f(x) = \begin{cases} 1 & \text{for } x_{ij} \ge 0 \\ -1 & \text{otherwise} \end{cases} \qquad (11.4)$$

and that it obeys, for any positive value $a$ to $f(x) = f(x/a)$, we can divide the argument of $f$ in eq. (11.3) by the nonnegative number $-2\sum_k w'_{jk}$. The division gives

$$\eta_j = g\left(\sum_i w''_{ij}\right)\eta_i \qquad (11.5)$$

where $g(x) = 1/2(1-f(x-1/2))$ and $w''_{ij} = w'_{ij}/\sum_k w'_{jk}$. This dynamics rule is relatively simple to implement optically because the interconnections $w_{ij}$ are all nonnegative, and the neural response function $g(x)$ is a steplike function centered around a fixed an uniform value of $1/2$. Each neurons output is therefore either zero or one.

In their approach using the modifications described above, I. Shariv and A.A. Friesem [396] realized such a system with a liquid-crystal light valve for implementing a two-dimensional array of inhibitory neurons and an array of subholograms for the interconnections. With that system, which is schematically scetched in fig. 11.4, two stable states can be tretrieved from noisy optical inputs, using an interconnection matrix having 16 neurons. These are represented by 16 equally intense plane waves, derived from an Argon ion laser, that are incident upon the read side of a liquid crystal light valve (LCLV). The backreflected plane waves from the LCLV represent the neuron outputs. Each of the reflected plane waves is fed back to illuminate its corresponding subhologram, which in turn fanned out the light in different directions. The diffracted intensities from all the holograms are incoherently summed at an array of $4 \times 4$ circles on the write side of the LCLV. These circles were located precisely across from the 16 plane waves incident on the read side and therefore determined the local reflectivity of the LCLV at each of these points.

**Figure 11.4:** Optical neural network of the Hopfield-type using only inhibitory interconnections M = mirror, H = subhologram array, L = imaging lens, PCBS = polarizing cubic beam splitter. Only three representative neurons are shown (after [396]).

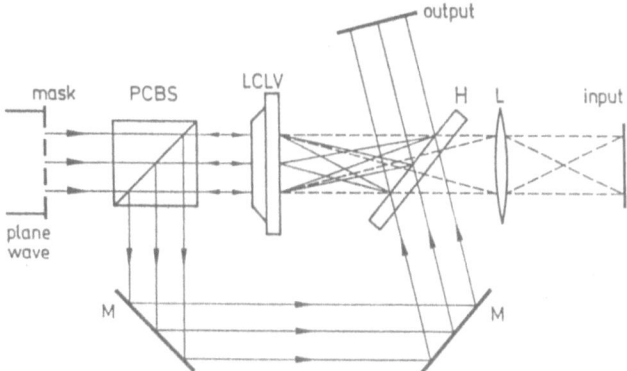

During operation, an input transparency is illuminated with white light and imaged onto the write side of the LCLV to determine the inital state of the system. When the input is blocked, the network evolves towards the closest stable state and remains there. The state of the network at any time is determined from the light transmitted directly through the holograms. Note that in order to set the network at any initial state, the complementary state has to be presented at the input, owing to the use of inverted neurons. Preliminary results showed that it always converges to the stored pattern that was the closest to the complement of the input. No oscillations or spurious states were observed as well as the same robustness known from bipolar Hopfield networks. This sort of reduced Hopfield network using only inhibitory interconnections represents an easy solution to realize incoherent optical Hopfield networks without using electronic calculation of the feedback unit of the system.

### 11.1.3 Inhibitory Interconnections by Photorefractive Grating Decay

Another interesting implementation idea of Hopfield networks was proposed by J.S. Jang and coworkers [397], implementing the bipolar weights in an all-optical way. They realized the network using a photorefractive crystal and a holographic lenslet array. The crystal is used for recording (excitation) and erasing (adaptation or inhibition) the synaptic weight values, and the lenslet aray is used for interconnection between the neurons. As I have discussed in detail in sections 4.3 and 4.4, photorefractive crystals are able to perform interconnection weights in a natural way due to the modulation of the refractive index proportional to the outer product of the incoming beam amplitudes and phases. The realization of bipolar neurons is in this case a coherent one, realized through definite erasure instead of exposure of the photorefractive grating modulation depth. In their system, J.S. Jang still used electrooptic feedback and thresholding for the feedback unit.

---

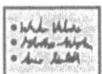 ## Summary

---

Adding a threshold and an (electro-)optic feedback loop to an interconnection device is sufficient to realize simple incoherent optical networks based on the Hopfield model. The major task to be fulfilled is the realization of bipolar weights. They can be built using

- a compact pocket-sized liquid crystal televisions (LCTV) as the input and the interconnection device. Bipolar weights are realized in a dual channel configuration.
- a liquid crystal light valve, creating only inhibitory interconnections. Although the network is composed of both positive and negative elements, this is possible because the stored information is not equally divided between the positive and negative elements. For small networks (e.g. 100 neurons) the performance of the system is essentially unchanged by discarding (setting to zero) all positive interconnections.
- a photorefractive crystal for recording (excitation) and erasing (inhibition) the synaptic weight values and a holographic lenslet array for interconnection between the neurons. The bipolar neurons are here realized in a coherent way, using definite erasure for inhibitory and definite exposure for excitatory weight modulation.

## 11.2    Coherent Realizations of Hopfield-like Networks

Among the neural network models that have been discussed as candidates that are well suited to be realized using coherent optical devices, Hopfield networks play an important role, because they can be simply realized using optical storage and feedback devices. The equality of Hopfield's learning algorithm and the principles of storage of optical interference terms in holographic media is one point which makes the Hopfield network especially attractive for coherent photorefractive realizations. Moreover, feedback can be achieved with more ease if self-aligning components as optical phase conjugators can be implemented in the setup. The most important problem, however, in coherent realizations of a lot of different network algorithms is to realize the threshold function. From the examples and possiblities discussed in chapter 5 Fabry-Perot etalons and liquid crystal thresholding devices seem to be the most promising candidates for coherent processing. In the following, I have chosen among the number of realizations an early one that clearly describes the possibilities and limitations of all-optical coherent realizations of Hopfield networks and an exemplarily one for each of the two main photorefractive realizations, two- and four-wave mixing.

### 11.2.1    Digital Hopfield Neural Network

 As in the case of optoelectronic realizations, it is possible to modify the network in such a way that no negative numbers or subtractions appear in the Hopfield model. For this case, also coherent implementations may be realized without referring to the phase in order to realize bipolar or especially inhibitory weights. Here, I will present a rearrangement of the Hopfield model in order to remove negative numbers without modifying the basic structure of the Hopfield model and without requiring subtraction stages or a time varying thresholding level. The conventional Hopfield model algorithm defines the following rule. If an input to the system is defined as a vector $x^{(\text{in})}$, the output is defined as a vector $x^{(\text{out})}$ with

$$x_i^{(\text{out})} = \begin{cases} 1 & \text{for } \sum_j^N w_{ij} x_j^{(\text{in})} \geq 0 \\ -1 & \text{for } \sum_j^N w_{ij} x_j^{(\text{in})} < 0 \end{cases} \tag{11.6}$$

where

**Figure 11.5:** Optical implementation of the Hopfield model, incorporating computer-generated holographic interconnections and a nonlinear bistable etalon (after [334]).

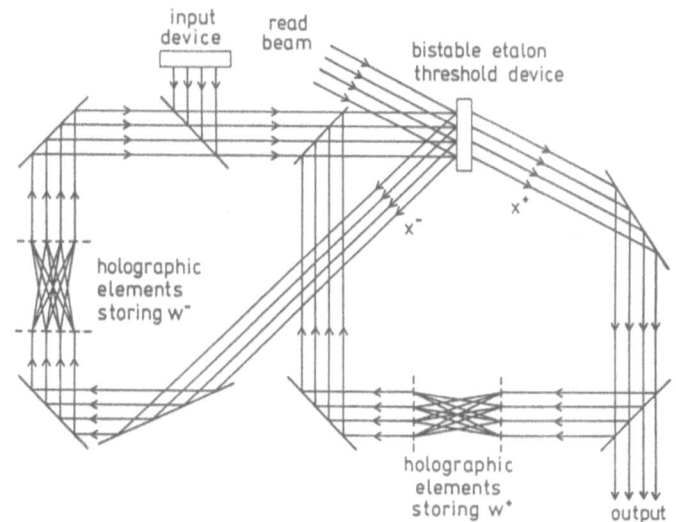

$$w_{ij} = \sum_{s=0}^{M-1} x_i^{(s)} x_j^{(s)} \tag{11.7}$$

represents the interconnection weight matrix for the holographically stored exemplar patterns. H.J. White et al. [334] defined a novel set of output vectors that contains no negative numbers or subtraction stages by using the following definitions

$$
\begin{aligned}
\vec{x} &= \vec{x}^+ - \vec{x}^- \\
w &= w^+ - w^- \\
x_j^+ + x_j^- &= 1,
\end{aligned}
\tag{11.8}
$$

where the elements of $\vec{x}^+$, $\vec{x}^-$, $w^+$ and $w^-$ are positive real numbers.

Then, eq. 11.6 can be rewritten as

$$
\begin{cases} x_i^{(\text{out})+} = 1 \\ x_i^{(\text{out})-} = 0 \end{cases}
\quad \text{for} \quad
\sum_j^N (w_{ij}^+ x_j^{(\text{in})+} + w_{ij}^- x_j^{(\text{in})-}) \geq \frac{1}{2} \sum_j^N (w_{ij}^+ + w_{ij}^-) \tag{11.9}
$$

$$
\begin{cases} x_i^{(\text{out})+} = 0 \\ x_i^{(\text{out})-} = 1 \end{cases}
\quad \text{for} \quad
\sum_j^N (w_{ij}^+ x_j^{(\text{in})+} + w_{ij}^- x_j^{(\text{in})-}) < \frac{1}{2} \sum_j^N (w_{ij}^+ + w_{ij}^-) \tag{11.10}
$$

With that set, the Hopfield model is rearranged to contain only positive numbers and is therefore ideally suited for an optical implementation.

An early system that realizes this modified Hopfield model is shown in fig. 11.5. It includes an input device, which enters the vector $\vec{x}^{(in)}$ into the system, the holographic interconnections, which store the matrices $w^+$ and $w^-$ and perform the vector-matrix multiplications and, finally, a threshold device, which operates upon the resultant vector of the vector matrix multiplications, producing two new binary vectors that are fed back into the matrix multipliers as the algorithm dictates. As the input device, a spatial light modulator — either electronically or optically addressed — can be used. For our model described above, the input SLM would be used to enter the input image represented by the input vector $\vec{x}^{(in)+}$ or $\vec{x}^{(in)-}$.

The holographic interconnections are used to perform the vector-matrix multiplications in eq. 11.10. The subproducts required for this are $w_{ij}^- x_j^-$ and $w_{ij}^+ x_j^+$ for $j \to N$, where $x_j$ is a given element of the input vector. The method proposed from H.J. White et al. to perform these calculations is to use computer-generated holograms to precisely encode the values of the $w^+$ and $w^-$ matrices.

A holographic element can be created that stores all of the elements of $w_{ij}^+$ for a particular value of $j$. The operation of such a hologram is that when addressed by a light intensity $x_j^+$, $N$ spatially separated outputs are produced that have intensities equal to the products $w_{ij}^+ x_j^+$, one for each value of $i$. Consequently, $N$ holograms will be required, one for each vaule of $j$, such that the outputs from the holograms overlap and incoherently add to produce $N$ spots. This incoherent addition is performed by integrating the intensities over the defined output area. The intensity of the spots gives the resultant vector $\vec{x}'^+$, where

$$x_i'^+ = \sum_j^N w_{ij}^+ x_j^+. \tag{11.11}$$

The desired output intensities of the holograms are formed in the first order diffraction pattern. It is important to note that the matrix $w$ needs to be calculated before the hologram is constructed. The matrix is inherently fixed once the hologram is constructed.

Each element of the input vector $\vec{x}^{(in)+}$ addresses only one hologram. Therefore, the vector is no longer constrained to be a linear array of light intensities. It can be folded to form a two-dimensional array or a binary image. This has the practical advantage of making full use of the available space and reducing the required diffraction angles of the hologram. The interconnections for $\vec{x}^{(in)-}$ are formed in an identical manner. Note also that this is an efficient method of forming the interconnections, as opposed to reducing a set intensity to the required amount for each interconnection, since only the required amount of light is diffracted for each interconnection.

In the system shown above, especially for purposes of pattern recognition, space-variant holographic interconnections are required. Considering planar holograms, computer generation of these holograms is the most flexible method of production. The number of holographic elements and hence the number of processing elements in the Hopfield model will be governed by the space bandwidth product of the holograms. The required space-bandwidth product of each of the holographic arrays increases with the number of sub-holographic elements, the required fan-out of each element, and the required range of diffraction efficiencies. The number of subholograms will equal the number of elements in the vector $\vec{x}$ — $N$. The required fan-out will on the average equal half the number of elements in the vector $\vec{s}$ — $N/2$. This is because a processing element is just as likely to have an interconnection from the holographic array representing the negative weights as from the positive ones but will not have an interconnection from both. The range of diffraction efficiencies will be half the number of stored memories ($M/2$). Hence, the space-bandwidth product of each of the holographic arrays is of the order of $1/4 \cdot N^2 M$.

However, to incorporate learning into the digital associative memory based on the Hopfield model would require a way of updating the interconnection. This can be done using photorefractive crystals and a controlling SLM. As proposed by H.J. White et al. , the SLM can be divided into $N$ subarrays, each containing $N$ elements. The output from each subarray is either focused or imaged onto a separate area of the photorefractive crystal, where a real-time hologram is formed with its corresponding beam from the input array.

These holograms are read out to produce phase-conjugate beams that re-form the image of the SLM. Using imaging optics, the subarrays can be superimposed as one array of $N$ elements on the output threshold array. With this configuration, the intensities of the SLM outputs control the weights of the interconnections between the input and output arrays. Each interconnection is formed only if both, the input beam and the relevant interconnection beam, are present. The interconnection weights can be changed at any point by altering the output from the SLM, which can be either optically or electronically addressed. With volume holographic recording media, the storage capacity compared to these holograms can be increased significantly. Moreover, the space-bandwidth product can be increased up to two orders of magnitude using angular multiplexing (see section 4.4.3).

The thresholding will be performed by a nonlinear device onto which the outputs from the holographic interconnections ($\vec{x}'^+$ and $\vec{x}'^-$) are imaged. It must be able to produce both $\vec{x}^{(out)+}$ and $\vec{x}^{(out)-}$ and therefore work by modulating a uniform read beam. Since $\vec{x}^{(out)+}$ and $\vec{x}^{(out)-}$ are the complement of each other, they are formed simultaneously. Both nonlinear Fabry-Perot etalons and liquid crystal thresholding devices can achieve this (see chapter 5). The response of the device must be highly nonlinear, as e.g. a Heaviside step function or a sigmoidal function.

Although in this short description I have only assumed binary data, the technique can also be adapted to process gray-level images. Thus, a digital Hopfield model for optical associative memory purposes can be realized using holographic interconnection elements and nonlinear optical thresholding devices.

### 11.2.2 Coherent Hopfield Net Based on a Photorefractive Ring Resonator

H.M. Stoll and L.S. Lee and coworkers [218] realized with the concept of an optical thresholding neuron based on two-beam coupling (see section 3.4), a complete continuous-time optical neural network capable of executing a broad class of energy-minimizing neural net algorithms. The easiest net among them is the Hopfield network, allowing to realize complex bipolar neurons directly without special adaptation algorithms. The network shown in fig. 11.6 is a ring resonator which contains a saturable two-beam amplifier (BaTiO$_3$-crystal) acting as the optical thresholding neuron array element and two volume holograms (LiNbO$_3$) in combination with a Fourier-plane linear two-beam amplifier (BaTiO$_3$). The two volume holograms provide global network interconnectivity and the linear amplifier supplies sufficient cavity gain to permit a resonant, convergent operation of the network.

The interconnection architecture can be understood by first recalling the basic operation of a continous-time, recurrent neural network as I have introduced it in chapter 2. Within this type of Hopfield-like networks, the instantaneous neural state vector $x(t)$ is continuously updated through multiplication by the network interconnection matrix $w$ — in order to calculate new neural inputs — and subsequent nonlinear processing by the neuron array — in order to calculate new neural outputs. For a network that is structured according to Hebb's law (see section 2.2), the following must therefore be repeatedly calculated:

$$w \cdot x(t) = \sum_k \vec{v}^k (\vec{v}^k)^T \cdot x(t)$$

$$= \sum_k [x(t), \vec{v}^k] \cdot \vec{v}^k, \qquad (11.12)$$

where $\vec{v}^k$ is a e.g. lexicographically ordered training-set pattern vector whose ordering is consistent with that of $x(t)$, $(\cdot)^T$ denotes a vector transpose, and $[\cot, \cdot]$ denotes a vector inner product. The architecture shown in fig. 11.6 calculates $wx(t)$ by using a modified Vander Lugt filter to compute the inner product terms of eq. (11.12) and angle-multiplexed, object-space holograms of the $\vec{v}^k$ to perform the summation. The modified Vander Lugt filter (spherical lenses $L_1$ and $L_3$ and angle-multiplexed Fourier-space holograms of phase-encoded versions of the $\vec{v}^k$) works by first correlating a phase-encoded $x(t)$ with each of the identically phase-encoded $\vec{v}^k$ and then using spatial filtering techniques within the correlation plane to reject all but the dc component of the resulting correlation integrals. Spatial filtering perpendicular to the plane of the network is achieved by using cylinder lens $L_2$ in combination with an appropriately shaped pump beam to create a ribbon-shaped interaction region within the linear amplifier. Efficient filtering in this direction occurs when the narrow dimension of the interaction region is equal to the coherence length of the phase encoder. This length is chosen to be significantly less than a typical pattern vector pixel dimension. Spatial filtering within the plane of the network occurs as a natural consequence of the Bragg selectivity within the Fourier-space holograms. Finally, summation over the $\vec{v}^k$ is accomplished by illuminating the right-hand LiNbO$_3$ crystal with the spatially filtered plane waves that emerge from the linear amplifier. Proper algebraic summation of the terms that constitute $wx(t)$ is ensured by the fact that each hologram pair corresponding to a given $\vec{v}^k$ is generated by using a common, plane reference beam $A^k$. Within the saturable two-beam amplifier, the pump-beam and the signal-beam angles of incidence with respect to the BaTiO$_3$ crystalline c-axis are selected in such a way that the small signal neural gain is maximized without permitting the pump beam to intersect more than one neural volume before it leaves the BaTiO$_3$ crystal.

The authors could show in numerical simulations of the network's equations of motion [398], that for real-valued neural state vectors the network operates in almost the same way as Hopfields continuous-time model. For complex-valued neural state vectors, the network always converges to the dominant network attractor, thereby suggesting a paradigm for solving optimization problems in which entrapment by local minima is avoided. In their experimental investigations [218], the network has been used as an associative memory to store and recall multiple, nonorthogonal 2d-images.

As a second example of coherent neural networks, let us examine a complex neural network based on a Hopfield-like algorithm realized with photorefractive phase-conjugation.

### 11.2.3 Complex Hopfield-like Networks Using Phase Conjugation

To understand the networks presented in that section, I want to remind you that coherent optical implementations of neural networks in general have the feature that the states of the neurons take naturally not only bipolar values but also complex values whose amplitude and phase correspond to those of the optical fields. I have discussed an coherent optical network implementation for complex multilayer perceptrons in section 9.4. The same is true with the synaptic weights, which are represented by complex amplitude transmittance or reflectance. Moreover, negative synaptic weights can be realized simply without loosing storage capacity by a change in the phase of the amplitude transmittance (or reflectance).

In 1988, A.J. Noest [379] proposed a Hopfield-like model for these complex neural networks, which he referred to as phasor neural networks. He showed that a Hopfield-like

**Figure 11.6:** Diagram of a continous-time optical neural network. Optical neurons are generated within the saturable, two-beam amplifier, and a global neural interconnect is realized by using Fourier-and object-space holograms to implement an innter-product, matrix-vector multiplication algorithm (after [218]).

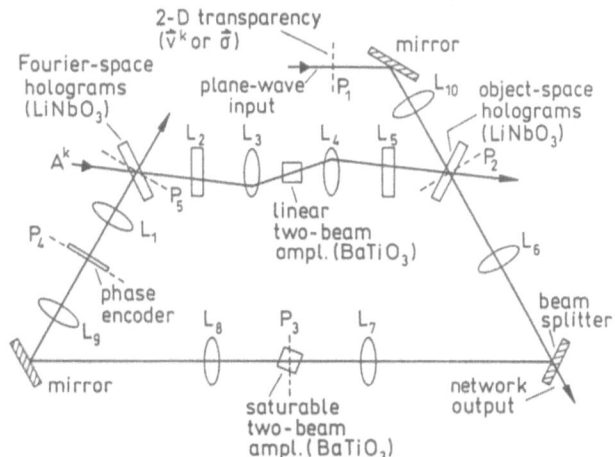

energy function exists when synaptic weights of the network have the form of a Hermitian matrix $w_{ij} = w_{ji}^*$, where * denotes a complex conjugate. These complex neural networks are of great relevance in optical computing when implemented in resonator memories.

An extension of this model has been given in 1992 by M. Takeda and T. Kishigami [399], who included amplitude variation in their model, which is one of the most fundamental physical characteristics of actual optical fields that are building up inside an optical cavity with a gain medium. This model permits the change of both amplitude and phase. Its dynamics has a close analogy with that of self-oscillation generated by degenerate four-wave mixing. Thus, a theoretical and experimental demonstration of the Hopfield-like energy function of the complex optical neural fields generated by phase-conjugate gain and feedback have been realized by M. Takeda and T. Kishigami.

*Energy Function for Complex Hopfield Neural Networks*

Here, I will discuss two different models of complex neural networks that are composed of discrete neurons with feedback in analogy to the Hopfield model.

Let us first consider a Noest-type complex neural network in which neurons change their states according to the following equations of dynamics (compare to 2.49):

$$\tau \frac{d\mu_j}{dt} = -\alpha\mu_j + \sum_i w_{ij} x_i \tag{11.13}$$

$$x_i = g(|\mu_i|)\frac{\mu_i}{|\mu_i|}, \tag{11.14}$$

where $x_i$ and $\mu_i$ denote complex external and internal states of the neurons, respectively, as I have denoted them already in section 2.5 and shown in fig. 2.19. $w_{ij}$ is now a complex synaptic interconnection weight for the complex signal flowing from neuron $i$ to neuron $j$, $\tau$ and $\alpha$ are constant real parameters, and $g(\cdot)$ is a nondecreasing nonnegative real function. Eq. (11.13) states that, while the amplitude of $|\mu_i|$ of the internal state is transformed by the nondecreasing function $g(\cdot)$ as a conventional real network, the value of the phase is preserved. This is the reason why these neurons arenamed *phase-preserving* *networks* (PPN). The phasor model of A.J. Noest is included in that model as a special case. It can be obtained when setting $g(\cdot) = 1$ and $\alpha = 0$. When using synapses with Hermitian symmetry, $w_{ij} = w_{ji}^*$, these phase-preserving neurons change their states in such a way that the energy function

$$E = -\frac{1}{2}\sum_i \sum_j w_{ij} x_i^* x_j + \alpha \sum_i \int_0^{|x_i|} g^{-1}(s)ds \qquad (11.15)$$

reduces its value monotonically with the time evolution of the system. In eq. (11.15), $g^{-1}(\cdot)$ is an inverse function of $g(\cdot)$ and $s$ is a real parameter for integration. Note that, as a result of Hermitian symmetry, $E$ becomes a real function.

The second model for a complex neural network has *phase-conjugate neurons* (PCN) that are connected by symmetric synapses $w_{ij} = w_{ji}$. Equations of the dynamics for this model are given by [399]

$$\tau\frac{d\mu_j}{dt} = -\alpha\mu_j + \sum_i w_{ij} x_i \qquad (11.16)$$

$$x_i = g(|\mu_i|)\frac{\mu_i^*}{|\mu_i|}, \qquad (11.17)$$

Although these equations look at a first glance equal to eqs. (11.13) and (11.14), the only, but important difference to the previous model is that internal states are transformed into external states with the sign of their phases reversed (or phase-conjugate). If the complex synaptic weights have symmetry $w_{ij} = w_{ji}$, the network has a Hopfield-like energy function (see eq. (2.50)) [399]

$$E = -\frac{1}{2}\mathcal{R}\left\{\sum_i \sum_j w_{ij} x_i x_j\right\} + \alpha \sum_i \int_0^{|x_i|} g^{-1}(s)ds, \qquad (11.18)$$

which reduces its value again monotonically with the time evolution of the system. Here, $\mathcal{R}$ denotes the real part. On the one hand, neurons in the PPN model do not have phase-conjugate properties, but Hermitian symmetric synapses have phase-conjugate relations $w_{ij} = w_{ji}*$. On the other hand, symmetric synapses in the PCN model do not have such phase-conjugate relations, but neurons have the phase-conjugate properties $x_i\,\mu_n^*$. If we consider a neural network as a graph with complex neurons at its nodes and complex synapses on its bidirectional edges, the two models may be considered to have a dual relationship in the sense that phase-conjugating functions are embedded either in the nodes or in the edges. Both of these models require a certain mechanism for phase conjugation in order for the complex network to have an energy function that decreases with the time evolution of the system.

*Analogy Between Complex Hopfield Networks and Phase-Conjugate Oscillator Systems*

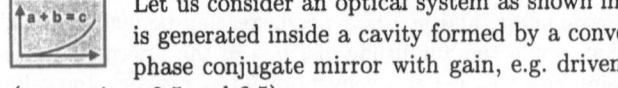 Let us consider an optical system as shown in fig. 11.7, where self-oscillation is generated inside a cavity formed by a conventional reflective mirror and a phase conjugate mirror with gain, e.g. driven by external four-wave mixing (see sections 3.5 and 6.5).

Self-oscillations in these resonators are possible if the gain is larger than the losses in the feedback loop. The author's group has demonstrated several different configurations of these phase conjugating oscillators [291, 294, 391, 400, 401] using Fabry-Perot-like or Sagnac-like feedback structures. Although actually dynamics of the optical fields in these physical systems are too complex to be expressed by simple equations of motion, the analogy of the phase-conjugate neuron model and systems with phase-conjugate gain can be made plausible by showing that a rough model of the dynamics of the optical fields inside the cavity can be derived from the phase-conjugate neuron model. The equations of dynamics of phase-conjugate neurons (see eqs. (11.16) and (11.17)) can be easily modified to describe the dynamics of spatially continuous neural fields [399]

$$\tau \frac{\delta y(\vec{r},t)}{\delta t} = -\alpha y(\vec{r},t) + \int_{-\infty}^{+\infty} w(\vec{r},\vec{r}')x(\vec{r}',t)d\vec{r}' \qquad (11.19)$$

$$x(\vec{r},t) = g(|u(\vec{r},t)|)\frac{u^*(\vec{r},t)}{|u(\vec{r},t)|}. \qquad (11.20)$$

In a similar manner, these complex neural fields change their states in such a way that the energy function defined by [399]

$$E = -\frac{1}{2} \int_{-\infty}^{+\infty}\int_{-\infty}^{+\infty} \mathcal{R}\left\{w(\vec{r},\vec{r}')x(\vec{r},t)x(\vec{r}',t)\right\}d\vec{r}d\vec{r}' + \alpha \int_{-\infty}^{+\infty}\int_{0}^{|x(\vec{r},t)|} g^{-1}(s)dsd\vec{r}, \qquad (11.21)$$

reduces its value monotonically with the time evolution of the system. Referring to fig. 11.7, the state of a complex neuron $x(\vec{r},t)$ can be interpreted as the complex amplitude of the optical field emitted from a point at $\vec{r}$ in the crystal. The internal state of the complex neuron $\mu(\vec{r},t)$ can be regarded as a complex amplitude of the grating formed at $\vec{r}$ by the interference between one of the pump beams and the beams originating from other sources $x(\vec{r}',t)$ at points $\vec{r}'$ in the crystal and that reach the point at $\vec{r}$ through reflection at the reflecting surface. M. Takeda et al. interprete the complex synaptic weight $w(\vec{r},\vec{r}')$ as a transmission function that describe the propagation of the beam from the point at $\vec{r}'$ to the point at $\vec{r}$. Based on this interpretation, eq. (11.19) may be considered to represent a writing process of the grating where the complex grating amplitude $\mu(\vec{r},t)$increases in proportion to the sum of the complex fields of the writing beams $\int_{-\infty}^{+\infty} w(\vec{r},\vec{r}')x(\vec{r}',t)d\vec{r}'$. The decay term $-\alpha\mu(\vec{r},t)$ may be considered to express a rate of erasing which the grating undergoes during the writing process. Likewise, eq. (11.20) may be considered to represent a readout process in which a beam is read out that is phase-conjugated to the writing beam and whose amplitude is transformed by a nondecreasing real function $g(\cdot)$. It is even possible to incorporate the effects of gain saturation and/or thresholding into this amplitude transformation function. These two equations of dynamics, eq. (11.19) and (11.20), together form a set of simultaneous equations, so they may reflect the fact that the writing and the readout processes occur simultaneously in the degenerate four-wave mixing.

Since the complex synaptic weight $w(\vec{r},\vec{r}')$ has been interpreted as a transmission function of the optical field, we may expect that the vast majority of optical feedback systems will have symmetric synapses because of the reciprocity theorem of Helmholtz , $w(\vec{r},\vec{r}') = w(\vec{r}',\vec{r})$. This in turn implies that the phase-preserving network model is less suitable for optical implementations because its synaptic weights need to have Hermitian symmetry — a condition that is generally not satisfied by the physical law of

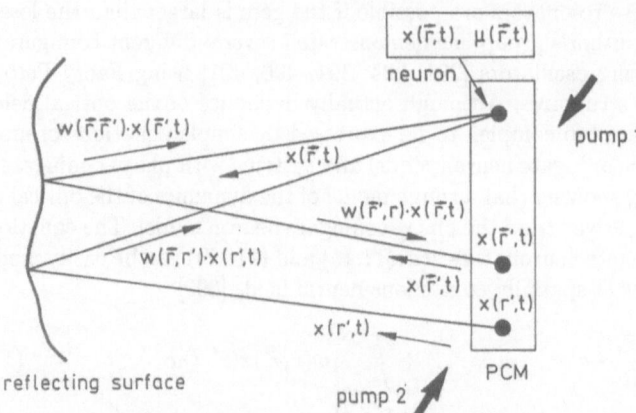

**Figure 11.7:** Scheme of the analogy between the phase-conjugate neuron model and the dynamics of optical fields generated in a phase-conjugate resonator (after [399]).

wave propagation. What does this analogy mean for the phase-conjugate neuron model? Suppose that the phase conjugation mechanism not necessary and the phase-conjugate neurons could be replaced by ordinary phase-preserving neurons. Then any optical feedback system with gain should have an energy function, since we have already seen that the condition of symmetric synaptic weights is automatically fulfilled by the Helmholtz reciprocity theorem. Because of their energy-minimization characteristics, optical fields inside any such optical feedback system with sufficiently large gain should always converge into some stable oscillating modes, irrespective of the shape of the reflecting surface.

This is obviously not what laser physicists usually experience with laser oscillations, in which stable oscillations are generated only when the reflectors form a stable cavity. Thus, the role of the phase-conjugating neurons can be interpreted as making the cavity stable in order to guarantee the convergence of the optical fields into a stable mode. The fact that a phase-conjugating mirror can always form a stable cavity has already been shown in section 6.5. Here, I now have discussed the same behaviour from the viewpoint of its relation to the existence of the Hopfield-like energy function for the optical fields inside the cavity.

The question that immediatedly arises is whether we are able to observe the predicted energy function of the complex neural fields in real experiments. It is generally not possible to specify the synaptic weights $w(\vec{r}, \vec{r}')$ for all the possible optical ray paths between the distributed neurons. Therefore, M. Takeda and T. Kishigami have eliminated them by substituting eqs. (11.19) and (11.20) into eq. (11.21) and obtain [399]

$$E = -\frac{1}{2} \int\limits_{-\infty}^{+\infty} \left\{ |x(\vec{r},t)| \left[ \tau \frac{\delta}{\delta t} g^{-1}(|x(\vec{r},t)|) + \alpha g^{-1}(|x(\vec{r},t)|) \right] - 2\alpha \int\limits_{0}^{|x(\vec{r},t)|} g^{-1}(s)ds \right\} d\vec{r}$$

(11.22)

This relates the energy function to the modulus of the field's amplitude $|x(\vec{r},t)|$.

In most cases, in which the oscillation grows rather slowly, both the field amplitude and the grating amplitude remain small for some time period after the start of the oscillation. In such a weak field limit ($|x(\vec{r},t)|, |y(\vec{r},t)| \ll 1$) for $0 \leq t \leq t_w$, the gain function can be approximated by a linear function $g(|y(\vec{r},t)|) \approx a \cdot |y(\vec{r},t)|$, so that

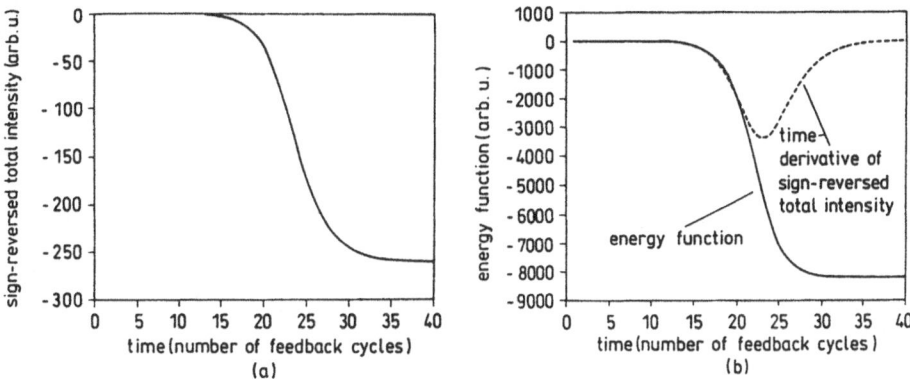

**Figure 11.8:** Sign-reversed total intensity (a), the energy function (b, slid curve) and the time-derivative of the sign-reversed total intensity (b, dashed curve), showing similar behaviour in the weak-field limit (after [399]).

$g^{-1}(|x(\vec{r},t)|) \approx a^{-1} \cdot |x(\vec{r},t)|$, with $a$ being a positive constant. Substituting this case in eq. (11.22) gives

$$E = -\frac{\tau}{4a}\frac{dI}{dt} \qquad (11.23)$$

with $I(t) = \int_{\infty}^{\infty} |x(\vec{r},t)|^2 d\vec{r}$. Thus in the weak field limit, the energy function is proportional to the time derivative of the sign-reversed total intensity of the fields. In other words, the energy function (with its sign reversed), is proportional to the rate of the intensity growth of the oscillating beam. Because $dE/dt \leq 0$, we have

$$\frac{d^2 I}{dt^2} = -\frac{4a}{\tau}\frac{dE}{dt} \geq 0. \qquad (11.24)$$

This indicates that the total intensity $I(t)$ grows as a downward convex function of time. Thus, our question from the beginning of this section can be answered: since the total intensity of the oscillating beam $I(t)$ is a measurable quantity, the behaviour of the energy function in the weak-field limit can be observed by experiments. In fig. 11.8, a numerical simulation of the sign-reversed total intensity behaviour and its time derivative — compared to the energy function — are shown. In the weak-field limit, the analogy in the behaviour is clearly visible. The appropriate experiments have been performed by M. Takeda and T. Kishigami [399] using a photorefractive $BaTiO_3$ phase-conjugate mirror driven by external pump beams. The experimental setup is sketched in fig. 11.9. Experimental proofs of self-oscillations have been given by numerous authors, among them the authors and her coworkers (see e.g. [391, 400, 401]). The aim of these experiments was to observe the analogy to the Hopfield-like energy function through the measurements of the total intensity of the oscillating fields and the computation of its sign-reversed derivative. Therefore, the total intensity of the fields propagating from mirror $M4$ towards the phase-conjugate mirror, given by $\hat{I}(t) = \int_{-\infty}^{\infty} |\int_{-\infty}^{\infty} w(r_x, \hat{r}_x)x(\hat{r}_x,t)|^2 d\hat{r}_x$ is measured. Generally, $\hat{I}(t)$ is not exactly equal to $I(t)$. However, if the transmission function $w(r_x, \hat{r}_x)$ representing the synaptic weights has the form of the Fourier transform kernel, we can regard $\hat{I}(t)$ as being equal to $I(t)$ because of Parseval's theorem of the Fourier transform. From the calculations above it becomes obvious that the derivative of

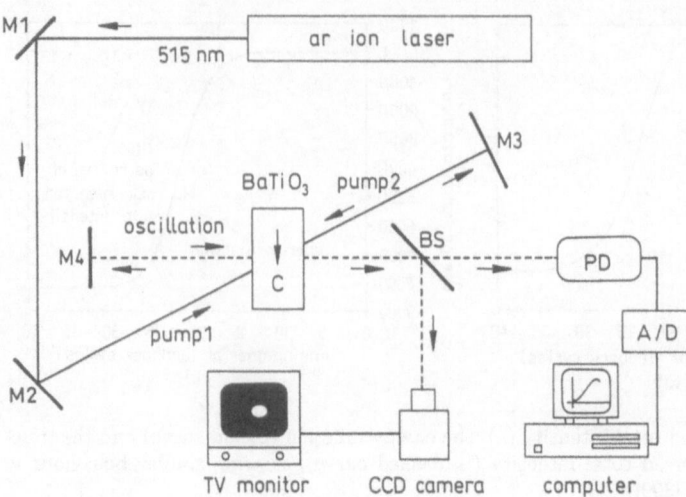

**Figure 11.9:** Experimental configuration for the observation of the Hopfield-like energy function of the optical fields generated in a phase-conjugate resonator. M1 - M4: conventional mirrors; BS: beamsplitter, PD: photodiode; A/D: analog-to-digital converter (after [399]).

the sign-reversed total intensity must be equal to the energy function for low intensities. In fact, M. Takeda and T. Kishigami could show this equivalence impressively.

To summarize, the analogy between the dynamics of our complex phase-conjugate neuron model and the dynamics of self-oscillation generated in a phase-conjugate resonator has been shown. From the physical interpretation of the model, the optical gain medium should have a phase-conjugate property in order for the generated optical fields to have a Hopfield-like energy function that decreases monotonically with the time evolution of the fields. In the weak-field limit, the energy function can be approximated by the time derivative of the sign-reversed total intensity of the fields and is observable by the experiments.

These results are an outstanding example for the importance of finding analogies between neural networks and different physical systems. J.J. Hopfield [10, 11] historically proposed his neural network model and the concept of the energy function based on the analogy to the physics of spin glasses (see section 2.6.4), J. Kirkpatrick et al. [402] as well as T.J. Sejnowski et al. [403] have been motivated by statistical physics to introduce simulated annealing and the Boltzmann machine (see section 2.6). Thus, it is obvious to propose this new network model based on the knowledge of optical physics, rather than to discuss the optical technology issues for the implementation of the existing neural network models. Although applications for complex neural networks are still at the beginning of the development, the optical implementation of the model would be easy and natural, because the model was derived from the analogy to optical physics.

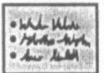 ## Summary

The equality of Hopfield's learning algorithm and the principles of storage of optical interference terms in holographic media is one point which makes the Hopfield network especially attractive for coherent realizations. Moreover, feedback can be achieved with more ease if self-aligning components as optical phase conjugators can be implemented in the setup. Therefore, coherent optical Hopfield networks are built by photorefractive interconnection devices. Special features of these systems are:

- In order to realize bipolarity for the interconnection weights, the Hopfield model can be rearranged to contain only positive numbers and is therefore ideally suited for an optical implementation.
- Thresholding can be accomplished using Fabry-Perot etalons, liquid crystal devices or photorefractive beam coupling devices, driven in the highly nonlinear regime to simulate the sigmoid threshold function.
- Feedback can be performed using unidirectional feedback systems with amplification by photorefractive two-beam coupling or photorefractive phase-conjugating elements. In the latter case, one can exploit the analogy between complex Hopfield networks and oscillator systems with phase conjugation.

## 11.3 Optical Realizations of Interpattern Association Networks

Interpattern association has been discussed in section 2.5 as an extension of the Hopfield model, which pays special attention to the fact that overlap regions between different patterns may be weightened differently. Thus, the information obtained from special areas (e.g. areas that are occupied by only one pattern) is more important than the one from common or overlapped areas. As a result, the associations between reference patterns are emphasized, thus allowing to discriminate even association for patterns that are not independent of each other.

Optically, this network structure can be implemented as easily as Hopfield models, using matrix-vector interconnections and electronic feedback devices. F.T.S. Yu and coworkers modified their compact optical neural network presented in section 11.1 (see fig. 11.2) only slightly in order to implement the interpattern association. Instead of a spatial light modulator used to realize the interconnection weight matrix, a video monitor is used. Again, a lenslet array is used to establish the optical interconnections, bipolar weights are achieved by using two different subsets of interconnection weight matrices that are displayed in a time sequential modus and are subtracted at the end of the learning cycle, and a pocket-sized programmable spatial light modulator is used as the input device. The light beam emanating from each block of the video monitor (i.e. a submatrix of the interconnection weight matrix) is passed through a specific lens of the lenslet array and is imaged on the input SLM. The beams passing the SLM are then imaged onto the output plane by an imaging lens to form an $N \times N$ array output image which represents the product of a four-dimensional interconnection weight matrix with a two-dimensional input array. The output signals can be picked up by an $N \times N$ array photodetector. To construct a closed loop neural network, again the output signals from the detector array are fed back to the input SLM via an electronic threshold circuit. By using the computer to calculate and modify the interconnection weight matrix, the optical neural network is no longer a coherent one, but can be made rather easily both adaptive and programmable.

The results of the interpattern association model show that the model has — compared to the results of the Hopfield model (see section 11.1) — two major advantages, namely fewer interconnection and fewer gray levels. The latter is significant because the

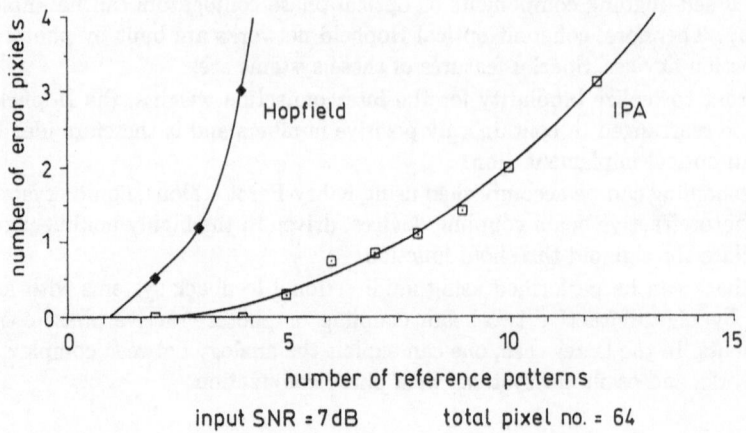

**Figure 11.10:** Comparison of the interpattern association and the Hopfield model (after [12]).

interpattern association model requires only three gray levels to represent the interconnection weight matrix, whereas the Hopfield model needs $2M + 1$ gray levels, where $M$ denotes the number of stored reference patterns. For an input pattern embedded in 30% noise (SNR = 7 db), fig. 11.10 shows clearly that the interpattern association is superior to the Hopfield model for similar patterns. Interpattern association opens also the possibility to include shift- and rotation invariance in the pattern recognition process. The technique replaces the reference exemplar's training set by an encoded binary data set. This has been performed by F.T.S. Yu and coworkers [404] using a two-level network. The first level encodes the spatial features of the input pattern, thus creating a shift- and rotation invariant data set. The second level of the system constructs the interpattern hetero-associative memory.

The second level has been discussed above, so let us here focus our attention on the first level of spatial feature encoding. For this purpose, two methods can be considered — a feature sampling method (i.e. a ring detector), which is rather easy to implement, but has a low discrimination rate, or the use of the circular harmonic expansion, which in turn requires a complex calculation, but achieves higher degree of discrimination.

The method of feature sampling using a ring detector or computer-generated ring detecting data allows to sample the modulus of the Fourier spectrum. Since the Fourier spectrum rotates as the input object rotates, the ring detecting data would remain unchanged due to object rotation. By virtue of the symmetry of the power density spectrum, only half of the spectral domain is needed for the ring detector. However, the ring sampling method does not provide adequate discrimination among similar patterns [404].

The second method, circular harmonic expansion, is based on the expansion of the input pattern into a series of circular harmonic components. If $f(x, y)$ is the input pattern to be expanded into a series of circular harmonic components, it can be written as

$$f(r, \Theta) = \sum_{m=-\infty}^{\infty} f_m(r) \exp(im\Theta) \tag{11.25}$$

where $f_m(r) = (2\pi)^{-1} \int_0^{2\pi} f(r, \Theta) \exp(-im\Theta) d\Theta = | f_m(r) | \cdot \exp(i\phi_m(r))$ is the $m$th order circular harmonic component.

The problem in that method lies in the fact that the circular harmonic expansion coefficient $f_m(r)$ is dependent upon the center of the expansion, which is difficult to calculate. Thus both methods have several drawbacks, but result in a sufficient encoding of the spatial features to allow for space- and rotation invariant pattern recognition. F.T.S. Yu and coworkers [404] could show experimentally in a setup very similar to the one used for interpattern association and for Hopfield networks that shifted and rotated input patterns which are embedded in 15% additive noise and partial inputs could be recognized without problems and associated with the corresponding exemplars.

However, one of the obvious disadvantages of the system is that the actual position and orientation — or in other words the spatial information content — of the input objects is lost.

---

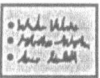 ## Summary

---

The interpattern association model allows to discriminate the common and special features of the reference patterns.

- Using optoelectronic devices, the interconnection weight matrix can be easily realized, requiring merely an SLM with three gray levels and a video monitor.
- Computer simulations as well as experiments revealed that the model is more effective in performing pattern recognition among similar patterns than the models as the Hopfield model that deal only with the intrapattern association.
- The model can be extended to realize shift- and rotation invariant pattern recognition.

## 11.4 Optical Realizations of Boltzmann Networks

The essential feature of Boltzmann machines compared to Hopfield nets is their ability to let the system escape from local minima and converge to a global one by "shaking" the energy function first in a strong and later on in a more and more decreased way — a feature that is more commonly known as simulated annealing. Consequently, optical realizations of Boltzmann machines are all based on optical realizations of simulated annealing. In the following I will discuss two of the possibilities to realize optical structures that are able to implement simulated annealing processes.

### 11.4.1 Simulated Annealing in Photorefractive Feedback Systems

 First, let us consider a basic system as I have discussed it already in the context of matrix-vector multiplications (see section 4.3.3) and Hopfield networks (see sections 11.1 and 11.2). The network, realized by N. Farhat and D. Psaltis [393] and shown in fig. 11.11, consists of $N$ neurons and is partitioned into three groups. Two groups, $V_1$ and $V_2$, represent visible units that can be viewed as input and output groups, respectively, and the third group $H$ are hidden or internal units. The

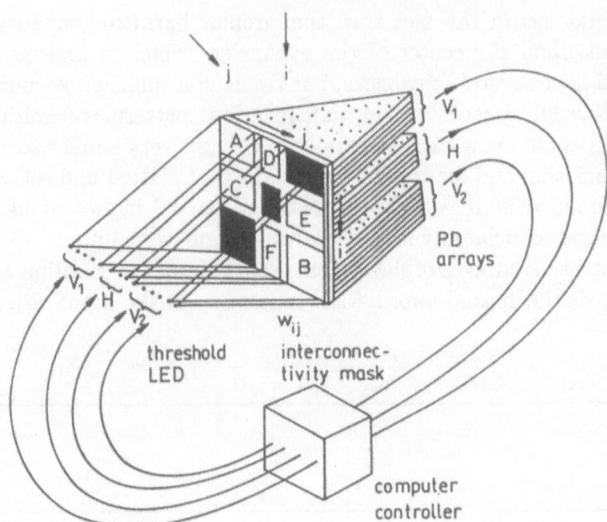

**Figure 11.11:** Optoelectronic analogue of a self-organizing neural network partitioned into three layers in order to realize stochastic self-programming and learning (after [393]).

partition is such that $N_1 + N_2 + N_3 = N$, where $N_1, N_2, N_3$ refer to the number of neurons in the $V_1, V_2$ and $H$ groups, respectively. The interconnectivity matrix $w_{ij}$ is appropriately partitioned into six submatrices shown as blackened or opaque regions in the $w_{ij}$ mask. The LED array represents the state of the neurons, assumed to be unipolar binary (on/off of the LED's), the $w_{ij}$ mask represents the strength of interconnection between neurons in a manner similar to the arrangement presented in section 11.1. Light from each LED is smeared vertically over the corresponding column of the $w_{ij}$ mask with the aid of an anamorphic lens system. Light emerging asymmetrically from each row of the mask is focused with the aid of another anamorphic lens system onto the corresponding elements of the photodetector array. For convenience, both lens systems are not shown in fig. 11.11. Bipolar values are realized using the dual rail concept of two subrows that will be subtracted electronically (see section 4.2.4 for further details).

The partitioning into three groups of the interconnection unit results in several submatrices, that are responsible for different interconnection areas. Submatrix $A$ ($N_1 \times N_1$ elements) provides the interconnection weights between units or neurons within group $V_1$. Submatrix $B$ ($N_2 \times N_2$ elements) does the same for group $V_2$. Submatrices $C$ ($N_1 \times N_3$ elements) and $D$ ($N_3 \times N_1$ elements) provide the interconnection weights between units of $V_1$ and $H$ and submatrices $E$ ($N_2 \times N_3$ elements) and $F$ ($N_3 \times N_2$ elements) finally provide the interconnection weights of units $V_2$ and $H$. By blackening certain regions of the interconnection weight matrix $w_{ij}$, units in $V_1$ and $V_2$ can not communicate among each other directly. In the same way, similar units within $H$ are prevented from communication by blocking the center square of $w_{ij}$.

Thresholding is accomplished in this setup in a global way. The LED element is of graded response. Its output represents the state of an auxiliary neuron in the net that is always on to provide a global threshold level to all units by contributing only to the light focused onto negative photosites of the photodetector arrays from pixels in the $G$ column in the interconnectivity mask. This is achieved by suitable modulation of the transmittance of pixels in the $G$ column.

By introducing a computer-controlled nonvolatile spatial light modulator to implement the $w_{ij}$ mask in the network and including a computer controller as shown in fig. 11.11, the system can be made self-programming with the ability of modifying the weights of synaptic links between its neurons. This is done by fixing or clamping the states of the input ($V_1$) and output ($V_2$) groups to each of the associations we want the net to learn and by repeated application of the simulated annealing procedure with the Boltzmann (or other stochastic update) rules and collection of the statistics on the states of the neurons at the end of each run when the net reaches thermodynamic equilibrium.

How is stochastic learning by simulated annealing performed in such a partitioned, optical net? Let us consider the steps in that net in a similar way as I have done it in chapter 2.6. Note that these steps are performed optically, but that the sequence of steps and the decisions are still realized by a computer [393].

1. Clamp $V_1$ and $V_2$ to the desired association and keep $H$ free running with an initially arbitrary interconnection matrix $w_{ij}$.

2. Randomly select a neuron in $H$ (neuron $k$) and flip its state (remember that we are dealing with binary neurons).

3. Calculate the change $\Delta E_k$ in the global energy $E$ of the net caused by this change in the state of the $k$th neuron. If

$$\Delta E_k \begin{cases} < 0 & \text{adopt the change} \\ \geq 0 & \text{calculate the Boltzmann probability factor: } P_k = \exp\left(-\Delta E_k / T\right). \end{cases}$$
(11.26)

Compare the outcome to a random number $N_r \in [0, 1]$. If

$$P_k \begin{cases} < N_r & \text{adopt the change of states of the } k\text{th neuron,} \\ & \text{even if it leads to an energy increase} \\ \geq N_r & \text{discard the change and return the } k\text{th neuron to its original state} \end{cases}$$
(11.27)

4. Select once again a neuron in $H$ randomly and repeat the previous steps.

5. Repeat steps (1. - 4.) reducing at every cycle the temperature $T$ gradually (e.g. $T = T_0 / \log(1 + m)$) ($m$ = cycle number) until a situation is reached where changing the states of the neurons in $H$ does not alter the energy $E$ any more. In other words, find $\Delta E_k \to 0$. Then the thermodynamic equilibrium or the state of the global energy minimum has been reached. Recall that the temperature $T$ determines the fineness of search for a global minimum and is the critical parameter for a successful search. A high $T$ produces coarse search and low $T$ a finer grained search.

6. Record the state vector at thermodynamic equilibrium, i.e. the states of all neurons in the net or those in $H$ and those in $V_1$ and $V_2$ that are clamped.

7. Repeat steps (1. - 6.) for all other associations on $V_1$ and $V_2$ the net should learn and collect statistics on the states of all neurons by storing the states at thermodynamic equilibrium in the computer memory. This completes the first phase of exposing the net to its environment.

8. Generate the probabilities $P_{ij}$ of finding the $i$th neuron and the $j$th neuron in the same state. This completes the first phase of learning in a Boltzmann net, which we called the incremental phase in section 2.6 and which reinforces the interconnections to be learned. In this phase, the nonvisible, free-running units settle according to

$$\mu_i = f_h \left[ \sum_j \frac{w_{ij} \mu_j}{T} \right]$$
(11.28)

where $\mu_i$ are the internal states, $f_h$ is the threshold function as the sigmoidal one, and $T$ is the temperature which decreases during the process of settling down. For the application of the mean-field theory, the variables $\mu_i$ can be interpreted as the thermal averages of the binary variables: $\mu_i \to < \mu_i >$.

9. Unclamp neurons in $V_2$ and let them run free as with neurons in $H$.

10. Repeat steps (1. - 8.) for all input vectos $V_1$ and collect statistics on the states of all neurons in the net.

11. Generate the probabilities $P'_{ij}$ of finding neuron $i$ and neuron $j$ in the same state.

12. Increment the current connectivity matrix $w_{ij}$ by $\Delta w_{ij} = \epsilon(P_{ij} - P'_{ij})$, where $\epsilon$ is a constant representing and controlling the speed of learning. This completes the second phase of the learning cycle — the decremental phase of unlearning as we have called it in section 2.6. In this phase, the internal and the input units settle according to

$$\mu'_i = f_h \left[ \sum_j \frac{w_{ij}\mu'_j}{T} \right] \qquad (11.29)$$

13. Repeat steps (1. - 12.) until the increments $\Delta w_{ij}$ tend to zero or — in real situations — become smaller than some prescribed small number. When this is accomplished, the net has captured the underlying structure or formed its own representations of its environment defined by the associations presented to it.

The search for the state of the global energy minimum is — as I have already discussed in detail in section 2.6 — basically a gradient descent procedure that allows for probabilistic hill climbing to avoid entrapment in a state of local energy minimum. The relative probability of two global states $\alpha$ and $\beta$ is given by the Boltzmann distribution $P_\alpha / P_\beta = \exp(E_\beta - E_\alpha)/T$. Therefore the lowest energy state is the most probable at any temperature and is sought by the procedure.

One of the advantages of the Boltzmann net that distinguishes it from other nets is that there is no need for a lack in correlation among the stored vectors in order to realize ideal storage and recall. In optical implementations, this is the big difference that distinguishes optical Boltzmann nets from all nets with resonant optical configurations, where orthogonal modes are used to represent the stored vectors. In fact, learning by simulated annealing in a Boltzmann machine looks for underlying similarities or correlations in the training set to generate weights that can make the net generalize. Generalization in turn is a property where the net recognizes an entity presented to it even though it was not among those specifically used in the learning session.

Evidently, the stochastic nature of the algorithm (involving probabilistic state transition rules and simulated annealing) makes the learning procedure lengthy, which makes the system taking hours in a digital simulation for nets of a few tens to a few hundred neurons. Hence, speeding up the process by using analog optical or optoelectronic implementations is one of the major contributions of optics in order to improve the implementations of Boltzmann networks. Therefore, in the following I will discuss some of the modifications of the Boltzmann net algorithm described above to take advantage of the available parallelism of optics and speed up stochastic learning by several orders of magnitude compared to serial digital implementations. The ideas go back to the early publication of N. Farhat in 1987 [393].

In learning by simulated annealing, the energy $E$ of the net has to be calculated, as I have described in section 2.6.1. A simple optoelectronic analog circuit for calculating the contribution $E_i$ to the global energy $E$ of the net requires the use of an electronically addressed nonvolatile binary (on-off) spatial light modulator consisting of a single column of $N$ pixels — e.g. a parallel addressed magnetooptic spatial light modulator (MOSLM). The function of this modulator will then be as follows. A fraction of the focused light emerging from each row of the $w_{ij}$-mask is deflected by a beam splitter onto individual pixels of the vertical column MOSLM such that light from adjacent pairs of subrows of $w_{ij}$ falls on one pixel of the MOSLM. The MOSLM pixels are overlayed by a checkered binary mask, in which opaque and transparent pixels are staggered in such a fashion that light emerging from the left subcolumn will originate from the positive subrows $w_{ij}^+$ only and light emerging from the right subcolumn will originate from the negative subrows $w_{ij}^-$. By separately focussing the light from the left and right subcolumns onto two photodetectors and subtracting and halfing their outputs — thus realizing the standard dual channel implementation of bipolar weights (see section 4.2.4) — one obtains

$$E = -\frac{1}{2} \sum_i \left[ \sum_{j \neq i} \left( w_{ij}^+ - w_{ij}^- \right) \vec{x}_j \right] \vec{x}_i = -\frac{1}{2} \sum_i \sum_{j \neq i} w_{ij} \vec{x}_i \vec{x}_j \qquad (11.30)$$

which is the required global energy.

The learning procedure described above requires fast random number generation for use in random drawing and switching of the state of the neurons from $H$ (during the first phase of learning) and from $H$ and $V_2$ (during the second phase of learning). Another random number is also needed to execute the stochastic state update rule when $\Delta E_k > 0$. Although fast digital pseudorandom number generation up to $10^9 \mathrm{s}^{-1}$ is feasible, it is more efficient to use an optoelectronic random number generation when the total number of neurons in the net is large, despite the slower rate of $10^5$s. An optoelectronic method for generating the Boltzmann probability factor $p(\Delta E_k)$ is to employ speckle statistics, an optical one is to use photon counting image acquisition systems or clipped laser speckles.Photon counting image acquisition systems have the advantage of being able to generate normalized random numbers with any probability density function. A more important advantage of optical generation of random number arrays however is the ability to exploit the parallelism of optics to modify the simulated annealing and the Boltzmann machine formalism detailed above to achieve a significant improvement in speed. With parallel optical random number generation, a spatially and temporally uncorrelated linear array of perculating light spots of suitable size can be generated and imaged on the photodetector array of fig. 11.11 such that both the positive and negative photosites of the photodetector array are subjected to random irradiance. This introduces a random noise component in the threshold, which can be interpreted as a bipolar noisy threshold. The noisy threshold produces in turn a noisy component in the energy. The magnitude of the noise components can be controlled by varying the standard deviation of the random light intensity array irradiating the photodetector array. The noisy threshold therefore produces random controlled perturbation or shaking of the energy landscape of the net. This helps shake the net loose whenever it gets trapped in a local energy minimum. The procedure can be viewed as generating a controlled deformation of tremor in the energy landscape of the net to prevent entrapment in a local energy minimum and thereby ensure convergence to a state of the global energy minimum. Both, the random drawing of neurons and the stochastic state update of the net are now done in parallel at the same time. This leads to significant acceleration of the simulated annealing process. Moreover, the speed enhancement will be dependent on the characteristics of the light emitting array, the photodetector array, the spatial light modulator and the speed of the computer-controlled interface used. In summation, the enhancement over digital serial computation can be significant, approaching 5-6 orders of magnitude, especially for large multilayer networks consisting of a few hundred neurons [393].

## 11.4.2   Photorefraction for Mean-Field Learning

 My second description of the optical realization of simulated annealing deals with the potential of photorefractive crystals for mean-field theory learning (see section 2.6.3). To achieve this aim, the Boltzmann algorithm has to be subjected to two minor, but important modifications. Both have been described by C. Peterson et al. [369] and are concerned with the method of updating. In my derivation in the second chapter, updating or learning takes place with the learning rule

$$\Delta w_{ij} = \beta(\mu_i \mu_j - \mu_i' \mu_j') \tag{11.31}$$

where $\beta$ is the learning rate. In contrast to the original mean-field theory learning, where asynchronous updating took place [405], synchronous updating is natural in optical implementations. However, iterative techniques as eq. (11.28) and eq. (11.29) are less efficient (a factor two more iterations are needed [406]) and are known to tend to cyclic behaviour when synchronous updating is used [407]. Second, eq. (11.31) requires storing the solutions of eqs. (11.28) and (11.29) and a subtraction, which should also be realized in optical implementations. However, there are no fundamental obstacles to realize intermediate updating with the following sequence

$$\Delta w_{ij} = \beta \mu_i \mu_j \tag{11.32}$$
$$\Delta w_{ij} = -\beta \mu_i' \mu_j' \tag{11.33}$$

Again, we have to think how to realize bipolar weightening in this case. Although the methods of implementation of bipolar weights are the same as presented in section 4.2.4, I want to discuss here the consequences these techniques have on the realization and performance of a mean-field theory learning algorithm.

C. Peterson and coworkers [369] suggested first to use the standard model of two sets of positive gratings, one for positive enforcement and one for the negative one. The negative sign is then added electronically with a subtraction, giving the local update rule

$$\mu_i = f_h \left[ (\sum_j w_{ij}^+ \mu_j - w_{ij}^- \mu_i) \right] \tag{11.34}$$

In the modified mean-field theory learning algorithm, the adjustment of eq. (11.33) is always positive, the one of eq. (11.33) always negative. So the clamped, incremental phase needs to use only positive weights and the decremental phase needs only negative weights. In a photorefractive realization, each connection strength is represented by two gratings, $w_{ij}^+$ and $w_{ij}^-$. Thus, as I have shown in section 4.2.4, $n$ neurons require $2n^2$ connections for full interconnectivity. In the read cycle, the reference beam is iteratively processed through the crystal and its stored gratings, until is has settled according to eq. (11.28) or eq. (11.29). The write cycle is slightly more complicated and differs between clamped and free phases. In the clamped phase, the beam again settles according to eq. (11.28). It then splits into the reference and the object beam. The two beams then impinge simultaneously onto the crystal, with the object beam hitting only to $w_{ij}^+$ columns. The decremental phase proceeds in the same way except that the object beam hits only the $w_{ij}^-$ columns.

The second model of of bipolar weight realization by C. Peterson et al. is a coherent one, where negative weights are realized without doubling the number of gratings per connection. In this well-known method, the weight subtraction of the free phase (which is equivalent to eq. (11.33)) takes place by shifting the phase of the holographic gratings by $\pi$. Interconnection strengths of negative sign are still accomplished by an electronic subtraction. Then, the local updating rule takes the form

$$\mu_i = f_h \left[ \sum_j (w_{ij} + \alpha)\mu_j - \alpha \sum_k \mu_k \right], \qquad (11.35)$$

or in the formulation of the mean-field algorithm

$$\mu_i = \frac{1}{2} \left\{ 1 + \tanh \left[ \sum_j \frac{(w_{ij} + \alpha)\mu_j - \alpha \sum_k \mu_k}{T} \right] \right\} \qquad (11.36)$$

where $w_{ij}$ is in the range of $-\alpha$ to $\alpha$. The argument of the *tanh* function is evaluated in the photorefractive crystal by the matrix multiplication with the diffraction efficiency $\sqrt{\eta} = w_{ij} + \alpha$. In addition, a row of gratings with strength $\alpha$ performs the sum over $\mu_j$. Subsequently, subtraction in eq. (11.36) is performed electronically, giving negative signs of $w_{ij}$ as in the double column case above. The write cycle for the mean-field theory algorithm works in this case as follows:

- In the clamped, decremental phase, the $\mu_i$ settle according to eq. (11.36) and then recording is performed with a single phase column. The partially destructive nature of a readout can be exploited to accomplish negative weight modifications. The change in the weight strength, $\Delta w_{ij}$, during such a modification period is given by

$$\Delta w_{ij} = (\Delta \eta)^{1/2} = \text{"read} = \text{decay"} + \text{"write} = \text{reconstruction"} \qquad (11.37)$$

For short readout times, the photorefractive writing and read out process can be written (see section 4.4) as

$$\begin{aligned} \eta &= \eta_{\max} \cdot t_1^2 / \tau_w^2 &= \eta_{\max} t_1 c_w I_w \\ \eta &= \eta_0 (1 - t_2/\tau_r)^2 &= \eta_0 (1 - t_2 c_r I_r) \end{aligned} \qquad (11.38)$$

where the time constant $\tau_w$ of the writing process is inversely proportional to the write intensity $I_w$, $\tau_w = 1/c_s I_w$ and the time constant $\tau_r$ of the decay process is inversely proportional to the read intensity $\tau_r = 1/c_r I_r$ [369]. This gives

$$\begin{aligned} \Delta w_{ij} &= (\Delta \eta_0)^{1/2} (1 - t_2 c_r I_r) + t_1 c_w I_w \qquad \text{or} \\ \Delta w_{ij} &= (\Delta \eta)^{1/2} = t_2 c_r I_r - t_1 c_w I_w. \end{aligned} \qquad (11.39)$$

- The free, decremental phase is slightly more complicated since all gratings are by definition positive. However, it can be accomplished by manipulating the interacting beams in the spatial light modulator in their intensities according to four different, sequentially performed writing steps. First, beam 1 has the value $\sqrt{k}$, where $k$ is a constant, and beam 2 is given by $\mu_i'$, giving the grating $\sqrt{k}\mu_i'$. Next, beam 1 is equal to $\sqrt{k} - \mu_j'$, beam 2 is equal to $\sqrt{k}$, resulting in the grating $k - \sqrt{k}\mu_j'$. In the third step, beam 1 is given by $\sqrt{k} + \mu_j'$, beam 2 is $\sqrt{k} - \mu_j'$, giving the grating $k - \sqrt{k}\mu_j' + \sqrt{k}\mu_j' - \mu_i'\mu_j'$. Finally, beam 1 is a plane wave and beam two is set to $2k$, giving the grating $-2k$.

The experimental setup to realize the mean-field learning algorithm with photorefractive crystals has already been shown in section 9.1.4, fig. 9.5. This setup is versatile because it is not tied to a specific learning algorithm, but is able to implement a wide range of supervised learning algorithms as mean-field theory, backpropagation and Kanerva-style networks. Again, the architecture is based on a single photorefractive crystal with spatial multiplexing that is able to handle hidden units and places no restrictions on connectivity.

## Summary

The essential feature of Boltzmann machines is their ability to let the system escape from local minima by simulated annealing. Consequently, optical realizations of Boltzmann machines are based on optical realizations of simulated annealing, mainly driven by computer-algorithms. This can be done using

- a conventional matrix-vector interconnection scheme, e.g. built by photorefractive gratings. In learning by simulated annealing, the energy $E$ of the net has to be calculated. This can be accomplished by an electronically addressed nonvolatile binary (on-off) spatial light modulator with a single column of $N$ pixels — e.g. a parallel addressed magnetooptic one (MOSLM), driven by a computer.
- photorefractive interconnection schemes in combination with computer-driven bipolar, coherent weight modifications, input by liquid-crystal spatial light modulators.

## Further Reading

1. B.K. Jenkins, A.R. Tanguay, *Photonic Implementations of Neural Networks*, Chapter 9 in *Neural Networks for Signal Processing*, B. Kosko, Ed., Prentice-Hall (1992). A description of optical neural network realizations including realizations of Hopfield and Boltzmann networks.
2. N.H. Farhat, D. Psaltis, *Optical Impelemtation of Associative Memory Based on Models of Neural Networks*, in *Optical Signal Processing*, Ed. J.L. Horner, Academic Press (1987). This chapter discusses the early realizations of different optical associative memories using the Hopfield model.

# 12  Optical Realizations of Adaptive Resonance Theory Networks

Since its conception in 1987 by G. Carpenter and S. Grossberg [15, 9], the adaptive resonance theory (ART) has become more and more attractive in solving a growing number of problems, especially in group-technology problems and in motion control. This is mainly due to its stable unsupervised learning properties and its ability to process large input-pattern fields. Successful electronic applications of ART used input fields in excess of $10^7$ nodes.It seems that ART's scaling properties are limited only by hardware and software implementations — a significant motivation for many researchers in the past years to exploit the features of ART in many different applications using optical realizations with their inherent potential of parallel processing.

Because at the same time ART is one of the most complex neural network algorithms, let me review the outline of the operation of ART — in order to recall the description of the network in section 2.7. My aim ist merely to point out once again here some properties of ART networks that are essential to their operation, yet presenting hardware challenges: rest and massive connectivity. Again, as in section 2.7, I have substituted most of S. Grossberg's terminology by terms that are more instructive to somebody who wants to get a first insight in the features of ART. I will here refer to ART1 as the first of the series of ART networks that S. Grossberg and his coworkers have developed.

An ART network has several important layers as shown in fig. 12.1. The recognition layer $R$, the comparison layer $C$, the input layer $I$, the vigilance layer $V$, and $Re$ the reset layer. From the left to the right in fig. 12.1, the ART units action is shown, First, the input is registered at the comparison layer and fed up into the recognition layer (fig. 12.1a). The recognition layer's winner-takes-all property finds the node corresponding to the initial best guess (fig. 12.1b). This guess is tested by replaying the winning nodes' previously learned template (or weight values) onto the comparison layer. The input layer, which still contains the original input, and the comparison layer, containing the winning template, compare these two patterns by sending competing signals to the vigilance node (fig. 12.1c). When the match is not good, the vigilance node may activate the reset layer, which suppresses only nodes at the output that have been recently active (e.g. only the prior one), and has no effect on the rest (fig. 12.1d). In this example, only the prior winner is affected. With this node removed, the network reclassifies the pattern and continues to do so until it finds a good match. When such a good match is found, the winning node's activation is permitted to continue. The simultaneous activation of the input nodes and a winning recognition node creates a resonance. This resonance is frustrated by the reset mechanism until a good match is found. The key is that the pattern classification takes place in a feedback loop and that learning does not set in until resonance is permitted to persist. Until then, the reset mechanism permits a search for a better pattern match, removes all classifications considered previously, and suspends learning until the best answer is found.

This short description already shows, that the communication between the comparison layer and the recognition layer requires a massive number of interconnections. Moreover,

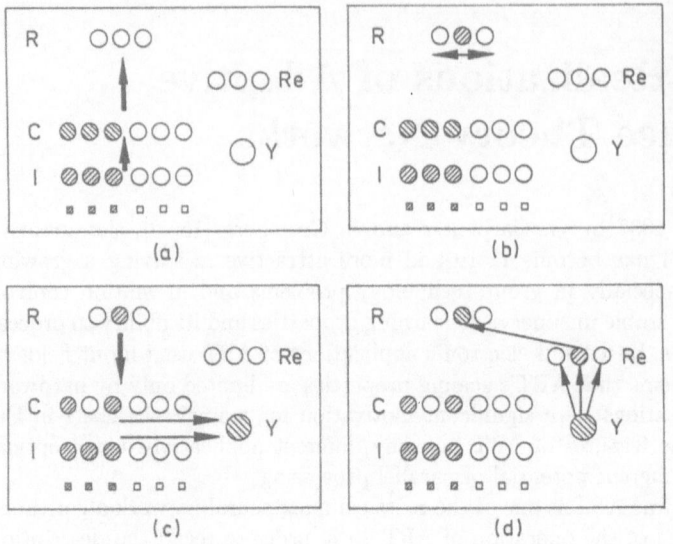

**Figure 12.1:** Network dynamics of an ART1 network: bottom-up pattern matching (a,b) is balanced by top-down feedback expectancy (c,d) (after [408]).

these connections must be able of real-time adaptation. This is the most compelling argument for the importance of optical implementations of ART neural networks.

In the following sections, I will discuss some of the optical and optoelectronic implementations that have been recently discussed. Most of them are still in a premature state, but seen in relation with the young age of the ART algorithm itself, already impressing results have been achieved in order to realize optical ART networks that are able to solve pattern recognition problems. I will first discuss a proposal for a hybrid optoelectronic adaptive theory processor, where parallel computations are relegated to free-space optics, while serial operations are performed in very-large-scale integrated electronics. Next, I will describe a holographic adaptive resonance theory system which self-searches through the data stored in a highly logical means, making especially use of the symmetric bottom-up and top-down (feedforward and feedback) structure of the algorithm. This is a concept that is completely different from the previous one, because it exploits the similarity of Fourier-optical pattern recognition and neural network structures in order to realize a concept that resembles the ART algorithm. Finally, I will go back to the resonant structure using a cavity with two photorefractive phase conjugators — the same I have discussed in section 6.4 — and exploit its structure to realize the different layers necessary in ART. This third approach is based on the assumption that cavity modes can be used as the energy minimum storage functions in a neural net in order to store input data as linear combinations of these eigenmodes. These three rather different approaches to the concept of adaptive resonance theory will give an insight in the vast possibilities, optics provide in order to realize highly complex networks as the adaptive resonance theory.

## 12.1 Optoelectronic Possibilities to Realize Adaptive Resonance Theory

When we want to realize an optoelectronic architecture of the ART1 neural network, we have to realize the operations we have described in the previous section. Several operations among them are matrix-vector multiplications, arising from the comparison and the vigilance phase. In fact, the major computational load arises from the computation of the vector dot product and norms. Fortunately, we have already seen in section 4.3, that optical systems can perform these types of operations efficiently. In contrast, all number comparisons (e.g. whether the norm $S$ is larger than $\rho$), simple divisions and additions can be realized easily using electronic devices.

Here, I want to show in relation to the publication of T.P. Caudell [408] for a single example, that all vector and matrix-vector operations in an ART1 neural network can be performed in parallel by using a modification of a well-known optoelectronic processor. Already in 1985, N. Farhat and D. Psaltis proposed and demonstrated the use of an optical marix-vector multiplication processor in a Hopfield neural network system (see section 11.1, fig. 11.1). The conceptually simple system performs a multiplication between a vector whose components are represented by the intensity of a light emitting diode array and a matrix whose components are represented by the transmission coefficients of the pixels in a two-dimensional spatial light modulator. The system is controlled by a digital computer, which contains in its memory images of the vector and the matrix to be multiplied as well as the controlling program. The vector and the matrix are loaded into the optical processor through electronic interface devices, and the resulting product is read back into memory from a controller for the detector array. How can this system now be used as a coprocessor in the execution of the ART1 algorithm? First, a two-dimensional binary image can be mapped onto the one-dimensional emitter array by lining up the lines one after the other. Moreover, the first memory template or exemplar can be mapped onto the two-dimensional SLM device. Thus, the system may produce the dot product of $T \cdot X$ for the vigilance test (see section 2.7). These dot products are read from the detector into the memory, and the remainder of the algorithm is executed within the digital computer. This mode of operation has also been presented by D. Wunsch et al. [409], realizing thus a 4f optical correlator implementation of an ART1 network.

A system using VLSI analog processing additionally to the above concept has been presented by P. Caudell [410]. Here, a smart input array substitutes the LED array, a smart SLM is used instead of a large ferroelectric SLM, and a smart output array is used. Especially the smart output array chip is an important modification compared to the previous setup, because it replaces most of the digital computer functionality. The chip incorporates a linear array of detectors, a linear array of emitters, an array of analog storage registers, analog scalar arithmetic operators and a winner-takes-all network. The latter implements the maximum operation required in the ART algorithm. The registers store the input norm $\| X \|$ computed through the first transparent row in the SLM and the template exemplar norms $\| T_k \|$ that are computed by loading the unit vector in the VLSI array. These values are used at various stages in the ART1 algorithm. They are combined with the dot products by using analog scalar arithmetic operators to feed the winner-takes-all network, which is also implemented in analog circuits. Specialized control circuitry regulates the reset and search processes and controls the logic of the comparison operations. In addition, a single mode control signal, is sent back to the smart input array along a small cable. Control of the smart SLM is performed by the smart output array indirectly through modulation of its emitters, which shine light back through the second cylindrical lens to illuminate a row on the modulator. The reverse illumination helps to perform the conjunction of the input pattern and the winning template and it helps the copying of an input pattern into a new template. These operations occur in the cells of the SLM when their reverse illumination highlights a template row and the smart input array is in the appropriate mode of operation. The four modes of operation that are possible in the smart input array are controlled by the signal from the smart output array. During the computation of the input-pattern-template dot products and the norm of the input pattern, the smart input array is operated in mode 0. In this mode, the emitter array illuminates the columns of the smart SLM with the binary input pattern previously clocked into the array's local buffers by means of a rapid shift register action. When operated in the third mode, all emitters are turned on to compute simultaneously the norms of the templates. During these two modes of operation, the smart output array is not reverse illuminating the SSLM.

During the copy and update operations, the smart output array will have selected a template row on the smart SLM for modification, owing to either a match with an existing template or a novel pattern detection. This selection manifests itself through the reverse illumination of a template row on the SLM. The state of each pixel on the SLM is determined by the state of a local flip-flop memory element (see also section 6.3). Thus, at power up time, each state is set to logical ON by a reset circuit. The state of the pixel is toggled when a threshold amount of light is received on the pixel's photodetector. The threshold is adjusted so that no state change occurs when only the forward or the reverse illumination is present alone. Both must be simultaneously active before a toggle of a SLM cell state can occur.

To copy an input pattern into a row on the SSLM, the smart output array illuminates the correct row, while the smart input array in mode 1 illuminates the columns to be toggled with the negation of the input pattern. Since the initial state of each pixel in an unused row is ON, the template pixels in the selected row and selected columns are toggled to the OFF state if the template pixel was initially ON and the input pixel OFF. To perform the conjunctive update operation, the smart output array illuminates the correct row, while the smart input array, again in mode 1, illuminates the columns to be toggled with the negation of the input pattern. This operation is an in-plane conjunction of the template and the input pattern. Note that the binary information stored on the SSLM does not change when the reverse illumination is removed.

The design of these three devices — the smart output, the smart input and the SSLM — combined with the two cylindrical lens systems, constitute a processor that could efficiently implement the entire ART1 algorithm. A promising concept of a physical implementation of this system by using ferroelectric liquid crystal spatial light modulators is presented in [410] and is shown in fig. 12.2.

The light source floods here the ferroelectric liquid-crystal SLM's on the smart input and smart output arrays with polarized light. This way the devices may act as emitters by reflecting the light towards the smart SLM while modulating the polarization. Polarizing beam splitters pass rotated polarization beams and deflect the unroted beams back towards the light source. A third polarizing beam splitter in front of the smart SLM combines the polarized beams from the two emitting arrays, which permits threshold logic to be performed during the template copy and update operations. This optical system does not require a coherent light source since no interference phenomenon is used in the processing. A narrow wavelength band may be required to achieve easonable contrast ra-

**Figure      12.2:**
One possible physical
implementation of an
ART1 neural network
by using FLC-SLM
CMOS devices. P.Bs.
are polarizing beam
splitters (after [410]).

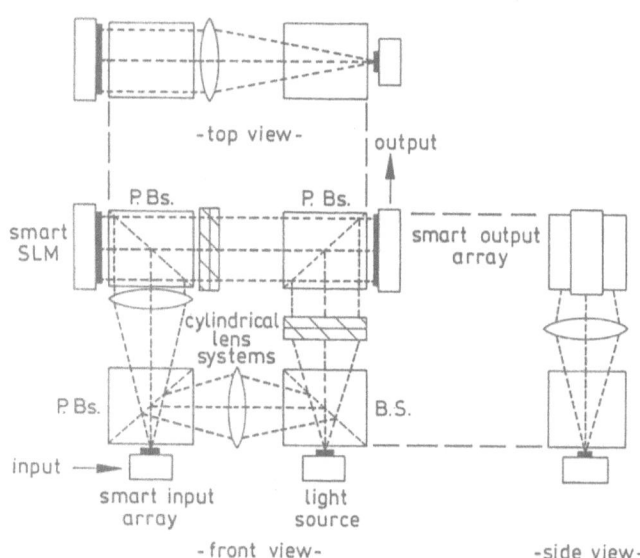

tios with the ferroelectric liquid-crystal SLM's. Consequently, high-power light-emitting diode arrays may supply the input illumination.

One of the most intriguing advantages of that system is that it can potentially occupy a small volume. If the cylindrical lenses are replaced by solid gradient-index elements or computer-generated holographic elements, the implementation concept could occupy a volume of less than 10 cm$^3$. The alignment could be frozen by cementing the elements into a solid structure. In principle, no computers would be required for operation, although external circuitry would be necessary to clock in the input array and to make use of the outputs. It is also important to know that the system is inherently quick, reaching up to a process and classification time of about 25 input vectors per second if a standard workstation computer is used. Thus the design has the potential to classify up to 10$^5$ small images per second into many types of classes, while remaining resilient to new or novel input patterns.

The major disadvantage of this design is the mapping of a two-dimensional input image into a one dimensional array. In many applications, input images are about or larger than 256 × 256 pixels. This size would require a SLM with 64.000 pixels along one axis — resulting e.g. in pixel sizes of 10$\mu$m × 10$\mu$m in a slab of 65 cm, a number that is clearly not within the reach of technical realization.

Moreover, the system at its present state is not suitable for optical cascading of these modules. To achieve this aim, a modification of the smart input array is necessary that allows an optical parallel load (or a write-with- light mode) beneath the electronic serial loading of the input vector. This could be done using at each pixel an additional photodetector that can be clocked to read the pattern of light on the array and to load this information into the input buffers. Additionally optics are necessary to map the square subimage onto a linear vector and to arrange the outputs of the first layer for the input to the second layer. Such a hierarchy of ART1 modules could be realized if the image is partitioned into small nonoverlapping patches, each patch being fed to individual ART1 modules. The outputs of this layer of separate ART1 modules are then funneled into a

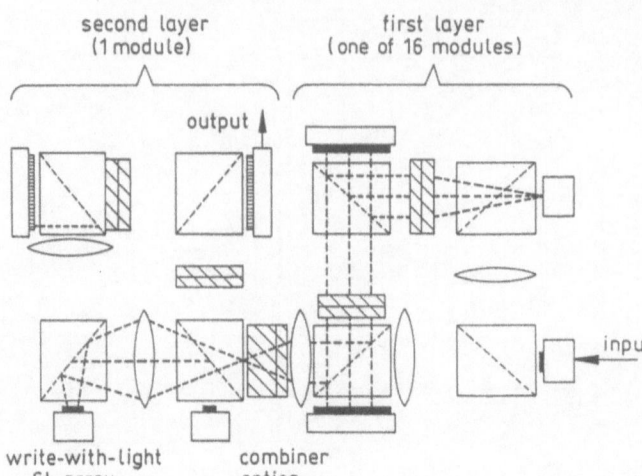

**Figure 12.3:** Optical processors cascaded together to form a macrocircuit of ART1 networks. Note that the SI array has been modified to include detectors that permit optical writing of the input vector. This extra circuitry also provides a mechanism for reading the learned templates stored on the smart SLM (after [410]).

single ART1 module that combines the lower level results. The vigilance parameters are adjusted so that this architecture acts as a voting machine, with the highest ART1 module producing an output code for the most popular interpretation of the input object. Fig. 12.3 shows how this architecture can be implemented with the optoelectronic processor described above. The temporal performance of such a hierarchical implementation is reduced by a factor proportional to the number of layers.

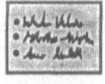 ## Summary

Realizing an optoelectronic architecture of the ART1 neural network means realizing matrix-vector multiplications in parallel in optics, arising from the comparison and the vigilance phase. In contrast, all number comparisons (e.g. whether the norm $S$ is larger than $\rho$), simple divisions and additions can be realized easily using electronic devices. For a compact realization, these units can be realized using smart input and output devices in VLSI technique.

## 12.2 Pattern Recognition by Adaptive Resonance Theory

Since the beginning of the formulation of neural network concepts, there have been investigations showing that the distinction between Fourier optical pattern recognition and neural networks is only marginal. In particular, it has been shown that adding a nonlinear feedback loop to a Fourier optical pattern recognizer can convert it easily into a simple associative memory. I have discussed these concepts of associative memories arising from concepts of optical information processing in detail in chapter 7.

In view of these successful transformations, why not trying to convert such a pattern recognizer into something more powerful — an associative memory acting like a fully trained adaptive resonance theory network? As in the case of simple associative memories, such a system can not be expected to be identical to the ART concept invented by S. Grossberg and G.A. Carpenter, but it may represent a good match to the concept. Here, I will discuss two important ideas of optical realizations. The first focusses its attention on the fact that in ART neural networks feedforward and feedback weights are operating simultaneously. The second takes advantage of phase conjugation to realize a resonant setup that is able to self-adapt.

### 12.2.1   Holographic Adaptive Resonance Theory System

The basic idea that has been presented in 1989 by H.J. Caulfield and D. Armitage [411] is based on the structure of the ART1 algorithm that enables symmetric information flow from top to bottom layers and back from bottom to top layers. It is to sort through stored patterns systematically (top down predispositions) comparing in detail the top down model to the received information from the outside world (bottom up correlation). When a match or — in the terminology of ART a resonance — occurs, the process stops. If no resonance occurs, the process continues after elimination of the rejected top down hypothesis.

If we consider as an example the operation of pattern recognition, realistic problems involve tremendous variations among a vast number of possible prototypes or examples. Consequently, the idea that one or even a bank of thousand filters could be adequate to perform such a task seems to be highly improbable. We have to vastly increase the information available in order to do the filtering. If we think about the precedent chapters, naturally page-oriented volume holography comes into ones mind as the method that is able to store up to a million masks (see section 4.3). This gives great complexity, but on the other hand exhaustive search of such a set is impractical for most of the actual application purposes. Therefore, a useful step would be to have a crude first estimate which of a large bank of filters holds the most promising results — and it would be best to perform this estimation in a parallel way. This is exactly the step which is done in bottom-up processing in the ART algorithm. Then, we need a way to select out the indicated filter and test it. This again is a step that is present in ART1, the top-down step. Thus we have achieved a formal equivalence between the steps necessary in a reliable pattern recognition system and in an ART1 neural network.

The remaining step is now to find the components that are able to realize these two different processing steps in pattern recognition configurations. For both steps, the two components which are key are a page oriented holographic memory and an optically-addressed spatial light modulator (SLM). In the reaization of H.J. Caulfield and J. Armitage [411], the page-oriented memory is a two-dimensional array of spatially discrete holograms, all of which produce their images in a single two-dimensional output plane. Generally, light is directed to one of these holograms through a beam deflector. The corresponding output image (data page) is then available in parallel for whatever purpose we choose. If we simply want to operate on these pages, we can do that in parallel by illuminating all the holograms simultaneously and placing the operator in what has previoulsy been regarded as the output (data page) plane. The image of the memory after such an operation shows the effect of data plane filtering on all the holograms. Both operations are shown in fig. 12.4 and 12.5.

The optically-addressed SLM has an important function for several reasons. First, it accepts optical information on the write side in parallel and is thus a suitable object to be

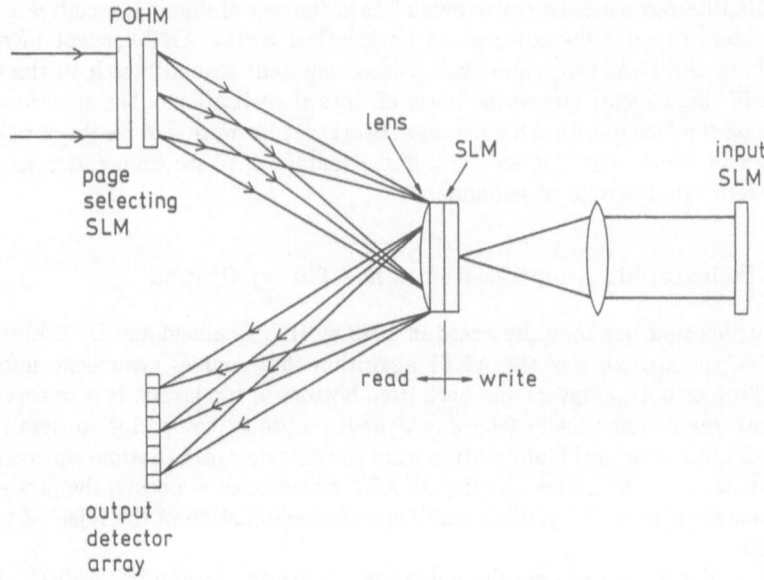

**Figure 12.4:** Bottom-up function of a whole page-oriented holographic memory storage. Each hologram produces its own Fourier pattern on the read side of the SLM where it is modulated by the power spectrum of the input scene. Each power spectrum correlation for the scene is measured by its own detector in the output plane. The highest reading suggests a filter for the top-down step (after [411]).

placed in the data plane of a page-oriented holographic memory. Second, it can operate in parallel on many inputs on the read side and is thus suitable for masking many pages in parallel. The basic scheme is to use a page-oriented holographic memory to store many (about 1000) recognition masks or templates. The mask design is such that it encodes the phase information holographically with a spatial carrier, whose frequency must be resolvable by the SLM. Alternatively, amplitude patterns on the write side may be used to produce phase patterns on the read side.

For the bottom-up operation, the Fourier transform of the input image is placed on the write side of the SLM. Then all the masks are addressed in parallel to place their mask patterns on the read side. The image of the memory through or off the SLM is a measure of the power spectrum overlap between the input scene and the scene to which the mask is matched. It indicates the preferred mask which has a good overall correlation with the scene. If we think of adding a nonlinear image processor or a nonlinear operator which favours bright points over weak ones — one can think e.g. about a photorefractive multiple beam amplifier (see section 3.4.4) — we are able to generate a single best-match mask.

The top-down step uses the page-oriented holographic memory in the conventional manner with a single beam deflected to the proper position so that the best mask hits the write side of the SLM. The Fourier transform of the input scene is filtered by the mask so created. The output of a two-dimensional scene is examined for a correlation point that has an intensity above a certain, fixed threshold. If such a point is not found, the search for a new mask begins, but the mask that has just failed the text is now excluded by an SLM inserted in the system just for that function. This selects a new best match

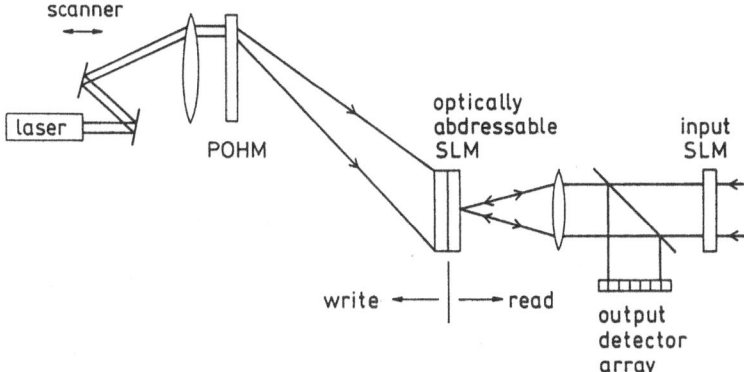

**Figure 12.5:** Top-down function using a single hologram from the page-oriented holographic memory. The filter is created by the light from the memory striking the write side of the SLM. The correlation pattern is examined spatially. If an acceptable correlation occurs, a match and the target location are declared. Otherwise, the system has to return to the bottom-up step but exclude failed filters from consideration by blocking them by a page-selecting SLM (after [411]).

and the cycle continues. In that way, we may even be able to recognize moving scenes. Temporal concern may in this case limit the number of searches if the scene is changing rapidly. This implies a limited number of cycles of "blackout" for rejected signals.

In in terms of classical neural network language, beneath the similarity to adaptive resonance, the system can be regarded as a bidirectional associative memory (BAM). In fact, adaptive resonance and the simple bidirectional association (see section 7) can be viewed as special cases of a more generalized bidirectional associative memory. Another way to put the system in a neural network context is to view it as a multilayer neural network. In the forward direction, layer 1 postulates a target and layer 2 tests the postulate. In the backward direction, the output of layer 2 can reconfigure layer 1 by excluding the prior postulate. The equilibrium solution is the correct output of the system.

The above discussion is an important example of how the neural network perspectives are more or less helpful to describe an optical system that is able to self-organize its structure to find an optimal match in pattern recognition. The same system can be described from several different, even divergent viewpoints, indicating thus that optical realizations allow different neural net functions to be mixed up in a single system and that pure neural net algorithms are not always sufficient to describe the optical processing of a pattern recognition system.

## 12.2.2 Photorefractive Adaptive Resonance Theory Network

The second system I want to discuss here is based on a resonant configuration using two phase-conjugate elements. This basic structure has already been described for the realization of optical image refreshment (see section 6.4.3, fig. 6.23) and several associative memories (see section 7.2, e.g. fig. 7.4). The key of the system is a photorefractive memory crystal ($BaTiO_3$), which is capable of recording multiple holograms in real-time

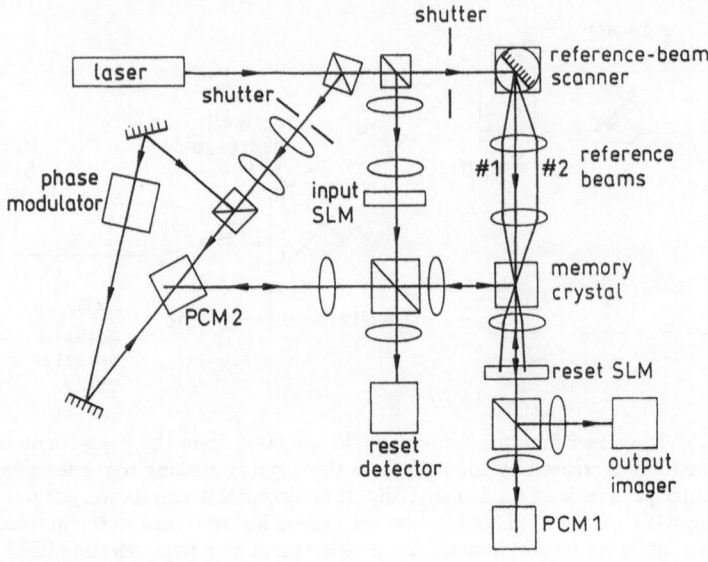

**Figure 12.6:** Schematic diagram of an optical adaptive resonance neural network using a photorefractive volume memory and a phase conjugating resonator (after [408]).

using one of the multiplexing techniques described in section 4.4 and which enables adaptivity of the interconnection weights by learning. Moreover, in difference to these setups, two supplementary components are introduced between the storage crystal and the first phase-conjugate mirror (PCM1): a reset spatial light modulator (the reset SLM) and a beamsplitter to permit imaging of the (angularly) multiplexed reference beams to identify the active mode. The operation of the system is — in its conventional configuration — as follows. The holographic resonator is activated with a partial input pattern introduced into the cavity by a beamsplitter. It is reflected towards the memory crystal, where it excites several reference beams if fixed holograms have been written before in the storage crystal. These reference beams are retroreflected by a thresholding phase-conjugate mirror (PCM1) back towards the hologram, setting up a resonant loop between PCM1 and PCM2. The loop is biased by the presence of the hologram and by the injected signal, thus suppressing all stored patterns and their reference beams except for the most closely matching pattern. This causes a readout of the stored image closest to the input image. The device can be considered in this mode of operation as a pattern classifier by considering the output reference beams to be angularly multiplexed classification codes. Also, light that contains information from both, the input and the stored matched image (or template), are overlapped at the opposite entrance face of the beamsplitter, allowing for a comparison of these two images.

To realize an ART system, D.C. Wunsch et al. [408] — and in a similar setup also A.V. Pavlov [329] — proposed to place a detector at the overlap position of the two images (see fig. 12.6). Moreover, the overall phase in the optically resonant loop is controlled and modulated using a pure phase modulator (see section 4.4.4) in one of the two pump beams in PCM2. The reset detector is an integrating photodetector. During the initial setup, with a single template stored in the memory crystal and with the same (matching) pattern input to the resonator, the phase of one of the pump waves of PCM2 is adjusted, while the total power at the reset detector is monitored to establish the phase at which maximum constructive interference between the input and the matching, stored patterns occurs. This implementation of phase adjustment naturally requires that PCM2 is operated in the four-wave mixing mode, while the nonlinear mode competition requires that PCM1 is operating in the photorefractive self-pumped mode having a sigmoidal intensity threshold characteristic (see section 5.2).

Subsequently, during the pattern-search mode (when one or more templates are stored in the memory crystal and a new pattern is presented), the adjustable PCM2 pump-beam phase is left at the pre-established setting, and the system is permitted to resonate until a single mode dominates. During this pattern-search mode the power used in the readout beams is insufficient (for the time scales involved) to rewrite the memory crystal. When a single template is dominant in the resonator (as indicated by a constant signal at the reset detector), the reset decision is made. To make this decision, one sweeps the phase of the pump beam of PCM2 through several cycles while looking for the presence of power modulation in ech pixel of the reset detector's field of view. Each pixel for which an overlap between input and template exists exhibits this modulation. The quality of the pattern match is determined from the number of overlap pixels detected by the reset detector. Thus, the reset detector measures the overlap of the input pattern and the template by the use of constructive or destructive interference — giving the inner product between the input and the template.

However, in most cases, these measurements have to be normalized by the input pattern size. This can be accomplished in two ways. The simpler, but significantly slower way is to shut off the pump beams to the four-wave mixing PCM2. This is done only when the input pattern is first presented. The reset detector then measures the total input pattern intensity and saves this number electronically. The pump beams are then switched back on and the reset detector reads the constructive interference pattern representing the overlap between the input and the template. The reset decision is made from these two pieces of information. The faster way to do this is to add an extra beamsplitter and to extract a copy of the input pattern before injecting it into the resonator. This copy is then measured by an extra detector and is used for the reset decision.

If the reset detector's match measurement is above a user-determined threshold, reset is not triggered and the system is permitted to resonate in its preferred mode, causing leraning of the new pattern. This learning can be accelerated by increasing the power in the readout beam for PCM2. If a reset condition is indicated (the match measurement is below the user-defined threshold), the following action takes place. The optical output imager is used to identify the location of maximum intensity at the reset SLM plane, which corresponds to the location of the reference beam associated with the dominant system mode. The formerly inactive reset SLM is then activated to produce attenuation at the known location at which the appropriate reference beam focuses through the reset SLM. This significantly reduces the gain of the dominant system mode and permits continued pattern search, in which the formerly dominant mode is not permitted to compete. This process continues, with poorly matching modes excluded sequentially from competition by the reset SLM, until a dominant mode that sufficiently matches is found. If no such mode is found, the input pattern is stored as a totally new template using a new reference beam aligned along a direction previously unused.

The role of the reset detector becomes even more important if one thinks about normalizing patterns by template size. This is a requirement of the ART neural network and an important problem, not just for this specific neural network, but for holographic pattern recognizers in general. This normalization of the templates needs to be area-based in order for the pattern classification to be meaningful. The modification of the system proposed by D.C. Wunsch et al. [408] is the following. The reset detector in its present configuration is measuring the size of the pattern that becomes the new recorded template. Therefore, when the reset is not indicated and a new template is recorded, the constructive interference measurement is used to determine the desired grating strength associated with that recorded template. This is done by providing the detector with enough electronic memory to store all the template sizes up to the capacity of the sstem. This uses only a small amount of the whole memory, even for a system of large capacity, because only one number must be stored for each template. This information is used to modify the recording schedule for all templates. Small templates are recorded longer, while big ones are recorded for a shorter time. Alternatively, recording power can be increased by different amounts, depending on the template size. This makes the recording procedure slightly more complicated, but the complexity is compensated by improved performance of the system.

In this implementation, the output of the neural network is the classification provided by the reference beam intensity, which can be read out by use of the output imaging optics viewing the reset SLM off the beam splitter (see fig. 12.6). However, it is also possible to read out the stored template information in a manner similar to that for the complete object output. Thus, the device can be used as either a heteroassociative or an autoassociative ART-based memory. Furthermore, it may be possible to replace the angularly multiplexed reference beams with a third SLM. This could be used for associating input patterns with known output patterns for supervised ART-based learning. The image on the reset SLM would be the inverse of the image on this third training SLM.

In the proposed experimental realization of this configuration, D.C. Wunsch et al. [408] suggested to implement human observation of the reset signal and activation of the reset SLM. This role can be automated in further implementations without problems, however electronically. The proper operation of the three photorefractive crystals is the key for the operation of the whole network. The memory crystal has two modes of operation, template storage and readout. The template storage mode uses relatively intense waves in the reference beams and the object input wave in order to quickly generate or modify the refractive index phase gratings within the crystal. In order to ensure that multiple stored hologams have similar formation time constants and decay rates, all gratings need to be formed with nearly the same geometry and spatial frequency magnitude. Therefore, special recording algorithms as sequential or incremental recording have to be applied (see section 4.4.3).

The readout mode involves much lower intensity waves in order to permit a search before overwriting of the stored holograms occurs. However, over time, both the high-power writing of new templates and the integrated effects of low-power readout degrades the templates stored previously. This is the crucial problem of the configuration and can only be circumvented by rereferencing the storage crystal with the template images. This could be compared with the refresh cycles necessary in a random-dynamic access memory. Methods for doing so have been described in detail in section 6.4. Such rereferencing could take place each time a new template is written after a classification search has been terminated.

For the feedback PCM's, four-wave mixing is employed in PCM2 with controllable pump-wavepower in such a way that adequate effective gain is established to overcome transmission and holographic efficiency losses within the resonator and allows the system to oscillate. PCM1 is operated in the self-pumped

Moreover, it is important to note that the reference beam used to record a particular template must be focused on the plane of the reset SLM after propagation through the memory crystal. Each of the reference beams, which are angularly multiplexed at the memory crystal, is focused at a different location on the reset SLM plane. Thus, the optical ART unit will be capable of processing large patterns and potentially has a large template capacity — the capcity just given by the storage capacity of holographic systems. Furthermore, the device is all-optical in the whole information processing path. The reset detector's electronics are never used sequentially when an input pattern has already been learned or when it matches an existing template sufficiently.

With this description of an all-optical, coherent ART realization using photorefractive opticsal devices I want to close the description of optical implementations of ART neural network types. Although most of the systems described above are in a rather premature state and have not yet proved to function in complete experimental setups, they show promising perspectives for the development of ART networks. The next future will therefore surely bring exciting results in this area and we may expect a broad range of investigations that may reveal the application potential of these optical proposals.

 ## Summary

The basic for an optical analogon of the ART algorithm is based on the structure of the ART1 algorithm that enables symmetric information flow from top to bottom layers (predispositions) and back from bottom to top layers (correlations). For this purpose, two ways can be accomplished:

- For both steps, a page oriented holographic memory and an optically-addressed spatial light modulator (SLM) can be used. The page-oriented memory is a two-dimensional array of spatially discrete holograms, all of which produce their images in a single two-dimensional output plane. Here, storage is the top to bottom and correlation matching the bottom to top information flow In terms of classical neural network language, beneath the similarity to adaptive resonance, the system can be regarded as a bidirectional associative memory (BAM).

- Both steps can be realized by a resonant configuration using two phase-conjugate elements. A photorefractive memory crystal records multiple holograms in real-time and enables adaptivity of the interconnection weights by learning. A reset spatial light modulator (the reset SLM) and a beamsplitter in the oscillator permit to threshold the reference beams to identify the active mode.

# 13 Outlook

When I began to develop the idea of that book in 1993, nonlinear optical realizations of neural networks were just at a beginning state, having produced several interesting and promising, but not fully mature systems. Now, a couple of years later, optical implementations of neural networks have spread out in a lot a research laboratories, optimizations of the first models are investigated in different countries and I do not exaggerate if I state that the development of optical neural networks has not yet reached its maximum.

Consequently, a lot of very recent novel realizations or technical optimizations of nonlinear optical neural network implementations are described only in a very short way or are often omitted in my description. Here, I want to give only a sudden insight in some of these new results, give several references that are well suited to follow the actual development and leave it to the reader to go more into the details of these investigations.

The most impressive advance has perhaps been reached in the area of holographic storage of iformation-bearing pages and optical interconnections. Compact, automatically working recording systems have been developped, combined with mature applications that are ready to be installed in industrial environments as analog gray-image [79] or digital video storage systems [412], analog recognition systems or associative memories with databases of several thousand exemplars [312, 413].

The quick development in that area is manifested by several special issues of different Journals, as e.g. the Special Issue of Optics and Quantum Electronics, 25(9), 1993, of Optical Engineering, 35 (8), 1996, and of Applied Optics 32(8), 1993 or 35(14), 1996. Moreover, the Journal of the Optical Society of America published biannually a special issue on recent innovations in Photorefractive Materials, Effects and Devices, last time appeared in September 1994, Vol. 11(9). Also, the SPIE Selected Paper Series have published the "Selected Papers on Holographic Storage" (Ed. G. Sincerbox), SPIE Selected Paper Series MS95, as well as the "Selected Papers on Optical Neural Networks" (Ed. S. Jutamulia) Series MS96. Finally, the journal of "Optical Memory and Neural Network" (Allerton Press) underline clearly the importance, the field has gained in the last years. Moreover, several conferences incoorporated special sessions on holographic storage and optical neural networks. These conferences, originally coming from the fields of nonlinear optics or neural networks, thus built a bridge to connect both fields close together. Here, I want to mention the biannual "Topical Meeting on Photorefractive Materials, Effects and Devices", held last time in Chiba, Japan, in June 117, the annual OSA Meeting, held this time in Long Beach, California, in October 1997, and the IEEE International Conference on Neural Networks, held last time in Texas, June 1997. Finally, new conferences has been installed as the "Optical Data Storage Topical Meeting", held recently in Tucson, Arizona, in April 1997, or the "International Symposium on Holographic Memories" held for the first time in Athens, May 1996. For those that are interested in the actual development in the field, the conference proceedings of these meetings supply a comprehensive source of information and hints for detailed literature.

These activities show also the constant interest and development of the field of optical neural networks, being lively proofs that the optical systems presented throughout this book are capable of performing certain complex operations in an easy and attractive way for engineering applications.

# Bibliography

[1] T.J. Sejnowski and C.R. Rosenberg. Parallel networks that learn to pronounce english text. *Compl. Syst.*, page 145, 1987.

[2] Y.S. Abu-Mostafa and D. Psaltis. Optical neural computers. *Sc. Am. 256*, page 88, March 1987.

[3] D. Psaltis, C.H. Park, and J. Hong. Higher-order associative memories and their optical implementation. *Neural Networks 2*, page 149, 1988.

[4] S. Kakizaki, P. Horan, A. Arimoto, H. Sako, A. Saito, and F. Kugiya. Optical implementation of a translation-invariant second-order neural network for multiple-pattern classification. *Appl. Opt. 33*, page 8270, 1994.

[5] Y. Qiao and D. Psaltis. Local learning algorithm for optical neural networks. *Appl. Opt. 31*, page 3285, 1992.

[6] K. Wagner and T.M. Slagle. Optical competitive learning with VLSI/liquid crystal winner-take-all-modulators. *Appl. Opt. 32*, page 1408, 1993.

[7] C. von der Malsburg. Self-organization of orientation sensitive cells in the striate cortex. *Kybernetik 14*, page 85, 1987.

[8] D.E. Rumelhardt and D. Zisper. Feature discovery by competitive learning. *Cognit. Sci. 9*, page 76, 1985.

[9] S. Grossberg. Competitive learning, from interactive activation to adaptive resonance. *Cognit. Sci. 11*, page 23, 1987.

[10] J.J. Hopfield. Neural networks and physical systems with emergent collective computational abilities. *Proc. Natl. Acad. Sci. USA 79*, page 2554, 1982.

[11] J.J. Hopfield. Neurons with graded response have collective computational properties like those of two-state neurons. *Proc. Natl. Acad. Sci. USA 81*, page 3088, 1984.

[12] T. Lu, X. Xu, S. Wu, and F.T.S. Yu. Neural network model using interpattern association. *Appl. Opt. 29*, page 284, 1990.

[13] E.G. Paek, P.F. Liao, and H. Gharavi. Derivation of neural network models and their computational circuits for associative memories. *Opt. Eng. 31*, page 986, 1992.

[14] M. Mezard, G. Parisi, and M.A. Virasoro. Spin glass theory and beyond. *World Scientific*, 1987.

[15] G.A. Carpenter and S. Grossberg. The ART of adaptive pattern recognition. *IEEE Computer 21*, 1988.

[16] M.K. Kim and R. Kachru. Storage and phase conjugation of multiple images using backward stimulated echoes in $Pr^+$:LaF$_3$. *Opt. Lett. 12*, page 593, 1987.

[17] M.K. Kim and R. Kachru. Long-term image storage and phase conjugation by a backward stimulated echo in $Pr^+$:LaF$_3$. *J. Opt. Soc. Am B 4*, page 305, 1987.

[18] M.K. Kim and R. Kachru. Multiple-bit long-term data by backward stimulated echo in Eu:YAlO$_3$. *Opt. Lett. 14*, page 423, 1989.

[19] N.W. Carlson, W.R. Babbitt, and T. W. Mossberg. Storage and phase conjugation of light pulses using stimulated photon echoes. *Opt. Lett. 8*, page 623, 1983.

[20] A. Ashkin, G.D. Boyd, J.M. Dziedzic, R.G. Smith, A.A. Ballmann, H.J. Levinstein, and K. Nassau. Optically induced refractive index inhomogeneities in LiNbO$_3$ and LiTaO$_3$. *Appl. Phys. Lett. 9*, page 72, 1960.

[21] F.S. Chen, J.T. La Macchia, and D.B. Frazer. Holographic storage in Lithium Niobate. *Appl. Phys. Lett. 13*, page 223, 1968.

[22] N.V. Kuktharev, V.B. Markov, S.G. Odulov, M.S. Soskin, and L. Vinetskii. Holographic storage in electrooptic crystals I. *Ferroel. 22*, page 949, 1979.

[23] R. Jaura, T.J. Hall, and P.D. Foote. Simplified band transport model of the photorefractive effect. *Opt. Eng. 25*, page 1068, 1986.

[24] J. Feinberg, D. Heiman, A.R. Tanguay, and R.W. Hellwarth. Photorefractive effects and light-induced charge migration in BaTiO$_3$. *J. Appl. Phys. 51*, page 1297, 1980.

[25] N.V. Kukhtarev, G.E. Dovgalenko, and V.N. Starkow. Influence of optical activity on hologram formation in photorefractive crystals. *Appl. Phys. A 33*, page 227, 1984.

[26] R. Orlowski and E. Krätzig. Hologram storage in electrooptic crystals. *Sol. State Comm. 27*, page 1351, 1978.

[27] M.B. Klein and G.C. Valley. Beam coupling in BaTiO$_3$ at 442 nm. *J. Appl. Phys 57*, page 4901, 1985.

[28] G.C. Valley. Erasure rates in photorefractive materials with two photoactive species. *Appl. Opt. 22*, page 3160, 1983.

[29] Y. Fainman, E. Klancnik, and S.H. Lee. Optimal coherent image amplification by two-wave coupling in photorefractive BaTiO$_3$. *Opt. Eng. 25*, page 228, 1986.

[30] H. Kogelnik. Coupled wave theory for thick hologram gratings. *Bell Syst. Tech. J. 48*, page 2909, 1969.

[31] A. Marrakchi, J.P. Huignard, and P. Günter. Diffraction efficiency and energy transfer in two-wave mixing experiments with Bi$_{12}$SiO$_{20}$ crystals. *Appl. Phys. 24*, page 131, 1981.

[32] C. Denz, L. Klees, and T. Tschudi. Multibeam coupling in photorefractive BaTiO$_3$. *ECO 2, Nonlinear Optical Materials II*, 1989.

[33] L. Klees, C. Denz, and T. Tschudi. Intensity crosstalk and angular selectivity of multibeam coupling in photorefractive BaTiO$_3$. *Opt. Comm. 77*, page 65, 1990.

[34] J. Ma, L. Liu, S. Wu, Z. Wang, L. Xu, and B. Shu. Multibeam coupling in photorefractive SBN:Ce. *Opt. Lett. 13*, page 1020, 1988.

[35] P.E. Anderson, P.M. Petersen, and P. Buchhave. Nonlinear combinations of gratings in drift-dominated recording in Bi$_{12}$SiO$_{20}$. *J. Opt. Soc. Am. B 12*, page 1422, 1995.

[36] P.E. Andersen, P.M. Petersen, and P. Buchhave. Crosstalk in dynamic optical interconnects in photorefractive crystals. *Appl. Phys. Lett. 65*, page 271, 1994.

[37] J.H. Hong, A.E. Chiou, and P. Yeh. Image amplification by two-wave mixing in photorefractive crystals. *Appl. Opt. 29*, page 3026, 1990.

[38] R. Saxena, F. Vachss, I. McMichael, and P. Yeh. Diffraction properties of multiple-beam photorefractive gratings. *Journ. Opt. Soc. Am. B 7*, page 1210, 1990.

[39] M. Segev, Y. Ophir, and B. Fischer. Nonlinear multi two-wave mixing, the fanning process and its bleaching in photorefractive media. *Opt. Comm. 77*, page 265, 1990.

[40] N.A. Vainos and R.W. Eason. Spatially multiplexed phase conjugate imaging and processing in photorefractive BSO. *Opt. Comm. 62*, page 311, 1987.

[41] D.Z. Anderson, M. Saffman, and A. Hermanns. Manipulating the information carried by an optical beam with reflexive photorefractive beam coupling. *Journ. Opt. Soc. Am. B 12*, page 118, 1995.

[42] M. Cronin-Golomb, B. Fischer, J.O. White, and A. Yariv. Theory and applications of four-wave mixing in photorefractive media. *IEEE Journ. Quant. El. 20*, page 12, 1984.

[43] J.F. Lam. Origin of phase-conjugate waves in self-pumped photorefractive mirrors. *Appl. Phys. Lett. 46*, page 909, 1985.

[44] J. Feinberg. Self-pumped continuous-wave phase conjugator using internal reflections. *Opt. Lett. 7*, page 486, 1982.

[45] A.A.Zozulya, G. Montemezzani, and D.Z. Anderson. Analysis of total internal reflection phase conjugate mirror. *Proc. of the 3rd Topical Meeting on Photorefractive Materials, Effects and Devices, Estes Park, Colorado, 11.-14.Juni 1995*, page 375, 1995.

[46] C. Denz, T. Rauch, and T. Tschudi. Analysis of irregular and chaotic fluctuations in a self-pumped BaTiO$_3$ phase-conjugate mirror. *Proc. SPIE 1281, Optical Interconnections and Networks, Den Haag*, 1990.

[47] T. Rauch, C. Denz, and T. Tschudi. Analysis of irregular fluctuations in a self-pumped BaTiO$_3$ phase-conjugate mirror. *Opt. Comm. 88*, pages 160–166, 1992.

[48] A.V. Nowak, T.R. Moore, and R.H. Fischer. Observation of internal beam production in BaTiO$_3$ phase conjugators. *J. Opt. Soc. Am B 5*, page 1864, 1988.

[49] A.M.C. Smout and R.W. Eason. Analysis of mutually incoherent beam coupling in BaTiO₃. *Opt. Comm. 59*, page 77, 1986.

[50] A.K. Mayumdar and J.L. Kobesky. Oscillations and possible transitions to optical chaos in phase-conjugate waves in self-pumped BaTiO₃ at 514.5 nm. *Opt. Comm. 75*, page 339, 1990.

[51] D.J. Gauthier, P. Narum, and R.W Boyd. Observation of deterministic chaos in a phase-conjugate mirror. *Phys. Rev. Lett. 58*, page 1640, 1987.

[52] M. Cronin-Golomb, B. Fischer, J.O. White, and A. Yariv. Exact solution of a nonlinear model of four-wave mixing and phase conjugation. *Opt. Lett. 7*, page 313, 1982.

[53] W. Królikowski, M. Belić, M. Cronin-Golomb, and A. Bledowski. Chaos in photorefractive four-wave mixing with a single grating and a single interaction region. *J. Opt. Soc. Am. B 7*, page 1204, 1990.

[54] M.R. Belić and W. Królikowski. Multigrating optical phase conjugation: Numerical results. *J. Opt. Soc. Am. B 6*, page 901, 1989.

[55] M.C. Gower and P. Hribek. Mechanism for internally self-pumped phase-conjugate emission from batio₃ crystals. *Journ. Opt. Soc. Am. B*, page 1750, 1988.

[56] A.D. McAulay. Optical computer architectures. *Wiley, New York*, 1991.

[57] S. Weiss, M. Segev, S. Sternklar, and B. Fischer. Photorefractive dynamic optical interconnects. *Appl. Opt. 27*, page 3422, 1988.

[58] C. Ward and J. Thackara. Operating modes of the microchannel spatial light modulator. *Opt. Eng. 22*, page 695, 1983.

[59] J. Grinberg, A. Jacobson, W. Bleha, L. Miller, L. Fraas, D. Boswell, and G. Myer. A new real-time non-coherent light image converter. The hybrid field effect liquid crystal light valve. *Opt. Eng. 14*, page 217, 1975.

[60] S. Morozumi and K. Oguchi. Current status of LCD-TV. *Mol. Cryst. and Liq. Cryst. 94*, page 43, 1983.

[61] H.K. Liu, J.A. Davis, and R.A. Lilly. Optical data processing properties of a liquid crystal television spatial light modulator. *Opt. Lett. 10*, page 635, 1985.

[62] F.T.S. Yu and S. Jutamulia. Optical parallel logic gates using inexpensive liquid-crystal televisions. *Opt. Lett. 12*, page 1050, 1987.

[63] T.H. Barnes, T. Eiju, K. Matsuda, H. Ichikawa, M.R. Taghizadeh, and J. Turunen. Reconfigurable free-space optical interconnections with a phase-only liquid crystal spatial light modulator. *Appl. Opt. 31*, page 5527, 1992.

[64] T.H. Barnes, K. Matsumoto, T. Eiju, K. Matsuda, and N. Ooyama. The application of phase-only filters to optical interconnects and pattern recognition. *J. Mod. Opt. 37*, page 1849, 1990.

[65] A.D. McAulay, J. Wang, and C.T. Ma. Optical heteroassociative memory using spatial light rebroadcasters. *Appl. Opt. 29*, page 2067, 1990.

[66] D.A. Gregory et al. Optical characteristics of a deformable-mirror spatial light modulator. *Opt. Lett. 13*, page 10, 1988.

[67] U.Efron. Spatial light modulator technology. *M. Dekker, New York*, 1995.

[68] U. Efron. Spatial light modulators and applications for optical information processing. *Proc. SPIE. 960*, 1988.

[69] C. Warde and A.D. Fisher. Spatial light modulators: applications and functional capabilites. *In J.L. Horner, "Optical Signal Processing", Academic Press*, 1987.

[70] R.A. Athale and C.W. Stirk. Compact architectures for adaptive neural nets. *Opt. Eng. 28*, page 447, 1989.

[71] K.Y. Hsu, C.J. Cheng, and T.C. Hsieh. Trainable pattern classifier using a computer generated hologram. *Jpn. J. Appl. Phys A 33*, page 1910, 1994.

[72] S.F. Habiby and S.A. Collins. Implementation of a fast digital matrix-vector multiplier using a holographic look-up table and residue arithmetics. *Appl. Opt. 26*, page 4639, 1987.

[73] Y. Kajiki, K. Matsushita, and E. Shimizu. An optical pattern classification using computer generated holograms. *Jap. J. Appl. Phys. 29*, page L 1274, 1990.

[74] S. Jutamulia, G.M. Storti, J. Lindmayer, and W. Seiderman. Use of electron trapping materials in optical signal processing, 1-4. *Appl. Opt. 29, 30 and 31*, 1990, 1991, 1992.

[75] H. Kurz. Photorefractive recording dynamics and multiple storage of volume holograms in photorefractive LiNbO₃. *Opt. Acta 24*, page 463, 1977.

[76] C. Denz, G. Pauliat, G. Roosen, and T. Tschudi. Phase coded hologram multiplexing for high-capacity optical data storage. *Proc. of the 3rd Topical Meeting on Photorefractive Materials, Effects and Devices, Boston*, August 1991.

[77] F.H. Mok, M.C. Tackitt, and H.M Stoll. Storage of 500 high-resolution holograms in a LiNbO₃-crystal. *Opt. Lett. 16*, page 605, 1991.

[78] M.-P. Bernal, H. Coufal, R.K. Grygier, J.A. Hoffnagle, C.M. Jefferson, R.M. Macfarlane, R. M. Shelby, G. T. Sincerbox, P. Wimmer, and G. Wittmann. A precision tester for studies of holographic optical storage materials and recording physics. *Appl. Opt. 35*, page 2360, 1996.

[79] I. McMichael, W. Christian, D. Pletche, T.W. Chang, and J.H. Hong. Compact holographic storage demonstrator with rapid access. *Appl. Opt. 35*, page 2375, 1996.

[80] D. Psaltis and F. Mok. Holographic data storage. *Sci. Am. 273*, page 70, 1995.

[81] A.F. Meixner, A. Renn, and U.P. Wild. Spectral hole burning and holography I: transmission and holographic detection of spectral holes. *J. Chem. Phys. 91*, page 6728, 1989.

[82] B. Kohler, S. Bernet, A. Renn, and U.P. Wild. Storage of 2000 holograms in a photochemical hole-burning system. *Opt. Lett. 18*, page 2144, 1993.

[83] R.T. Weverka, K. Wagner, and M. Saffman. Fully interconnected, two-dimensional neural arrays using wavelength-multiplexed volume holograms. *Opt. Lett. 16*, page 826, 1991.

[84] O. Ollikanen, A. Rebane, and K.K. Rebane. Error-corrective optical neural network modelled by persistent spectral hole burning. *Opt. and Quant. El. 25*, page S569, 1993.

[85] X.A. Shen and R. Kachru. High-speed pattern recognition by using stimulated echoes. *Opt. Lett. 17*, page 520, 1992.

[86] X.A. Shen and R. Kachru. Time-domain optical memory for image storage and high-speed image processing. *Appl. Opt. 32*, page 5810, 1993.

[87] E.Y. Xu, S. Kröll, D. Huestis, R. Kachru, and M.K. Kim. Nanosecond image processing using stimulated photon echoes. *Opt. Lett. 15*, page 562, 1990.

[88] Y.S. Bai and R. Kachru. Coherent time-domain data storage with a spread spectrum generated by random biphase shifting. *Opt. Lett. 18*, page 1189, 1993.

[89] M. Mitsunaga, R. Yano, and N. Uesugi. Time- and frequency domain hybrid optical memory: 1.6 kbit data storage. *Opt. Lett. 16*, page 1890, 1991.

[90] Y.S. Bai, W.R. Babbitt, N.W. Carlson, and T.W. Mossberg. Real-time optical waveform convolver/cross-correlator. *Appl. Phys. Lett. 45*, page 714, 1984.

[91] X.A. Shen, R. Hartmann, and R. Kachru. Impulse-equivalent time-domain optical memory. *Opt. Lett. 21*, page 833, 1996.

[92] X.A. Shen and R. Kachru. Experimental demonstration of impulse-equivalent time-domain optical memory. *Opt. Lett. 21*, page 2020, 1996.

[93] B.S. Ham and M.K. Kim. Photon-echo amplification by an external cavity amplifier. *Appl. Opt. 33*, page 4472, 1994.

[94] S. Hunter, F. Kiamilev, S. Esener, D.A. Parthenopoulos, and P.M. Rentzepis. Potentials of two-photon based 3-d optical memories for high-performance computing. *Appl. Opt. 29*, page 2058, 1990.

[95] M.M Wang, S.C. Esener, F.B. McCormick, I. Cokgör, A.S. Dvornikov, and P.M. Rentzepis. Experimental characterization of a two-photon memory. *Opt. Lett. 22*, page 558, 1997.

[96] Y.S. Bai and R. Kachru. Nonvolatile holographic storage with two-step recording in Lithium Niobate using cw lasers. *Phys. Rev. Lett. 78*, page 2944, 1997.

[97] M. Kazmarek and R.W. Eason. Very high single-pass two-beam coupling gain at 647 nm under conditions of induced transparency in Rh:doped BaTiO₃. *Proc. of the Conference on Photorefractive Materials, Effects and Devices, Estes Park, Colorado, 11.-14.6.95*, page 132, 1995.

[98]  M. Kaczmarek, G.W. Ross, P.M. Jeffey, R.W. Eason, P. Hribek, M.J. Damzen, R. Ramos-Garcia, R. Troth, M. H. Garrett, and D. Rytz. Dual wavelength characterisation of shallow traps in blue BaTiO₃. *Opt. Mat. 4*, page 158, 1995.

[99]  D. Psaltis, D. Brady, Xiang-Guang Gu, and S. Lin. Holography in artificial neural networks. *Nature 343*, page 325, January 1990.

[100]  P. van Heerden. Theory of optical information storage in solids. *Appl. Opt. 2*, page 393, 1963.

[101]  D. Psaltis, D. Brady, and K. Wagner. Adaptive optical networks using photorefractive crystals. *Appl. Opt. 27*, page 1752, 1988.

[102]  D. Gabor, G.W. Stroke, R. Restrick, A. Funkhouser, and D. Brumm. *Phys. Lett. Compl. 18*, page 116, 1965.

[103]  J.P. Huignard, J.P. Herriau, and F. Micheron. Coherent selective erasure of superimposed volume holograms in LiNbO₃. *Appl. Phys. Lett. 26*, page 256, 1975.

[104]  A. Marrakchi. Continuous coherent erasure of dynamic holographic interconnections in photorefractive crystals. *Opt. Lett. 14*, page 326, 1989.

[105]  J.H. Hong, S. Campbell, and P. Yeh. Optical pattern classifier with perceptron learning. *Appl. Opt. 29*, page 3019, 1990.

[106]  N. Konforti, E. Marom, and S.T. Wu. Phase-only modulation with twisted nematic liquid crystal spatial light modulators. *Opt. Lett. 13*, page 251, 1988.

[107]  M. Kranzdorf, B.J. Bigner, L. Zhang, and K.M. Johnson. Optical connectionist machine with polarization-based bipolar weight values. *Opt. Eng. 28*, page 844, 1989.

[108]  K. Noguchi. Large-scale two-dimensional optical hopfield associative memory using an incoherent optical free-space interconnection. *Opt. Lett. 16*, page 1110, 1991.

[109]  A.P. Ittycheriah, J.F. Walkup, T. F. Krile, and S.L. Lim. Outer product processor using polarization encoding. *Appl. Opt. 29*, page 275, 1990.

[110]  K. Wagner and D. Psaltis. Nonlinear etalons in competitive optical learning networks. *1st IEEE International Conference on Neural Networks, Vol. 3*, page 585, June 1987, San Diego.

[111]  D.Z. Anderson and D.M. Lininger. Dynamic optical interconnects: volume holograms as two-port operators. *Appl. Opt. 26*, page 5031, 1987.

[112]  A. Marrakchi and W. M. Hubbard. Fan-in issues in a holographic grating interconnect. *Opt. Lett. 16*, page 417, 1991.

[113]  J. Hong and P. Yeh. Photorefractive parallel matrix-matrix multiplier. *Opt. Lett. 16*, page 1343, 1991.

[114]  C. Gu, S. Campbell, and P. Yeh. Matrix-matrix multiplication by using grating degeneracy in photorefractive media. *Opt. Lett. 18*, page 146, 1993.

[115]  C.C. Sun, M.W. Chang, and K.Y. Hsu. Matrix-matrix multiplication by using anisotropic self-diffraction in BaTiO₃. *Appl. Opt. 33*, page 4501, 1994.

[116]  J.E. Ford, Y. Fainman, and S.H. Lee. Array interconnection by phase-coded optical correlation. *Opt. Eng. 15*, page 225, 1990.

[117]  J.E. Ford, Y. Fainman, and S.H. Lee. Reconfigurable array interconnection by photorefractive correlation. *Appl. Opt. 33*, page 5363, 1994.

[118]  P. Yeh, A.E. Chiou, J. Hong, P. Beckwith, T. Chang, and M. Koshnevisan. Photorefractive nonlinear optics and optical computing. *Opt. Eng. 28*, page 328, 1989.

[119]  J.O. White and A. Yariv. Real-time image processing via four-wave mixing in a photorefractive medium. *Appl. Phys. Lett. 37*, page 5, 1980.

[120]  G.N. Henderson, J.F. Walkup, and E.J. Bochove. Optical quadratic processor using four-wave mixing in BaTiO₃. *Opt. Lett. 14*, page 770, 1989.

[121]  D. Psaltis and N.H. Farhat. Optical information processing based on an associative memory model of neural nets with threshold and feedback. *Opt. Lett. 10*, page 98, 1985.

[122]  A.E. Chiou and P. Yeh. Energy efficiency of optical interconnections using photorefractive holograms. *Appl. Opt. 29*, page 1111, 1990.

[123]  A.E. Chiou and P. Yeh. 2 x 8 photorefractive reconfigurable interconnect with laser diodes. *Appl. Opt. 31*, page 5536, 1992.

[124] S. Wu, Q. Song, A. Mayers, D.A. Gregory, and F.T.S. Yu. Reconfigurable interconnections using photorefractive holograms. *Appl. Opt. 29*, page 1118, 1990.

[125] D.J. Brady and D. Psaltis. Holographic interconnections in photorefractive waveguides. *Appl. Opt. 30*, page 2324, 1991.

[126] A.A. Sawchuk, B.K. Jenkins, and C.S. Raghavendra. Optical crossbar networks. *IEEE Trans. Comput. C-20*, page 50, 1987.

[127] R.K. Kostuk, J.W. Goodman, and L. Hesselink. Design considerations for holographic optical interconnections. *Appl. Opt. 26*, page 3947, 1987.

[128] J.W. Goodman. Fan-in and fan-out with optical interconnections. *Opt. Acta 32*, page 1489, 1985.

[129] A. Marrakchi, W. M. Hubbard, S.M. Habiby, and J.S. Patel. Dynamic holographic interconnects with analog weights in photorefractive crystals. *Opt. Eng. 29*, page 215, 1990.

[130] E.S. Maniloff and K.M. Johnson. Dynamic holographic interconnections using static holograms. *Opt. Eng. 29*, page 225, 1990.

[131] Y. Owechko. Cascaded-grating holography for artificial neural networks. *Appl. Opt. 32*, page 1381, 1993.

[132] H. Lee. Volume holographic global and local interconnection patterns with maximal capacity and minimal crosstalk. *Appl. Opt. 28*, page 5312, 1989.

[133] H. Lee, X.G. Gu, and D. Psaltis. Volume holographic interconnection with maximal capacity and minimal crosstalk. *J. Appl. Phys. 65*, page 2191, 1989.

[134] J.R. Wullert II and Y. Lu. Limits of the capacity and density of holographic storage. *Appl. Opt. 33*, page 2192, 1994.

[135] J. Lembcke, C. Denz, T.H. Barnes, and T. Tschudi. Multiple image storage using phase encoding - latest results. *Digest of the Topical Meeting on Photorefractive Materials, Effects and Devices*, page SaCo2, Ukrainian Akademy of Science, 1993.

[136] D. Psaltis, C. Gu, and D. Brady. *Proc. SPIE 963*, page 605, 1988.

[137] W.S. Baek and H. Lee. Photorefractive intermode space-charge fields in volume holographic interconnections. *J. Appl. Phys. 67*, page 1194, 1990.

[138] C. Slinger. Analysis of the N-to-N volume holographic neural interconnect. *J. Opt. Soc. Am. A 8*, page 3074, 1991.

[139] K.J. Blotekjaer. Limitations on holographic storage capacity of photochromic and photorefractive media. *Appl. Opt. 18*, page 57, 1979.

[140] C. Gu, A. Chiou, and J. Hong. Cross-talk noise in photorefractive interconnection. *Appl. Opt. 32*, page 1437, 1993.

[141] Y. Owechko and B.H. Soffer. Optical interconnection method for neural networks using self-pumped phase-conjugate mirrors. *Opt. Lett. 16*, page 675, 1991.

[142] M.P. Schamschula and H.J. Caulfield. Photorefractive mirrors simplify adaptive interconnections. *Laser Focus World 5*, page 151, 1992.

[143] G.J. Dunning, Y. Owechko, and B.H. Soffer. Hybrid optoelectronic neural networks using a mutually pumped phase-conjugate mirror. *Opt. Lett. 16*, page 928, 1991.

[144] A.C. Strasser, E.S. Maniloff, K.M. Johnson, and S.D.D. Goggin. Procedure for recording multiple-exposure holograms with equal diffraction efficiency in photorefractive media. *Opt. Lett. 14*, page 6, 1989.

[145] L. d' Auria, J.P. Huignard, C. Slezak, and E. Spitz. Experimental holographic read-write memory using 3-d storage. *Appl. Opt. 13*, page 808, 1974.

[146] C. Denz. Photorefraktive Materialien als Komponenten in optisch neuronalen Netzstrukturen. *Dissertation*, 1992.

[147] J. Ma, T. Chang, J. Hong, R. R. Neurgaonkar, G. Barbastathis, and D. Psaltis. Electrical fixing of 1000 angle-multiplexed holograms in SBN:75. *Technical Digest of the Conference on "Nonlinear Optics: Materials, fundamentals and applications*, page 175, 1996.

[148] E.S. Maniloff and K.M. Johnson. Maximized photorefractive holographic storage. *J. Appl. Phys. 70*, page 4702, 1991.

[149] A.A. Goldstein, G.C. Petrisor, and K.B. Jenkins. Gain and exposure scheduling to compensate for photorefractive neural-network weight decay. *Opt. Lett. 20*, page 611, 1995.

[150] Y. Taketomi, J.E. Ford, H. Sasaki, J. Ma, Y. Fainman, and S.H. Lee. Incremental recording for photorefractive hologram multiplexing. *Opt. Lett. 16*, page 1774, 1991.

[151] S. Campbell, Y. Zhang, and P. Yeh. Writing and copying in volume holographic memories: approaches and analysis. *Opt. Comm. 123*, page 27, 1996.

[152] H. Sasaki, Y. Fainman, and S.H. Lee. Gray-scale fidelity in volume-multiplexed photorefractive memory. *Opt. Lett. 18*, page 1358, 1993.

[153] L. d' Auria, J.P. Huignard, and E. Spitz. Holographic read-write memory and capacity enhancement by 3-d storage. *IEEE. Transact. Magnetics MAG 9*, page 83, 1973.

[154] J.P. Huignard and E. Spitz. Holographic mass memories using volume holograms. *Laser 75 Optoelectronics Conf. Proc.*, page 248, 1975.

[155] J.P. Huignard. Holographic read-write memory, optical organisation andcapacity enhancement by three-dimensional storage. *1st European Electro-Optics Markets and Technology Conference Proc.*, page 241, 1975.

[156] J.F. Heanue, M.C. Bashaw, and L. Hesselink. Volume holographic storage and retrieval of digital data. *Science 19*, page 749, 1994.

[157] L. Hesselink, J.F. Heanue, and M.C. Bashaw. Holographic digital data storage systems. *Proc. SPIE Vol. 2778*, page 410, 1996.

[158] M.A. Neifeld and J.D. Hayes. Error-correction schemes for volume optical memories. *Appl. Opt. 34*, page 8183, 1995.

[159] M.A. Neifeld and M. McDonald. Error-correction for increasing the useable capacity of photorefractive memories. *Opt. Lett. 19*, page 1483, 1994.

[160] J.F. Heanue, M.C. Bashaw, and L. Hesselink. Channel codes for digital holographic data storage. *J. Opt. Soc. Am. A 12*, page 2432, 1995.

[161] J.F. Heanue, K. Gürkan, and L. Hesselink. Signal detection for page-access optical memories with intersymbolic interference. *Appl. Opt. 35*, page 2431, 1996.

[162] H. Sasaki, N. Maudit, J. Ma, Y. Fainman, S.H. Lee, and M.S. Gray. Dynamic digital photorefractive memory for optoelectronic neural network learning modules. *Appl. Opt. 35*, page 4641, 1996.

[163] A. Aharoni, M.C. Bashaw, and L. Hesselink. Distortion-free multiplexed holography in striated photorefractive media. *Appl. Opt. 32*, page 1973, 1993.

[164] J. Trisnadi and S. Redfield. Practical verification of hologram multiplexing without beam movement. *Proc. SPIE 1773, Photonic Neural Networks*, page 362, 1992.

[165] J.F. Heanue, M.C. Bashaw, and L. Hesselink. Recall of linear combinations of stored data pages based on phase-code multiplexing in volume holography. *Opt. Lett. 19*, page 1079, 1994.

[166] C. Denz, T. Dellwig, T. Rauch, and T. Tschudi. Parallel optical image addition and subtraction in a dynamic photorefractive optical memory using phase-code multiplexing. *Opt. Lett. 21*, page 278, 1996.

[167] J.F. Heanue, M.C. Bashaw, and L. Hesselink. Encrypted holographic data storge based on orthogonal phase-code multiplexing. *Appl. Opt. 34*, page 6012, 1995.

[168] Y.N. Denisyuk. Three dimensional and pseudodeep holograms. *Journ. Opt. Soc. Am. A*, page 1141, 1992.

[169] T.F. Krile, M.O. Hagler, W.D. Redus, and J.F. Walkup. Multiplex holography with chirp-modulated binary phase-coded reference-beam masks. *Appl. Opt. 18*, page 52, 1979.

[170] H. Lee and S.K. Jin. Experimental study of volume holographic interconnects using random patterns. *Appl. Phys. Lett. 62*, page 2191, 1993.

[171] K. Kamra, A. Kumar, and K. Singh. Novel optical photorefractive storage-retrieval sytem using speckle coding technique in beam-fanning geometry. *J. Mod. Opt. 43*, page 365, 1996.

[172] Y. Taketomi, J.E. Ford, H. Sasaki, J. Ma, Y. Fainman, and S.H. Lee. Hologram multiplexing using orthogonal phase codes and incremental recording. *Proc. of the Meeting on Optical Computing*, page 268, 1991.

[173] A.K. Jain. Fundamentals of digital image processing. *Prentice-Hall, New Jersey, USA*, page 155, 1988.

[174] X. Yang, Y. Xu, and Z. When. Generation of Hadamard matrices for phase-code multiplexed holographic memories. *Opt. Lett. 21*, page 1067, 1996.

[175] C. Denz, G. Pauliat, G. Roosen, and T. Tschudi. Potentialities and limitations of hologram multiplexing using the phase encoding technique. *Appl. Opt. 31*, page 5700, 1992.

[176] C. Alves, G. Pauliat, and G. Roosen. Dynamic phase-encoding storage of 64 images in a BaTiO$_3$ photorefractive crystal. *Opt. Lett. 19*, page 1894, 1994.

[177] C. Denz, T. Rauch, K.-O. Müller, T. Heimann, J. Trumpfheller, and T. Tschudi. Realization of a high-capacity holographic memory for analog and digital data storage based on phase-encoded multiplexing. *Proc. of the 1997 Topical Meeting on Photorefractive Materials, Effects and Devices, Chiba, Japan, 11.-13.6.1997*, page 232, 1997.

[178] K.-O. Müller, C. Denz, T. Rauch, T. Heimann, J. Trumpfheller, and T. Tschudi. Analog data storage for high-resolution images in a phase-encoded holographic memory. *Proc. SPIE of the Int. Symposium on Optical Information Science and Technology, OIST 97, Moscow, Russia, 26.-30.8.1997*, 1997.

[179] V.V. Orlov and A.R. Bulygin. Volume holograms of co-orthogonal waves for optical channel switching and expansion of light waves in walsh basis functions. *Opt. Comm. 133*, page 415, 1997.

[180] G.A. Sefler, T.K. Gustafson, E. Yin, and Y. Spiridon. Orthogonal photorefractive grating storage. *Opt. Lett. 21*, page 293, 1996.

[181] F.M. Schellenberg, W. Lenth, and G.C. Bjorklund. Technological aspects of frequency domain data storage using persistent spectral hole burning. *Appl. Opt. 25*, page 3207, 1986.

[182] S. Campbell and P. Yeh. Sparse-wavelength angle-multiplexed volume holographic memory system: analysis and advances. *Appl. Opt. 35*, page 2380, 1996.

[183] S. Campbell, X. Yi, and P. Yeh. Hybrid sparse-wavelength angle-multiplexed optical data storage system. *Opt. Lett. 19*, page 2161, 1994.

[184] R. McRuer, J. Wilde, L. Hesselink, and J. Goodman. Two-wavelength photorefractive dynamic optical interconnect. *Opt. Lett. 14*, page 1174, 1989.

[185] G.A. Rakuljic, V. Leyva, and A. Yariv. Optical data storage by using orthogonal wavelength-multiplexed volume holograms. *Opt. Lett. 17*, page 1471, 1992.

[186] S. Yin, H. Zhou, M. Wen, Z. Yang, J. Zhang, and F.T.S. Yu. Wavelength multiplexed holographic storage in a sensitive photorefractive crystal using a visible-light tunable diode laser. *Opt. Comm. 131*, page 317, 1993.

[187] M.C. Bashaw, R.C. Singer, J.F. Heanue, and L. Hesselink. Coded-wavelength multiplex volume holography. *Opt. Lett. 20*, page 1916, 1995.

[188] D. Lande, J.F. Heanue, M.C. Bashaw, and L. Hesselink. A digital wavelength-multiplexed holographic data storage system. *Technical Digest of the Conference on "Nonlinear Optics: Materials, fundamentals and applications*, page 189, 1996.

[189] A. Kewitsch, M. Segev, and A. Yariv. Electric-field multiplexing/demultiplexing of volume holograms in photorefractive media. *Opt. Lett. 18*, page 534, 1993.

[190] M. Balberg, M. Razvag, S. Vidro, E. Refaeli, and A.J. Agranat. Electroholographic neurons implemented on potassium lithium tantalate niobate crystals. *Opt. Lett. 21*, page 1544, 1996.

[191] A.J. Agranat, R. Hofmeister, and A. Yariv. *Opt. Lett. 17*, page 17, 1992.

[192] J. Lembcke, C. Denz, and T. Tschudi. General formalism for angular and phase-encoding multiplexing in holographic image storage. *Opt. Mat. 4*, page 428, 1995.

[193] G.W. Burr, F.H. Mok, and D. Psaltis. Angle and space multiplexed holographic storage using the 90° geometry. *Opt. Comm. 117*, page 49, 1995.

[194] S. Tao, R. Selviah, and J.E. Midwinter. Spatioangular multiplexed storage of 750 holograms in a Fe:LiNbO$_3$-crystal. *Opt. Lett. 18*, page 912, 1993.

[195] S. Tao, Z.H. Song, R. Selviah, and J.E. Midwinter. Spatioangular-multiplexing scheme for dense holographic storage. *Appl. Opt. 34*, page 912, 1993.

[196] L. Solymar and D.J. Cooke. Volume Holography and Volume Gratings. *Academic Press, London*, page 362, 1981.

[197] D. Psaltis, M. Levene, A. Pu, G. Barbastathis, and K. Curtis. Holographic storage using shift multiplexing. *Opt. Lett. 20*, page 782, 1995.

[198] G. Barbastathis, M. Levene, and D. Psaltis. Shift multiplexing with spherical reference waves. *Appl. Opt. 35*, page 2403, 1996.

[199] K. Curtis, A. Pu, and D. Psaltis. Method for holographic storage using peristrophic multiplexing. *Opt. Lett. 19*, page 993, 1994.

[200] X. Yi, P. Yeh, and C. Gu. Statistical analysis of cross-talk noise and storage capacity in volume holographic memory. *Opt. Lett. 19*, page 1580, 1994.

[201] E.G. Ramberg. Holographic information storage. *RCA Rev. 33*, page 5, 1972.

[202] C. Gu, J. Hong, I. McMichael, R. Saxena, and F. Mok. Cross-talk-limited storage capacity of volume holographic memory. *J. Opt. Soc. Am. A 9*, page 1978, 1992.

[203] E.N. Leith, A. Kozma, J. Upathniecks, J. Marks, and N. Massey. Holographic data storage in three-dimensional media. *Appl. Opt. 5*, page 1303, 1966.

[204] M.C. Bashaw, A. Aharoni, J.F. Walkup, and L. Hesselink. Cross talk considerations for angular and phase-encoded multiplexing in volume holography. *J. Opt. Soc. Am. B 11*, page 1820, 1994.

[205] K. Curtis and D. Psaltis. Cross talk in phase-coded holographic memories. *J. Opt. Soc. Am. A 10*, page 2547, 1993.

[206] K. Curtis, C. Gu, and D. Psaltis. Cross talk in wavelength-multiplexed holographic memories. *Opt. Lett. 18*, page 1001, 1993.

[207] X. Yi, C. Yang, S.H. Lin, P. Yeh, and C. Gu. Spectral hole- and angle multiplexed volume holographic memory. *Technical Digest of the Conference on "Nonlinear Optics: Materials, fundamentals and applications*, page 194, 1996.

[208] H. Sasaki and K. Karaki. Direct pattern recongition of a motion picture by hole-burning holographiy of $Eu^{3+}$: $Y_2Sio_3$. *Appl. Opt. 36*, page 1742, 1997.

[209] A. Bergeron, H.H. Arsenault, E. Eustache, and D. Gingras. Optoelectronic thresholding module for winner-take-all operations in optical neural networks. *Appl. Opt. 33*, page 1463, 1994.

[210] A. von Lehmen, E.G. Paek, P.F. Liao, A. Marrakchi, and J. S Patel. Influence of interconnection weight discretization and noise in an optoelectronic neural network. *Opt. Lett. 14*, page 928, 1989.

[211] H.J. Kim, J. Katz, S. Lin, and D. Psaltis. Monolithically integrated two-dimensional arrays of optoelectronic threshold devices for neural network applications. *Proc. SPIE 1043*, page 44, 1989.

[212] M. Ziari, W.H. Steier, and R.L.S. Devine. Infrared nonlinear neurons using the field shielding effect in CdTe. *Appl. Opt. 29*, page 2074, 1990.

[213] H. Yoshinaga, K. Kitayama, and T. Hara. All-optical error signal generation for backpropagation learning in optical multilayer neural networks. *Opt. Lett. 14*, page 202, 1989.

[214] W. Kawakami, H. Yshinaga, and K. Kitayama. Demonstration of an optical inhibitory neural network. *Opt. Lett. 18*, page 984, 1989.

[215] B.K. Jenkins and C.H. Wang. Model for an incoherent neuron that subtracts. *Opt. Lett. 13*, page 892, 1988.

[216] A.D. Fisher, W.L. Lippincott, and J.N. Lee. Optical implementations of associative networks with versatile adaptive learning capabilities. *Appl. Opt. 26*, page 5039, 1987.

[217] D. Haronian and A. Lewis. Elements of a unique bacteriorhodopsin neural network architecture. *Appl. Opt. 30*, page 597, 1992.

[218] L.S. Lee, H.M. Stoll, and M.C. Tackitt. Continuous-time optical network associative memory. *Opt. Lett. 14*, page 162, 1989.

[219] V. Hornung-Lequeux, P. Lalanne, and G. Roosen. Photorefractive wave mixing in $BaTiO_3$ for graded neurons with gain and signal binarization. *Opt. Comm. 73*, page 12, 1989.

[220] G. Balzer, S. Matamontero, M. Vasnetsov, and T. Tschudi. Phase defects in a phase-conjugate photorefractive gain oscillator. *Digest of the Topical Meeting on Photorefractive Materials, Effects and Devices*, page ThD11, Ukrainian Akademy of Science, 1993.

[221] K. Sayano, G.A. Rakuljic, and Y. Yariv. Thresholding semilinear phase-conjugate mirror. *Opt. Lett. 13*, page 143, 1988.

[222] R. Yahalom and Y. Yariv. Optical thresholding and switiching using a fiber-coupled phase-conjugate mirror. *Opt. Lett. 13*, page 889, 1988.

[223] K. Wagner and D. Psaltis. Multilayer optical learning networks. *Appl. Opt. 26*, page 5061, 1987.

[224] G.J. Dunning and C.R. Guiliano. Optical computing using phase conjugation. *In: Nonlinear Optics and Optical Computing, Eds. S. Martellucci and A.N. Chester, Plenum Press*, page 173, 1990.

[225] M.B. Klein, G.J. Dunning, G.C. Valley, and T.R. O'Meara. Imaging threshold detector using a phase-conjugate resonator in $BaTiO_3$. *Opt. Lett. 11*, page 575, 1986.

[226] S.W. McCahon, G.J. Dunning, G.C. Valley K.W. Kirby, and M.B. Klein. Bistable operation of a phase-conjugate resonator containing an intracavity saturable absorber. *Opt.Lett. 17*, page 517, 1992.

[227] C. Gu, S. Campbell, J. Hong, Q.B. He, D. Zhang, and P. Yeh. Optical thresholding and maximum operations. *Appl. Opt. 31*, page 5661, 1992.

[228] A.E. Chiou, P. Yeh, and M. Khoshnevisan. Nonlinear optical image substraction for potential industrial applications. *Opt. Eng. 27*, page 385, 1988.

[229] N.G. Basov et al. Laser interferometry with wavefront-reversing mirrors. *Sov. Phys. JETP 52*, page 847, 1980.

[230] M.D. Ewbank, P. Yeh, M. Khoshnevisan, and J. Feinberg. Time reversal by an interferometer with coupled phase-conjugate reflectors. *Opt. Lett. 10*, page 282, 1985.

[231] M. Serwe, A. Herden, and T. Tschudi. Application of an interferometer containing a self-pumped phase-conjugating mirror. *Proc. SPIE 1017, Nonlinear Optical Materials*, page 180, 1988.

[232] J.P. Huignard and A. Marrakchi. Two-wave mixing and energy transfer in $Bi_{12}SiO_{20}$ crystals: application to image amplification and vibration analysis. *Opt. Lett. 6*, page 622, 1981.

[233] J. Feinberg. Real-time edge enhancement using the photorefractive effect. *Opt. Lett. 5*, page 33, 1980.

[234] E. Ochoa, W. Goodman, and L. Hesselink. Real-time enhancement of defects in a periodic mask using photorefractive $B_{12}SiO_{20}$. *Opt. Lett. 10*, page 430, 1985.

[235] T. Rauch. Die selbstgepumpte Phasenkonjugation in $BaTiO_3$: Dynamik und Anwendungen. *Diplomarbeit, TH Darmstadt*, Oktober 1991.

[236] J. Goltz, C. Denz, and T. Tschudi. Dynamics of holographic readout in photorefractive crystals for broken Bragg-conditions. *Opt. Comm. 68*, page 228, 1988.

[237] S.K. Kwong, G.A. Rakuljic, and A. Yariv. Real time image subtraction and exclusive or operation using a self-pumped phase conjugate mirror. *Appl. Phys. Lett. 48*, page 201, 1985.

[238] A.E. Chiou and P. Yeh. Parallel image subtraction using a phase-conjugate Michelson interferometer. *Opt. Lett. 11*, page 306, 1986.

[239] P. Yeh and A.E. Chiou. Real-time contrast reversal via four-wave mixing in nonlinear media. *Opt. Comm. 64*, page 160, 1987.

[240] Y. Tomita, R. Yahalom, and A. Yariv. Real-time image subtraction with the use of wave polarization and phase conjugation. *Appl. Phys. Lett. 52*, page 425, 1988.

[241] H. Fujiware, K. Nakagawa, and T. Suzuki. Real-time image subtraction and addition using two cross-polarized phase-conjugated waves. *Opt. Comm. 79*, page 6, 1990.

[242] N.A. Vainos and R.W. Eason. Strictly real-time image differentiation in photorefractive BSO. *J. Mod. Opt. 35*, page 505, 1988.

[243] N.A. Vainos, J.A. Khoury, and R.W. Eason. Real-time parallel optical logic in photorefractive bismuth silicon oxide. *J. Mod. Opt. 35*, page 505, 1988.

[244] J.A. Khoury and R.W. Eason. Photorefractive deconvolution techniques for optical differentiation of images. *J. Mod. Opt. 36*, page 369, 1989.

[245] J. Zhang, H. Xu, Y. Yuan, and K. Xu. Real-time coherent image differentiation using a self-pumped phase conjugator with cu:knsbn. *Appl. Opt. 32*, page 1470, 1993.

[246] G.T. Hengst and W.B. Roh. Switchable optical image adder/subtractor. *Opt. Lett. 16*, page 165, 1991.

[247] M. Ogusu, S.I. Tanaka, and K. Kuroda. Optical logic operations using three-beam phase-conjugate interferometry. *Proc. of the 3rd Topical Meeting on Photorefractive Materials, Effects and Devices, OSA 1991 Technical Digest Series Vol. 14*, page 77, July 1991.

[248] U.P. Wild, C. De Caro, S. Bernet, M. Traber, and A. Renn. Molecular computing. *Journ. Luminesc. 48/49*, page 335, 1991.

[249] D.Z. Anderson and M.C. Erie. Resonator memories and optical novelty filters. *Opt. Eng. 26*, page 434, 1987.

[250] D.Z. Anderson and J. Feinberg. Optical novelty filters. *IEEE J. Quant. El. QE-25*, page 535, 1989.

[251] M. Cronin-Golomb, A.M. Biernacki, C. Lin, and H. Kong. Photorefractive time differentiation of coherent optical images. *Opt. Lett. 12*, page 1029, 1987.

[252] M. Sedlatschek, T. Rauch, C. Denz, and T. Tschudi. Demonstrator concepts and performance of a photorefractive optical novelty filter. *Opt. Mat. 4*, page 376, 1995.

[253] M. Sedlatschek, T. Rauch, C. Denz, and T. Tschudi. Generalized theory of the resolution of object tracking novelty filter. *Opt. Comm. 116*, page 25, 1995.

[254] J.E. Ford, Y. Fainman, and S.H. Lee. Time-integrating interferometry using photorefractive fanout. *Opt. Lett. 13*, page 856, 1988.

[255] T. Tschudi, A. Herden, J. Goltz, H. Klumb, F. Laeri, and J. Albers. Image amplification by two- and four-wave mixing in BaTiO$_3$ photorefractive crystals. *IEEE J. Quant. El. QE-22*, page 1493, 1986.

[256] D.Z. Anderson, C. Benkert, B. Chorbajian, and A. Hermanns. Photorefractive flip-flop. *Opt. Lett. 16*, page 250, 1991.

[257] D.L. Staebler and J.J. Amodei. Thermally fixed holograms in LiNbO$_3$. *Ferroel. 3*, page 107, 1972.

[258] J.J. Amodei and D.L. Staebler. Holographic pattern fixing in electrooptic crystals. *Appl. Phys. Lett. 18*, page 540, 1971.

[259] S.W. McCahon, D. Rytz, G.C. Valley, M.B. Klein, and B.A. Wechsler. Hologram fixing in Bi$_{12}$TiO$_{20}$ using heating and an AC electric field. *Appl. Opt. 28*, page 1967, 1989.

[260] Y. Qiao, S. Orlov, D. Psaltis, and R.R. Neurgoankar. Electrical fixing of holograms in SBN:75,. *Opt. Lett. 18*, page 1004, 1993.

[261] R.S. Cudney, J. Fousek, M. Zgonik, P. Bernasconi, and P. Günter. Ferroelectric domain fixing and space-charge field enhancement in Barium Titanate and Potassium Niobate. *Proceedings of Photorefractive Materials Effects and Devices (PR '95), Optical Society of America, Technical Digest TPC5*, page 288, 1995.

[262] S. Redfield and L. Hesselink. Enhanced nondestructive holographic readout in strontium barium niobate. *Opt. Lett. 13*, page 880, 1988.

[263] E.K. Gulanyan, I.R. Dorosh, V.D. Iskin, A.L. Mikaelyan, and M.A. Maiorchuk. Nondestructive readout of holograms in iron-doped Lithium niobate crystals. *Sov. J. Quant. Electron. 9*, page 647, 1979.

[264] M.P. Petrov, S.I. Stepanov, and A.A. Kamshilin. Holographic storage of information and pecularities of light diffraction in birefringent electro-optic crystals. *Opt. Laser Technol. 6*, page 149, 1979.

[265] H. C. Külich. A new approach to read volume holograms at different wavelengths. *Opt. Comm. 64*, page 407, 1987.

[266] D. Psaltis, F. Mok, and H.Y. Sidney Li. Nonvolatile storage in photorefractive crystals. *Opt. Lett. 19*, page 210, 1994.

[267] E.J. Bjornson, M.C. Bashaw, and L. Hesselink. Digital quasi-phase-matched two-color nonvolatile holographic storage. *Appl. Opt. 36*, page 3090, 1997.

[268] Y. Qiao and D. Psaltis. Sampled dynamic holography. *Opt. Lett. 17*, page 1376, 1992.

[269] H. Sasaki, J. Ma, Y. Fainman, S.H. Lee, and Y. Taketomi. Fast update of dynamic photorefractive optical memory. *Opt. Lett. 17*, page 1468, 1992.

[270] J.V. Alvarez-Bravo and L. Arizmendi. Coherent erasure and updating of holograms in LiNbO$_3$. *Opt. Mat. 4*, page 419, 1995.

[271] H. Rajbenbach, S. Bann, and J.P. Huignard. Long-term readout of photorefractive memories by using a storage / amplification two-crystal configuration. *Opt. Lett. 17*, page 1712, 1992.

[272] J.C. Palais and J.A. Wise. Improving the efficiency of very low efficiency holograms by copying. *Appl. Opt. 10*, page 667, 1971.

[273] K.M. Johnson, M. Armstrong, L. Hesselink, and J.W. Goodman. Multiple multiple-exposure holograms. *Appl. Opt. 24*, page 4467, 1985.

[274] D. Brady, K. Hsu, and D. Psaltis. Periodically refreshed multiply exposed photorefractive holograms. *Opt. Lett. 15*, page 819, 1990.

[275] D. Psaltis, D. Brady, X.-G. Gu, and K. Hsu. *Optical Processing and Computing, H. Arsenault, Ed. (Academic Press, Orlando, Florida*, 1989.

[276] H. Sasaki, Y. Fainman, J.E. Ford, Y. Taketomi, and S.H. Lee. Dynamic photorefractive optical memory. *Opt. Lett. 16*, page 1874, 1991.

[277] T. Dellwig, C. Denz, T. Rauch, and T. Tschudi. Coherent refreshment and updating for dynamic photorefractive optical memories using phase conjugation. *Opt. Comm. 119*, page 333, 1995.

[278] D.Z. Anderson. Coherent optical eigenstate memory. *Opt. Lett. 11*, page 56, 1986.

[279] J.P. Jiang and J. Feinberg. Dancing modes and frequency shifts in a phase conjugator. *Opt. Lett. 12*, page 266, 1987.

[280] E.J. Bochove. Transverse-mode instability and chaos in an optical cavity with phase-conjugate mirror. *Opt. Lett. 11*, page 727, 1986.

[281] G.J. Dunning, S.W. McCahon, G.J. Dunning, M.B. Klein, and D.M. Pepper. Improved spatial and temporal performance of a phase-conjugate resonator with an intracavity mode homogenizer. *J. Opt. Soc. Am. B 11*, page 339, 1994.

[282] Y. Qiao, D. Psaltis, J. Hong, P. Yeh, and R.R. Neurgoankar. Phase-locked sustainment of photorefractive holograms using phase conjugation. *J. Appl. Phys. 70*, page 4646, 1991.

[283] P. Yeh, C. Gu, C.-J. Cheng, and K.-Y. Hsu. Hologram enhancement in photorefractive media. *Opt. Eng. 34*, page 2204, 1995.

[284] P. Yeh, C. Gu, C.-J. Cheng, and K.-Y. Hsu. Optical restoration of photorefractive volume holograms. *Appl. Phys. B 61*, page 511, 1995.

[285] S. Campbell, P. Yeh, C. Gu, S.-H. Lin, C.-J. Cheng, and K.-Y. Hsu. Optical restoration of photorefractive holograms via self-enhanced diffraction. *Opt. Lett. 20*, page 330, 1995.

[286] K. Rastani and W.M. Hubbard. Large interconnects in photorefractives: grating erasure problem and a proposed solution. *Appl. Opt.31*, page 598, 1992.

[287] P. Vanitt, P. Pagani, L.A. Lugiato, and M.V. Pinna. All-optical associative memory based on the use of the laser as a nonlinear element. *Opt. Lett. 17*, page 1526, 1992.

[288] D. Hennequin, L. Dambly, D. Dangoisee, and P. Glorieux. Basic transverse dynamics of a photorefractive oscillator. *J. Opt. Soc. Am. B 11*, page 676, 1994.

[289] C.O. Weiss. Spatio-temporal structures, Part II. Vortices and defects in lasers. *Phys. Rep. 219*, page 311, 1992.

[290] G.D'Alessandro. Spatiotemporal dynamics of an unidirectional ring oscillator with photorefractive gain. *Phys. Rev. A 46*, page 2791, 1992.

[291] F.E. Ammann, A. Herden, and T. Tschudi. Coupling channels in a photorefractive ring-resonator containing gain by two-wave mixing. *Ferroel. 92*, page 301, 1989.

[292] C. Denz, G. Balzer, O. Knaup, and T. Tschudi. Origin of pattern dynamics and circling vortices in a unidirectional photorefractive ring resonator. *Digest of the Topical Meeting on Photorefractive Materials, Effects and Devices*, page 415, Tokyo, June 1997.

[293] D.Z. Anderson and R. Saxena. Theory of multimode operation of a unidirectional ring oscillator having photorefractive gain: weak-field limit. *J. Opt. Soc. Am. B 4*, page 164, 1987.

[294] G. Balzer, S. Matamontero, M. Vasnetsov, and T. Tschudi. Phase defects in a phase-conjugate photorefractive gain-oscillator. *J. Mod. Optics 41*, page 807, 1994.

[295] F.T. Arrechi, G. Giacomelli, P.L. Ramazza, and S. Residori. Experimental evidence of chaotic itinerancy and spatiotemporal chaos in optics. *Phys. Rev. Lett. 20*, page 2531, 1990.

[296] P. Davis. Application of optical chaos to temporal pattern search in a nonlinear optical resonator. *Int. Top. Meet. on Opt. Comp. Technical Digest, Jap. Soc. Appl. Phys.*, page 272, 1990.

[297] H. Kogelnik. Laser beams and resonators. *Proc. IEEE 54*, page 1312, 1966.

[298] J. Auyeung, D. Fekete, D.M. Pepper, and A. Yariv. A theoretical and experimental investigation of the modes of optical resonators with phase-conjugate mirrors. *IEEE J. Quant. El. QE-15*, page 1180, 1979.

[299] J. Goltz, C.Denz, and T. Tschudi. Four-wave mixing in photorefractive $BaTiO_3$ by use of tilted pump waves. *Opt. Comm. 72*, page 129, 1989.

[300] M.S. Cohen. Self-organizing categorization and abstraction in a phase conjugating resonator. *Proc. SPIE 625*, page 214, 1986.

[301] J.F. Lam and W.P. Brown. Optical resonators with phase-conjugate mirrors. *Opt. Lett. 5*, page 61, 1980.

[302] F. Laeri. Is a resonator that contains a phase-conjugating mirror a resonator? *J. Opt. Soc. Am. B 77*, page 2169, 1990.

[303] E.G. Paek, P.F. Liao, and H. Gharavi. Derivation of neural network models and their computational circuits for associative memories. *Opt. Eng. 31*, page 986, 1992.

[304] B. Kosko. Bidirectional associative memories. *IEEE transactions on systems, man and cybernetics 18*, 1988.

[305] D. Gabor. Associative holographic memories. *IBM J. Res. Dev. 156*, 1969.

[306] G.W. Stroke, R. Restrick, A. Funkhouse, and D. Brumm. Resolution-retrieving compensation of source effects by correlative reconstruction in high-resolution holography. *Phys. Lett. 18*, page 274, 1965.

[307] E.N. Leith and J. Upathniecks. Reconstructed wavefronts and communication theory. *J. Opt. Soc .Am. 52*, page 1123, 1962.

[308] R.J. Collier and K.S. Pennington. Ghost imaging by holograms formed in the near field. *Appl. Phys. Lett. 8*, page 44, 1966.

[309] G.R. Knight. Page-oriented associative holographic memory. *Appl. Opt. 13*, page 904, 1974.

[310] D.M. Pepper, J.AuYeung, D. Fekete, and Y. Yariv. Spatial convolution and correlation of optical fields via degenerate four-wave mixing. *Opt. Lett. 3*, page 7, 1978.

[311] H. Rajbenbach, S. Bann, Ph. Refregier, P. Joffre, J.P. Huignard, H.-S. Buchkremer, A.S. Jensen, E. Rasmussen, K.-H. Brenner, and G. Lohman. Compact photorefractive correlator for robotic applications. *Appl. Opt. 31*, page 5666, 1992.

[312] A. Pu and D. Psaltis. High-density recording in photopolymer-based holographic three-dimensional disks. *Appl. Opt. 35*, page 2389, 1996.

[313] H.-Y. Sidney LI, Y. Qiao, and D. Psaltis. Optical network for real-time face recognition. *Appl. Opt. 32*, page 5026, 1993.

[314] H. Yoshinaga, K. Kitayama, and H. Oguri. Application of fiber holography to associative memory. *Opt. Lett. 16*, page 669, 1991.

[315] Y.K. Lee. Optical angular mapping for pattern recognition using self-pumped phase conjugation. *Appl. Opt. 33*, page 6228, 1994.

[316] D. Psaltis and J.H. Hong. Shift-invariant optical associative memory. *Opt. Eng. 26*, page 010, 1987.

[317] F.E. Ammann. Herstellung holographisch-optischer Elemente und deren Anwendung in optischen Rückkopplungssystemen. *Diploma Thesis*, page 62 ff, 1987.

[318] Y. Owechko. Nonlinear holographic associative memories. *IEEE J. Quant. El. 25*, page 619, 1989.

[319] E.L. Kral, J.F. Walkup, and M.O. Hagler. Correlation properties of random phase diffusers for multiplex holography. *Appl. Opt. 21*, page 1281, 1982.

[320] Y.S. Abu-Mostafa and J.N. St.Jacques. Information capacity of the Hopfield model. *IEEE Trans.Inform.Theory IT-31*, page 461, 1985.

[321] R.J. McEliece, E.C. Posner, E.R. Rodemich, and S.S. Venkatesh. The capacity of the Hopfield associative memory. *IEEE Trans. Inform. Theory IT-33*, page 461, 1987.

[322] A.D. Bruce, E. Gardner, and D.J. Wallace. Dynamics and statistical mechanics of the Hopfield model. *J. Phys. A: Math. Gen. 20*, page 2909, 1987.

[323] Y. Owechko, G.J. Dunning, E. Marom, and B.H. Soffer. Holographic associative memory with nonlinearities in the correlation domain. *Appl. Opt. 26*, page 1900, 1987.

[324] A. Yariv and S.K. Kwong. Associative memories based on message-bearing optical modes in phase-conjugate resonators. *Opt. Lett. 11*, page 186, 1986.

[325] D.M. Pepper. Optical phase conjugator with spatially resolvable thresholding utilizing a liquid crystal light valve. *U.S. Patent no. 4 762 397*, 1988.

[326] X. Sun, Z. Zhou, Y. Jiang, K.Xu, and J. Zhang. Holographic associative memory using a coherently induced double phase conjugate mirror. *Opt. Eng. 35*, page 2135, 1996.

[327] H. Kang, C.X. Yang, G.G. Mu, and Z.K. Wu. Real-time holographic associative memory using doped LiNbO3 in a phase-conjugate resonator. *Opt. Lett. 15*, page 637, 1990.

[328] Y. Owechko. Optoelectronic resonator neural network. *Appl. Opt. 26*, page 5104, 1987.

[329] A.V. Pavlov. Adaptive-optical neural network for classifying paterns on structured background. *Opt. Spectrosc. 75*, page 391, 1993.

[330] E.G. Paek and D. Psaltis. Optical associative memory using Fourier transform holograms. *Opt. Lett. 10*, page 98, 1985.

[331] E.G. Paek and A.Von Lehmen. Holographic associative memory for word-break recognition. *Opt. Lett. 14*, page 205, 1989.

[332] Oh Changsuk and Park Park Hankyu. Real-time Fourier holographic associative memory with photorefractive material. *Proc. SPIE 963 Optical Computing*, page 554, 1988.

[333] H. Xu, Y. Yuan, K. Xu, and Y. Xu. Performance of real-time associative memory using a photorefractive crystal and liquid crystal electrooptic switches. *Appl. Opt. 29*, page 3375, 1990.

[334] H.J. White, N.B. Aldridge, and I. Lindsay. Digital and analog holographic associative memories. *Opt. Eng. 27*, page 30, 1988.

[335] A. Yariv, S.K. Kwong, and K. Kyuma. Demonstration of an all-optical holographic associative memory. *Appl. Phys. Lett. 48*, page 1114, 1986.

[336] C.J. Kuo. On the feedback holographic associative memory. *Opt. and Quant. El. 25*, page 761, 1993.

[337] L. Wang, V. Esch, R. Feinleib, L. Zahng, R. Jin, H.M Chou, R.W. Sprague, H.A. MacLeod, G. Khitrova, H.M. Gibbs, K. Wagner, and D. Psaltis. Interference filters as nonlinear decision-making elements for three-spot recognition and associative memories. *Appl. Opt. 27*, page 1715, 1988.

[338] E.G. Paek and E.C. Jung. Simplified holographic associative memory using enhanced nonlinear processing with a thermoplastic plate. *Opt. Lett. 15*, page 637, 1990.

[339] M. Ingold, M. Duelli, P. Günter, and M. Schadt. All-optical associative memory based on a nonresonant cavity with image-bearing beams. *J. Opt. Soc. Am B 9*, page 1327, 1992.

[340] M.S. Cohen. Multistability and associative memory in a phase-conjugating resonator. *J. Opt. Soc. Am. B 8*, page 106, 1991.

[341] M. Ishikawa, N. Mukohzaka, H. Toyoda, and Y. Suzuki. Optical associatron: a simple model for optical associative memory. *Appl. Opt. 28*, page 291, 1989.

[342] H. Toyoda, N. Mukohzaka, M. Ishikawa, and Y. Suzuki. Adaptive optical processing system with optical associative memory. *Appl. Opt. 32*, page 1355, 1993.

[343] J.M. Kinser, H.J. Caulfield, and J. Shamir. Design for a massive all-optical bidirectional associative memory: the big BAM. *Appl. Opt. 27*, page 3442, 1988.

[344] C.C. Guest and R. TeKolste. Design and devices for optical bidirectional associative memories. *Appl. Opt. 26*, page 5055, 1987.

[345] F.T.S. Yu, S. Yin, and C.-M. Uang. A content-adressable polychromatic neural net using a (Ce:Fe)-doped LiNbO3 photorefractive crystal. *Opt. Comm. 29*, page 300, 1994.

[346] M. Takeda and J.W. Goodman. Neural networks for computation: number representations and programming complexity. *Appl. Opt. 25*, page 3033, 1986.

[347] G.Y. Sirat, A.D. Maruani, and R.C. Chevallier. Gray level neural networks. *Appl. Opt. 28*, page 414, 1989.

[348] W. Zhang, K. Itoh, J. Tanida, and Y. Ichioka. Hopfield model with multistate neurons and its optoelectronic implementation. *Appl. Opt. 30*, page 1195, 1991.

[349] F.T.S. Yu, C.M. Uang, and S. Yin. Gray-level discrete associative memory. *Appl. Opt. 32*, page 1322, 1993.

[350] O.V. Dubrovskaya and E.I. Shubnikov. Holographic correlators as optical neural networks. *Opt. Spectrosc. 72*, page 524, 1992.

[351] S. Abramson, D. Saad, E. Marom, and N. Konforti. Four-quadrant optical matrix-vector multiplication machine as a neural-network processor. *App. Opt. 32*, page 1330, 1993.

[352] E.G. Paek, J.R. Wullert, and J.S. Patel. Holographic implementation of a learning machine based on a multicategory perceptron algorithm. *Opt. Lett 14*, page 1303, 1989.

[353] E.G. Paek, J.R. Wullert, and J.S. Patel. Holographic on-line learning machine for multicategory classification. *Jap. Journ. Appl. Phys. 29*, page L 1332, 1990.

[354] H. Yoshinaga nad K. Kitayama and T. Hori. Experimental learning in an optical perceptronlike neural network. *Opt. Lett 14*, page 716, 1989.

[355] C. Denz. Zwei- und Vierwellenmischen in photorefraktivem BaTiO3: parameterabhängige Untersuchungen. *Diplomarbeit, TH Darmstadt*, Februar 1988.

[356] J.H. Hong, S. Campbell, and P. Yeh. Optical implementation of perceptrons using photorefractive media. *Proc. of the Topical Meeting on Photorefractive Materials, Effects and Devices, Aussois, France*, page 265, 1990.

[357] J.H. Hong. Applications of photorefractive crystals for optical neural networks. *Opt. and Quant. El. 25*, page S551, 1993.

[358] K.Y. Hsu, S.H. Lin, and P. Yeh. Conditional convergence of photorefractive perceptron learning. *Opt. Lett 24*, page 2135, 1993.

[359] C.J. Cheng, P. Yeh, and K.Y. Hsu. Generalized perceptron learning rule and its implications for photorefractive neural networks. *J. Opt. Soc. Am. B 11*, page 1619, 1994.

[360] G.C. Petrisor, A.A. Goldstein, B.K. Jenkins, E.J. Herbulock, and A.R. Tanguay. Convergence of backward error-propagation learning in photorefractive crystals. *Appl. Opt. 35*, page 1328, 1996.

[361] C.C. Mao and K.M. Johnson. Optoelectronic array that computes error and weight modification for a bipolar optical neural network. *Appl. Opt. 32*, page 1291, 1993.

[362] P.J. de Groot and R.J. Noll. Adaptive neural network in a hybrid optical/electronic architecture using lateral inhibition. *Appl. Opt. 28*, page 3852, 1989.

[363] F. Ioth, K. Kitayama, and Y. Tamura. Optical outer-product learning in a neural network using optically stimulable phosphor. *Opt. Lett. 15*, page 860, 1990.

[364] Y. Kuratomi, A. Takimoto, K. Akiyama, and H. Ogawa. Optical neural network using vector-feature extraction. *Appl. Opt. 32*, page 5750, 1993.

[365] Z. When, P. Yeh, and X. Yang. Modified two-dimensional hamming neural network and its optical implementation using dammann gratings. *Opt. Eng. 35*, page 2136, 1996.

[366] T. Lu, d. Mintzer, A. Kostrzewski, and F. Lin. Compact holographic optical neural network system for real-time pattern recognition. *Opt. Eng. 35*, page 2122, 1996.

[367] K. Fukushima, S. Miyake, and T. Ito. Neocognitron: a neural network model for a mechanism of visual pattern recognition. *IEEE Trans. Syst, Man Cybern. SMC-13*, page 826, 1983.

[368] T.H. Chao and W.W. Stoner. Optical implementation of a feature-based neural network with application to automatic target recognition. *Appl. Opt. 32*, page 1359, 1993.

[369] C. Petersen, S. Redfield, J.D. Keeler, and E. Hartman. Optoelectronic implementation of multi-layer neural networks in a single photorefractive crystal. *Opt. Eng. 29*, page 359, 1989.

[370] P.A. Shoemaker, M.J. Carlin, and R.L. Shimabukuro. Backpropagation learning with trinary quantization of weight updates. *Neural Net. 4*, page 231, 1991.

[371] C.L. Giles and T. Maxwell. Learning, invariance and generalization in higher-order neural networks. *Appl. Opt. 26*, page 4972, 1987.

[372] P. Horan, D. Uecker, and A. Arimoto. Holographic on-line learning machine for multicategory classification. *J. Journ. of Appl. Phys. 29*, page L 1328, 1990.

[373] P. Horan, A. Jennings, B. Kelly, and J. Hegarty. Optical implementation of a second-order translation-invariant network algorithm. *Appl. Opt. 32*, page 1311, 1993.

[374] A. Jennings, P. Horan, and J. Hegarty. Optical neural network with quantum well devices. *Appl. Opt. 33*, page 1469, 1994.

[375] A. von Lehmen, E.G. Paek, L.C. Carrion, J. S Patel, and A. Marrakchi. Optoelectronic chip implementation of a quadratic associative memory. *Opt. Lett. 15*, page 279, 1990.

[376] A.L. Mikaelian, B.S. Kiselyov, N.Y. Kulakov, V.A. Shkitin, and V.A. Ivanov. Optical implementation of high-order associative memory. *Int. J. of Opt. Comp.* , 1990.

[377] E.A. Manykin and M.N. Belov. Higher-order neural networks and photon-echo effect. *Neural Networks 4*, page 417, 1991.

[378] S. Lin and L. Liu. Opto-electronic implementation of a neural network with a third-order interconnection for quadratic associative memory. *Opt. comm. 73*, page 268, 1989.

[379] A.J. Noest. Discrete-state phasor neural networks. *Phys. Rev. A 38*, page 2196, 1988.

[380] A.J. Noest. Associative memory in sparse phasor neural networks. *Europhys. Lett. 6*, page 469, 1988.

[381] G.D. Little, S.C. Gustafson, and R.A. Senn. Generalization of the backpropagation neural network learning algorithm to permit complex weights. *Appl. Opt. 29*, page 1591, 1990.

[382] T. Kohonen. Self-Organization and Associative Memory. *Springer Verlag*, 1984.

[383] S.A. Collins, S.C. Ahalt, A.K. Krishnamurthy, and D.F. Stewart. Optical implementation and applications of closest-vector selection in neural networks. *Appl. Opt. 32*, page 1297, 1993.

[384] A.L. Lentine, H.S. Hinton, D. Miller, J.E. Henry, J.E. Cunningham, and L. Chirovsky. Symmetric self-electro-optic effect device: optical set-reset latch. *Appl. Phys. Lett. 52*, page 1419, 1988.

[385] M. Killinger, J.L. de Bougrenet, and P. Cambon. Controlling the grey-level capacity of a bistable FLC spatial light modulator. *Ferroel. 122*, page 89, 1991.

[386] J. Duvillier, M. Killinger, K. Heggarty, K. Yao, J.L. de Bougrenet, and P. Cambon. All-optical implementation of a self-organizing map: a preliminary approach. *Appl. Opt. 33*, page 258, 1994.

[387] T. Galstyan, Gilles Pauliat, Andre Villing, and Gerald Roosen. Adaptive photorefractive neurons for self-organizing networks. *Opt. Comm. 109*, page 35, 1994.

[388] Y. Frauel, T. Galstyan, Gilles Pauliat, Andre Villing, and Gerald Roosen. Topological map from a photorefractive self-organizing neural networks. *Opt. Comm. 135*, page 179, 1997.

[389] C. Benkert and D.Z. Anderson. Controlled competitive dynamics in a photorefractive ring oscillator: 'winner-takes-all' and the 'voting paradox' dynamics. *Phys. Rev. A 44*, page 4633, 1991.

[390] D.Z. Anderson, C. Benkert, V. Hebler, J.S. Jand, D. Montgomery, and M. Saffman. Optical implementation of a self-organized feature extractor. *Advances in neural-information processing systems IV; J.E. Moody et al., eds.*, page 821, 1992.

[391] R. Blumrich, T. Kobialka, and T. Tschudi. Behaviour of the self-oscillation pattern in a phase-conjugate ring resonator. *J. Opt. Soc. Am. B 7*, page 2299, 1990.

[392] M. Saffman, D. Montgomery, and D.Z. Anderson. Collapse of a transverse mode continuum in a self-imaging photorefractively pumped ring resonator. *Opt. Lett. 19*, page 518, 1994.

[393] N.H. Farhat. Optoelectronic analoges of self-programming neural nets: architecture and methodologics for implementing fast stochastic learning by simulated annealing. *Appl. Opt. 26*, page 5093, 1987.

[394] N.H. Farhat, D. Psaltis, A. Prata, and E. Paek. Optical implementation of the Hopfield model. *Appl. Opt. 24*, page 1469, 1985.

[395] F.T.S. Yu, T. Lu, X. Yang, and D.A. Gregory. Optical neural network with pocket-sized liquid crystal televisions. *Opt. Lett. 15*, page 863, 1990.

[396] I. Shariv and A.A. Friesem. All-optical neural network with inhibitory neurons. *Opt. Lett. 24*, page 485, 1989.

[397] J.S. Jang, S.G. Shin, S.Y. Shin, and S.Y. Lee. Dynamic Hopfield-like network using a holographic lenslet array and a photorefractive crystal. *Proc. Conference*, page 325, 1990.

[398] H.M. Stoll and L.S. Lee. A continuous-time optical neural network. *Proc. of the IEEE Int. Conf. on Neural Networks, San Diego, California, Vol. II*, 24.-27.7.1988.

[399] M. Takeda and T. Kishigami. Complex neural fields with a Hopfield-like energy function and an analogy to opticals fields generated in phase-conjugate resonators. *J. Opt. Soc. Am. A 9*, page 2182, 1992.

[400] T. Tschudi, H. Klumb, T. Kobialka, A. Herden, and F. Laeri. Image transmission and self-oscillation in an active conjugated ring-resonator using $BaTiO_3$. *Proc. SPIE. 813*, page 71, 1987.

[401] H. Klumb, A. Herden, T. Kobialka, F. Laeri, and T. Tschudi. Active coherent optical feedback system with phase- conjugating image amplifier. *J. Opt. Soc. Am. B 5*, page 2379, 1988.

[402] S. Kirkpatrick, C.D. Gelat, and M.P. Vecchi. Optimization by simulated annealing. *Science 220*, page 671, 1983.

[403] G.E. Hinton and T.J. Sejnowski. Learning and relearning in Boltzmann machines. *Parallel Distributed Processing I, ed.: D.E. Rumelhardt and J.L. McClelland, MIT Press, Cambridge, Massachusetts*, page Capt. 7, 1987.

[404] C.M. Uang, S. Yin, G. Lu, and F.T.S. Yu. Shift- and rotation invariant interpattern heteroassociation (IHA) model. *Opt. Comm. 104*, page 285, 1994.

[405] C. Peterson and J.R. Anderson. A mean field theory learning algorithm for neural networks. *Compl. Syst. 1*, page 995, 1987.

[406] C. Peterson and E. Hartman. Explorations of the mean field theory learning algorithm. *Neur. Netw. 2*, page 475, 1989.

[407] M.Y. Choi and B. Huberman. Digital dynamics and the simulation of magnetic systems. *Phys. Rev. B 28*, page 2547, 1983.

[408] D.C. Wunsch II, D.J. Morris, R.L. McGann, and T.P. Caudell. Photorefractive adaptive resonance neural network. *Appl. Opt. 32*, page 1399, 1993.

[409] D. Wunsch, T.P. Caudell, D. Capps, and A. Falk. Optoelectronic learning machine. *OSA Annual Meeting 15*, page MJJ5, 1990, OSA Technical Digest Series.

[410] T.P. Caudell. Hybrid optoelectronic adaptive resonance theory neural processor, art1. *Appl. Opt. 31*, page 6220, 1992.

[411] H.J. Caulfield and D. Armitage. Adaptive resonance theory of optical pattern recognition. *Appl. Opt. 28*, page 4060, 1989.

[412] J. Daiber, R. Snyder, J. Colvin, B. Okas, and L. Hesselink. Fully functional digital video holographic storage system. *OSA Annual Meeting, 12.-17.10.1997, California*, page Talk ThR3, 1997.

[413] M.A. Neifeld and D. Psaltis. Programmable image associative memory using an optical disk and a photorefractive crystal. *Appl. Opt. 32*, page 4398, 1993.

# Index